Conic Sections

All conic sections have an equation of the form $Ax^2 + Bx + Cy^2 + Dy + E = 0$ where either A or C is non zero. The particular conic sections are listed below.

Conic Section	Characteristic
Parabola	Either A or C is 0, but not both.
Circle	$A = C \neq 0$
Ellipse	$A \neq C$, $AC > 0$
Hyperbola	$AC < 0$

Function

A function is a correspondence or rule that assigns to each element of one set (the domain) exactly one element from another set (the range).

Composition of Functions

$(g \circ f)(x) = g[f(x)]$

One-to-One Function

f is a one-to-one function if $a \neq b$ implies $f(a) \neq f(b)$.

Inverse Function

Let f be a one-to-one function. The inverse function, f^{-1}, can be found by exchanging x and y in the equation for f; then solving for y if possible.

Value of e

$e \approx 2.718281828$

Definition of Logarithm

$y = \log_a x$ if and only if $x = a^y$.

Properties of Logarithms

(a) $\log_a xy = \log_a x + \log_a y$ (b) $\log_a \dfrac{x}{y} = \log_a x - \log_a y$

(c) $\log_a x^r = r \log_a x$ (d) $\log_a a = 1$

(e) $\log_a 1 = 0$

Arithmetic Sequences

nth term: $a_n = a_1 + (n - 1)d$

Sum of n terms: $S_n = \dfrac{n}{2}[a_1 + a_n]$ or $S_n = \dfrac{n}{2}[2a_1 + (n - 1)d]$

Geometric Sequences

nth term: $a_n = a_1 r^{n-1}$

Sum of n terms: $S_n = \dfrac{a_1(1 - r^n)}{1 - r}$ $(r \neq 1)$

Sum of an infinite geometric sequence: $S_\infty = \dfrac{a_1}{1 - r}$, $-1 < r < 1$

Permutations

The number of arrangements of n things taken r at a time is $P(n, r) = \dfrac{n!}{(n - r)!}$.

Combinations

The number of ways to choose r things from a group of n things is $\dbinom{n}{r} = \dfrac{n!}{(n - r)!r!}$.

Properties of Probability

For any events E and F:
1. $0 \leq P(E) \leq 1$
2. P (a certain event) $= 1$
3. P (an impossible event) $= 0$
4. $P(E') = 1 - P(E)$
5. $P(E \text{ or } F) = P(E \cup F) = P(E) + P(F) - P(E \cap F)$.

Binomial Theorem

For any positive integer n, $(x + y)^n = x^n + \dbinom{n}{n-1} x^{n-1} y + \dbinom{n}{n-2} x^{n-2}y^2 + \ldots + \dbinom{n}{1} xy^{n-1} + y^n$.

College Algebra

Fourth Edition

Margaret L. Lial

Charles D. Miller
American River College

Scott, Foresman and Company

Glenview, Illinois
London, England

To the Student

A *Study Guide* and *Student Solutions Manual* to accompany this book are available from your local college bookstore. These books can help you study and review the course material.

Cover photo: #8140 Totem, 1981, by McCrystle Wood. Collection of Central Bank Corporation, Cincinnati, Ohio. © McCrystle Wood.

Library of Congress Cataloging in Publication Data

Lial, Margaret L.
  College algebra.

  Includes index.
  1. Algebra. I. Miller, Charles David.
II. Title.
QA154.2.L5   1985   512.9   84-20228
ISBN 0-673-18024-7

1 2 3 4 5 6-RRC-89 88 87 86 85 84

Preface

The Fourth Edition of *College Algebra* is designed for a one-semester or one-quarter course that will prepare students for work in trigonometry, statistics, finite mathematics, calculus, discrete mathematics, or courses in computer science.

We have written the book assuming that students have had an earlier course in algebra. For students whose knowledge of past courses is not as solid as it might be, we have included two chapters of review topics. In fact, Chapter 3 on equations will be considered review for some students. A diagnostic pretest covering the first three chapters is provided in the *Instructor's Guide* and can be used to determine the necessary topics for review.

KEY FEATURES

Examples   Nearly 350 worked-out examples clearly illustrate concepts and techniques. Second color is used to identify pertinent steps within examples, as well as to highlight explanatory side comments.

Applications   The text features a rich selection of applications from many different fields of study. This edition includes several new applications from business and the mathematics of finance.

Intuitive Treatment   Throughout the book, properties are stated simply and intuitively. A formal theorem-proof format has been downplayed.

EXERCISES

Graded Exercises   The range of difficulty in the exercise sets affords students ample practice with drill problems. Then they are eased gradually through problems of increasing difficulty to problems that will challenge outstanding students. More than 4800 exercises, including approximately 4450 drill problems and 350 word problems, are provided in the text.

Calculator Exercises   A large number of exercise sets feature calculator problems. These optional exercises are identified with the symbol ▤. Brief discussions of the difficulties that can arise when students use calculators also are presented in the text as appropriate.

Chapter Review Exercises   A lengthy set of review exercises is given at the end of each chapter and reviews each section of the text thoroughly. A total of about 900 review exercises provides opportunity for mastery of the material before students take an examination. In addition, a list of key words and formulas is given at the end of each chapter.

SECOND COLOR

Second color is used pedagogically in the following ways.
- Screens set off key definitions, formulas, and procedures, helping students review easily.
- Color side comments within examples explain the structure of the problem.
- For clarity, the end of each example is indicated with a color symbol, ●.

CONTENT FEATURES

The text begins with a thorough **algebra review** in Chapters 1 and 2. Operations on the real numbers, absolute value, exponents, polynomials, factoring, rational expressions, and radicals are covered.

Graphing is a key idea of the text. Starting with number lines in Chapter 1, graphing is emphasized throughout. Chapter 4 teaches the techniques of graphing in detail, and the applications of graphing to functions are discussed in Chapter 5.

The concept of **functions** is crucial in this course, and we have included all the functions topics together in Chapter 5 so students can focus on learning this material without being introduced to other new concepts at the same time.

Chapter 6 on **exponential and logarithmic functions** can be approached with tables only, with calculators only, or with a combination of both.

Polynomial and rational functions are introduced early in the book through a discussion of graphing, and zeros of polynomials are thoroughly discussed in Chapter 9.

A new section on **probability theory** has been added to the last chapter of the text. It is preceded by two sections on permutations and combinations.

SUPPLEMENTS

The **Instructor's Guide** gives a diagnostic pretest with answers, answers to the even-numbered exercises in the text, three alternative forms of chapter tests in a format that is ready for duplication, a test bank for each chapter, and answers to all test items.

A **Study Guide,** in a semiprogrammed format, includes a pretest and posttest for each chapter, plus exercises that give additional practice and reinforcement for students.

A **Student Solutions Manual** has solutions to all the odd-numbered exercises in the text. Some students may want to use this book as an additional source of examples.

Audiotapes that cover all topics in the text are available at no charge to users of the book. Students who need help with a particular topic or who have missed class find that these tapes help them master the material.

College Algebra, Fourth Edition, is part of a series of texts, including *Beginning Algebra,* Fourth Edition; *Intermediate Algebra,* Fourth Edition; *Fundamentals of College Algebra; Trigonometry,* Third Edition; and *Algebra and Trigonometry,* Third Edition. The use of a series of texts increases student understanding by offering continuity of notation, definitions, and format.

We thank the many users of the previous editions of this book who were kind enough to share their experiences with us. This revision has benefited from their comments and suggestions.

We also thank the people who reviewed all or part of the revised manuscript and gave us many helpful suggestions: James F. Connelly, Monroe Community College; Jan Elrod, Troy State University; Laurel L. Finke, Oregon State University; Richard Fritsche, Northeast Louisiana University; Louis F. Hoelzle, Bucks County Community College; Daniel A. Hogan, Hinds Junior College; Ruth Hunt, University of Missouri; Bill E. Jordan, Seminole Community College; William J. Lewis, University of Nebraska; Mary R. Muse, University of Central Arkansas; Warren G. Strickland, Del Mar College; Lowell Stultz, Kalamazoo Valley Community College; Bruce Walek, Santa Fe Community College; Glenn Wallace, Texas Southmost College; and Thomas J. Woods, Central Connecticut State University.

Our appreciation also goes to the staff at Scott, Foresman and Company, who did an excellent job in working with us toward publication.

Margaret L. Lial
Charles D. Miller

Contents

fri. 201 (handwritten)

1

Number Systems

set theory (handwritten)

2

Algebraic Expressions

scientific notation (handwritten)

9-1 Synthetic division (handwritten)

3

Equations and Inequalities

4

Graphing   *135*

5

Functions   *193*

6

Exponential and Logarithmic Functions

246

7

Matrices

290

8

Systems of Equations and Inequalities

333

9

Zeros of Polynomials

10

Sequences

11

Counting and Probability

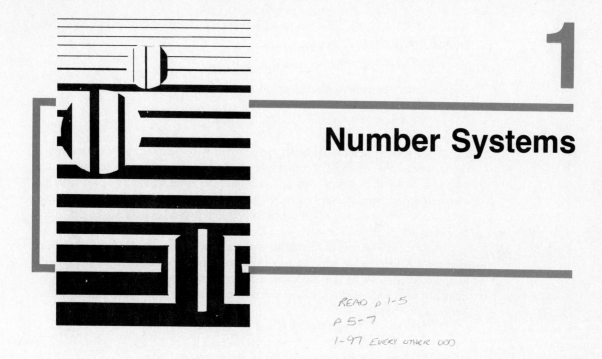

1

Number Systems

READ p 1-5
p 5-7
1-97 EVERY OTHER ODD

Today, people use algebra in fields ranging from accounting to zoology. Algebra is required in all these fields because it can be used to solve a variety of problems. Equations, graphs, and functions are important tools for people in business, science, and other areas. As preparation to discuss these topics of algebra, we will look at the properties of sets.

1.1 Sets

Think of a **set** as a collection of objects. The objects that belong to a set are called the **elements** or **members** of the set. In algebra, the elements of a set are usually numbers. Sets are commonly written using **set braces,** { }. For example, the set containing the elements 1, 2, 3, and 4 is written

$$\{1, 2, 3, 4\}.$$

Since the order in which the elements are listed is not important, this same set can also be written as $\{4, 3, 2, 1\}$ or with any other arrangement of the four numbers.

To show that 4 is an element of the set $\{1, 2, 3, 4\}$, use the symbol $\in$ and write

$$4 \in \{1, 2, 3, 4\}.$$

1

Also, $2 \in \{1, 2, 3, 4\}$. To show that 5 is *not* an element of this set, place a slash through the symbol:

$$5 \notin \{1, 2, 3, 4\}.$$

It is customary to name sets with capital letters. If S is used to name the set above, then

$$S = \{1, 2, 3, 4\}.$$

Set S was written by listing its elements. It is sometimes easier to describe a set in words. For example, set S might be described as ''the set containing the first four counting numbers.'' In this example, the notation $\{1, 2, 3, 4\}$, with the elements listed between set braces, is briefer than the verbal description. However, the set F, consisting of all fractions between 0 and 1, could not be described by listing its elements. (Try it.)

Set F is an example of an *infinite set,* one that has an unending list of distinct elements. A *finite set* is one that has a limited number of elements. Some infinite sets, unlike F, can be described by a listing process. For example, the set of numbers used for counting, called the **natural numbers,** or the **counting numbers,** can be written as

$$N = \{1, 2, 3, 4, \ldots\},$$

where the three dots show that the list of the elements of the set continues in the same way.

Sets are often written using a **variable,** a letter that is used to represent one element from a set of numbers. The set of numbers is called the **domain** of the variable. For example,

— SUCH THAT

$$\{x \mid x \text{ is a natural number between 2 and 7}\}$$

(read ''the set of all elements x such that x is a natural number between 2 and 7'') represents the set $\{3, 4, 5, 6\}$. The numbers 2 and 7 are *not* between 2 and 7. The notation used here, $\{x \mid x \text{ is a natural number between 2 and 7}\}$, is called **set-builder notation.**

EXAMPLE 1   Write the elements belonging to each of the following sets.

(a) $\{x \mid x \text{ is a counting number less than 5}\}$
The counting numbers less than 5 make up the set $\{1, 2, 3, 4\}$.

(b) $\{x \mid x \text{ is a state that touches Florida}\}$
The states touching Florida make up the set $\{\text{Alabama, Georgia}\}$.   ● *

When discussing a particular situation or problem, we can usually identify a **universal set** (whether expressed or implied) that contains all the elements ap-

---

*The symbol ● represents the end of an example.

pearing in any set used in the given problem. The letter U is used to represent the universal set.

At the other extreme from the universal set is the **null set,** or **empty set,** the set containing no elements. The set of all people twelve feet tall is an example of the null set. Write the null set in either of two ways: use the special symbol $\emptyset$ or else write set braces enclosing no elements, { }. (Be careful not to combine these symbols; $\{\emptyset\}$ is *not* the null set.)

Every element of the set $S = \{1, 2, 3, 4\}$ is a natural number. Because of this, set S is a *subset* of the set N of natural numbers, written $S \subseteq N$. By definition, set A is a **subset** of set B if every element of set A is also an element of set B. For example, if $A = \{2, 5, 9\}$ and $B = \{2, 3, 5, 6, 9, 10\}$, then $A \subseteq B$. However, there are some elements of B that are not in A, so B is not a subset of A, written $B \nsubseteq A$. By the definition, every set is a subset of itself.

Figure 1.1 shows a set A that is a subset of set B. The rectangle of the drawing represents the universal set U. Such diagrams are called **Venn diagrams.** These diagrams are used as an aid in clarifying and discussing the relationships among sets.

By the definition of subset, the null set is a subset of every set. That is,

if A is any set, then $\emptyset \subseteq A$.

Two sets A and B are **equal** whenever $A \subseteq B$ and $B \subseteq A$. In other words, $A = B$ if the two sets contain exactly the same elements. For example,

$$\{1, 2, 3\} = \{3, 1, 2\},$$

since both sets contain exactly the same elements. However,

$$\{1, 2, 3\} \neq \{0, 1, 2, 3\},$$

since the set $\{0, 1, 2, 3\}$ contains the element 0, which is not an element of $\{1, 2, 3\}$.

VENN
DIAGRAMS

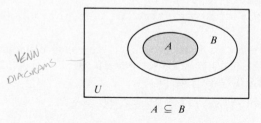

$A \subseteq B$

Figure 1.1

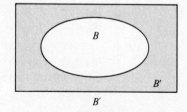

B'

Figure 1.2

Operations on Sets   Given a set A and a universal set U, the set of all elements of U that do not belong to set A is called the **complement** of set A. For example, if set A is the set of all the female students in your class, and U is the set of all students in the class, then the complement of A would be the set of all the male students in the class. The complement of set A is written A' (read "A-prime"). The Venn diagram in Figure 1.2 shows a set B. Its complement, B', is colored.

EXAMPLE 2   Let $U = \{1, 2, 3, 4, 5, 6, 7\}$, $A = \{1, 3, 5, 7\}$, and $B = \{3, 4, 6\}$. Find each of the following sets.

(a) A'

Set A' contains the elements of U that are not in A.

$$A' = \{2, 4, 6\}$$

(b) $B' = \{1, 2, 5, 7\}$

(c) $\emptyset' = U$ and $U' = \emptyset$ ●

Given two sets A and B, the set of all elements belonging both to set A and to set B is called the **intersection** of the two sets, written $A \cap B$. For example, the elements that belong to $A = \{1, 2, 4, 5, 7\}$ and $B = \{2, 4, 5, 7, 9, 11\}$ are 2, 4, 5, and 7, so

$$A \cap B = \{1, 2, 4, 5, 7\} \cap \{2, 4, 5, 7, 9, 11\} = \{2, 4, 5, 7\}.$$

The Venn diagram in Figure 1.3 shows two sets A and B; their intersection, $A \cap B$, is shown in color.

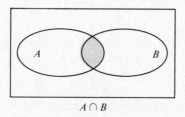

$A \cap B$

Figure 1.3

EXAMPLE 3   **(a)** $\{9, 15, 25, 36\} \cap \{15, 20, 25, 30, 35\} = \{15, 25\}$

The elements 15 and 25 are the only ones belonging to both sets.

(b) $\{2, 3, 4, 5, 6\} \cap \{1, 2, 3, 4\} = \{2, 3, 4\}$ ●

Two sets that have no elements in common are called **disjoint sets.** For example, there are no elements common to both $\{50, 51, 54\}$ and $\{52, 53, 55, 56\}$, so these two sets are disjoint, and

$$\{50, 51, 54\} \cap \{52, 53, 55, 56\} = \emptyset.$$

This result can be generalized:

if A and B are any two disjoint sets, then $A \cap B = \emptyset$.

The set of all elements belonging to set A or to set B is called the **union** of the two sets, written $A \cup B$. For example,

$$\{1, 3, 5\} \cup \{3, 5, 7, 9\} = \{1, 3, 5, 7, 9\}.$$

The Venn diagram in Figure 1.4 shows two sets A and B; the union of the sets, $A \cup B$, is shown in color.

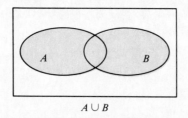

$A \cup B$

Figure 1.4

EXAMPLE 4   **(a)** Find the union of $\{1, 2, 5, 9, 14\}$ and $\{1, 3, 4, 8\}$.

Begin by listing the elements of the first set, $\{1, 2, 5, 9, 14\}$. Then include any elements from the second set that are not already listed. Doing this gives

$$\{1, 2, 5, 9, 14\} \cup \{1, 3, 4, 8\} = \{1, 2, 3, 4, 5, 8, 9, 14\}.$$

(b) $\{1, 3, 5, 7\} \cup \{2, 4, 6\} = \{1, 2, 3, 4, 5, 6, 7\}$ ●

The processes of finding the complement of a set, the intersection of two sets, and the union of two sets are called **set operations.** These operations are similar to operations on numbers, such as addition, subtraction, multiplication, or division.

The various operations on sets are summarized below.

Set Operations

> *For all sets A and B, and universal sets U:*
>
> The **complement** of set A is the set A' of all elements in the universal set that do not belong to set A.
>
> $$A' = \{x | x \in U, \quad x \notin A\}$$
>
> The **intersection** of sets A and B, written $A \cap B$, is made up of all the elements belonging to set A and set B at the same time.
>
> $$A \cap B = \{x | x \in A \text{ and } x \in B\}$$
>
> The **union** of sets A and B, written $A \cup B$, is made up of all the elements belonging to set A or to set B.
>
> $$A \cup B = \{x | x \in A \text{ or } x \in B\}$$

1.1   EXERCISES

List all the elements of each of the following sets.

1. $\{12, 13, 14, \ldots, 20\}$ **2.** $\{8, 9, 10, \ldots, 17\}$ **3.** $\{1, 1/2, 1/4, \ldots, 1/32\}$

4. $\{3, 9, 27, \ldots, 729\}$ **5.** $\{17, 22, 27, \ldots, 47\}$ **6.** $\{74, 68, 62, \ldots, 38\}$

7. $\{$all natural numbers greater than 7 and less than 15$\}$

8. $\{$all natural numbers not greater than 4$\}$

Identify the sets in Exercises 9–16 as finite *or* infinite.

9. {4, 5, 6, . . . , 15}

10. {4, 5, 6, . . .}

11. {1, 1/2, 1/4, 1/8, . . .}

12. {0, 1, 2, 3, 4, 5, . . . , 75}

13. {x|x is a natural number larger than 5}

14. {x|x is a person alive on the earth now}

15. {x|x is a fraction between 0 and 1}

16. {x|x is a positive even natural number}

Complete the blanks with either ∈ *or* ∉ *so that the resulting statement is true.*

17. 6 _____ {3, 4, 5, 6}

18. 9 _____ {3, 2, 5, 9, 8}

19. −4 _____ {4, 6, 8, 10}

20. −12 _____ {3, 5, 12, 14}

21. 0 _____ {2, 0, 3, 4}

22. 0 _____ {5, 6, 7, 8, 10}

23. {3} _____ {2, 3, 4, 5}

24. {5} _____ {3, 4, 5, 6, 7}

25. {0} _____ {0, 1, 2, 5}

26. {2} _____ {2, 4, 6, 8}

27. 0 _____ ∅

28. ∅ _____ ∅

Tell whether each statement is true *or* false.

29. 3 ∈ {2, 5, 6, 8}

30. 6 ∈ {−2, 5, 8, 9}

31. 1 ∈ {3, 4, 5, 11, 1}

32. 12 ∈ {18, 17, 15, 13, 12}

33. 9 ∉ {2, 1, 5, 8}

34. 3 ∉ {7, 6, 5, 4}

35. {2, 5, 8, 9} = {2, 5, 9, 8}

36. {3, 0, 9, 6, 2} = {2, 9, 0, 3, 6}

37. {5, 8, 9} = {5, 8, 9, 0}

38. {3, 7, 12, 14} = {3, 7, 12, 14, 0}

39. {x|x is a natural number less than 3} = {1, 2}

40. {x|x is a natural number greater than 10} = {11, 12, 13, ...}

41. {5, 7, 9, 19} ∩ {7, 9, 11, 15} = {7, 9}

42. {8, 11, 15} ∩ {8, 11, 19, 20} = {8, 11}

43. {2, 1, 7} ∪ {1, 5, 9} = {1}

44. {6, 12, 14, 16} ∪ {6, 14, 19} = {6, 14}

45. {3, 2, 5, 9} ∩ {2, 7, 8, 10} = {2}

46. {8, 9, 6} ∪ {9, 8, 6} = {8, 9}

47. {3, 5, 9, 10} ∩ ∅ = {3, 5, 9, 10}

48. {3, 5, 9, 10} ∪ ∅ = {3, 5, 9, 10}

49. {1, 2, 4} ∪ {1, 2, 4} = {1, 2, 4}

50. {1, 2, 4} ∩ {1, 2, 4} = ∅

51. ∅ ∪ ∅ = ∅

52. ∅ ∩ ∅ = ∅

Let A = {2, 4, 6, 8, 10, 12}
B = {2, 4, 8, 10}
C = {4, 10, 12}
D = {2, 10}
U = {2, 4, 6, 8, 10, 12, 14}.

Tell whether each statement is true *or* false.

53. A ⊆ U

54. C ⊆ U

55. D ⊆ B

56. D ⊆ A

57. A ⊆ B

58. B ⊆ C

59. ∅ ⊆ A

60. ∅ ⊆ ∅

61. {4, 8, 10} ⊆ B

62. {0, 2} ⊆ D

63. B ⊆ D

64. A ⊄ C

Insert ⊆ *or* ⊄ *in each blank to make the resulting statement true.*

65. {2, 4, 6} ＿＿＿ {3, 2, 5, 4, 6}

66. {1, 5} ＿＿＿ {0, −1, 2, 3, 1, 5}

67. {0, 1, 2} ＿＿＿ {1, 2, 3, 4, 5}

68. {5, 6, 7, 8} ＿＿＿ {1, 2, 3, 4, 5, 6, 7}

69. ∅ ＿＿＿ {1, 4, 6, 8}

70. ∅ ＿＿＿ ∅

Let U = {0, 1, 2, 3, 4, 5, 6, 7, 8, 9, 10, 11, 12, 13}
  M = {0, 2, 4, 6, 8}
  N = {1, 3, 5, 7, 9, 11, 13}
  Q = {0, 2, 4, 6, 8, 10, 12}
  R = {0, 1, 2, 3, 4}.

Use these sets to find each of the following. Identify any disjoint sets. See Examples 2–4.

71. $M \cap R$

72. $M \cup R$

73. $M \cup N$

74. $M \cap N$

75. $M \cap U$

76. $M \cup Q$

77. $N \cup R$

78. $U \cap N$

79. N'

80. Q'

81. $M' \cap Q$

82. $Q \cap R'$

83. $\emptyset \cap R$

84. $\emptyset \cap Q$

85. $N \cup \emptyset$

86. $R \cup \emptyset$

87. $(M \cap N) \cup R$

88. $(N \cup R) \cap M$

89. $(Q \cap M) \cup R$

90. $(N \cup R) \cap M$

91. $(M' \cup Q) \cap R$

92. $Q \cap (M \cup N)$

93. $Q' \cap (N' \cap U)$

94. $(U \cap \emptyset') \cup R$

Let U = {all students in this school}
  M = {all students taking this course}
  N = {all students taking calculus}
  P = {all students taking history}.

Describe each of the following sets in words.

95. M'

96. $M \cup N$

97. $N \cap P$

98. $N' \cap P'$

99. $M \cup P$

100. $P' \cup M'$

In the study of probability, the set of all possible outcomes for an experiment is called the sample space for the experiment. For example, the sample space for tossing an honest coin is the set {h, t}. Find the sample space for each of these exercises.

101. Rolling an honest die

102. Tossing a coin twice

103. Tossing a coin three times

104. Having three children

State the conditions on sets X and Y for which the following are true.

105. $X \cap Y = \emptyset$

106. $X \cup Y = \emptyset$

107. $X \cap Y = X$

108. $X \cup Y = X$

109. $X \cap \emptyset = X$

110. $X \cup \emptyset = X$

1.2   Real Numbers and Their Operations

In the last section we mentioned the set of natural numbers,

$$\{1, 2, 3, 4, 5, \ . \ . \ .\}.$$

The idea of counting, as suggested by the natural numbers, goes back into the mists of antiquity. Much more recent is the idea of counting *no* objects, that is, the idea of the number 0. Including 0 with the set of natural numbers gives the set of **whole numbers,**

$$\{0, 1, 2, 3, 4, 5, \ . \ . \ .\}.$$

(All these sets of numbers are summarized later in this section.)

About 500 years ago, people came up with the idea of counting backwards, from 4 to 3, to 2, to 1, to 0. There seemed no reason not to continue this process, calling the new numbers -1, -2, -3, and so on. Including these numbers with the set of whole numbers gives the very useful set of **integers,**

$$\{. \ . \ . \ , \ -4, \ -3, \ -2, \ -1, \ 0, \ 1, \ 2, \ 3, \ . \ . \ .\}.$$

The integer -7 might represent a temperature 7 degrees below zero, a debt of 7 dollars, or a loss of 7 yards in a football game. When thinking of time, the integer -7 might represent a time 7 seconds ago.

Integers can be shown pictorially with a **number line.** (A number line is similar to a thermometer on its side.) To construct a number line, draw a horizontal line and choose any point on the line to represent 0. Then choose any point to the right of 0 and label it 1. The distance from 0 to 1 sets up a unit measure that can be used to locate other points to the right of 1, which are labeled 2, 3, 4, 5, and so on, and corresponding points to the left of 0, labeled -1, -2, -3, -4, and so on. As an example, the elements of the set $\{-3, \ -1, \ 0, \ 1, \ 3, \ 5\}$ are located on the number line in Figure 1.5.

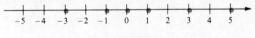

Figure 1.5

The integers 1, 2, 3, 4, and so on (those to the right of 0 on the number line) are called **positive integers,** while -1, -2, -3, and so on (to the left of 0) are **negative integers.** The integer 0 is neither positive nor negative. Positive and negative numbers are sometimes called **signed numbers.**

As you have seen in a previous algebra course, if two integers are added, subtracted, or multiplied, the answer is an integer. However, the result of dividing two integers need not be an integer. For example,

$$\frac{15}{4}, \qquad \frac{11}{7}, \qquad \text{and} \qquad \frac{28}{9}$$

are not integers. The result of dividing two integers, with the second integer not zero, is called a *rational number*.

By definition, the **rational numbers** are the elements of the set

$$\left\{ \frac{p}{q} \mid p, \ q \text{ are integers,} \quad q \neq 0 \right\}.$$

Examples of rational numbers include 3/4, −5/8, 7/2, −14/9, and so on. All integers are rational numbers, since any integer can be written as the quotient of itself and 1. For example, $6 = 6/1$, while $-8 = -8/1$, and $0 = 0/1$. The set of integers is a subset of the set of rational numbers.

Rational numbers can be located on a number line by a process of subdivision. For example, 5/8 can be located by dividing the interval from 0 to 1 into 8 equal parts, then labeling the fifth part 5/8. Several rational numbers are located on the number line in Figure 1.6.

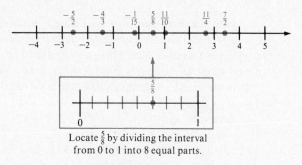

Locate $\frac{5}{8}$ by dividing the interval from 0 to 1 into 8 equal parts.

Figure 1.6

The set of all numbers that correspond to points on a number line is called the set of **real numbers.** See Figure 1.7.

While all rational numbers are real numbers, there are numbers that can be located on a number line that are not rational. Figure 1.8 shows the diagonal of a square, 1 foot on a side. By the Pythagorean formula from geometry, the length of this diagonal is

$$\sqrt{1^2 + 1^2}, \quad \text{or} \quad \sqrt{2} \ \text{ft.}$$

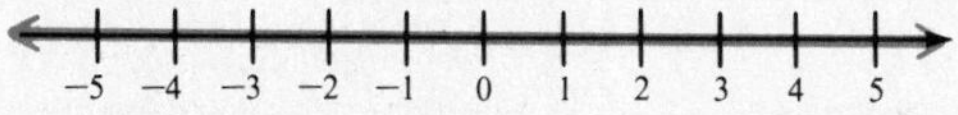

Figure 1.7

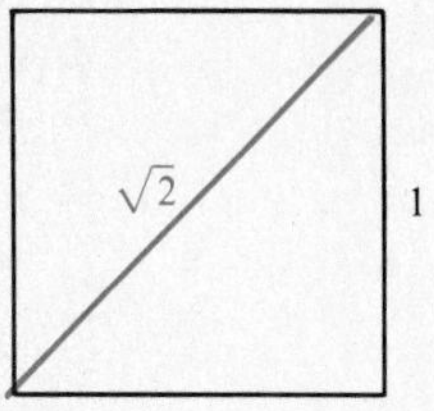

Figure 1.8

As shown in Exercise 81 on pages 14–15, $\sqrt{2}$ cannot be written as the quotient of two integers, so $\sqrt{2}$ is a real number that is not rational. A real number that is not rational is called an **irrational number.**

Other examples of irrational numbers include $\sqrt{3}$, $\sqrt{5}$ (but not $\sqrt{4}$, which equals 2), and π, which is approximately equal to 3.14159. Using a calculator or a square root table shows that $\sqrt{2} \approx 1.414$ (where $\approx$ is read "is approximately equal to") and $\sqrt{5} \approx 2.236$. Using these approximations, the numbers in the set $\{-2/3, 0, \sqrt{2}, \sqrt{5}, \pi, 4\}$ can be located on a number line as shown in Figure 1.9 below.

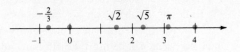

Figure 1.9

Real numbers can also be defined in another way, in terms of decimals. Using repeated subdivisions, any number could be located (at least in theory) as a point on a number line. This process would let us define the set of real numbers as the set of all decimals. Further work would show that the set of rational numbers is the set of all decimals that repeat or terminate. For example,

$$.25 \qquad\qquad = \quad 1/4$$
$$.833333. . . \qquad = \quad 5/6$$
$$.076923076923. . . \quad = \quad 1/13,$$

and so on. The set of irrational numbers is then the set of decimals that do *not* repeat or terminate. For example,

$$\sqrt{2} = 1.414213562373. . .$$
$$\pi = 3.141592654359. . . .$$

The sets of numbers discussed so far are summarized below.

Sets of Numbers

Real numbers	$\{x \mid x \text{ corresponds to a point on a number line}\}$
Integers	$\{. . . , -3, -2, -1, 0, 1, 2, 3, . . .\}$
Rational numbers	$\left\{ \dfrac{p}{q} \middle\vert p \text{ and } q \text{ are integers,} \quad q \neq 0 \right\}$
Irrational numbers	$\{x \mid x \text{ is real but not rational}\}$
Whole numbers	$\{0, 1, 2, 3, 4, . . .\}$
Natural numbers	$\{1, 2, 3, 4, . . .\}$

EXAMPLE 1   Let set $A = \{-8, -6, -3/4, 0, 3/8, 1/2, 1, \sqrt{2}, \sqrt{5}, 6, 9/0\}$.
List the elements from set A that belong to each of the following sets.

(a) The natural numbers in set A are 1 and 6.

(b) The whole numbers are 0, 1, and 6.

(c) The integers are -8, -6, 0, 1, and 6.

(d) The rational numbers are -8, -6, $-3/4$, 0, 3/8, 1/2, 1, and 6.

(e) The irrational numbers are $\sqrt{2}$ and $\sqrt{5}$.

(f) All elements of A are real numbers except 9/0. Division by 0 is not permitted, so 9/0 is not a number. (The reason that division by 0 is not permitted is explained later in this chapter.)   ●

The relationships among the various sets of real numbers are shown in Figure 1.10 below.

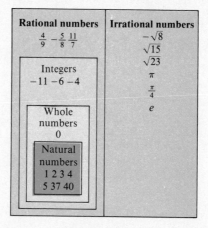

Figure 1.10

Operations on Real Numbers   There are four fundamental operations on real numbers: addition, subtraction, multiplication, and division. Addition is indicated with the symbol $+$. Subtraction is written with the symbol $-$, as in $8 - 2 = 6$. Multiplication is written in a variety of ways. All the symbols 2×8, $2 \cdot 8$, $2(8)$, and $(2)(8)$ represent the product of 2 and 8, or 16. When writing products involving variables, no operation symbols may be necessary: $2x$ represents the product of 2 and x, while xy indicates the product of x and y. Division of real numbers a and b is written $a \div b$, or more commonly, a/b.

When a problem involves more than one operation symbol, use the following **order of operations.**

Order of Operations

1 IF EXPONENTS OR ROOTS ARE INVOLVED, SIMPLIFY THEM FIRST

> *If parentheses or square brackets are present:*
>
> **1.** Work separately above and below any fraction bar.
>
> **2.** Use the rules below within each set of parentheses or square brackets. Start with the innermost and work outward.
>
> *If no parentheses are present:*
>
> **1.** Do any multiplications or divisions in the order in which they occur, working from left to right.
>
> **2.** Do any additions or subtractions in the order in which they occur, working from left to right.

1. If exponents or roots are involved, simplify the exponents or roots first, and then use the order of operations.

EXAMPLE 2 Use the order of operations given above to simplify each of the following.

(a) $6 \div 3 + 2 \cdot 4 = 2 + 2 \cdot 4 = 2 + 8 = 10$

(b)
$$(8 + 6) \div 7 \cdot 3 + (-6) = 14 \div 7 \cdot 3 + (-6)$$
$$= 2 \cdot 3 + (-6)$$
$$= 6 + (-6)$$
$$= 0$$

(c)
$$\frac{-9(-3) + (-5)}{2(-8) - 5(3)} = \frac{27 + (-5)}{-16 - 15}$$
$$= \frac{22}{-31} = -\frac{22}{31} \quad \bullet$$

EXAMPLE 3 Use the order of operations to evaluate each expression if $x = -2$, $y = 5$, and $z = -3$.

(a) $8x - 7y + 4z$

Replace x with -2, y with 5, and z with -3.
$$8x - 7y + 4z = 8(-2) - 7(5) + 4(-3)$$
$$= -16 - 35 - 12$$
$$= -63$$

(b)
$$\frac{9(x - 5) + 4y}{z + 4} = \frac{9(-2 - 5) + 4(5)}{-3 + 4}$$
$$= \frac{9(-7) + 20}{1}$$
$$= -63 + 20$$
$$= -43 \quad \bullet$$

1.2   EXERCISES

Let set $B = \{-6, -12/4, -5/8, -\sqrt{3}, 0, 1/4, 1, \pi/2, 3, \sqrt{12}\}$. List all the elements of B that are elements of the following sets.

1. Natural numbers   **2.** Whole numbers   **3.** Integers

4. Rational numbers   **5.** Irrational numbers   **6.** Real numbers

For Exercises 7–18, choose all words from the following list that apply: (a) natural number, (b) whole number, (c) integer, (d) rational number, (e) irrational number, (f) real number, (g) meaningless. See Example 1.

7. 12   **8.** 0   **9.** -9

10. 3/4   **11.** $-5/9$ **12.** $\sqrt{8}$

13. $-\sqrt{2}$ **14.** $\sqrt{25}$ **15.** $-\sqrt{36}$

16. 8/0   **17.** $-9/(0-0)$ **18.** $0/(3+3)$

Write each of the following sets by listing their elements between set braces.

19. $\{x | x$ is a natural number between 3 and 9$\}$ **20.** $\{x | x$ is a natural number less than 7$\}$

21. $\{x | x$ is a whole number between -3 and 4$\}$ **22.** $\{x | x$ is an integer between -3 and 5$\}$

23. $\{x | x$ is a positive odd integer greater than 10$\}$ **24.** $\{x | x$ is a negative integer less than $-6\}$

Use the order of operations to evaluate each of the following. See Example 2.

25. $-9 + (-4) - (-12)$ **26.** $8 - (-4) + 11$

27. $16 - (-9) + (-4)$ **28.** $-15 - 3 - (-8)$

29. $-1 - [-8 - (-12)]$ **30.** $-4 - [-7 - 9]$

31. $[(-2) + (-3)] - [-4 - (-1)]$ **32.** $[(-6) - (15)] + [-8 - (-2)]$

33. $-4(9) - 8(-7) + 2$ **34.** $6(-5) - (-3)(2) - 7$

35. $(-1 - 7)(2) - 4(-1 - 3)$ **36.** $(-6 + 11)(-4) - 3(8 - 9)$

37. $[4 + (-8)][-2 - (-5)](-2)$ **38.** $[-7 - (-9)][4 - (-3)](-1)$

39. $\dfrac{1}{2} - \left(\dfrac{2}{3} - \dfrac{5}{6}\right)$ **40.** $\dfrac{3}{4} - \left(\dfrac{7}{8} - \dfrac{5}{12}\right)$

41. $\left(-\dfrac{2}{9} - \dfrac{1}{4}\right) - \left[-\dfrac{5}{18} - \left(-\dfrac{1}{2}\right)\right]$ **42.** $\left[-\dfrac{5}{8} - \left(-\dfrac{2}{5}\right)\right] - \left(\dfrac{3}{2} - \dfrac{11}{10}\right)$

43. $\dfrac{-8 + (-4)(-6) \div 12}{4 - (-3)}$ **44.** $\dfrac{15 \div 5 \cdot 4 \div 6 - 8}{-6 - (-5) - 8 \div 2}$

45. $\dfrac{17 \div (3 \cdot 5 + 2) \div 8}{-6 \cdot 5 - 3 - 3(-11)}$ **46.** $\dfrac{-12(-3) + (-8)(-5) + (-6)}{17(-3) + 4 \cdot 8 + (-7)(-2) - (-5)}$

47. $\dfrac{6 - 5\left(\dfrac{-1 - 2}{3}\right) + (-2)}{-8 + 7 - 6 + (-2)}$ **48.** $\dfrac{-9\left(\dfrac{-3 - 5 - 2}{-1 + (-2)}\right) - 6(-2)}{-3 - [-5 - (-2)] - 7}$

49. $\dfrac{-9.23(5.87) + 6.993}{1.225(-8.601) - 148(.0723)}$

50. $\dfrac{189.4(3.221) - 9.447(-8.772)}{4.889[3.117 - 8.291(3.427)]}$

Evaluate each of the following if $p = -2$, $q = 4$, and $r = -5$. See Example 3.

51. $7p - 4q + 10$

52. $8q - r + 3p$

53. $-3(p + 5q)$

54. $2(q - r)$

55. $\dfrac{q + r}{q + p}$

56. $\dfrac{3q}{3p - 2r}$

57. $\dfrac{8q - 3r}{q + r + 1}$

58. $\dfrac{8r + q}{q + 2p}$

59. $\dfrac{\dfrac{q}{4} - \dfrac{r}{5}}{\dfrac{p}{2} + \dfrac{q}{2}}$

60. $\dfrac{\dfrac{3r}{10} - \dfrac{5p}{2}}{q + \dfrac{2r}{5}}$

61. $\dfrac{\dfrac{1}{p} + \dfrac{1}{r}}{\dfrac{1}{p} - \dfrac{1}{q}}$

62. $\dfrac{\dfrac{2}{p + 1} + \dfrac{1}{q - 2}}{\dfrac{1}{r + 2} - \dfrac{2}{q + 2}}$

Evaluate each of the following if $a = -4.706$, $b = 1.983$, and $c = -.942$. Round all answers to the nearest thousandth. See Example 3.

63. $.072a + .123b$

64. $.946c - .212a$

65. $.498b + \dfrac{c}{.273}$

66. $1.426(3a - 8b)$

67. $\dfrac{3.091c + .769a}{.842b + 5}$

68. $\dfrac{.309b + .574c}{.9a - 2}$

Let N represent the set of natural numbers, W the set of whole numbers, I the integers, F the rational numbers, H the irrational numbers, and let the universal set be R, the real numbers. Complete each of the following statements with an appropriate symbol: $\subseteq$, $=$, $\cap$, or $\cup$. There may be more than one correct answer.

69. N _____ I

70. N _____ F

71. N _____ $W = W$

72. N _____ $R = R$

73. I _____ $R = I$

74. F _____ $H = \emptyset$

75. F _____ $H = R$

76. $\emptyset$ _____ N

77. F' _____ H

78. I _____ $(N \cup I)$

79. W _____ $W = W$

80. R _____ $R = R$

81. This exercise is designed to show that $\sqrt{2}$ is irrational. Give a reason for each of steps (a) through (i).

There are two possibilities:

(i) A rational number $\dfrac{a}{b}$ exists such that $\left(\dfrac{a}{b}\right)^2 = 2$.

(ii) There is no such rational number.

We work with assumption (i). If it leads to a contradiction, then we will know that (ii) must be correct. Start by assuming that a rational number $\dfrac{a}{b}$ exists such that $\left(\dfrac{a}{b}\right)^2 = 2$. Assume also that $\dfrac{a}{b}$ is written in lowest terms.

(a) Since $\left(\dfrac{a}{b}\right)^2 = 2$, we must have $\dfrac{a^2}{b^2} = 2$, or $a^2 = 2b^2$.

(b) $2b^2$ is an even number.

---

*The symbol ▦ identifies problems that can be solved more quickly with the help of a calculator.

(c) Therefore, a^2, and a itself, must be even numbers.

(d) Since a is an even number, it must be a multiple of 2. That is, we can find a natural number c such that $a = 2c$. This changes $a^2 = 2b^2$ into $(2c)^2 = 2b^2$.

(e) Therefore, $4c^2 = 2b^2$ or $2c^2 = b^2$.

(f) $2c^2$ is an even number.

(g) This makes b^2 an even number, so that b must be even.

(h) We have reached a contradiction. Show where the contradiction occurs.

(i) Since assumption (i) lead to a contradiction, we are forced to accept assumption (ii), which says that $\sqrt{2}$ is irrational.

82. Go through steps similar to those in Exercise 81 to show that $\sqrt{3}$ is irrational.

1.3 Properties of the Real Numbers

We have discussed various sets of numbers and the operations on these numbers. In this section we will discuss some of the properties of these numbers and operations. A summary of all the properties is given at the end of the section. The first of these properties, the *commutative property,* says that two numbers may be added or multiplied in any order:

$$4 + (-12) = -12 + 4, \qquad 8(-5) = -5(8),$$
$$-9 + (-1) = -1 + (-9), \qquad (-6)(-3) = (-3)(-6),$$

and so on. Generalizing, the **commutative property** says that for all real numbers a and b,

$$a + b = b + a \qquad \text{and} \qquad ab = ba.$$

EXAMPLE 1   The following statements illustrate the commutative property. Notice that the order of the numbers changes from one side of the equals sign to the other.

(a) $6 + x = x + 6$

(b) $(6 + x) + 9 = (x + 6) + 9$

(c) $(6 + x) + 9 = 9 + (6 + x)$

(d) $5 \cdot (9 \cdot 8) = (9 \cdot 8) \cdot 5$

(e) $5 \cdot (9 \cdot 8) = 5 \cdot (8 \cdot 9)$ ●

By the **associative property,** if three numbers are to be added or multiplied, either the first two numbers or the last two may be "associated." For example, the sum of the three numbers -9, 8, and 7 may be found in either of two ways:

$$-9 + (8 + 7) = -9 + 15 = 6,$$
or
$$(-9 + 8) + 7 = -1 + 7 = 6.$$

In summary, the associative property says that for all real numbers a, b, and c,

$$(a + b) + c = a + (b + c) \qquad \text{and} \qquad (ab)c = a(bc).$$

EXAMPLE 2   The following statements illustrate the associative property. Here the order of the numbers does not change, but the placement of parentheses (the association of the numbers) does change.

(a) $4 + (9 + 8) = (4 + 9) + 8$

(b) $3(9x) = (3 \cdot 9) x$

(c) $(\sqrt{3} + \sqrt{7}) + 2\sqrt{6} = \sqrt{3} + (\sqrt{7} + 2\sqrt{6})$ ●

It is a common error to confuse the associative and commutative properties. To avoid this error, check the order of the terms: the order changes from one side of the equals sign to the other with the commutative property, while the order does not change with the associative property.

EXAMPLE 3   This example shows a list of statements using the same symbols. Notice the difference between the commutative and associative properties.

Commutative property	*Associative property*
$(x + 4) + 9 = (4 + x) + 9$	$(x + 4) + 9 = x + (4 + 9)$
$7 \cdot (5 \cdot 2) = (5 \cdot 2) \cdot 7$	$7 \cdot (5 \cdot 2) = (7 \cdot 5) \cdot 2$ ●

EXAMPLE 4   Simplify each expression, using the commutative and associative properties as needed.

(a) $6 + (9 + x) = (6 + 9) + x = 15 + x$ Associative property

(b) $\dfrac{5}{8}(16y) = \left(\dfrac{5}{8} \cdot 16\right)y = 10y$ Associative property

(c) $(-10p)\left(\dfrac{6}{5}\right) = \dfrac{6}{5}(-10p)$ Commutative property

$= \left[\dfrac{6}{5}(-10)\right]p$ Associative property

$= -12p$ ●

The **identity property** shows special properties of the numbers 0 and 1. The sum of 0 and any real number a is a itself:

$$0 + 4 = 4, \qquad -5 + 0 = -5,$$

and so on. The number 0 preserves the identity of a number under addition, making 0 the **identity element for addition.**

The number 1, the **identity element for multiplication,** preserves the identity of a number under multiplication, since the product of 1 and any number a is a:

$$5 \cdot 1 = 5, \qquad 1\left(-\dfrac{2}{3}\right) = -\dfrac{2}{3},$$

and so on. In summary, the identity property says that for every real number a, there exists a unique real number 0 such that

$$a + 0 = a \quad \text{and} \quad 0 + a = a,$$

and there exists a unique real number 1 such that

$$a \cdot 1 = a \quad \text{and} \quad 1 \cdot a = a.$$

EXAMPLE 5 The following statements illustrate the identity property.

(a) $-12 \cdot 1 = -12$

(b) $\dfrac{8}{5} + 0 = \dfrac{8}{5}$

(c) $\pi \cdot 1 = \pi \quad \text{and} \quad \pi + 0 = \pi$ ●

The sum of the numbers 5 and -5 is the identity element for addition, 0, just as the sum of $-2/3$ and $2/3$ is 0. In fact, for any real number a there is a real number, written $-a$, such that the sum of a and $-a$ is the identity element 0, or

$$a + (-a) = 0.$$

The number $-a$ is the **additive inverse** or **negative** of a.

Do not confuse the *negative of a number* with a *negative number*. Since a is a variable, it can represent either a positive or a negative number. The negative of a, written $-a$, can also be either a negative or a positive number (or zero). Do not make the common mistake of thinking that $-a$ *must* be a negative number. For example, if a is -3, then $-a$ is $-(-3) = 3$.

For each real number a (except 0) there is a real number $1/a$ such that the product of a and $1/a$ is the identity element for multiplication, 1, or

$$a \cdot \frac{1}{a} = 1 \quad \text{and} \quad \frac{1}{a} \cdot a = 1 \quad (a \neq 0).$$

The number $1/a$ is called the **multiplicative inverse** or **reciprocal** of the number a. Every real number except 0 has a reciprocal.

The existence of $-a$ and of $1/a$ comes from the **inverse property:** for any real number a, there exists a unique real number $-a$ such that

$$a + (-a) = 0 \quad \text{and} \quad -a + a = 0,$$

and for any nonzero real number a, there exists a unique real number $1/a$ such that

$$a \cdot \frac{1}{a} = 1 \quad \text{and} \quad \frac{1}{a} \cdot a = 1.$$

EXAMPLE 6   The following statements illustrate the inverse property.

(a) $9 + (-9) = 0$

(b) $-15 + 15 = 0$

(c) $\dfrac{4}{3} \cdot \dfrac{3}{4} = 1$

(d) $-8 \cdot \left(\dfrac{1}{-8}\right) = 1$

(e) $\dfrac{1}{\sqrt{5}} \cdot \sqrt{5} = 1$

(f) There is no real number x such that $0 \cdot x = 1$. In fact, 0 is the only real number without an inverse for multiplication.   ●

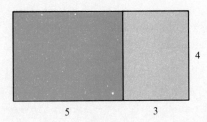

Figure 1.11

Distributive Property   The area of the entire region shown in Figure 1.11 can be found in two ways. One way is to multiply the length of the base of the entire region, or $5 + 3 = 8$, by the width of the region:

$$4(5 + 3) = 4(8) = 32.$$

Another way to find the area of the region is to add the areas of the smaller rectangles on the left and right:

$$4(5) = 20 \qquad \text{and} \qquad 4(3) = 12,$$

with total area

$$4(5) + 4(3) = 20 + 12 = 32,$$

the same result. The equal results from finding the area in two ways,

$$4(5 + 3) = 4(5) + 4(3),$$

illustrate the **distributive property,** which says that for all real numbers a, b, and c,

$$a(b + c) = ab + ac.$$

The distributive property is one of the key properties of the real numbers. This property is used to change products to sums and sums to products.

EXAMPLE 7   The following statements illustrate the distributive property.

(a) $9(6 + 4) = 9 \cdot 6 + 9 \cdot 4$

(b) $3(x + y) = 3x + 3y$

(c) $-(m - 4n) = -1 \cdot (m - 4n) = -m + 4n$

(d) $7p + 21 = 7p + 7 \cdot 3 = 7(p + 3)$ ●

The distributive property can be extended to include more than two numbers in the sum. For example,

$$9(5x + y + 4z) = 9(5x) + 9y + 9(4z)$$
$$= 45x + 9y + 36z.$$

Now we can summarize the properties of the real numbers.

Properties of the Real Numbers

For all real numbers a, b, and c:

Commutative property
$a + b = b + a$
$ab = ba$

Associative property
$(a + b) + c = a + (b + c)$
$(ab)c = a(bc)$

Identity property
There exists a unique real number 0 such that
$$a + 0 = a \quad \text{and} \quad 0 + a = a.$$
There exists a unique real number 1 such that
$$a \cdot 1 = a \quad \text{and} \quad 1 \cdot a = a.$$

Inverse property
There exists a unique real number $-a$ such that
$$a + (-a) = 0 \quad \text{and} \quad (-a) + a = 0.$$
If $a \neq 0$, there exists a unique real number $1/a$ such that
$$a \cdot \frac{1}{a} = 1 \quad \text{and} \quad \frac{1}{a} \cdot a = 1.$$

Distributive property
$a(b + c) = ab + ac$

1.3   EXERCISES

Tell whether each statement is true *or* false. *If false, tell why.*

1. $8 + (-2) = -2 + 8$ illustrates the commutative property.

2. $6 + 4 = 4 + 6$ illustrates the associative property.

3. $(-9 + 6) + 8 = -9 + (6 + 8)$ illustrates the associative property.

4. There exists a real number b such that $15 + b = 0$.

5. By the identity property, $8 \cdot 0 = 0$.

6. By the identity property, $8 + 1 = 9$.

7. The set of natural numbers contains an identity element for addition.

8. The set of even integers satisfies the inverse property for addition.

9. By the associative property, $6 + (5 + y) = (6 + 5) + y$.

10. By the associative property, $(8 \cdot 5)z = z(5 \cdot 8)$.

11. Using the distributive property, $4 + (7 \cdot 9) = (4 + 7) \cdot (4 + 9)$.

12. Using the commutative property, $8 - 15 = 15 - 8$.

13. Using the commutative property, $-7 + 4 = 4 - 7$.

14. Using the associative property, $(19 - 3) - 1 = 19 - (3 - 1)$.

15. By the identity property, $2 + 1/2 = 5/2$.

16. By the inverse property, $2 \cdot 1/2 = 1$.

Identify the properties that are illustrated in each of the following statements. Some will require more than one property. Assume that all variables represent real numbers. See Examples 1–3 and 5–7.

17. $8 \cdot 9 = 9 \cdot 8$

18. $3 + (-3) = 0$

19. $0 + (-7) = (-7) + 0$

20. $-7 + 0 = -7$

21. $8 + (12 + 6) = (8 + 12) + 6$

22. $[9(-3)] \cdot 2 = 9[(-3) \cdot 2]$

23. $6 \cdot 12 + 6 \cdot 15 = 6(12 + 15)$

24. $8(m + 4) = 8m + 8 \cdot 4$

25. $(x + 6) \cdot \left(\dfrac{1}{x + 6} \right) = 1$ (if $x + 6 \neq 0$)

26. $\dfrac{2 + m}{2 - m} \cdot \dfrac{2 - m}{2 + m} = 1$ (if $m \neq 2$, and $m \neq -2$)

27. $(7 - y) + 0 = 7 - y$

28. $[9 + (-9)] \cdot 5 = 5 \cdot 0$

29. $8(6 + 4) = 8 \cdot 4 + 8 \cdot 6$

30. $(-5 + 7)(3 + 4) = (-5 + 7) \cdot 3 + (-5 + 7) \cdot 4$

31. $x \cdot \dfrac{1}{x} + x \cdot \dfrac{1}{x} = x \left(\dfrac{1}{x} + \dfrac{1}{x} \right)$ (if $x \neq 0$)

32. $(-5 + 7) \cdot 3 + (-5 + 7) \cdot 4 = (-5) \cdot 3 + 7 \cdot 3 + (-5) \cdot 4 + 7 \cdot 4$

Use the distributive property to rewrite sums as products and products as sums. See Example 7.

33. $6(x + y)$

34. $9(r - s)$

35. $-3(z - y)$

36. $-2(m + n)$

37. $-(8r - k)$

38. $-(-2 + 5m)$

39. $a(r + s - t)$

40. $p(q - w + x)$

41. $8r - 16$

42. $12z - 36$

43. $mn + mp$

44. $xy - xq$

Use the various properties of real numbers to simplify each of the following expressions.
See Example 4.

45. $9(4m)$

46. $-6(-3r)$

47. $\dfrac{5}{3}(12y)$

48. $\dfrac{10}{11}(22z)$

49. $\left(\dfrac{3}{4}r\right)(-12)$

50. $\left(-\dfrac{5}{8}p\right)(-24)$

51. $\dfrac{2}{3}(12y - 6z + 18q)$

52. $-\dfrac{1}{4}(20m + 8y - 32z)$

53. $-\dfrac{7}{10}(40p - 10m + 50)$

54. $-\dfrac{5}{7}(14 - 28r + 56z)$

55. $\dfrac{3}{8}\left(\dfrac{16}{9}y + \dfrac{32}{27}z - \dfrac{40}{9}\right)$

56. $\dfrac{2}{3}\left(\dfrac{9}{4}a + \dfrac{15}{14}b - \dfrac{27}{20}\right)$

Give numerical examples to show that each of the following is false.

57. "Commutative property of subtraction," $a - b = b - a$

58. "Commutative property of division," $a/b = b/a$

59. "Associative property of subtraction," $(a - b) - c = a - (b - c)$

60. "Associative property of division," $(a/b)/c = a/(b/c)$

61. We know that $a(b + c) = ab + ac$. This property is the distributive property of multiplication over addition. Is there a distributive property of addition over multiplication? That is, does

$$a + (b \cdot c) = (a + b)(a + c)$$

for all real numbers a, b, and c? To find out, try various sample values of a, b, and c.

62. Is there an identity property for subtraction? That is, does a real number k exist such that

$$a - k = a \quad \text{and} \quad k - a = a$$

for every real number a? How about an identity property for division?

1.4   Order and Absolute Value

In this section we will study the order relationship between real numbers. (This lets us decide which of two given numbers is smaller.) Figure 1.12 shows a number line with the points corresponding to several different numbers marked on the line. A number that corresponds to a particular point on a line is called the **coordinate** of the point. For example, the leftmost marked point in Figure 1.12 has coordinate -4. The correspondence between points on a line and the real numbers is called a **coordinate system** for the line. (From now on, the phrase "the point on a number line with coordinate a" will be abbreviated as "the point with coordinate a," or simply "the point a.")

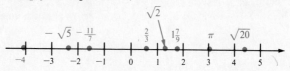

Figure 1.12

If the real number a is to the left of the real number b on a number line, then *a is less than b,* written $a < b$. If a is to the right of b, then *a is greater than b,* written $a > b$. (Remember that the "point" of the symbol goes toward the smaller number.)

As an alternative to this geometric definition of "is less than" or "is greater than," we could give an algebraic definition: if a and b are two real numbers and if the difference $a - b$ is positive, then $a > b$. If $a - b$ is negative, then $a < b$. We can summarize the geometric and algebraic statements of order as follows.

Statement	Geometric form	Algebraic form
$a > b$	a is to the right of b	$a - b$ is positive
$a < b$	a is to the left of b	$a - b$ is negative

[handwritten: $-a =$ NEGAIVE OF A OR OPPOSITE OF A]

EXAMPLE 1

Part (a) of this example shows how to identify the smaller of two numbers with the geometric approach, while part (b) uses the algebraic approach.

(a) In Figure 1.12, $-\sqrt{5}$ is to the left of 2/3, so

$$-\sqrt{5} < \frac{2}{3}.$$

Since 2/3 is to the right of $-\sqrt{5}$,

$$\frac{2}{3} > -\sqrt{5}.$$

Figure 1.12 also suggests that $2 < \pi$, while $\pi < \sqrt{20}$.

(b) The difference $2/3 - (-11/7)$ is positive, showing that

$$\frac{2}{3} > -\frac{11}{7}.$$

The difference $-11/7 - 2/3$ is negative, showing that

$$-\frac{11}{7} < \frac{2}{3}. \quad \bullet$$

The following variations on $<$ and $>$ are often used.

Symbol	Meaning
$\leq$	is less than or equal to
$\geq$	is greater than or equal to
$\not<$	is not less than
$\not>$	is not greater than

Statements involving these symbols, as well as $<$ and $>$, are called **inequalities.**

EXAMPLE 2   The list below shows several statements and the reason that each is true.

Statement	Reason
$8 \le 10$	$8 < 10$
$8 \le 8$	$8 = 8$
$-9 \ge -14$	$-9 > -14$
$-8 \not> -2$	$-8 < -2$
$4 \not< 2$	$4 > 2$ ●

The expression $a < b < c$ says that b is *between* a and $c,$ since

$$a < b < c$$

means $a < b$ and $b < c.$

In the same way, $a \le b \le c$

means $a \le b$ and $b \le c.$

When writing these "between" statements, make sure that both inequality symbols point in the same direction. For example, both $2 < 7 < 11$ and $5 > 4 > -1$ are true statements, but $3 < 5 > 2$ is meaningless.

The following **properties of order** give the basic properties of $<$ and $>$.

Properties of Order

For all real numbers $a,$ $b,$ and $c:$

Transitive property   If $a < b$ and $b < c,$ then $a < c.$

Addition property   If $a < b,$ then $a + c < b + c.$

Multiplication property   If $a < b,$ and if $c > 0,$ then $ac < bc.$
      If $a < b,$ and if $c < 0,$ then $ac > bc.$

EXAMPLE 3   **(a)** By the transitive property, if $3z < k$ and $k < p,$ then $3z < p.$

(b) By the addition property, any real number can be added to both sides of an inequality. For example, adding -2 to both sides of

$$x + 2 < 5$$

gives

$$x + 2 + (-2) < 5 + (-2)$$
$$x < 3.$$

This process is explained in more detail in Chapter 3.

(c) While any number may be *added* to both sides of an inequality, more care must be used when *multiplying* both sides by a number. For example, multiplying both sides of

$$\frac{1}{2}x < 5$$

by 2 gives

$$2 \cdot \frac{1}{2}x < 2 \cdot 5$$

$$x < 10,$$

by the first part of the multiplication property. On the other hand, multiplying both sides of

$$-\frac{3}{5}r \geq 9$$

by $-\frac{5}{3}$ gives

$$-\frac{5}{3}\left(-\frac{3}{5}r\right) \leq -\frac{5}{3} \cdot 9$$

$$r \leq -15.$$

Here the $\geq$ symbol was changed to $\leq$, using the second part of the multiplication property. •

Absolute Value The distance on the number line from a number to 0 is called the **absolute value** of that number. The absolute value of the number a is written $|a|$. For example, the distance on the number line from 9 to 0 is 9, as is the distance from -9 to 0. (See Figure 1.13.) Therefore,

$$|9| = 9 \qquad \text{and} \qquad |-9| = 9.$$

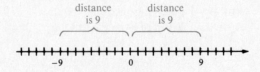

Figure 1.13

EXAMPLE 4   **(a)** $|-4| = 4$

(b) $|2\pi| = 2\pi$

(c) $\left| -\dfrac{5}{8} \right| = \dfrac{5}{8}$

(d) $-|8| = -(8) = -8$

(e) $-|-2| = -(2) = -2$ •

The definition of the absolute value of the real number a can be stated as follows.

Absolute Value

$$|a| = \begin{cases} a \text{ if } a \geq 0 \\ -a \text{ if } a < 0 \end{cases}$$

The second part of this definition requires some thought. If a is a negative number, that is, if $a < 0$, then $-a$ is positive. Thus, for a *negative* number a,

$$|a| = -a.$$

For example, if $a = -5$, then $|a| = |-5| = -(-5) = 5$.

EXAMPLE 5

Write each of the following without absolute value bars.

(a) $|-8 + 2|$

Since $-8 + 2 = -6$, $|-8 + 2| = |-6| = 6$.

(b) $|\sqrt{5} - 2|$

Since $\sqrt{5} > 2$ (see the square root table or use a calculator), $\sqrt{5} - 2 > 0$, and $|\sqrt{5} - 2| = \sqrt{5} - 2$.

(c) $|\pi - 4|$

We know $\pi < 4$, so that $\pi - 4 < 0$, and

$$\begin{aligned} |\pi - 4| &= -(\pi - 4) \\ &= -\pi + 4 \\ &= 4 - \pi. \end{aligned}$$

(d) $|m - 2|$ if $m < 2$

If $m < 2$, then $m - 2 < 0$, so

$$\begin{aligned} |m - 2| &= -(m - 2) \\ &= -m + 2 \quad \text{or} \quad 2 - m. \end{aligned}$$ •

The definition of absolute value can be used to prove the properties of absolute value listed at the top of page 26.

Properties of Absolute Value

> For all real numbers a and b,
>
> $$|a| \geq 0 \qquad\qquad |-a| = |a| \qquad\qquad |a| \cdot |b| = |ab|$$
>
> $$\left|\frac{a}{b}\right| = \frac{|a|}{|b|} \quad (b \neq 0)$$
>
> $$|a + b| \leq |a| + |b| \quad \text{the triangle inequality.}$$

1.4 EXERCISES

Write the following numbers in numerical order, from smallest to largest. Use the square root table in the Appendix or a calculator, as necessary.

1. $-9, -2, 3, -4, 8$

2. $7, -6, 0, -2, -3$

3. $|-8|, -|9|, -|-6|$

4. $-|-9|, -|7|, -|-2|$

5. $\sqrt{8}, -4, -\sqrt{3}, -2, -5, \sqrt{6}, 3$

6. $\sqrt{2}, -1, 4, 3, \sqrt{8}, -\sqrt{6}, \sqrt{7}$

7. $3/4, \sqrt{2}, 7/5, 8/5, 22/15$

8. $-9/8, -3, -\sqrt{3}, -\sqrt{5}, -9/5, -8/5$

9. $|-8 + 2|, -|3|, -|-2|, -|-2| + (-3), -|-8| - |-6|$

10. $-2 -|-4|, -3 + |2|, -4, -5 + |-3|$

Let $x = -4$ and $y = 2$. Evaluate each of the following.

11. $|2x|$

12. $|-3y|$

13. $|x - y|$

14. $|2x + 5y|$

15. $|3x + 4y|$

16. $|-5y + x|$

17. $|-4x + y| - |y|$

18. $|-8y + x| - |x|$

19. $\dfrac{|x| + 2|y|}{5 + x}$

20. $\dfrac{2|x| - |y + 2|}{|x| \cdot |y|}$

21. $\dfrac{x|-2 + y + |x||}{|x + 3|}$

22. $\dfrac{(y - 3)|5 - |x| + 2|y + 4||}{2 - |y - 1|}$

Write each of the following without absolute value bars. See Example 5.

23. $|-6|$

24. $-|-2|$

25. $-|-8| + |-2|$

26. $3 - |-4|$

27. $|8 - \sqrt{50}|$

28. $|2 - \sqrt{3}|$

29. $|\sqrt{7} - 5|$

30. $|\sqrt{2} - 3|$

31. $|\pi - 3|$

32. $|\pi - 5|$

33. $|x - 4|$, if $x > 4$

34. $|y - 3|$, if $y < 3$

35. $|2k - 8|$, if $k < 4$

36. $|3r - 15|$, if $r > 5$

37. $|7m - 56|$, if $m < 4$

38. $|2k - 7|$, if $k > 4$

39. $|-8 - 4m|$, if $m > -2$

40. $|6 - 5r|$, if $r < -2$

41. $|x - y|$, if $x < y$

42. $|x - y|$, if $x > y$

43. $|3 + x^2|$

44. $|x^2 + 4|$

45. $|-1 - p^2|$

46. $|-r^4 - 16|$

Multiply both sides of the given inequalities by the indicated number. See Example 3(c).

47. $-2 < 5, \quad 3$

48. $-7 < 1, \quad 2$

49. $-6 < -4, \quad -5$

50. $-9 < -8, \quad -4$

51. $-11 \geq -22, \quad -1/11$

52. $-2 \geq -14, \quad -1/2$

53. $3x < 9$, $1/3$ **54.** $5x > 45$, $1/5$ **55.** $-9k > 63$, $-1/9$

56. $-3p \leq 24$, $-1/3$ **57.** $(-2/5)\,k < -10$, $-5/2$ **58.** $(-3/5)\,a \geq -9$, $-5/3$

Justify each of the following statements by giving the correct property from this section. Assume that all variables represent real numbers and that no denominators are zero. See Example 3.

59. If $2k < 8$, then $k < 4$.

60. If $x + 8 < 15$, then $x < 7$.

61. If $-4x < 24$, then $x > -6$.

62. If $x < 5$ and $5 < m$, then $x < m$.

63. If $m > 0$, then $9m > 0$.

64. If $k > 0$, then $8 + k > 8$.

65. $|8 + m| \leq |8| + |m|$

66. $|k - m| \leq |k| + |-m|$

67. $|8| \cdot |-4| = |-32|$

68. $|12 + 11r| \geq 0$

69. $\left|\dfrac{-12}{5}\right| = \dfrac{|-12|}{|5|}$

70. $\left|\dfrac{6}{5}\right| = \dfrac{|6|}{|5|}$

Under what conditions are the following statements true?

71. $|x| = |y|$

72. $|x + y| = |x| + |y|$

73. $|x + y| = |x| - |y|$

74. $|x| \leq 0$

75. $|x - y| = |x| - |y|$

76. $||x + y|| = |x + y|$

Evaluate each of the following if x is a nonzero real number.

77. $\dfrac{|x|}{x}$

78. $\left|\dfrac{|x|}{x}\right|$

Chapter 1   Summary

Key Words			
	set	elements	members
	set braces	natural numbers	counting numbers
	variable	domain	set-builder notation
	universal set	null set	Venn diagram
	empty set	subset	disjoint sets
	complement	intersection	coordinate
	union	number line	inequalities
	absolute value		

Sets of Numbers		
Real numbers	$\{x \mid x$ corresponds to a point on a number line$\}$
Integers	$\{\ldots,\, -3,\, -2,\, -1,\, 0,\, 1,\, 2,\, 3,\, \ldots\}$
Rational numbers	$\left\{\dfrac{p}{q} \;\middle	\; p \text{ and } q \text{ are integers}, \quad q \neq 0\right\}$
Irrational numbers	$\{x \mid x$ is real but not rational$\}$
Whole numbers	$\{0,\, 1,\, 2,\, 3,\, 4,\, \ldots\}$
Natural numbers	$\{1,\, 2,\, 3,\, 4,\, \ldots\}$

Properties of the Real Numbers

For all real numbers a, b, and c:

Commutative property

$$a + b = b + a$$
$$ab = ba$$

Associative property

$$(a + b) + c = a + (b + c)$$
$$(ab)c = a(bc)$$

Identity property

There exists a unique real number 0 such that

$$a + 0 = a \quad \text{and} \quad 0 + a = a.$$

There exists a unique real number 1 such that

$$a \cdot 1 = a \quad \text{and} \quad 1 \cdot a = a.$$

Inverse property

There exists a unique real number $-a$ such that

$$a + (-a) = 0 \quad \text{and} \quad (-a) + a = 0.$$

If $a \neq 0$, there exists a unique real number $1/a$ such that

$$a \cdot \frac{1}{a} = 1 \quad \text{and} \quad \frac{1}{a} \cdot a = 1.$$

Distributive property

$$a(b + c) = ab + ac$$

Chapter 1   Review Exercises

Tell whether each statement is true *or* false.

1. $-6 \in \{8, 4, -3, -9, 6\}$

2. $8 \notin \{3, 9, 7\}$

3. $2 \notin \{0, 1, 2, 3, 4\}$

4. $0 \in \{0, 1, 2, 3, 4\}$

5. $\{3, 4, 5\} \subseteq \{2, 3, 4, 5, 6\}$

6. $\{1, 2, 5, 8\} \subseteq \{1, 2, 5, 10, 11\}$

7. $\{3, 6, 9, 10\} \subseteq \{3, 9, 11, 13\}$

8. $\emptyset \subseteq \{1\}$

9. $\{2, 8\} \nsubseteq \{2, 4, 6, 8\}$

10. $0 \subseteq \emptyset$

List the elements in the following sets.

11. $\{x | x$ is a counting number more than 8 and less than 5$\}$

12. $\{x | x$ is an integer, $-3 \leq x < 1\}$

13. $\{$all counting numbers less than 5$\}$

14. $\{$all whole numbers not greater than 2$\}$

Let $U = \{$all students at State University$\}$
  $A = \{$all chemistry majors$\}$
  $B = \{$all freshmen$\}$

$C = \{$all female students$\}$
$D = \{$all students out for track$\}$.

Describe the following sets in words.

15. $A \cap C$

16. $B \cap D$

17. $A \cup D$

18. $A' \cap D$

19. $B' \cap C'$

20. $(B \cup C)'$

Let $U = \{-4, -3, -2, -1, 0, 1, 2, 3, 4, 5\}$
$\quad A = \{-3, 0, 1, 4\}$
$\quad B = \{-4, -3, -2, -1, 0\}$
$\quad C = \{3, 5\}$
$\quad D = \{0, 1, 2, 3, 4, 5\}.$

Find each of the following.

21. $A \cap B$ **22.** $B \cup C$ **23.** A'

24. $B' \cup C$ **25.** $D \cup \emptyset$ **26.** $C \cap \emptyset$

27. $(A \cap B) \cup C$ **28.** $(D' \cup C) \cap B$ **29.** $(A' \cap B') \cap C'$

Let set $K = \{-12, -6, -9/10, -\sqrt{7}, -\sqrt{4}, 0, 1/8, \pi/4, 6, \sqrt{11}\}$. List the elements of K that are members of the following sets.

30. Natural numbers   **31.** Whole numbers   **32.** Integers

33. Rational numbers   **34.** Irrational numbers   **35.** Real numbers

For Exercises 36–44 choose all words from the following list that apply.

(a) natural numbers   *(b) whole numbers*   *(c) integers*   *(d) rational numbers*
(e) irrational numbers   *(f) real numbers*   *(g) meaningless*

36. 22   **37.** 0   **38.** $\sqrt{36}$

39. $-\sqrt{25}$ **40.** -1 **41.** 5/8

42. $\sqrt{15}$ **43.** $-6/0$ **44.** $3\pi/4$

Use the order of operations to simplify each of the following.

45. $-4 - [2 - (-9)]$ **46.** $[8 - (-5)] - 4$

47. $2(-9) - 4(-6)$ **48.** $-7(-5) - 3(-9)$

49. $(-4 - 1)(-3 - 5) - 8$ **50.** $(6 - 9)(-2 - 7) - (-4)$

51. $\left(-\dfrac{5}{9} - \dfrac{2}{3}\right) - \dfrac{5}{6}$ **52.** $\left(-\dfrac{8}{5} - \dfrac{3}{4}\right) - \left(-\dfrac{1}{2}\right)$

53. $\dfrac{6(-4) - 9(-3)}{-5[-2 - (-6)]}$ **54.** $\dfrac{(-7)(-3) - (-8)(-5)}{(-4 - 2)(-1 - 6)}$

Evaluate each of the following when $a = -1$, $b = -2$, and $c = 4$.

55. $9a - 5b + 4c$ **56.** $6b - 11a - 3c$ **57.** $-4(2a - 5b)$

58. $6(3b + 2c)$ **59.** $\dfrac{9a + 2b}{a + b + c}$ **60.** $\dfrac{4a - 5c}{2a + 3b}$

Identify the properties illustrated in each of the following.

61. $6 \cdot 4 = 4 \cdot 6$ **62.** $8(5 + 9) = (5 + 9)8$

63. $3 + (-3) = 0$ **64.** If $x \neq 0$, then $x \cdot \dfrac{1}{x} = 1.$

65. $4 \cdot 6 + 4 \cdot 12 = 4(6 + 12)$ **66.** $3 \cdot (4 \cdot 2) = (3 \cdot 4) \cdot 2$

67. $-(r - 2) = -r + 2$ **68.** $[-7 + 7] \cdot 2 = 0 \cdot 2$

69. $(9 + p) + 0 = 9 + p$

70. $\dfrac{1}{r} \cdot r + \dfrac{1}{r} \cdot r = \left(\dfrac{1}{r} + \dfrac{1}{r} \right) \cdot r \quad (r \neq 0)$

71. $[6 \cdot 5 + 8 \cdot 5] \cdot 2 = [(6 + 8) \cdot 5] \cdot 2$

72. $-6(r - 5) = -6r + 30$

Use the distributive property to rewrite sums as products and products as sums.

73. $9(p + q)$

74. $-11(m - n)$

75. $-3(5 - r)$

76. $k(r + s - t)$

77. $ab + ac$

78. $mn - 9m$

Write the following numbers in numerical order from smallest to largest.

79. $-7, -3, 8, \pi, -2, 0$

80. $5/6, 1/2, -2/3, -5/4, -3/8$

81. $|6 - 4|, -|-2|, |8 + 1|, -|3 - (-2)|$

82. $\sqrt{7}, -\sqrt{8}, -|\sqrt{16}|, |-\sqrt{12}|$

Write without absolute value bars.

83. $-|-6| + |3|$

84. $|-5| + |-9|$

85. $7 - |-8|$

86. $|3 - \sqrt{7}|$

87. $|\sqrt{8} - 3|$

88. $|\sqrt{52} - 8|$

89. $|m - 3|$ if $m > 3$

90. $|-6 - x^2|$

91. $|\pi - 4|$

92. $|3 + 5k|$ if $k < -3/5$

Justify each of the following by giving the correct property. Assume that all variables represent real numbers.

93. If $-7r < -14$, then $r > 2$.

94. If $p > 0$, then $12p > 0$.

95. If $4k - 9 \leq 6$, then $4k \leq 15$.

96. $|9 + 7p| \leq |9| + |7p|$

97. $\left| \dfrac{10}{7} \right| = \dfrac{|10|}{|7|}$

98. $|-5 + 14z| \geq 0$

99. $|-2| \cdot |-5| = |(-2)(-5)|$

100. If $-\dfrac{2}{3}y > 12$, then $y < -18$.

Under what conditions are the following statements true?

101. $|x| = x$

102. $x + |x| = 0$

103. $|x + y| = -x - y$

104. $|x + y| = -|x| - |y|$

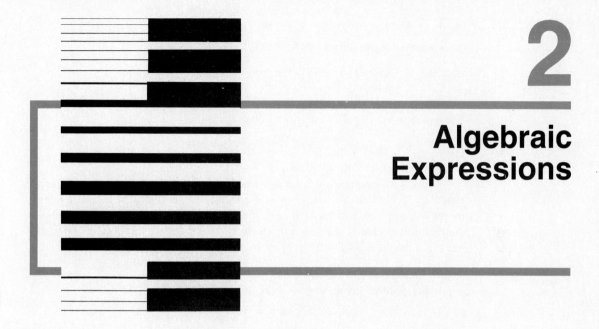

2
Algebraic Expressions

An **algebraic expression** is any combination of variables or constants joined by the basic operations of addition, subtraction, multiplication, division (except by zero), or extraction of roots. The simplest algebraic expression is a polynomial, which we discuss in the second section of this chapter, after a look at exponents.

2.1 Integer Exponents

Exponents are used to write the products of repeated factors. For example, the product $2 \cdot 2 \cdot 2$ can be written as 2^3, where the 3 shows that three factors of 2 appear in the product. Generalizing, the symbol a^n is defined as follows.

Definition of a^n

> If n is any positive integer and a is any real number,
>
> $$a^n = a \cdot a \cdot a \cdots a, \quad \text{where } a \text{ appears } n \text{ times.}$$

The integer n is called the **exponent,** and a is the **base.** (Read a^n as "a to the nth power," or just "a to the nth.")

EXAMPLE 1   (a) $4^3 = 4 \cdot 4 \cdot 4 = 64$ The base is 4 and the exponent is 3

(b) $(-6)^2 = (-6)(-6) = 36$ Base: -6; exponent: 2

(c) $-6^2 = -(6 \cdot 6) = -36$ Base: 6; exponent: 2   ●

One common error with exponents occurs with expressions such as $4 \cdot 3^2$. The exponent of 2 applies only to the base 3, so that

$$4 \cdot 3^2 = 4 \cdot 3 \cdot 3 = 36.$$

It is *not correct* to write $4 \cdot 3^2$ as $4 \cdot 3 \cdot 4 \cdot 3$.

Work with exponents can be simplified by using various properties of exponents. By definition, a^m (where m is a positive integer and a is a real number) means a appears as a factor m times. In the same way, a^n (where n is a positive integer) means that a appears as a factor n times. In the product $a^m \cdot a^n$, the number a would appear as a factor $m + n$ times. Therefore,

Product Rule

> for any positive integers m and n and any real number a,
>
> $$a^m \cdot a^n = a^{m+n}.$$

EXAMPLE 2   (a) $y^4 \cdot y^7 = y^{4+7} = y^{11}$

(b) $(6z^5)(9z^3)(2z^2) = 6 \cdot 9 \cdot 2 \cdot z^{5+3+2} = 108z^{10}$

(c) $(2k^m)(k^{1+m}) = 2k^{m+(1+m)} = 2k^{1+2m}$ (if m is a natural number)   ●

To get a similar rule for division, we must first define a negative exponent. Start with the quotient

$$\frac{a^3}{a^7}.$$

By the definition of exponent,

$$\frac{a^3}{a^7} = \frac{a \cdot a \cdot a}{a \cdot a \cdot a \cdot a \cdot a \cdot a \cdot a} = \frac{1}{a \cdot a \cdot a \cdot a} = \frac{1}{a^4}.$$

In the rule for products given above, $a^m \cdot a^n = a^{m+n}$, we *added* exponents. This would suggest that when dividing, we would subtract exponents. Subtracting exponents gives

$$\frac{a^3}{a^7} = a^{3-7} = a^{-4}.$$

The only way to keep these results consistent is to define a^{-4} as $1/a^4$. This definition can be generalized as shown at the top of the next page.

Definition of a^{-n}

If a is a nonzero real number and n is any integer,

$$a^{-n} = \frac{1}{a^n}.$$

EXAMPLE 3

(a) $4^{-2} = \frac{1}{4^2} = \frac{1}{16}$

(b) $2^{-3} = \frac{1}{2^3} = \frac{1}{8}$

(c) $\left(\frac{2}{5}\right)^{-3} = \frac{1}{\left(\frac{2}{5}\right)^3} = \frac{1}{\frac{8}{125}} = \frac{125}{8}$

(d) $-4^{-2} = -\frac{1}{4^2} = -\frac{1}{16}$ ●

Using the definition of a negative exponent, we can prove the following property of exponents.

Quotient Rule

If a is a nonzero real number and if m and n are any integers,

$$\frac{a^m}{a^n} = a^{m-n}.$$

In the following examples, assume that all variables represent nonzero real numbers.

EXAMPLE 4

(a) $\frac{12^5}{12^2} = 12^{5-2} = 12^3$

(b) $\frac{a^5}{a^8} = a^{5-8} = a^{-3} = \frac{1}{a^3}$

(c) $\frac{16m^9}{12m^{11}} = \frac{4}{3m^2}$

(d) $\frac{25r^7z^5}{10r^9z} = \frac{5z^4}{2r^2}$

(e) $\frac{x^{5y}}{x^{3y}} = x^{5y-3y} = x^{2y}$ (if y is an integer) ●

What happens to the property $a^m/a^n = a^{m-n}$ if m and n are the same? For example,

$$\frac{k^5}{k^5} = k^{5-5} = k^0.$$

If k is a nonzero real number, then $k^5/k^5 = 1$. To be consistent, k^0 must be defined to equal 1. That is,

Definition of a^0

> for any nonzero real number a,
> $$a^0 = 1.$$

The symbol 0^0 is not defined. (An attempt to give meaning to 0^0 leads to a contradiction, such as dividing by 0.)

EXAMPLE 5

(a) $3^0 = 1$

(b) $(-4)^0 = 1$

(c) $-4^0 = -1$

(d) $-(-4)^0 = -1$

(e) $(7r)^0 = 1$ if $r \neq 0$ ●

The expression $(2^5)^3$ can be written as

$$(2^5)^3 = 2^5 \cdot 2^5 \cdot 2^5.$$

By a generalization of the first property of exponents, this product is

$$(2^5)^3 = 2^{5+5+5} = 2^{15}.$$

The same exponent could have been obtained by multiplying 3 and 5. This example suggests the first of the **power rules** given below.

Power Rules

> If a and b are real numbers, and m and n are integers,
> $$(a^m)^n = a^{mn}$$
> $$(ab)^m = a^m b^m$$
> $$\left(\frac{a}{b}\right)^m = \frac{a^m}{b^m} \quad (b \neq 0).$$

EXAMPLE 6

(a) $(5^3)^{-2} = 5^{3(-2)} = 5^{-6} = \dfrac{1}{5^6}$

(b) $(3^4 x^2)^3 = (3^4)^3 (x^2)^3 = 3^{4(3)} x^{2(3)} = 3^{12} x^6$

(c) $\left(\dfrac{2^5}{b^4}\right)^3 = \dfrac{(2^5)^3}{(b^4)^3} = \dfrac{2^{15}}{b^{12}}$ if $b \neq 0$

(d) $\left(\dfrac{5^{-2}}{q^3}\right)^{-4} = \dfrac{5^8}{q^{-12}} = \dfrac{5^8}{\dfrac{1}{q^{12}}} = 5^8 \cdot \dfrac{q^{12}}{1} = 5^8 \cdot q^{12}$ if $q \neq 0$ ●

Now we can summarize the properties of exponents.

Properties of Exponents

For all real numbers a and b and integers m and n:

Product rule $a^m \cdot a^n = a^{m+n}$

Quotient rule $\dfrac{a^m}{a^n} = a^{m-n}$ $(a \neq 0)$

Power rules $(a^m)^n = a^{mn}$

$(ab)^m = a^m b^m$

$\left(\dfrac{a}{b}\right)^m = \dfrac{a^m}{b^m}$ $(b \neq 0).$

EXAMPLE 7   Use the properties of exponents to simplify each expression. Write answers without negative exponents. Assume that all variables represent nonzero real numbers.

(a) $3x^{-2}(4^{-1}x^{-5})^2 = 3x^{-2}(4^{-2}x^{-10})$

$\qquad = 3 \cdot 4^{-2} \cdot x^{-2+(-10)}$

$\qquad = 3 \cdot 4^{-2} \cdot x^{-12}$

$\qquad = \dfrac{3}{16x^{12}}$

(b) $\dfrac{5m^{-3}}{10m^{-5}} = \dfrac{5}{10} m^{-3-(-5)}$

$\qquad = \dfrac{1}{2} m^2,$ or $\dfrac{m^2}{2}$

(c) $\dfrac{12p^3 q^{-1}}{8p^{-2}q} = \dfrac{12}{8} \cdot \dfrac{p^3}{p^{-2}} \cdot \dfrac{q^{-1}}{q}$

$\qquad = \dfrac{3}{2} \cdot p^{3-(-2)}q^{-1-1}$ $(q = q^1)$

$\qquad = \dfrac{3}{2} p^5 q^{-2}$

$\qquad = \dfrac{3p^5}{2q^2}$

(d) $\dfrac{(3x^2)^{-1}(3x^5)^{-2}}{(3^{-1}x^{-2})^2} = \dfrac{3^{-1}x^{-2}3^{-2}x^{-10}}{3^{-2}x^{-4}}$

$= \dfrac{3^{-1+(-2)}x^{-2+(-10)}}{3^{-2}x^{-4}}$

$= \dfrac{3^{-3}x^{-12}}{3^{-2}x^{-4}}$

$= 3^{-3-(-2)}x^{-12-(-4)}$

$= 3^{-1}x^{-8}$

$= \dfrac{1}{3x^8}$ ●

Scientific Notation One application of exponents comes from science, where calculations often involve very large or very small numbers. To avoid working with many zeros, scientists write such numbers in *scientific notation*.

Scientific
Notation

> In **scientific notation,** a number is written as the product of a number *a* whose absolute value is at least 1, but less than 10, and a power of 10, or
>
> $$a \times 10^n,$$
>
> for an appropriate integer power *n*.

EXAMPLE 8

Write each number in scientific notation.

(a) 125,000

The number 1.25 is between 1 and 10. In scientific notation, 125,000 will be written as the product of 1.25 and some power of 10. To go from 1.25 to 125,000 requires moving the decimal point 5 places to the right, so

$$125,000 = 1.25 \times 10^5.$$

(b) $-7900 = -7.9 \times 10^3$

(c) $.003 = 3 \times 10^{-3}$

(d) $.0000056 = 5.6 \times 10^{-6}$

(e) $8.739 = 8.739 \times 10^0$ ●

One advantage of scientific notation is that some calculations can be simplified using the properties of exponents. This process is shown in the next example.

EXAMPLE 9

Find each of the following.

(a) (28,000,000)(.000012)

Write each number in scientific notation, as shown on the next page.

$$(28{,}000{,}000)(.000012) = (2.8 \times 10^7)(1.2 \times 10^{-5})$$
$$= (2.8 \times 1.2) \times 10^{7+(-5)}$$
$$= 3.36 \times 10^2$$
$$= 336$$

(b) $\dfrac{.082}{.0000097} = \dfrac{8.2 \times 10^{-2}}{9.7 \times 10^{-6}}$

$$= \frac{8.2}{9.7} \times 10^{-2-(-6)}$$

$$= .85 \times 10^4 \qquad \text{The quotient is rounded to the nearest}$$

$$= 8500 \quad \bullet \qquad \text{hundredth}$$

Another use of scientific notation comes when writing **significant digits.** While we do not go into the fine points of significant digits in this book, we will show some examples.

Number	Significant digits
9	1
.00271	3
2.00094	6

Confusion comes from a number like 8000. Is this number measured to the nearest one, the nearest ten, the nearest hundred, or the nearest thousand? To decide, write the number as follows.

Number	Significant digits	Measured to nearest
8.000×10^3	4	One
8.00×10^3	3	Ten
8.0×10^3	2	Hundred
8×10^3	1	Thousand

2.1 EXERCISES

Simplify each of the following. Write results with no negative exponents. Assume that all variables represent nonzero real numbers. See Examples 1–7.

1. 4^5

2. 3^4

3. 3^{-5}

4. 2^{-3}

5. $(-4)^{-3}$

6. $(-5)^{-2}$

7. $\left(\dfrac{1}{2}\right)^{-3}$

8. $\left(\dfrac{2}{3}\right)^{-2}$

9. $\left(\dfrac{5}{x}\right)^{-4}$

10. $\left(\dfrac{n}{3}\right)^{-2}$

11. $2^{-3} - 2^{-2}$

12. $3^{-3} + 3^{-1}$

13. 7^0

14. $\left(\dfrac{2}{3}\right)^0$

15. -7^0

16. $\left(-\dfrac{5}{8}\right)^0$

17. $-(-7^0)$

18. $-\left(\dfrac{2}{3}\right)^0$

19. $(2^2)^5$

20. $(6^4)^3$

21. $(x^5y^4)^3$

22. $(m^3n^9)^2$

23. $\left(\dfrac{p^4}{q}\right)^2$

24. $\left(\dfrac{r^8}{s^2}\right)^3$

25. $(z^{-3}y^3)^{-3}$

26. $(y^{-3}x^{-5})^{-1}$

27. $(2a^3b^{-3})^3$

28. $(5p^4q^{-1})^{-2}$

29. $\left(\dfrac{r^{-2}}{s^{-5}}\right)^{-3}$

30. $\left(\dfrac{p^{-1}}{q^{-5}}\right)^{-2}$

31. $\left(\dfrac{a}{b^{-3}}\right)^{-1}$

32. $\left(\dfrac{m^{-1}}{n}\right)^{-2}$

33. $[(m^2n)^{-1}]^4$

34. $(x^5t^{-2})^{-2}$

35. $\left(\dfrac{2c^2}{d^3}\right)^{-2}$

36. $\left(\dfrac{a^{-1}}{b^2}\right)^{-3}$

37. $(-2x^{-3}y^2)^{-2}$

38. $(-3p^2q^{-2}s^{-1})^7$

39. $(6^{-2})(6^{-5})(6^3)$

40. $(4^8)(4^{-10})(4^2)$

41. $(m^{10})(m^{-4})(m^{-5})$

42. $(r^{-3})(r^{-2})(r^4)$

43. $\dfrac{6m^2n^4}{2mn^5}$

44. $\dfrac{3x^5y^{10}}{9xy^5z^4}$

45. $\left(\dfrac{3m^4n}{2x^3}\right)^5$

46. $\left(\dfrac{2a^2b^3}{5c}\right)^4$

47. $(3m)^2(-2m)^3(-3m^4)$

48. $(-p^3)(2p^2)^3(-2p^4)$

49. $\dfrac{2^{-1}x^3y^{-3}}{xy^{-2}}$

50. $\dfrac{5^{-2}m^2y^{-5}}{5^2m^{-1}y^{-2}}$

51. $\dfrac{(4+s)^3(4+s)^{-2}}{(4+s)^{-5}}$

52. $\dfrac{(m+n)^4(m+n)^{-2}}{(m+n)^{-5}}$

53. $\dfrac{(2x^3)^{-2}(2^2x^5)^{-1}}{(2x^4)^{-3}}$

54. $\dfrac{(5p^2q)^{-1}(5p^3q^{-2})^2}{5(pq)^{-3}(p^4q^{-2})}$

55. $\dfrac{(r^2s^3t^4)^{-2}(r^3s^2t)^{-1}}{(rst)^{-3}(r^2st^2)^3}$

56. $\dfrac{(ab^2c)^{-3}(a^2bc)^{-2}}{(abc^2)^4}$

57. $\dfrac{[k^2(p+q)^4]^{-1}}{k^{-4}(p+q)^3}$

58. $\dfrac{t^5(x+y)^4t^{-8}}{(x+y)^{-2}t^{-5}}$

Evaluate each expression, letting $a = 2$, $b = -3$, and $c = 0$.

59. $a^3 + b$

60. $b^3 + a^2$

61. $-b^2 + 3(c + 5)$

62. $2a^2 - b^4$

63. $a^7b^9c^5$

64. $12a^{15}b^6c^8$

65. $a^b + b^a$

66. $b^c + a^c$

67. $b^aa^ba^c$

68. $c^aa^bb^a$

69. $(3a^3 + 5b^2)^c$

70. $-(4b^5 - 2a^3)^c$

71. $a^{-1} + b^{-1}$

72. $b^{-2} - a$

73. $\dfrac{2b^{-1} - 3a^{-1}}{a + b^2}$

74. $\dfrac{3a^2 - b^2}{b^{-3} + 2a^{-1}}$

75. $\left(\dfrac{a}{3}\right)^{-1} + \left(\dfrac{b}{2}\right)^{-2}$

76. $\left(\dfrac{2b}{5}\right)^2 - 3\left(\dfrac{a^{-1}}{4}\right)$

Express each of the following numbers in scientific notation. See Example 8.

77. 69,300

78. 5000

79. 12,700,000

80. 6,000,000,000

81. .0001

82. .0000023

83. .00792

84. .054

85. .965

Express each of the following numbers without exponents. See Example 8.

86. 8.2×10^5

87. 3.7×10^3

88. 2.98×10^2

89. 1.7×10^8

90. 3.61×10^{-2}

91. 4.02×10^{-5}

92. 6.15×10^{-3}

93. 9.3×10^{-6}

94. 8.76×10^{-1}

Use scientific notation to evaluate each of the following. See Example 9.

95. $(4600)(.00092)$ **96.** $(.87)(.0004)$ **97.** $(.00002)(.00009)$ **98.** $(.0004)(.0000015)$

99. $\dfrac{.000034}{.017}$ **100.** $\dfrac{.0000042}{.0006}$ **101.** $\dfrac{28}{.0004}$ **102.** $\dfrac{50}{.0000025}$

Calculators with scientific notation capability display the number 1.46×10^{12} as follows.

$$1.46 \qquad 12$$

Use a calculator to evaluate each of the following. Write all answers in scientific notation rounded to three significant digits.

103. $(1.66 \times 10^4)(2.93 \times 10^3)$

104. $(6.92 \times 10^7)(8.14 \times 10^5)$

105. $(4.08 \times 10^{-9})(3.172 \times 10^4)$

106. $(9.113 \times 10^{12})(8.72 \times 10^{-11})$

107. $\dfrac{-(4.389 \times 10^4)(2.421 \times 10^{-2})}{1.76 \times 10^{-9}}$

108. $\dfrac{(1.392 \times 10^{10})(5.746 \times 10^{-8})}{4.93 \times 10^{-15}}$

109. $\dfrac{-3.9801 \times 10^{-6}}{(7.4993 \times 10^{-8})(2.117 \times 10^{-4})}$

110. $\dfrac{-5.421 \times 10^{-7}}{(4.2803 \times 10^{-4})(9.982 \times 10^{-8})}$

Use scientific notation to work the following exercises.

111. The distance to the sun is 9.3×10^7 mi. How long would it take a rocket, traveling at 2.9×10^3 mph, to reach the sun?

112. A *light-year* is the distance that light travels in one year. Find the number of miles in a light-year if light travels 1.86×10^5 mi per sec.

113. Use the information given in the previous two exercises to find the number of minutes necessary for light from the sun to reach the earth.

114. A computer can do one addition in 1.4×10^{-7} sec. How long would it take the computer to do a trillion (10^{12}) calculations? Give the answer in seconds and then in hours.

115. Assume that the volume of the earth is 5×10^{14} cu m and that the volume of a bacterium is 4×10^{-16} cu m. If the earth could be packed full of bacteria, how many would it contain?

116. Our galaxy is approximately 1.2×10^{17} km across. Suppose a spaceship could travel at 1.5×10^5 km per sec (half the speed of light). Find the approximate number of years needed for the spaceship to cross the galaxy.

Find each of the following products or quotients. Assume that all variables appearing in exponents represent integers and all variables are nonzero.

117. $(2k^m)(k^{1-m})$

118. $(y^m)(y^{1+m})$

119. $5p^r(6p^{3-2r})$

120. $(z^{p-1})(z^{3-p})$

121. $(5x^p)(3x^2)$

122. $(-z^{2r})(4z^{r+3})$

123. $\dfrac{k^{y-5}}{k^{y+2}}$

124. $\dfrac{4a^{m+2}}{2a^{m-1}}$

125. $\dfrac{(b^2)^y}{(2b^y)^3}$

126. $\left(\dfrac{3p^a}{2m^b}\right)^c$

127. $\dfrac{(2m^n)^2(-4m^{2+n})}{8m^{4n}}$

128. $\dfrac{(-5z^p)^3(2z^{1-4p})}{-10z^{-p}}$

Use a calculator with an x^y key to suggest that each statement is correct.

129. $(9.0086)^0 = 1$

130. $(3.27599)^0 = 1$

131. $(2.7)^3(1.9)^3 = [2.7(1.9)]^3$

132. $[(5.8)(3.1)]^3 = (5.8)^3(3.1)^3$

133. $(8.24)^2 + (3.5)^2 \neq (8.24 + 3.5)^2$

134. $[(2.3)^2]^3 = (2.3)^6$

2.2 Polynomials

The product of a real number and one or more variables raised to powers is called a **term.** The real number is called the **numerical coefficient,** or just the **coefficient.** The coefficient in $-3m^4$ is -3, while the coefficient in $-p^2$ is -1.

A **polynomial** is defined as a finite sum of terms, with only nonnegative integer exponents permitted on the variables. If the terms of a polynomial contain only the variable x, then the polynomial is called a **polynomial in x.** (Polynomials in other variables are defined similarly.)

Examples of polynomials include

$$5x^3 - 8x^2 + 7x - 4, \qquad 9p^5 - 3, \qquad 8r^2, \qquad \text{and} \qquad 6.$$

The expression $9x^2 - 4x - 6/x$ is not a polynomial because of $-6/x$. Since $-6/x = -6x^{-1}$, the exponent is not a nonnegative integer.

The highest exponent in a polynomial in one variable is the **degree** of the polynomial. A nonzero constant is said to have degree 0. (The polynomial 0 has no degree.) For example, $3x^6 - 5x^2 + 2x + 3$ is a polynomial of degree 6.

A polynomial can have more than one variable. A term containing more than one variable has degree equal to the sum of all the exponents appearing on the variables in the term. For example, $-3x^4y^3z^5$ is of degree $4 + 3 + 5 = 12$.

The degree of a polynomial in more than one variable is equal to the highest degree of any term appearing in the polynomial. By this definition, the polynomial

$$2x^4y^3 - 3x^5y + x^6y^2$$

is of degree 8 because of the x^6y^2 term.

A polynomial containing exactly three terms is called a **trinomial;** one containing exactly two terms is a **binomial;** and a single-term polynomial is called a **monomial.** For example, $7x^9 - 8x^4 + 1$ is a trinomial of degree 9.

EXAMPLE 1

The list below shows several polynomials, gives the degree of each, and identifies each as a monomial, binomial, or trinomial.

Polynomial	Degree	Type
$9p^7 - 4p^3 + 8p^2$	7	Trinomial
$29x^{11} + 8x^{15}$	15	Binomial
$-10r^6s^8$	14	Monomial
$5a^3b^7 - 3a^5b^5 + 4a^2b^9 - a^{10}$	11	None of the above ●

Since the variables used in polynomials represent real numbers, a polynomial represents a real number. This means that all the properties of the real numbers mentioned in Chapter 1 hold for polynomials. In particular, the distributive property holds, so

$$3m^5 - 7m^5 = (3 - 7)m^5 = -4m^5.$$

To *add* polynomials, add coefficients of like terms; to *subtract* polynomials, subtract coefficients of the same powers. (Notice how the distributive property was used to suggest these definitions.)

EXAMPLE 2   Add or subtract, as indicated.

(a) $(2y^4 - 3y^2 + y) + (4y^4 + 7y^2 + 6y)$

$$= (2 + 4)y^4 + (-3 + 7)y^2 + (1 + 6)y$$
$$= 6y^4 + 4y^2 + 7y$$

(b) $(6r^4 + 2r^2) + (3r^3 + 9r) = 6r^4 + 3r^3 + 2r^2 + 9r$

(c) $(-3m^3 - 8m^2 + 4) - (m^3 + 7m^2 - 3)$

$$= (-3 - 1)m^3 + (-8 - 7)m^2 + [4 - (-3)]$$
$$= -4m^3 - 15m^2 + 7$$

(d) $8m^4p^5 - 9m^3p^5 + (11m^4p^5 + 15m^3p^5) = 19m^4p^5 + 6m^3p^5$ ●

The distributive property, together with the properties of exponents, can also be used to find the product of two polynomials. For example, to find the product of $3x - 4$ and $2x^2 - 3x + 5$, treat $3x - 4$ as a single expression and use the distributive property as follows.

$$(3x - 4)(2x^2 - 3x + 5) = (3x - 4)(2x^2) - (3x - 4)(3x) + (3x - 4)(5)$$

Now use the distributive property three separate times on the right of the equals sign to get

$$(3x - 4)(2x^2 - 3x + 5)$$
$$= (3x)(2x^2) - 4(2x^2) - (3x)(3x) - (-4)(3x) + (3x)5 - 4(5)$$
$$= 6x^3 - 8x^2 - 9x^2 + 12x + 15x - 20$$
$$= 6x^3 - 17x^2 + 27x - 20.$$

It is sometimes more convenient to write such a product vertically, as shown below.

$$
\begin{array}{r}
2x^2 - 3x + 5 \\
3x - 4 \\
\hline
-8x^2 + 12x - 20 \\
6x^3 - 9x^2 + 15x \\
\hline
6x^3 - 17x^2 + 27x - 20
\end{array}
\qquad \text{Add in columns}
$$

EXAMPLE 3   Find each product.

(a) $(6m + 1)(4m - 3) = (6m)(4m) + (6m)(-3) + 1(4m) + 1(-3)$
$$= 24m^2 - 14m - 3$$

(b) $(2x + 7)(2x - 7) = 4x^2 - 14x + 14x - 49$
$$= 4x^2 - 49$$

(c) $(2k^n - 5)(k^n + 3) = 2k^{2n} - 5k^n + 6k^n - 15$
$$= 2k^{2n} + k^n - 15 \quad \text{(if } n \text{ is an integer)} \quad \bullet$$

In parts (a) and (c) of Example 3, the product of two binomials was a trinomial, while in part (b) the product of two binomials was a binomial. The product of two binomials of the form $x + y$ and $x - y$ is always a binomial. Check by multiplying that

Difference of Two Squares

$$(x + y)(x - y) = x^2 - y^2.$$

This product is called the **difference of two squares.**

EXAMPLE 4   (a) $(3p + 11)(3p - 11)$

Using the pattern discussed above, replace x with $3p$ and y with 11. This gives

$$(3p + 11)(3p - 11) = (3p)^2 - 11^2 = 9p^2 - 121.$$

(b) $(5m^3 - 3)(5m^3 + 3) = (5m^3)^2 - 3^2 = 25m^6 - 9$

(c) $(9k - 11r^3)(9k + 11r^3) = (9k)^2 - (11r^3)^2 = 81k^2 - 121r^6 \quad \bullet$

The **square of a binomial** is another special product. The products $(x + y)^2$ and $(x - y)^2$ are shown below.

Square of a Binomial

$$(x + y)^2 = x^2 + 2xy + y^2$$
$$(x - y)^2 = x^2 - 2xy + y^2$$

EXAMPLE 5   (a) $(2m + 5)^2 = (2m)^2 + 2(2m)(5) + (5)^2$
$$= 4m^2 + 20m + 25$$

(b) $(3x - 7y^4)^2 = (3x)^2 - 2(3x)(7y^4) + (7y^4)^2$
$$= 9x^2 - 42xy^4 + 49y^8 \quad \bullet$$

One last special product gives the cube of a binomial.

Cube of a Binomial

$$(x + y)^3 = x^3 + 3x^2y + 3xy^2 + y^3$$
$$(x - y)^3 = x^3 - 3x^2y + 3xy^2 - y^3$$

EXAMPLE 6

(a) $(m + 4n)^3$

Replace x with m and y with $4n$ in the pattern for $(x + y)^3$ to get

$$(m + 4n)^3 = m^3 + 3m^2(4n) + 3m(4n)^2 + (4n)^3$$
$$= m^3 + 12m^2n + 48mn^2 + 64n^3.$$

(b) $(5k - 2z^5)^3 = (5k)^3 - 3(5k)^2(2z^5) + 3(5k)(2z^5)^2 - (2z^5)^3$
$$= 125k^3 - 150k^2z^5 + 60kz^{10} - 8z^{15} \quad \bullet$$

It is useful to memorize and be able to apply these special products. (Both the square of a binomial and the cube of a binomial are special cases of the binomial theorem, discussed in the last chapter of this book.)

Quotients of Polynomials To find the quotient of a polynomial P and a monomial Q, multiply P by the reciprocal of Q.

$$\frac{P}{Q} = P \cdot \frac{1}{Q} \quad \text{(if } Q \neq 0\text{)}$$

EXAMPLE 7

(a) $\dfrac{2m^5 - 6m^3}{2m^3} = (2m^5 - 6m^3) \cdot \dfrac{1}{2m^3} = \dfrac{2m^5}{2m^3} - \dfrac{6m^3}{2m^3} = m^2 - 3$

Since the quotient $m^2 - 3$ is a polynomial, $2m^5 - 6m^3$ is divisible by $2m^3$. (The word *divisible* means "with remainder 0.")

(b) $\dfrac{3y^6x^3 - 6y^3x^6 + 8y^5x}{3y^3x^3} = \dfrac{3y^6x^3}{3y^3x^3} - \dfrac{6y^3x^6}{3y^3x^3} + \dfrac{8y^5x}{3y^3x^3} = y^3 - 2x^3 + \dfrac{8y^2}{3x^2}$

Here the quotient is not a polynomial, so $3y^6x^3 - 6y^3x^6 + 8y^5x$ is *not* divisible by $3y^3x^3$. $\quad \bullet$

To divide one polynomial by another, use a **division algorithm.** (An **algorithm** is an orderly procedure for performing an operation.) The division algorithm for polynomials is very similar to that used for dividing whole numbers.

EXAMPLE 8

Divide $6q^3 - 17q^2 + 22q - 23$ by $2q - 3$.

Begin by writing both polynomials with exponents in descending order.

$$2q - 3 \overline{)6q^3 - 17q^2 + 22q - 23}$$

Step 1 The quotient of $6q^3$ and $2q$ is $3q^2$. Multiply $3q^2$ and $2q - 3$, getting $6q^3 - 9q^2$.

$$
\begin{array}{r}
3q^2 \\
2q - 3 \overline{)6q^3 - 17q^2 + 22q - 23} \\
6q^3 - 9q^2
\end{array}
$$

$\dfrac{6q^3}{2q} = 3q^2$

$3q^2(2q - 3)$

Step 2 Subtract $6q^3 - 9q^2$ from $6q^3 - 17q^2$ by changing the signs on $6q^3 - 9q^2$ and adding. Then bring down $22q$.

$$
\begin{array}{r}
3q^2 \\
2q - 3 \overline{) 6q^3 - 17q^2 + 22q - 23} \\
\underline{-6q^3 + 9q^2} \\
-8q^2 + 22q
\end{array}
$$

Change signs and add

Step 3 The quotient of $-8q^2$ and $2q$ is $-4q$. Multiply $-4q$ and $2q - 3$.

$$
\begin{array}{r}
3q^2 - 4q \longleftarrow \\
2q - 3 \overline{) 6q^3 - 17q^2 + 22q - 23} \\
\underline{6q^3 - 9q^2} \\
-8q^2 + 22q \\
\underline{-8q^2 + 12q} \longleftarrow
\end{array}
$$

$\dfrac{-8q^2}{2q} = -4q$

$-4q(2q - 3)$

Step 4 Subtract $-8q^2 + 12q$ from $-8q^2 + 22q$. Bring down -23.

$$
\begin{array}{r}
3q^2 - 4q \\
2q - 3 \overline{) 6q^3 - 17q^2 + 22q - 23} \\
\underline{6q^3 - 9q^2} \\
-8q^2 + 22q \\
\underline{8q^2 - 12q} \\
10q - 23
\end{array}
$$

Change signs and add

Step 5 The quotient of $10q$ and $2q$ is 5. Multiply 5 and $2q - 3$.

$$
\begin{array}{r}
3q^2 - 4q + 5 \longleftarrow \\
2q - 3 \overline{) 6q^3 - 17q^2 + 22q - 23} \\
\underline{6q^3 - 9q^2} \\
-8q^2 + 22q \\
\underline{-8q^2 + 12q} \\
10q - 23 \\
\underline{10q - 15} \longleftarrow
\end{array}
$$

$\dfrac{10q}{2q} = 5$

$5(2q - 3)$

Step 6 Subtract $10q - 15$ from $10q - 23$, getting -8.

$$
\begin{array}{r}
3q^2 - 4q + 5 \\
2q - 3 \overline{) 6q^3 - 17q^2 + 22q - 23} \\
\underline{6q^3 - 9q^2} \\
-8q^2 + 22q \\
\underline{-8q^2 + 12q} \\
10q - 23 \\
\underline{-10q + 15} \\
-8
\end{array}
$$

Change signs and add

The quotient is $3q^2 - 4q + 5$, with remainder -8. Write this result as

$$
\frac{6q^3 - 17q^2 + 22q - 23}{2q - 3} = 3q^2 - 4q + 5 + \frac{-8}{2q - 3}. \quad \bullet
$$

EXAMPLE 9   Divide $3x^3 - 2x^2 - 150$ by $x - 4$.

The polynomial $3x^3 - 2x^2 - 150$ has a missing term, the term where the variable is x. When dividing a polynomial with a missing term, it is convenient to allow for that term by leaving a space, as shown below.

$$
\begin{array}{r}
3x^2 + 10x + 40 \\
x - 4\overline{\smash{)}\,3x^3 - 2x^2 \qquad\quad - 150} \\
\underline{3x^3 - 12x^2} \\
10x^2 \\
\underline{10x^2 - 40x} \\
40x - 150 \\
\underline{40x - 160} \\
10
\end{array}
$$

This result is written as follows:

$$\frac{3x^3 - 2x^2 - 150}{x - 4} = 3x^2 + 10x + 40 + \frac{10}{x - 4}. \quad \bullet$$

By the process of division shown above, if a polynomial P is divided by a polynomial D of lower degree, the result is a **quotient** polynomial Q and a **remainder** R such that

$$P = D \cdot Q + R.$$

The degree of R is less than the degree of D.

2.2 EXERCISES

Identify each of the following as a polynomial or not a polynomial. For each polynomial, give the degree and identify as monomial, binomial, trinomial, *or* none of these. *See Example 1.*

1. $5x^{11}$

2. $-9y^{12} + y^2$

3. $8p^5 + 6p$

4. $2a^6 - 5a^2 + 4a$

5. $\sqrt{2}x + \sqrt{3}x^6$

6. $-\sqrt{7}m^5 + 2\sqrt{3}m^4$

7. $\frac{1}{3}r^2 - \frac{3}{5}r^4 + r$

8. $\frac{3}{10}p^7 - \frac{2}{7}p^5$

9. $\frac{5}{p} + \frac{2}{p^2} - \frac{5}{p^3}$

10. $\frac{q^2}{6} - \frac{q^5}{9} + \frac{q^8}{10}$

11. $5\sqrt{z} + 2\sqrt{z^3} - 5\sqrt{z^5}$
   NOT POLYNOMIAL

12. $\frac{8}{\sqrt{y}} - \frac{6}{\sqrt{y^3}} + \frac{3}{\sqrt{y^5}}$

Find each of the following sums and differences. See Example 2.

13. $(3x^2 - 4x + 5) + (-2x^2 + 3x - 2)$

14. $(4m^3 - 3m^2 + 5) + (-3m^3 - m^2 + 5)$

15. $(x^2 + 4x) + (3x^3 - 4x^2 + 2x + 2)$

16. $(r^5 - r^3 + r) + (3r^5 - 4r^4 + r^3 + 2r)$

17. $(12y^2 - 8y + 6) - (3y^2 - 4y + 2)$

18. $(8p^2 - 5p) - (3p^2 - 2p + 4)$

19. $(p^3 - 4p^2 + p) - (3p^2 + 2p + 7)$

20. $(4y^2 - 2y + 7) - (6y - 9)$

21. $(3a^2 - 2a) + (4a^2 + 3a + 1) - (a^2 + 2)$

22. $(6m^4 - 3m^2 + m) - (2m^3 + 5m^2 + 4m) + (m^2 - m)$

23. $(5b^2 - 4b + 3) - (2b^2 + b) - (3b + 4)$

24. $-(8x^3 + x - 3) + (2x^3 + x^2) - (4x^2 + 3x - 1)$

25. $-3(4q^2 - 3q + 2) + 2(-q^2 + q - 4)$

26. $2(3r^2 + 4r + 2) - 3(-r^2 + 4r - 5)$

Find each of the following products. See Examples 3–6.

27. $p(4p - 6) + 2(3p - 8)$

28. $m(5m - 2) + 9(5 - m)$

29. $-y(y^2 - 4) + 6y^2(2y - 3)$

30. $-z^3(9 - z) + 4z(2 + 3z)$

31. $(4r - 1)(7r + 2)$

32. $(5m - 6)(3m + 4)$

33. $(6p + 5q)(3p - 7q)$

34. $(2z + y)(3z - 4y)$

35. $\left(3x - \dfrac{2}{3}\right)\left(5x + \dfrac{1}{3}\right)$

36. $\left(2m - \dfrac{1}{4}\right)\left(3m + \dfrac{1}{2}\right)$

37. $\left(\dfrac{2}{5}y + \dfrac{1}{8}z\right)\left(\dfrac{3}{5}y + \dfrac{1}{2}z\right)$

38. $\left(\dfrac{3}{4}r - \dfrac{2}{3}s\right)\left(\dfrac{5}{4}r + \dfrac{1}{3}s\right)$

39. $(5r + 2)(5r - 2)$

40. $(6z + 5)(6z - 5)$

41. $(4x + 3y)(4x - 3y)$

42. $(7m + 2n)(7m - 2n)$

43. $(6k - 3)^2$

44. $(3p + 5)^2$

45. $(4m + 2n)^2$

46. $(a - 6b)^2$

47. $[(2p - 3) + q]^2$

48. $[(4y - 1) + z]^2$

49. $[(3q + 5) - p][(3q + 5) + p]$

50. $[(9r - s) + 2][(9r - s) - 2]$

51. $(2z - 1)^3$

52. $(3m + 2)^3$

53. $(5r + 3t^2)^3$

54. $(2z^4 - 3y)^3$

55. $4x^2(3x^3 + 2x^2 - 5x + 1)$

56. $2b^3(b^2 - 4b + 3)$

57. $5m(3m^3 - 2m^2 + m - 1)$

58. $4y^3(y^3 + 2y^2 - 6y + 3)$

59. $(2z - 1)(-z^2 + 3z - 4)$

60. $(x - 1)(x^2 - 1)$

61. $(3p - 1)(9p^2 + 3p + 1)$

62. $(2p - 1)(3p^2 - 4p + 5)$

63. $(2m + 1)(4m^2 - 2m + 1)$

64. $(k + 2)(12k^3 - 3k^2 + k + 1)$

65. $(m - n + k)(m + 2n - 3k)$

66. $(r - 3s + t)(2r - s + t)$

67. $(a - b + 2c)^2$

68. $(k - y + 3m)^2$

69. $(x^{-1} - 2)^2$

70. $(1 + 2y^{-1})^2$

71. $(3m^{-1} - 2n^{-1})(4m^{-1} + n^{-1})$

72. $(-k^{-1} + 3q^{-1})(4k^{-1} - 3q^{-1})$

Perform each of the following divisions. See Examples 7–9.

73. $\dfrac{4m^3 - 8m^2 + 16m}{2m}$

74. $\dfrac{30k^5 - 12k^3 + 18k^2}{6k^2}$

75. $\dfrac{15x^4 + 30x^3 + 12x^2}{3x}$

76. $\dfrac{16a^6 + 24a^5 - 48a^4}{8a^2}$

77. $\dfrac{25x^2y^4 - 15x^3y^3 + 40x^4y^2}{5x^2y^2}$

78. $\dfrac{-8r^3s - 12r^2s^2 + 20rs^3}{4rs}$

79. $\dfrac{6y^2 + y - 2}{2y - 1}$

80. $\dfrac{16r^2 + 2r - 3}{2r + 1}$

81. $\dfrac{20p^2 + 11p - 3}{4p + 3}$

82. $\dfrac{32m^2 + 20m - 3}{8m - 1}$

83. $\dfrac{12x^2 - 7x - 12}{4x - 5}$

84. $\dfrac{12p^2 - 13p - 16}{3p - 2}$

85. $\dfrac{8z^2 + 14z - 20}{2z + 5}$

86. $\dfrac{6a^2 - 13a - 18}{2a + 1}$

87. $\dfrac{2x^3 - 11x^2 + 19x - 10}{2x - 5}$

88. $\dfrac{3p^3 - 11p^2 + 5p + 3}{3p + 1}$

89. $\dfrac{15x^3 + 11x^2 + 20}{3x + 4}$

90. $\dfrac{4r^3 + 2r^2 - 14r + 15}{2r + 5}$

91. $\dfrac{3m^4 - 14m^3 + 22m^2 - 22m + 6}{m - 3}$

92. $\dfrac{5a^4 + 4a^3 - a^2 + 19a - 10}{a + 2}$

93. $\dfrac{x^4 + 2x^3 + 2x^2 - 2x - 3}{x^2 - 1}$

94. $\dfrac{2y^5 + y^3 - 2y^2 - 1}{2y^2 + 1}$

95. $\dfrac{4z^5 - 4z^2 - 5z + 3}{2z^2 + z + 1}$

96. $\dfrac{12z^4 - 25z^3 + 35z^2 - 26z + 10}{4z^2 - 3z + 5}$

97. $\dfrac{6p^6 - 13p^5 - 8p^4 - 2p^3 + 3p^2 - 6p + 2}{3p^3 + p^2 - 1}$

98. $\dfrac{6r^6 - 31r^5 + 22r^4 + 40r^3 + 14r^2 + 27r - 6}{2r^3 - 5r^2 - 3}$

99. $\dfrac{6x^3 - 17x^2y + 14xy^2 - 3y^3}{2x - 3y}$

100. $\dfrac{20p^3 - 43p^2q + 45pq^2 - 24q^3}{4p - 3q}$

101. $\dfrac{2y^4 - 8y^3z + 13y^2z^2 - 10yz^3 + 3z^4}{2y^2 - 4yz + 3z^2}$

102. $\dfrac{10a^4 + 3a^3b + 8a^2b^2 + ab^3 + 2b^4}{5a^2 - ab + 2b^2}$

In each of the following, find the coefficient of x^3 without finding the entire product.

103. $(x^2 + 4x)(-3x^2 + 4x - 1)$ **104.** $(4x^3 - 2x + 5)(x^3 + 2)$ **105.** $(1 + x^2)(1 + x)$

106. $(3 - x)(2 - x^2)$ **107.** $x^2(4 - 3x)^2$ **108.** $-4x^2(2 - x)(2 + x)$

Find each of the following products. Assume that all variables used in exponents represent integers.

109. $(k^m + 2)(k^m - 2)$ **110.** $(y^x - 4)(y^x + 4)$ **111.** $(b^r + 3)(b^r - 2)$

112. $(q^y + 4)(q^y + 3)$ **113.** $(3p^x + 1)(p^x - 2)$ **114.** $(2^a + 5)(2^a + 3)$

115. $(m^x - 2)^2$ **116.** $(z^r + 5)^2$ **117.** $(q^p - 5p^q)^2$

118. $(3y^x - 2x^y)^2$ **119.** $(3k^a - 2)^3$ **120.** $(r^x - 4)^3$

Suppose one polynomial has degree m and another has degree n, where m and n are
natural numbers with n < m. Find the degree of the following for the polynomials.

121. Sum   **122.** Difference   **123.** Product

124. What would be the degree of the square of the polynomial of degree m?

125. Derive a formula for $(x + y + z)^2$.

126. Derive a formula for $(x + y + z)(x + y - z)$.

2.3 Factoring Polynomials

The process of finding polynomials whose product equals a given polynomial is called **factoring.** For example, since $4x + 12 = 4(x + 3)$, both 4 and $x + 3$ are called **factors** of $4x + 12$. Also, $4(x + 3)$ is the **factored form** of $4x + 12$. A polynomial is **factored completely** when it is written as a product of polynomials, none of which can be written as the product of polynomials of positive degree. A polynomial that cannot be written as a product of two polynomials of positive degree is a **prime** or **irreducible** polynomial.

One way to factor a polynomial is to use the distributive property. For example, to factor $6x^2y^3 + 9xy^4 + 18y^5$, look for a monomial that is the greatest common factor of each term. This polynomial has $3y^3$ as greatest common factor. By the distributive property,

$$6x^2y^3 + 9xy^4 + 18y^5 = (3y^3)(2x^2) + (3y^3)(3xy) + (3y^3)(6y^2)$$
$$= 3y^3(2x^2 + 3xy + 6y^2).$$

EXAMPLE 1   Factor the greatest common factor from each polynomial.

(a) $9y^5 + y^2$

The greatest common factor is y^2.

$$9y^5 + y^2 = y^2 \cdot 9y^3 + y^2 \cdot 1$$
$$= y^2 (9y^3 + 1)$$

Do not make the common error of forgetting to include the 1. Since $y^2(9y^3) \neq 9y^5 + y^2$, the 1 is essential in the answer.

(b) $6x^2t + 8xt + 12t = 2t(3x^2 + 4x + 6)$

(c) $14m^4(m + 1) - 28m^3(m + 1) - 7m^2(m + 1)$

The greatest common factor is $7m^2(m + 1)$. Use the distributive property.

$$14m^4(m + 1) - 28m^3(m + 1) - 7m^2(m + 1)$$
$$= [7m^2(m + 1)](2m^2 - 4m - 1)$$
$$= 7m^2(m + 1)(2m^2 - 4m - 1) \quad \bullet$$

Now we need to look at methods of factoring a trinomial of degree 2, such as $kx^2 + mx + n$, where k, m, and n are integers. Any factorization will be of the form $(ax + b)(cx + d)$, with a, b, c, and d integers. Multiplying out the product $(ax + b)(cx + d)$ gives

$$(ax + b)(cx + d) = acx^2 + (ad + bc)x + bd,$$

which equals $kx^2 + mx + n$ only if

$$ac = k, \qquad ad + bc = m, \qquad \text{and} \qquad bd = n. \qquad (*)$$

By this result, a trinomial $kx^2 + mx + n$ can be factored by finding four integers a, b, c, and d satisfying the conditions given in equations (*). If no such numbers exist, the trinomial is prime.

EXAMPLE 2   Factor each polynomial.

(a) $k^2 - 5k + 6$

To factor this polynomial, we need integers a, b, c, and d such that

$$k^2 - 5k + 6 = (ak + b)(ck + d).$$

By equations (*) above, these integers will satisfy the conditions $ac = 1$, $ad + bc = -5$, and $bc = 6$. The condition $ac = 1$ requires that $a = 1$ and $c = 1$. If both a and c equal 1, then $ad + bc = -5$ becomes $d + b = -5$. Now we must find integers b and d such that $b + d = -5$ and $bd = 6$. Inspection leads to the integers -2 and -3. Either b or d can equal either -2 or -3. Finally,

$$k^2 - 5k + 6 = (k - 2)(k - 3).$$

(b) $6p^2 - 7p - 5$

To find integers a, b, c, and d so that

$$6p^2 - 7p - 5 = (ap + b)(cp + d),$$

look for integers a, b, c, and d such that $ac = 6$, $ad + bc = -7$, and $bd = -5$. To find these integers, try various possibilities. Since $ac = 6$, we might let $a = 2$ and $c = 3$. Since $bd = -5$, we might let $b = -5$ and $d = 1$, giving

$$(2p - 5)(3p + 1) = 6p^2 - 13p - 5. \qquad \text{Incorrect}$$

We need to make another attempt; let us try

$$(3p - 5)(2p + 1) = 6p^2 - 7p - 5. \qquad \text{Correct}$$

Finally, $6p^2 - 7p - 5$ factors as $(3p - 5)(2p + 1)$.

(c) $4x^3 + 6x^2r - 10xr^2$

The greatest common factor of this polynomial is $2x$.

$$4x^3 + 6x^2r - 10xr^2 = 2x(2x^2 + 3xr - 5r^2)$$

After removing the $2x$, factor the trinomial $(2x^2 + 3xr - 5r^2)$.

To factor $2x^2 + 3xr - 5r^2$, we need the factors of $2x^2$ and of $-5r^2$ that will yield the correct middle term of $3xr$. By inspection,

$$4x^3 + 6x^2r - 10xr^2 = 2x(2x + 5r)(x - r).$$

Remember to write the $2x$ as part of the answer.

(d) $r^2 + 6r + 7$

It is not possible to find integers, a, b, c, and d so that

$$r^2 + 6r + 7 = (ar + b)(cr + d);$$

therefore, $r^2 + 6r + 7$ is a prime polynomial. ●

Factoring by Grouping   When a polynomial has more than three terms, it can sometimes be factored by a method called **factoring by grouping.** For example, to factor

$$ax + ay + 6x + 6y,$$

collect the terms into two groups,

$$ax + ay + 6x + 6y = (ax + ay) + (6x + 6y),$$

and then factor each group, getting

$$ax + ay + 6x + 6y = a(x + y) + 6(x + y).$$

The quantity $(x + y)$ is now a common factor, which can be factored out, producing

$$ax + ay + 6x + 6y = (x + y)(a + 6).$$

It is not always obvious which terms should be grouped. Experience, plus repeated trials, is the best teacher for techniques of factoring.

EXAMPLE 3   Factor by grouping.

(a) $mp^2 + 7m + 3p^2 + 21$

Group the terms as follows.

$$mp^2 + 7m + 3p^2 + 21 = (mp^2 + 7m) + (3p^2 + 21)$$

Use the greatest common factor for each part.

$$(mp^2 + 7m) + (3p^2 + 21) = m(p^2 + 7) + 3(p^2 + 7)$$
$$= (p^2 + 7)(m + 3)$$

(b) $2y^2 - 2z - ay^2 + az$

Grouping terms as above gives

$$2y^2 - 2z - ay^2 + az = (2y^2 - 2z) + (-ay^2 + az)$$
$$= 2(y^2 - z) + a(-y^2 + z).$$

The expression $-y^2 + z$ is the negative of $y^2 - z$, so the terms should be grouped as follows.

$$2y^2 - 2z - ay^2 + az = (2y^2 - 2z) - (ay^2 - az)$$
$$= 2(y^2 - z) - a(y^2 - z)$$
$$= (y^2 - z)(2 - a)$$

(c) $6p^2 - 15p + 16p - 40 = (6p^2 - 15p) + (16p - 40)$
$$= 3p(2p - 5) + 8(2p - 5)$$
$$= (2p - 5)(3p + 8) \quad \bullet$$

2.3 EXERCISES

Factor out the greatest common factor from each polynomial. See Example 1.

1. $12m + 36$

2. $5y + 10$

3. $8k^2 + 4$

4. $9z^3 + 3$

5. $12mn - 8m$

6. $3pq - 18pqr$

7. $12r^3 + 6r^2 - 3r$

8. $5k^2g - 25kg - 30kg^2$

9. $6px^2 - 8px^3 - 12px$

10. $9m^2n^3 - 18m^3n^2 + 27m^2n^4$

11. $45a^4b^2 - 63a^3b^4 + 18a^4b^3$

12. $8x^4y^2 + 24x^3y^3 - 12x^2y^4$

13. $4k^2m^3 + 8k^4m^3 - 12k^2m^4$

14. $28r^4s^2 + 7r^3s - 35r^4s^3$

15. $2(a + b) + 4m(a + b)$

16. $4(y - 2)^2 + 3(y - 2)$

17. $a^3(r + s) + b^2(r + s)$

18. $p^4(m - 2n) + q(m - 2n)$

19. $(2y - 3)(y + 2) + (y + 5)(y + 2)$

20. $(6a - 1)(a + 2) + (6a - 1)(3a - 1)$

21. $(5r - 6)(r + 3) - (2r - 1)(r + 3)$

22. $(3z + 2)(z + 4) - (z + 6)(z + 4)$

23. $2(m - 1) - 3(m - 1)^2 + 2(m - 1)^3$

24. $5(a + 3)^3 - 2(a + 3) + (a + 3)^2$

Factor each of the following polynomials twice. First use a common factor with a positive coefficient, then use a common factor with a negative coefficient. For example, $9x^2 - 3x^5$ would be factored as $3x^2(3 - x^3)$ and $-3x^2(-3 + x^3)$.

25. $25r^4 - 50r^3$

26. $60m^3 - 90m^6$

27. $-12a^5 + 24a^3 - 36a^2$

28. $-20p^4 + 15p^5 - 35p^6$

29. $-36m^4n^3 + 72m^2n^5 - 144m^5$

30. $-32z^3y^4 + 80z^5y^2 - 48z^3y^3$

Factor each trinomial. See Example 2.

31. $m^2 + 9m + 14$

32. $p^2 - 2p - 15$

33. $x^2 + 4x - 5$

34. $y^2 + y - 72$

35. $z^2 + 9z + 20$

36. $k^2 + 8k + 15$

37. $a^2 + 5ab + 4b^2$

38. $y^2 - 6yx + 8x^2$

39. $s^2 + 2st - 35t^2$

40. $n^2 - 12np + 35p^2$

41. $y^2 - 4yz - 21z^2$

42. $r^2 + rs - 42s^2$

43. $6a^2 - 48a - 120$

44. $8h^2 - 24h - 320$

45. $3m^3 + 12m^2 + 9m$

46. $9y^4 - 54y^3 + 45y^2$

47. $4p^2 + 3p - 1$

48. $6x^2 + 7x - 3$

49. $2z^2 + 7z - 30$

50. $4m^2 + m - 3$

51. $12r^2 + 24r - 15$

52. $12p^2 + p - 20$

53. $18r^2 - 3rs - 10s^2$

54. $12m^2 + 16mn - 35n^2$

55. $20a^2 - 13ab + 2b^2$

56. $10k^2 + 13kg - 3g^2$

57. $6k^2 + 5kp - 6p^2$

58. $14m^2 + 11mr - 15r^2$

59. $2x^2 - 5x - 3$

60. $3r^2 - r - 2$

61. $2a^2 - 17a + 30$

62. $3k^2 + 2k - 8$

63. $15y^2 + y - 2$

64. $6x^2 + x - 1$

65. $3p^2 - 7p + 10$

66. $8r^2 + r + 6$

67. $5a^2 - 7ab - 6b^2$

68. $12s^2 + 11st - 5t^2$

69. $21m^2 + 13mn + 2n^2$

70. $20y^2 + 39yx - 11x^2$

71. $9x^2 - 6x^3 + x^4$

72. $30a^2 + am - m^2$

73. $2r^2z - rz - 3z$

74. $3m^4 + 7m^3 + 2m^2$

75. $24a^4 + 10a^3b - 4a^2b^2$

76. $18x^5 + 15x^4z - 75x^3z^2$

77. $32z^5 - 20z^4a - 12z^3a^2$

78. $15x^4 - 7x^3p - 4x^2p^2$

Factor each of the following by grouping. See Example 3.

79. $mp + 4p + 3m + 12$

80. $rs + 6s - 3r - 18$

81. $rq + 2r + 2qs + 4s$

82. $zx + 2yx + 3z + 6y$

83. $6st + 9t - 10s - 15$

84. $10ab - 6b + 35a - 21$

85. $4x + 4y^2 - mx - my^2$

86. $x^2 + xy - 5x - 5y$

87. $rt^3 + rs^2 - pt^3 - ps^2$

88. $2m^4 + 6 - am^4 - 3a$

89. $6p^2 - 14p + 15p - 35$

90. $8r^2 - 10r + 12r - 15$

91. $20z^2 - 8zx - 45zx + 18x^2$

92. $16a^2 + 10ab - 24ab - 15b^2$

93. $15 - 5m^2 - 3r^2 + m^2r^2$

94. $6y + 3yz^2 - 4x^2 - 2x^2z^2$

95. $3m^2 - 3m^2p + 2 - 2p$

96. $2z^2 - 4z^2x^2 + 7x - 14x^3$

Factor each of the following. Assume that all variables used in exponents represent nonnegative integers.

97. $m^2 - mn^p - 2n^{2p}$

98. $r^2 + rs^q - 6s^{2q}$

99. $6z^{2a} - z^ax^b - x^{2b}$

100. $5p^{2c} + 13p^cq^c - 6q^{2c}$

101. $2x^{2n} - 23x^ny^n - 39y^{2n}$

102. $3a^{2x} + 7a^xb^x + 2b^{2x}$

Factor the coefficient of smallest absolute value from each of the following.

103. $9.44x^3 + 48.144x^2 - 30.208x + 37.76$

104. $12.915m^2 + 2.05m - 23.575m^3 + 20.5$

105. $64.616z^4 - 23.64z^2 + 7.88z + 114.26$

106. $1.47y^5 - 8.526y^3 + 69.384y^2 - 36.75$

Factor the variable of smallest exponent, together with any numerical common factor, from each of the following expressions. [For example, factor $9x^{-2} - 6x^{-3}$ as $3x^{-3}(3x - 2)$.

107. $p^{-4} - p^{-2}$

108. $m^{-1} + 3m^{-5}$

109. $12k^{-3} + 4k^{-2} - 8k^{-1}$

110. $6a^{-5} - 10a^{-4} + 18a^{-2}$

111. $110p^{-6} - 50p^{-2} + 75p^2$

112. $32y^{-3} + 48y - 64y^2$

2.4   Further Techniques of Factoring

Each of the special patterns of multiplication given earlier can be used in reverse to get a pattern for factoring. One of the most common of these is shown below.

Difference of Two Squares

$$x^2 - y^2 = (x + y)(x - y)$$

CONCEPT MARS

EXAMPLE 1   Factor each of the following polynomials.

(a) $4m^2 - 9$

First, recognize that $4m^2 - 9$ is the difference of two squares, since $4m^2 = (2m)^2$ and $9 = 3^2$. Use the pattern for the difference of two squares with $2m$ replacing x and 3 replacing y. Doing this gives

$$4m^2 - 9 = (2m)^2 - 3^2$$
$$= (2m + 3)(2m - 3).$$

(b) $144r^2 - 25s^2 = (12r + 5s)(12r - 5s)$

(c) $256k^4 - 625m^4$

Use the difference of two squares pattern twice, as follows:

$$256k^4 - 625m^4 = (16k^2)^2 - (25m^2)^2$$
$$= (16k^2 + 25m^2)(16k^2 - 25m^2)$$
$$= (16k^2 + 25m^2)(4k + 5m)(4k - 5m).$$

(d) $(a + 2b)^2 - 4c^2 = (a + 2b)^2 - (2c)^2$
$$= [(a + 2b) + 2c][(a + 2b) - 2c]$$
$$= (a + 2b + 2c)(a + 2b - 2c)$$

(e) $y^{4q} - z^{2q} = (y^{2q} + z^q)(y^{2q} - z^q)$ (if q is an integer)   ●

A perfect square trinomial can be factored as shown below.

Perfect Square Trinomial

$$x^2 + 2xy + y^2 = (x + y)^2$$
$$x^2 - 2xy + y^2 = (x - y)^2$$

EXAMPLE 2   Factor each polynomial.

(a) $16p^2 - 40pq + 25q^2$

Since $16p^2 = (4p)^2$ and $25q^2 = (5q)^2$, use the second pattern shown above with $4p$ replacing x and $5q$ replacing y to get

$$16p^2 - 40pq + 25q^2 = (4p)^2 - 2(4p)(5q) + (5q)^2$$
$$= (4p - 5q)^2.$$

Make sure that the middle term of the trinomial being factored, $-40pq$ here, is twice the product of the two terms in the binomial $(4p - 5q)^2$.

$$-40pq = 2(4p)(-5q)$$

(b) $169x^2 + 104xy^2 + 16y^4 = (13x + 4y^2)^2$

(c) $x^2 - 6x + 9 - y^4 = (x^2 - 6x + 9) - y^4$

$$= (x - 3)^2 - (y^2)^2$$

$$= [(x - 3) + y^2][(x - 3) - y^2]$$

$$= (x - 3 + y^2)(x - 3 - y^2)$$

Example 2(c) above uses the methods for the difference of two squares and for a perfect square trinomial.

(d) $4p^{2a} - 12p^a q^{2a} + 9q^{4a} = (2p^a - 3q^{2a})^2$ (if a is an integer)   ●

Two other special results of factoring are listed below.

Difference and Sum of Cubes

MEMORIZE

Difference of two cubes	$x^3 - y^3 = (x - y)(x^2 + xy + y^2)$
Sum of two cubes	$x^3 + y^3 = (x + y)(x^2 - xy + y^2)$

EXAMPLE 3

Factor each polynomial.

(a) $m^3 - 64n^3$

Since $64n^3 = (4n)^3$, the given polynomial is a difference of two cubes. To factor, use the first pattern in the box above, replacing x with m and y with $4n$, to get

$$m^3 - 64n^3 = m^3 - (4n)^3$$

$$= (m - 4n)[m^2 + m(4n) + (4n)^2]$$

$$= (m - 4n)(m^2 + 4mn + 16n^2).$$

(b) $8q^6 + 125p^9$

Write $8q^6$ as $(2q^2)^3$ and $125p^9$ as $(5p^3)^3$, so that the given polynomial is a sum of two cubes. Factor it as

$$8q^6 + 125p^9 = (2q^2)^3 + (5p^3)^3$$

$$= (2q^2 + 5p^3)[(2q^2)^2 - (2q^2)(5p^3) + (5p^3)^2]$$

$$= (2q^2 + 5p^3)(4q^4 - 10q^2p^3 + 25p^6).$$

(c) $(2a - 1)^3 + 8$

Use the pattern for the sum of two cubes, with $x = 2a - 1$ and $y = 2$. Doing so gives

$$(2a - 1)^3 + 8 = [(2a - 1) + 2][(2a - 1)^2 - (2a - 1)2 + 2^2]$$

$$= (2a - 1 + 2)(4a^2 - 4a + 1 - 4a + 2 + 4)$$

$$= (2a + 1)(4a^2 - 8a + 7).$$ ●

Method of Substitution   Some more complicated polynomials can be factored by substituting one expression for another. The next example shows this **method of substitution.**

EXAMPLE 4   Factor each polynomial.

(a) $6z^4 - 13z^2 - 5$

Replace z^2 with y, so that $y^2 = (z^2)^2 = z^4$. This replacement gives

$$6z^4 - 13z^2 - 5 = 6y^2 - 13y - 5.$$

Factor $6y^2 - 13y - 5$ as

$$6y^2 - 13y - 5 = (2y - 5)(3y + 1).$$

Replacing y with z^2 gives

$$6z^4 - 13z^2 - 5 = (2z^2 - 5)(3z^2 + 1).$$

(b) $10(2a - 1)^2 - 19(2a - 1) - 15$

Replacing $2a - 1$ with x gives

$$10x^2 - 19x - 15 = (5x + 3)(2x - 5).$$

Now replace x with $2a - 1$ and simplify.

$$
\begin{aligned}
10(2a - 1)^2 - 19(2a - 1) - 15 &= [5(2a - 1) + 3][2(2a - 1) - 5] \\
&= (10a - 5 + 3)(4a - 2 - 5) \\
&= (10a - 2)(4a - 7) \\
&= 2(5a - 1)(4a - 7) \quad \bullet
\end{aligned}
$$

2.4 EXERCISES

Factor each difference of two squares. See Example 1.

1. $p^2 - 100$

2. $m^2 - 81$

3. $9a^2 - 16$

4. $16q^2 - 25$

5. $144z^2 - 121$

6. $100r^2 - 169s^2$

7. $25s^4 - 9t^2$

8. $36z^2 - 81y^4$

9. $x^2y^2 - 25a^2$

10. $9p^2 - q^2r^2$

11. $(a + b)^2 - 16$

12. $(p - 2q)^2 - 100$

13. $9 - (2m - n)^2$

14. $16 - (a - 3b)^2$

15. $36q^2 - (10a - b)^2$

16. $49r^2 - (3p - q)^2$

17. $(x + y)^2 - (x - y)^2$

18. $(m + 2p)^2 - (m - 2p)^2$

19. $p^4 - 625$

20. $m^4 - 81$

21. $9r^8 - 16s^{10}$

22. $25y^6 - 81z^{12}$

23. $100m^6 - 25p^4$

24. $9z^2 - 81x^4$

Factor each perfect square trinomial. See Example 2.

25. $9m^2 - 12m + 4$

26. $16p^2 - 40p + 25$

27. $36a^2 + 60a + 25$

28. $4z^2 + 28z + 49$

29. $25k^2 - 40kp + 16p^2$

30. $49y^2 + 70yz + 25z^2$

31. $32a^2 - 48ab + 18b^2$

32. $20p^2 - 100pq + 125q^2$

33. $4x^2y^2 + 28xy + 49$

34. $9m^2n^2 - 12mn + 4$

35. $(a - 3b)^2 - 6(a - 3b) + 9$

36. $(2p + q)^2 - 10(2p + q) + 25$

37. $q^2 + 6q + 9 - p^2$

38. $4b^2 + 4bc + c^2 - 16$

39. $a^2 + 2ab + b^2 - x^2 - 2xy - y^2$

40. $d^2 - 10d + 25 - c^2 + 4c - 4$

Factor each sum or difference of cubes. See Example 3.

41. $8 - a^3$

42. $r^3 + 27$

43. $125x^3 - 27$

44. $8m^3 - 27n^3$

45. $216p^3 + 125q^3$

46. $1000x^3 + 343y^3$

47. $64 - x^6$

48. $125m^6 - 216$

49. $1331p^3q^3 + 125k^6$

50. $512z^3x^3 + 27y^6$

51. $27y^9 + 125z^6$

52. $1000m^{12} - 27n^3$

53. $125p^6 - 1000q^3$

54. $27z^3 + 729y^3$

55. $(r + 6)^3 - 216$

56. $(b + 3)^3 - 27$

57. $27 - (m + 2n)^3$

58. $125 - (4a - b)^3$

59. $125m^6 - 8(z + x^2)^3$

60. $64p^6 - 27(p^2 - q)^3$

Factor each of the following, using the method of substitution. See Example 4.

61. $m^4 - 3m^2 - 10$

62. $a^4 - 2a^2 - 48$

63. $6p^4 + 7p^2 - 3$

64. $4z^4 - 7z^2 - 15$

65. $7(3k - 1)^2 + 26(3k - 1) - 8$

66. $6(4z - 3)^2 + 7(4z - 3) - 3$

67. $(2y - 1)^2 - 4(2y - 1) + 4$

68. $(3a + 5)^2 - 18(3a + 5) + 81$

69. $9(a - 4)^2 + 30(a - 4) + 25$

70. $16(2r - 3)^2 + 24(2r - 3) + 9$

71. $6(1 - 3m)^2 - (1 - 3m) - 35$

72. $20(4 - p)^2 - 3(4 - p) - 2$

Factor each of the following polynomials. Assume that all variables used in exponents represent positive integers.

73. $9a^{4k} - b^{8k}$

74. $16y^{2c} - 25x^{4c}$

75. $16 - (x^a + y^a)^2$

76. $(3p^r + 2q^r)^2 - 9$

77. $4y^{2a} - 12y^a + 9$

78. $25x^{4c} - 20x^{2c} + 4$

79. $x^{3n} - y^{6n}$

80. $a^{9c} - b^{3c}$

81. $125r^{6k} + 216s^{12k}$

82. $216p^{12r} + 125q^{15r}$

83. $6(m + p)^{2k} + (m + p)^k - 15$

84. $8(2k + q)^{4z} - 2(2k + q)^{2z} - 3$

Find a value of b or c that will make the following expressions perfect square trinomials.

85. $p^2 + bp + 49$

86. $k^2 + bk + 16$

87. $4z^2 + bz + 81$

88. $9p^2 + bp + 25$

89. $100r^2 - 60r + c$

90. $49x^2 + 70x + c$

2.5 Rational Expressions

An expression that is the quotient of two algebraic expressions (with denominator not 0) is called a **fractional expression.** The most common fractional expressions are those that are the quotients of two polynomials; these are called **rational expressions.** Since fractional expressions involve quotients, it is important to keep track of the values of the variable that satisfy the requirement that no denominator be 0. The set of such values is the **domain** of the expression.

For example, -2 is not in the domain of the rational expression

$$\frac{x + 6}{x + 2}$$

because replacing x with -2 makes the denominator equal 0. The domain of this rational expression can be written $\{x | x \neq -2\}$. In a similar way, the domain of

$$\frac{(x + 6)(x + 4)}{(x + 2)(x + 4)}$$

can be written $\{x | x \neq -2 \text{ and } x \neq -4\}$.

Just as the fraction 6/8 is written in lowest terms as 3/4, rational expressions must also be written in lowest terms. Do this with the **fundamental principle,** which says that

Fundamental Principle

$$\frac{ac}{bc} = \frac{a}{b} \quad (\text{if } c \neq 0).$$

EXAMPLE 1   Write each expression in lowest terms.

(a) $\dfrac{2p^2 + 7p - 4}{5p^2 + 20p}$

Factor the numerator and denominator to get

$$\frac{2p^2 + 7p - 4}{5p^2 + 20p} = \frac{(2p - 1)(p + 4)}{5p(p + 4)}.$$

By the fundamental principle,

$$\frac{2p^2 + 7p - 4}{5p^2 + 20p} = \frac{2p - 1}{5p}.$$

The domain of the original expression is $\{p | p \neq 0 \text{ and } p \neq -4\}$, so this result is valid only for values of p other than 0 and -4. From now on, we shall always assume such restrictions when reducing rational expressions.

(b) $\dfrac{6 - 3k}{k^2 - 4}$

Factor to get

$$\frac{6 - 3k}{k^2 - 4} = \frac{3(2 - k)}{(k + 2)(k - 2)}.$$

The factors $2 - k$ and $k - 2$ have opposite signs. Because of this, multiply numerator and denominator by -1, as follows.

$$\frac{6 - 3k}{k^2 - 4} = \frac{3(2 - k)(-1)}{(k + 2)(k - 2)(-1)}$$

Since $(k - 2)(-1) = -k + 2$, or $2 - k$,

$$\frac{6 - 3k}{k^2 - 4} = \frac{3(2 - k)(-1)}{(k + 2)(2 - k)},$$

finally giving

$$\frac{6 - 3k}{k^2 - 4} = \frac{-3}{k + 2}.$$

Working in an alternate way would lead to the equivalent result

$$\frac{3}{-k - 2}. \quad \bullet$$

To multiply or divide rational expressions, use properties and definitions from earlier work.

Multiplication and Division

$$\frac{a}{b} \cdot \frac{c}{d} = \frac{ac}{bd}$$

$$\frac{a}{b} \div \frac{c}{d} = \frac{a}{b} \cdot \frac{d}{c} \quad \left(\text{if } \frac{c}{d} \neq 0\right)$$

EXAMPLE 2

Multiply or divide, as indicated.

(a) $\dfrac{2y^2}{9} \cdot \dfrac{27}{8y^5} = \dfrac{2y^2 \cdot 27}{9 \cdot 8y^5}$

$$= \frac{2 \cdot 9 \cdot 3y^2}{2 \cdot 9 \cdot 4y^2 \cdot y^3}$$

$$= \frac{3}{4y^3}$$

The product was written in lowest terms in the last step.

(b) $\dfrac{3m^2 - 2m - 8}{3m^2 + 14m + 8} \cdot \dfrac{3m + 2}{3m + 4} = \dfrac{(m - 2)(3m + 4)}{(m + 4)(3m + 2)} \cdot \dfrac{3m + 2}{3m + 4}$

$$= \dfrac{(m - 2)(3m + 4)(3m + 2)}{(m + 4)(3m + 2)(3m + 4)}$$

$$= \dfrac{m - 2}{m + 4}$$

(c) $\dfrac{5}{8m + 16} \div \dfrac{7}{12m + 24} = \dfrac{5}{8(m + 2)} \div \dfrac{7}{12(m + 2)}$

$$= \dfrac{5}{8(m + 2)} \cdot \dfrac{12(m + 2)}{7}$$

$$= \dfrac{5 \cdot 12(m + 2)}{8 \cdot 7(m + 2)}$$

$$= \dfrac{15}{14}$$

(d) $\dfrac{3p^2 + 11p - 4}{24p^3 - 8p^2} \div \dfrac{9p + 36}{24p^4 - 36p^3} = \dfrac{(p + 4)(3p - 1)}{8p^2(3p - 1)} \div \dfrac{9(p + 4)}{12p^3(2p - 3)}$

$$= \dfrac{(p + 4)(3p - 1)(12p^3)(2p - 3)}{8p^2(3p - 1)(9)(p + 4)}$$

$$= \dfrac{12p^3(2p - 3)}{9 \cdot 8p^2}$$

$$= \dfrac{p(2p - 3)}{6} \quad \bullet$$

To add or subtract two rational expressions, use the appropriate properties from earlier work.

Addition and Subtraction

$$\dfrac{a}{b} + \dfrac{c}{d} = \dfrac{ad + bc}{bd}$$

$$\dfrac{a}{b} - \dfrac{c}{d} = \dfrac{ad - bc}{bd}$$

In practice, rational expressions are normally added or subtracted after rewriting all the rational expressions with a common denominator found with the steps given below.

Common Denominator

1. Write each denominator as a product of prime factors.
2. Form a product of all the different prime factors. Each factor should have as exponent the *highest* exponent which appears on that factor.

EXAMPLE 3   Add or subtract, as indicated.

(a) $\dfrac{5}{9x^2} + \dfrac{1}{6x}$

Write each denominator as a product of prime factors, as follows.

$$9x^2 = 3^2 \cdot x^2$$

$$6x = 2^1 \cdot 3^1 x^1$$

For the common denominator, form the product of all the prime factors, with each factor having the highest exponent that appears on it. Here the highest exponent on 2 is 1, while both 3 and x have a highest exponent of 2. The common denominator is

$$2^1 \cdot 3^2 \cdot x^2 = 18x^2.$$

Now write both of the given rational expressions with this denominator.

$$\frac{5}{9x^2} + \frac{1}{6x} = \frac{5 \cdot 2}{9x^2 \cdot 2} + \frac{1 \cdot 3x}{6x \cdot 3x}$$

$$= \frac{10}{18x^2} + \frac{3x}{18x^2}$$

$$= \frac{10 + 3x}{18x^2}$$

(b) $\dfrac{y + 2}{y^2 - y} - \dfrac{3y}{2y^2 - 4y + 2}$

Factor each denominator, giving

$$\frac{y + 2}{y^2 - y} - \frac{3y}{2y^2 - 4y + 2} = \frac{y + 2}{y(y - 1)} - \frac{3y}{2(y - 1)^2}.$$

The common denominator, by the method above, is $2y(y - 1)^2$. Write each rational expression with this denominator, as follows.

$$\frac{y + 2}{y(y - 1)} - \frac{3y}{2(y - 1)^2} = \frac{2(y - 1)(y + 2)}{2y(y - 1)^2} - \frac{y \cdot 3y}{2y(y - 1)^2}$$

$$= \frac{2(y^2 + y - 2)}{2y(y - 1)^2} - \frac{3y^2}{2y(y - 1)^2}$$

$$= \frac{2y^2 + 2y - 4 - 3y^2}{2y(y - 1)^2}$$

$$= \frac{-y^2 + 2y - 4}{2y(y - 1)^2}$$

(c) $\dfrac{3}{(x-1)(x+2)} - \dfrac{4}{(x+3)(x-4)}$

$$= \dfrac{3(x+3)(x-4)}{(x-1)(x+2)(x+3)(x-4)} - \dfrac{4(x-1)(x+2)}{(x+3)(x-4)(x-1)(x+2)}$$

$$= \dfrac{3(x^2-x-12) - 4(x^2+x-2)}{(x-1)(x+2)(x+3)(x-4)}$$

$$= \dfrac{3x^2 - 3x - 36 - 4x^2 - 4x + 8}{(x-1)(x+2)(x+3)(x-4)}$$

$$= \dfrac{-x^2 - 7x - 28}{(x-1)(x+2)(x+3)(x-4)} \quad \bullet$$

Complex Fractions   Any quotient of two rational expressions is called a **complex fraction.** Complex fractions can often be simplified by the methods shown in the following examples.

EXAMPLE 4   Simplify each complex fraction.

(a) $\dfrac{6 - \dfrac{5}{k}}{1 + \dfrac{5}{k}}$

Multiply both numerator and denominator by the common denominator k.

$$\dfrac{k\left(6 - \dfrac{5}{k}\right)}{k\left(1 + \dfrac{5}{k}\right)} = \dfrac{6k - k\left(\dfrac{5}{k}\right)}{k + k\left(\dfrac{5}{k}\right)} = \dfrac{6k - 5}{k + 5}$$

(b) $\dfrac{\dfrac{a}{a+1} + \dfrac{1}{a}}{\dfrac{1}{a} + \dfrac{1}{a+1}}$

Multiply both numerator and denominator by the common denominator of all the fractions, in this case $a(a+1)$. Doing so gives

$$\dfrac{\dfrac{a}{a+1} + \dfrac{1}{a}}{\dfrac{1}{a} + \dfrac{1}{a+1}} = \dfrac{\left(\dfrac{a}{a+1} + \dfrac{1}{a}\right)a(a+1)}{\left(\dfrac{1}{a} + \dfrac{1}{a+1}\right)a(a+1)}$$

$$= \dfrac{a^2 + (a+1)}{(a+1) + a} = \dfrac{a^2 + a + 1}{2a + 1}.$$

As an alternate method of solution, first perform the indicated additions in the numerator and denominator, and then divide.

$$\frac{\dfrac{a}{a + 1} + \dfrac{1}{a}}{\dfrac{1}{a} + \dfrac{1}{a + 1}} = \frac{\dfrac{a^2 + 1(a + 1)}{a(a + 1)}}{\dfrac{1(a + 1) + 1(a)}{a(a + 1)}}$$

$$= \frac{\dfrac{a^2 + a + 1}{a(a + 1)}}{\dfrac{2a + 1}{a(a + 1)}}$$

$$= \frac{a^2 + a + 1}{a(a + 1)} \cdot \frac{a(a + 1)}{2a + 1}$$

$$= \frac{a^2 + a + 1}{2a + 1} \quad \bullet$$

The next example shows how negative exponents can lead to rational expressions.

EXAMPLE 5 Simplify $\dfrac{(x + y)^{-1}}{x^{-1} + y^{-1}}$. Write the result with only positive exponents.

Use the definition of negative integer exponent to get

$$\frac{(x + y)^{-1}}{x^{-1} + y^{-1}} = \frac{\dfrac{1}{x + y}}{\dfrac{1}{x} + \dfrac{1}{y}}$$

$$= \frac{\dfrac{1}{x + y}}{\dfrac{y + x}{xy}}$$

$$= \frac{1}{x + y} \cdot \frac{xy}{x + y} = \frac{xy}{(x + y)^2}. \quad \bullet$$

2.5 EXERCISES

Give the domain of each of the following.

1. $\dfrac{x - 2}{x + 6}$

2. $\dfrac{x + 5}{x - 3}$

3. $\dfrac{2x}{5x - 3}$

4. $\dfrac{6x}{2x - 1}$

5. $\dfrac{-8}{x^2 + 1}$

6. $\dfrac{3x}{3x^2 + 7}$

Write each of the following in lowest terms. See Example 1.

7. $\dfrac{25p^3}{10p^2}$

8. $\dfrac{14z^3}{6z^2}$

9. $\dfrac{8k + 16}{9k + 18}$

10. $\dfrac{20r + 10}{30r + 15}$

11. $\dfrac{3(t + 5)}{(t + 5)(t - 3)}$

12. $\dfrac{-8(y - 4)}{(y + 2)(y - 4)}$

13. $\dfrac{8x^2 + 16x}{4x^2}$

14. $\dfrac{36y^2 + 72y}{9y}$

15. $\dfrac{m^2 - 4m + 4}{m^2 + m - 6}$

16. $\dfrac{r^2 - r - 6}{r^2 + r - 12}$

17. $\dfrac{8m^2 + 6m - 9}{16m^2 - 9}$

18. $\dfrac{6y^2 + 11y + 4}{3y^2 + 7y + 4}$

Find each of the following products or quotients. See Example 2.

19. $\dfrac{x(y + 2)}{4} \div \dfrac{x}{2}$

20. $\dfrac{mn}{m + n} \div \dfrac{p}{m + n}$

21. $\dfrac{15p^3}{9p^2} \div \dfrac{6p}{10p^2}$

22. $\dfrac{3r^2}{9r^3} \div \dfrac{8r^3}{6r}$

23. $\dfrac{2k + 8}{6} \div \dfrac{3k + 12}{2}$

24. $\dfrac{5m + 25}{10} \cdot \dfrac{12}{6m + 30}$

25. $\dfrac{9y - 18}{6y + 12} \cdot \dfrac{3y + 6}{15y - 30}$

26. $\dfrac{12r + 24}{36r - 36} \div \dfrac{6r + 12}{8r - 8}$

27. $\dfrac{x^2 + x}{5} \cdot \dfrac{25}{xy + y}$

28. $\dfrac{3m - 15}{4m - 20} \cdot \dfrac{m^2 - 10m + 25}{12m - 60}$

29. $\dfrac{4a + 12}{2a - 10} \div \dfrac{a^2 - 9}{a^2 - a - 20}$

30. $\dfrac{6r - 18}{9r^2 + 6r - 24} \cdot \dfrac{12r - 16}{4r - 12}$

31. $\dfrac{p^2 - p - 12}{p^2 - 2p - 15} \cdot \dfrac{p^2 - 9p + 20}{p^2 - 8p + 16}$

32. $\dfrac{x^2 + 2x - 15}{x^2 + 11x + 30} \cdot \dfrac{x^2 + 2x - 24}{x^2 - 8x + 15}$

33. $\dfrac{k^2 - k - 6}{k^2 + k - 12} \cdot \dfrac{k^2 + 3k - 4}{k^2 + 2k - 3}$

34. $\dfrac{n^2 - n - 6}{n^2 - 2n - 8} \div \dfrac{n^2 - 9}{n^2 + 7n + 12}$

35. $\dfrac{m^2 + 3m + 2}{m^2 + 5m + 4} \div \dfrac{m^2 + 5m + 6}{m^2 + 10m + 24}$

36. $\dfrac{y^2 + y - 2}{y^2 + 3y - 4} \div \dfrac{y^2 + 3y + 2}{y^2 + 4y + 3}$

37. $\dfrac{2m^2 - 5m - 12}{m^2 - 10m + 24} \div \dfrac{4m^2 - 9}{m^2 - 9m + 18}$

38. $\dfrac{6n^2 - 5n - 6}{6n^2 + 5n - 6} \cdot \dfrac{12n^2 - 17n + 6}{12n^2 - n - 6}$

39. $\left(1 + \dfrac{1}{x}\right)\left(1 - \dfrac{1}{x}\right)$

40. $\left(3 + \dfrac{2}{y}\right)\left(3 - \dfrac{2}{y}\right)$

41. $\dfrac{x^3 + y^3}{x^2 - y^2} \cdot \dfrac{x + y}{x^2 - xy + y^2}$

42. $\dfrac{8y^3 - 125}{4y^2 - 20y + 25} \cdot \dfrac{2y - 5}{y}$

43. $\dfrac{x^3 + y^3}{x^3 - y^3} \cdot \dfrac{x^2 - y^2}{x^2 + 2xy + y^2}$

44. $\dfrac{x^2 - y^2}{(x - y)^2} \cdot \dfrac{x^2 - xy + y^2}{x^2 - 2xy + y^2} \div \dfrac{x^3 + y^3}{(x - y)^4}$

Perform each of the following additions or subtractions. See Example 3.

45. $\dfrac{8}{r} + \dfrac{6}{r}$

46. $\dfrac{3}{y} + \dfrac{4}{y}$

47. $\dfrac{3}{2k} + \dfrac{5}{3k}$

48. $\dfrac{8}{5p} + \dfrac{3}{4p}$

49. $\dfrac{2}{3y} - \dfrac{1}{4y}$

50. $\dfrac{6}{11z} - \dfrac{5}{2z}$

51. $\dfrac{a + 1}{2} - \dfrac{a - 1}{2}$

52. $\dfrac{y + 6}{5} - \dfrac{y - 6}{5}$

53. $\dfrac{3}{p} + \dfrac{1}{2}$

54. $\dfrac{9}{r} - \dfrac{2}{3}$

55. $\dfrac{2}{y} - \dfrac{1}{4}$

56. $\dfrac{6}{11} + \dfrac{3}{a}$

57. $\dfrac{1}{6m} + \dfrac{2}{5m} + \dfrac{4}{m}$

58. $\dfrac{8}{3p} + \dfrac{5}{4p} + \dfrac{9}{2p}$

59. $\dfrac{1}{y} + \dfrac{1}{y + 1}$

60. $\dfrac{2}{3(x - 1)} + \dfrac{1}{4(x - 1)}$

61. $\dfrac{2}{a + b} - \dfrac{1}{2(a + b)}$

62. $\dfrac{3}{m} - \dfrac{1}{m - 1}$

63. $\dfrac{1}{a + 1} - \dfrac{1}{a - 1}$

64. $\dfrac{1}{x + z} + \dfrac{1}{x - z}$

65. $\dfrac{m + 1}{m - 1} + \dfrac{m - 1}{m + 1}$

66. $\dfrac{2}{x - 1} + \dfrac{1}{1 - x}$

67. $\dfrac{3}{a - 2} - \dfrac{1}{2 - a}$

68. $\dfrac{q}{p - q} - \dfrac{q}{q - p}$

69. $\dfrac{x + y}{2x - y} - \dfrac{2x}{y - 2x}$

70. $\dfrac{m - 4}{3m - 4} + \dfrac{3m + 2}{4 - 3m}$

71. $\dfrac{1}{a^2 - 5a + 6} - \dfrac{1}{a^2 - 4}$

72. $\dfrac{-3}{m^2 - m - 2} - \dfrac{1}{m^2 + 3m + 2}$

73. $\dfrac{1}{x^2 + x - 12} - \dfrac{1}{x^2 - 7x + 12} + \dfrac{1}{x^2 - 16}$

74. $\dfrac{2}{2p^2 - 9p - 5} + \dfrac{p}{3p^2 - 17p + 10} - \dfrac{2p}{6p^2 - p - 2}$

75. $\dfrac{3a}{a^2 + 5a - 6} - \dfrac{2a}{a^2 + 7a + 6}$

76. $\dfrac{2k}{k^2 + 4k + 3} + \dfrac{3k}{k^2 + 5k + 6}$

77. $\dfrac{4x - 1}{x^2 + 3x - 10} + \dfrac{2x + 3}{x^2 + 4x - 5}$

78. $\dfrac{3y + 5}{y^2 - 9y + 20} + \dfrac{2y - 7}{y^2 - 2y - 8}$

Perform the indicated operations. See Example 4.

79. $\left(\dfrac{3}{p - 1} - \dfrac{2}{p + 1} \right) \left(\dfrac{p - 1}{p} \right)$

80. $\left(\dfrac{y}{y^2 - 1} - \dfrac{y}{y^2 - 2y + 1} \right) \left(\dfrac{y - 1}{y + 1} \right)$

81. $\dfrac{\dfrac{1}{x + h} - \dfrac{1}{x}}{h}$

82. $\dfrac{1}{h} \left(\dfrac{1}{(x + h)^2 + 9} - \dfrac{1}{x^2 + 9} \right)$

83. $\dfrac{1 + \dfrac{1}{x}}{1 - \dfrac{1}{x}}$

84. $\dfrac{2 - \dfrac{2}{y}}{2 + \dfrac{2}{y}}$

85. $\dfrac{\dfrac{1}{x + 1} - \dfrac{1}{x}}{\dfrac{1}{x}}$

86. $\dfrac{\dfrac{1}{y + 3} - \dfrac{1}{y}}{\dfrac{1}{y}}$

87. $\dfrac{1 + \dfrac{1}{1 - b}}{1 - \dfrac{1}{1 + b}}$

88. $m - \dfrac{m}{m + \dfrac{1}{2}}$

89. $\dfrac{m - \dfrac{1}{m^2 - 4}}{\dfrac{1}{m + 2}}$

90. $\dfrac{\dfrac{3}{p^2 - 16} + p}{\dfrac{1}{p - 4}}$

91. $\dfrac{\dfrac{8}{z^2 - 25z} + 3}{\dfrac{5}{z(z + 1)}}$

92. $\dfrac{9 - \dfrac{1}{r^2 - r}}{\dfrac{3}{r(r + 2)}}$

Perform all indicated operations and write all answers with positive integer exponents.
See Example 5.

93. $\dfrac{a^{-1} + b^{-1}}{(ab)^{-1}}$

94. $\dfrac{p^{-1} - q^{-1}}{(pq)^{-1}}$

95. $\dfrac{r^{-1} + q^{-1}}{r^{-1} - q^{-1}} \cdot \dfrac{r - q}{r + q}$

96. $\dfrac{xy^{-1} + yx^{-1}}{x^2 + y^2}$

97. $(a + b)^{-1}(a^{-1} + b^{-1})$

98. $(m^{-1} + n^{-1})^{-1}$

99. $\dfrac{x - 9y^{-1}}{(x - 3y^{-1})(x + 3y^{-1})}$

100. $\dfrac{(m + n)^{-1}}{m^{-2} - n^{-2}}$

2.6 Rational Exponents

Earlier in this chapter we gave a definition of a^n for integer values of n. Now we shall extend this definition to include rational values of n. Let us begin with the symbol $a^{1/n}$. We would want any definition of $a^{1/n}$ to be consistent with past properties of exponents. One of these properties says that if a is a nonzero real number, and m and n are integers,

$$(a^m)^n = a^{mn}.$$

Assume $n \neq 0$ and replace m with $1/n$ to get

$$(a^{1/n})^n = a^{(1/n)n} = a^1 = a,$$

so that a is the nth power of $a^{1/n}$. For consistency with past work, $a^{1/n}$ must be defined as an **nth root** of a. That is, a definition of $a^{1/n}$ must satisfy the condition

$$(a^{1/n})^n = a.$$

The definition of $a^{1/n}$ depends on whether n is odd or even and whether a is positive or negative. For example, $16^{1/2}$ is the 2nd root, or square root, of 16. Since $4^2 = 16$ and $(-4)^2 = 16$, there are two square roots of 16, namely 4 and -4. The symbol $16^{1/2}$ is reserved for the positive root, so

$$16^{1/2} = 4.$$

Also, $-16^{1/2} = -4$, because $-16^{1/2}$ means $-(16^{1/2})$. There is no real number square root of -16, so $(-16)^{1/2}$ is not a real number.

The symbol $a^{1/3}$ is the 3rd root, or cube root, of a. For example,

$$216^{1/3} = 6 \quad \text{since} \quad 6^3 = 216$$

and $\qquad (-8)^{1/3} = -2 \quad \text{since} \quad (-2)^3 = -8.$

The symbol $a^{1/n}$ is defined as follows.

Definition of $a^{1/n}$

If a is a real number, and n is a positive integer, then $a^{1/n}$, the **nth root** of a, is defined as follows.

n is	$a > 0$	$a < 0$	$a = 0$
even	$a^{1/n}$ is the positive real number such that $(a^{1/n})^n = a.$	$a^{1/n}$ is not a real number.	$a^{1/n}$ is 0.
odd	$a^{1/n}$ is the real number such that $(a^{1/n})^n = a.$		$a^{1/n}$ is 0.

EXAMPLE 1

(a) $36^{1/2} = 6$ since $6^2 = 36$

(b) $-100^{1/2} = -10$

(c) $-(225)^{1/2} = -15$

(d) $625^{1/4} = 5$

(e) $(-1296)^{1/4}$ is not a real number, but $-1296^{1/4} = -6$

(f) $(-27)^{1/3} = -3$

(g) $-32^{1/5} = -2$ ●

What about more general rational exponents? We want to define the symbol $a^{m/n}$ so that all the past rules for exponents still hold. For the power rule to hold, we must have $(a^{1/n})^m = a^{m/n}$. Therefore, $a^{m/n}$ is defined as follows.

Definition of $a^{m/n}$

For all integers m and all positive integers n, and for all real numbers a for which $a^{1/n}$ exists,

$$a^{m/n} = (a^{1/n})^m.$$

EXAMPLE 2

(a) $125^{2/3} = (125^{1/3})^2 = 5^2 = 25$

(b) $32^{7/5} = (32^{1/5})^7 = 2^7 = 128$

(c) $81^{3/2} = (81^{1/2})^3 = 9^3 = 729$ ●

By starting with $(a^{1/n})^m$ and raising it to the nth power, we can show that $(a^{1/n})^m$ is also equal to $(a^m)^{1/n}$. This means that we could have defined $a^{m/n}$ in either of the following ways.

For all real numbers a and positive integers n for which $a^{1/n}$ exists,

$$a^{m/n} = (a^{1/n})^m \quad \text{or} \quad a^{m/n} = (a^m)^{1/n}.$$

Now $a^{m/n}$ can be evaluated in either of two ways: as $(a^{1/n})^m$ or as $(a^m)^{1/n}$. It is usually easier to find $(a^{1/n})^m$. For example, $27^{4/3}$ can be evaluated in either of two ways:

$$27^{4/3} = (27^{1/3})^4 = 3^4 = 81$$
$$27^{4/3} = (27^4)^{1/3} = 531{,}441^{1/3} = 81.$$

The form $(27^{1/3})^4$ is easier to evaluate.

It can be shown that all the earlier results concerning integer exponents also apply to rational exponents.

Rules for Exponents

> Let r and s be rational numbers. The results below are valid for all real numbers a and b for which all indicated powers exist, and as long as no denominators are 0.
>
> $$a^r \cdot a^s = a^{r+s} \qquad \left(\frac{a}{b}\right)^r = \frac{a^r}{b^r}$$
>
> $$\frac{a^r}{a^s} = a^{r-s} \qquad (a^r)^s = a^{rs}$$
>
> $$(ab)^r = a^r \cdot b^r \qquad a^{-r} = \frac{1}{a^r}$$

EXAMPLE 3

(a)
$$\frac{27^{1/3} \cdot 27^{5/3}}{27^3} = \frac{27^{1/3 + 5/3}}{27^3}$$
$$= \frac{27^2}{27^3} = 27^{2-3} = 27^{-1} = \frac{1}{27}$$

(b) $81^{5/4} \cdot 4^{-3/2} = (3^4)^{5/4}(2^2)^{-3/2} = 3^5 \cdot 2^{-3} = \dfrac{3^5}{2^3}$ or $\dfrac{243}{8}$.

(c) $6y^{2/3} \cdot 2y^{1/2} = 12y^{2/3 + 1/2} = 12y^{7/6}$ $(y \geq 0)$

(d) $\left(\dfrac{3m^{5/6}}{y^{3/4}}\right)^2 \cdot \left(\dfrac{8y^3}{m^6}\right)^{2/3} = \dfrac{9m^{5/3}}{y^{3/2}} \cdot \dfrac{4y^2}{m^4} = 36m^{5/3-4}y^{2-3/2} = \dfrac{36y^{1/2}}{m^{7/3}}$

$(m > 0, y > 0)$

(e) $m^{2/3}(m^{7/3} + 2m^{1/3}) = (m^{2/3 + 7/3} + 2m^{2/3 + 1/3}) = m^3 + 2m$ ●

The next example shows how to factor with rational exponents.

EXAMPLE 4   Factor out the lowest power of the variable. Assume that all variables represent positive real numbers.

(a) $4m^{1/2} + 3m^{3/2} = m^{1/2}(4 + 3m)$

To check this result, multiply $m^{1/2}$ by $4 + 3m$.

(b) $y^{-1/3} + y^{2/3} = y^{-1/3}(1 + y)$ ●

2.6 EXERCISES

Simplify each of the following. Assume that all variables represent nonnegative real numbers. See Examples 1 and 2.

1. $4^{1/2}$

2. $25^{1/2}$

3. $8^{2/3}$

4. $81^{3/4}$

5. $27^{-2/3}$

6. $32^{-4/5}$

7. $\left(\dfrac{4}{9}\right)^{-3/2}$

8. $\left(\dfrac{1}{8}\right)^{-5/3}$

9. $125^{-2/3}$

10. $64^{5/6}$

11. $\left(\dfrac{64}{27}\right)^{2/3}$

12. $\left(\dfrac{8}{125}\right)^{4/3}$

13. $\left(\dfrac{27}{64}\right)^{-4/3}$

14. $\left(\dfrac{121}{100}\right)^{-3/2}$

15. $(16p^4)^{1/2}$

16. $(36r^6)^{1/2}$

17. $(27x^6)^{2/3}$

18. $(64a^{12})^{5/6}$

19. $(125y^9z^{12})^{1/3}$

20. $(256z^{16}r^8)^{1/4}$

Write each of the following, using only positive exponents. Assume that all variables represent positive real numbers and that variables used as exponents represent rational numbers. See Example 3.

21. $2^{1/2} \cdot 2^{3/2}$

22. $5^{3/8} \cdot 5^{5/8}$

23. $27^{2/3} \cdot 27^{-1/3}$

24. $9^{-3/4} \cdot 9^{1/4}$

25. $\dfrac{4^{2/3} \cdot 4^{5/3}}{4^{1/3}}$

26. $\dfrac{3^{-5/2} \cdot 3^{3/2}}{3^{7/2} \cdot 3^{-9/2}}$

27. $(m^{2/3})(m^{5/3})$

28. $(x^{4/5})(x^{2/5})$

29. $(1 + n)^{1/2}(1 + n)^{3/4}$

30. $(m + 7)^{-1/6}(m + 7)^{-2/3}$

31. $(2y^{3/4}z)(3y^{1/4}z^{-1/3})$

32. $(4a^{-1/2}b^{2/3})(a^{3/2}b^{-1/3})$

33. $(r^{5/7}s^{3/5})(r^{-1/2}s^{-1/4})$

34. $(t^{3/2}w^{3/5})(t^{-1/3}w^{-1/5})$

35. $\dfrac{7^{-1/3} \cdot 7r^{-3}}{7^{2/3} \cdot (r^{-2})^2}$

36. $\dfrac{12^{3/4} \cdot 12^{5/4} \cdot y^{-2}}{12^{-1} \cdot (y^{-3})^{-2}}$

37. $\dfrac{6k^{-4} \cdot (3k^{-1})^{-2}}{2^3 \cdot k^{1/2}}$

38. $\dfrac{8p^{-3} \cdot (4p^2)^{-2}}{p^{-5}}$

39. $\dfrac{a^{4/3} \cdot b^{1/2}}{a^{2/3} \cdot b^{-3/2}}$

40. $\dfrac{x^{1/3} \cdot y^{2/3} \cdot z^{1/4}}{x^{5/3} \cdot y^{-1/3} \cdot z^{3/4}}$

41. $\dfrac{k^{-3/5} \cdot h^{-1/3} \cdot t^{2/5}}{k^{-1/5} \cdot h^{-2/3} \cdot t^{1/5}}$

42. $\dfrac{m^{7/3} \cdot n^{-2/5} \cdot p^{3/8}}{m^{-2/3} \cdot n^{3/5} \cdot p^{-5/8}}$

43. $\dfrac{k^{3/2} \cdot k^{-1/2}}{k^{1/4} \cdot k^{3/4}}$

44. $\dfrac{m^{2/5} \cdot m^{3/5} \cdot m^{-4/5}}{m^{1/5} \cdot m^{-6/5}}$

45. $\dfrac{x^{-2/3} \cdot x^{4/3}}{x^{1/2} \cdot x^{-3/4}}$

46. $\dfrac{-4a^{1/2} \cdot a^{2/3}}{a^{-5/6}}$

47. $\dfrac{8y^{2/3}y^{-1}}{2^{-1}y^{3/4} \cdot y^{-1/6}}$

With $a^{1/n}$ written as $\sqrt[n]{a}$, we can also write $a^{m/n}$ using radicals.

For all integers m, all positive integers n, and all real numbers a for which $\sqrt[n]{a}$ exists,

$$a^{m/n} = (\sqrt[n]{a})^m \qquad \text{or} \qquad a^{m/n} = \sqrt[n]{a^m}.$$

By the definition of $\sqrt[n]{a}$, for any positive integer n, if $\sqrt[n]{a}$ exists, then

$$(\sqrt[n]{a})^n = a.$$

If a is positive, or if a is negative and n is an odd positive integer,

$$\sqrt[n]{a^n} = a.$$

Because of the conditions just given, we *cannot* simply write $\sqrt{x^2} = x$. For example, if $x = -5$,

$$\sqrt{x^2} = \sqrt{(-5)^2} = \sqrt{25} = 5 \neq x.$$

To take care of the fact that a negative value of x can produce a positive result for the square root, use the following rule, which involves absolute value.

For any real number a,

$$\sqrt{a^2} = |a|.$$

For example,

$$\sqrt{(-9)^2} = |-9| = 9, \qquad \text{and} \qquad \sqrt{13^2} = |13| = 13.$$

To avoid difficulties when working with variable radicands, we will assume that all variables in radicands represent only nonnegative real numbers.

Three key rules for working with radicals are given in the box below.

Rules for Radicals

For all real numbers a and b and positive integers m and n for which the indicated roots exist,

$$\sqrt[n]{a} \cdot \sqrt[n]{b} = \sqrt[n]{ab}$$

$$\sqrt[n]{\frac{a}{b}} = \frac{\sqrt[n]{a}}{\sqrt[n]{b}} \qquad (b \neq 0)$$

$$\sqrt[m]{\sqrt[n]{a}} = \sqrt[mn]{a}.$$

EXAMPLE 2   Use the rules for radicals to simplify each of the following.

(a) $\sqrt{6} \cdot \sqrt{54} = \sqrt{6 \cdot 54} = \sqrt{324} = 18$

(b) $\sqrt{\dfrac{7}{64}} = \dfrac{\sqrt{7}}{\sqrt{64}} = \dfrac{\sqrt{7}}{8}$

(c) $\sqrt[7]{\sqrt[3]{2}} = \sqrt[21]{2}$

(d) $\sqrt[4]{\sqrt{3}} = \sqrt[8]{3}$ ●

Simplifying Radicals   In working with numbers, it is customary to write them in their simplest form. For example, 10/2 is written as 5, $-9/6$ is written as $-3/2$, and 4/16 is written as 1/4. Expressions with radicals too should be written in their simplest form.

Simplified Radicals

> An expression with radicals is *simplified* when all the following conditions are satisfied.
>
> **1.** All possible factors have been removed from under the radical sign.
>
> **2.** The index on the radical is as small as possible.
>
> **3.** All radicals are removed from any denominators (a process called *rationalizing the denominator*).
>
> **4.** All indicated operations have been performed (if possible).

EXAMPLE 3   Simplify each of the following. Assume that all variables represent nonnegative real numbers.

(a) $\sqrt{175} = \sqrt{25 \cdot 7} = \sqrt{25} \cdot \sqrt{7} = 5\sqrt{7}$

(b) $\sqrt[3]{81x^5y^7z^6} = \sqrt[3]{27 \cdot 3 \cdot x^3 \cdot x^2 \cdot y^6 \cdot y \cdot z^6}$

$\qquad = \sqrt[3]{(27x^3y^6z^6)(3x^2y)}$

$\qquad = 3xy^2z^2\sqrt[3]{3x^2y}$ ●

Radicals with the same radicand and the same index, such as $3\sqrt[4]{11pq}$ and $-7\sqrt[4]{11pq}$, are called **like radicals.** Only like radicals can be added or subtracted. As shown in parts (b) and (c) of the next example, it is sometimes necessary to simplify radicals before adding or subtracting.

EXAMPLE 4   (a) $3\sqrt[4]{11pq} + (-7\sqrt[4]{11pq}) = -4\sqrt[4]{11pq}$

(b) $\sqrt{98x^3y} + 3x\sqrt{32xy}$

First remove all perfect square factors from under the radical. Then use the distributive property, as follows.

$$\sqrt{98x^3y} + 3x\sqrt{32xy} = \sqrt{49 \cdot 2 \cdot x^2 \cdot x \cdot y} + 3x\sqrt{16 \cdot 2 \cdot x \cdot y}$$

$$= 7x\sqrt{2xy} + (3x)(4)\sqrt{2xy}$$

$$= 7x\sqrt{2xy} + 12x\sqrt{2xy}$$

$$= 19x\sqrt{2xy}$$

(c) $\sqrt[3]{64m^4n^5} - \sqrt[3]{-27m^{10}n^{14}} = \sqrt[3]{(64m^3n^3)(mn^2)} - \sqrt[3]{(-27m^9n^{12})(mn^2)}$

$$= 4mn\sqrt[3]{mn^2} - (-3)m^3n^4\sqrt[3]{mn^2}$$

$$= 4mn\sqrt[3]{mn^2} + 3m^3n^4\sqrt[3]{mn^2}$$

$$= (4 + 3m^2n^3)mn\sqrt[3]{mn^2} \quad \bullet$$

EXAMPLE 5   Simplify the following radicals by first rewriting with rational exponents.

(a) $\sqrt[6]{3^2} = 3^{2/6} = 3^{1/3} = \sqrt[3]{3}$

(b) $\sqrt[6]{x^{12}y^3} = (x^{12}y^3)^{1/6} = x^2y^{3/6} = x^2y^{1/2} = x^2\sqrt{y}$

(c) $\sqrt[9]{\sqrt{6^3}} = \sqrt[9]{6^{3/2}} = (6^{3/2})^{1/9} = 6^{1/6} = \sqrt[6]{6} \quad \bullet$

In Example 5(a), we simplified $\sqrt[6]{3^2}$ as $\sqrt[3]{3}$. However, to simplify $\sqrt[6]{x^2}$, the variable x must be nonnegative. For example,

$$(-8)^{2/6} = [(-8)^{1/6}]^2.$$

This result is not a real number, since $(-8)^{1/6}$ is not a real number. On the other hand,

$$(-8)^{1/3} = -2.$$

Here, even though $2/6 = 1/3$,

$$\sqrt[6]{x^2} \neq \sqrt[3]{x}.$$

If a is nonnegative, then it is always true that $a^{m/n} = a^{mp/np}$. Reducing rational exponents on negative bases must be considered case by case.

Multiplying radical expressions is much like multiplying polynomials.

EXAMPLE 6   (a) $(\sqrt{2} + 3)(\sqrt{8} - 5) = \sqrt{2}(\sqrt{8}) - \sqrt{2}(5) + 3\sqrt{8} - 3(5)$

$$= \sqrt{16} - 5\sqrt{2} + 3(2\sqrt{2}) - 15$$

$$= 4 - 5\sqrt{2} + 6\sqrt{2} - 15$$

$$= -11 + \sqrt{2}$$

(b) $(\sqrt{7} - \sqrt{10})(\sqrt{7} + \sqrt{10}) = (\sqrt{7})^2 - (\sqrt{10})^2$

$$= 7 - 10$$

$$= -3 \quad \bullet$$

Rationalizing the Denominator   The process of rationalizing the denominator, mentioned in condition 3 of the rules for simplifying radicals described above, is explained in the next few examples.

EXAMPLE 7   Rationalize each denominator.

(a) $\dfrac{4}{\sqrt{3}}$

To rationalize the denominator, multiply by $\sqrt{3}/\sqrt{3}$ (or 1) so that the denominator of the product is a rational number.

$$\frac{4}{\sqrt{3}} \cdot \frac{\sqrt{3}}{\sqrt{3}} = \frac{4\sqrt{3}}{3}$$

(b) $\sqrt[4]{\dfrac{3}{5}}$

Start by using the fact that the radical of a quotient can be written as the quotient of radicals.

$$\sqrt[4]{\frac{3}{5}} = \frac{\sqrt[4]{3}}{\sqrt[4]{5}}$$

To rationalize the denominator, multiply numerator and denominator by $\sqrt[4]{5^3}$. Use this number so that the denominator will be a rational number. This multiplication gives

$$\frac{\sqrt[4]{3}}{\sqrt[4]{5}} = \frac{\sqrt[4]{3} \cdot \sqrt[4]{5^3}}{\sqrt[4]{5} \cdot \sqrt[4]{5^3}} = \frac{\sqrt[4]{3 \cdot 5^3}}{\sqrt[4]{5^4}} = \frac{\sqrt[4]{375}}{5}. \quad \bullet$$

EXAMPLE 8   Rationalize the denominator of $\dfrac{1}{1 - \sqrt{2}}$.

The best approach here is to multiply both numerator and denominator by the **conjugate*** of the denominator, in this case $1 + \sqrt{2}$.

$$\frac{1}{1 - \sqrt{2}} = \frac{1(1 + \sqrt{2})}{(1 - \sqrt{2})(1 + \sqrt{2})} = \frac{1 + \sqrt{2}}{1 - 2} = -1 - \sqrt{2} \quad \bullet$$

2.7 EXERCISES

Simplify each of the following. Assume that all variables represent positive real numbers.
See Examples 1, 3, 5, and 7.

1. $\sqrt[3]{125}$ **2.** $\sqrt[4]{81}$ **3.** $\sqrt[4]{1296}$ **4.** $\sqrt[3]{343}$

5. $\sqrt[5]{-3125}$ **6.** $\sqrt[7]{-128}$ **7.** $\sqrt{50}$ **8.** $\sqrt{45}$

---

*The conjugate of $a\sqrt{m} + b\sqrt{n}$ is $a\sqrt{m} - b\sqrt{n}$.

9. $\sqrt{2000}$ **10.** $\sqrt{3200}$ **11.** $\sqrt{5000}$ **12.** $\sqrt{27{,}000}$

13. $\sqrt[3]{81}$ **14.** $\sqrt[3]{16}$ **15.** $\sqrt[3]{250}$ **16.** $\sqrt[3]{128}$

17. $-\sqrt[4]{32}$ **18.** $-\sqrt[4]{243}$ **19.** $-\sqrt{\dfrac{9}{5}}$ **20.** $\sqrt{\dfrac{3}{8}}$

21. $-\sqrt[3]{\dfrac{3}{2}}$ **22.** $-\sqrt[3]{\dfrac{4}{5}}$ **23.** $\sqrt[4]{\dfrac{3}{2}}$ **24.** $\sqrt[4]{\dfrac{32}{81}}$

25. $-\sqrt{24 \cdot 3^2 \cdot 2^4}$ **26.** $-\sqrt{18 \cdot 2^3 \cdot 4^2}$ **27.** $\sqrt[3]{16(-2)^4(2)^8}$ **28.** $\sqrt[3]{25(3)^4(5)^3}$

29. $\sqrt{8x^5z^8}$ **30.** $\sqrt{24m^6n^5}$ **31.** $\sqrt[3]{16z^5x^8y^4}$ **32.** $\sqrt[3]{81p^5x^2y^6}$

33. $\sqrt[4]{m^2n^7p^8}$ **34.** $\sqrt[4]{x^8y^6z^{10}}$ **35.** $\sqrt[6]{5^9r^8z^{10}}$ **36.** $\sqrt[4]{3^6m^6n^{14}}$

37. $\sqrt{\dfrac{2}{3x}}$ **38.** $\sqrt{\dfrac{5}{3p}}$ **39.** $\sqrt{\dfrac{x^5y^3}{z^2}}$ **40.** $\sqrt{\dfrac{g^3h^5}{r^3}}$

41. $\sqrt[3]{\dfrac{8}{x^2}}$ **42.** $\sqrt[3]{\dfrac{9}{16p^4}}$ **43.** $-\sqrt[3]{\dfrac{k^5m^3r^2}{r^8}}$ **44.** $-\sqrt[3]{\dfrac{9x^5y^6}{z^5w^2}}$

45. $\sqrt[4]{\dfrac{g^3h^5}{9r^6}}$ **46.** $\sqrt[4]{\dfrac{32x^{4/5}}{y^5}}$ **47.** $\dfrac{\sqrt[3]{mn}\cdot\sqrt[3]{m^2}}{\sqrt[3]{n^2}}$ **48.** $\dfrac{\sqrt[3]{8m^2n^3}\cdot\sqrt[3]{2m^2}}{\sqrt[3]{32m^4n^3}}$

49. $\dfrac{\sqrt[4]{32x^5y}\cdot\sqrt[4]{2xy^4}}{\sqrt[4]{4x^3y^2}}$ **50.** $\dfrac{\sqrt[4]{rs^2t^3}\cdot\sqrt[4]{r^3s^2t}}{\sqrt[4]{r^2t^3}}$ **51.** $\sqrt[3]{\sqrt{4}}$ **52.** $\sqrt[4]{\sqrt[3]{2}}$

53. $\sqrt[6]{\sqrt[3]{x}}$ **54.** $\sqrt[8]{\sqrt[4]{y}}$

Simplify each of the following, assuming that all variables represent nonnegative numbers. See Examples 4, 6, and 7.

55. $4\sqrt{3} - 5\sqrt{12} + 3\sqrt{75}$ **56.** $2\sqrt{5} - 3\sqrt{20} + 2\sqrt{45}$ **57.** $\sqrt{50} - 8\sqrt{8} + 4\sqrt{18}$

58. $6\sqrt{27} - 3\sqrt{12} + 5\sqrt{48}$ **59.** $3\sqrt{28p} - 4\sqrt{63p} + \sqrt{112p}$ **60.** $9\sqrt{8k} + 3\sqrt{18k} - \sqrt{32k}$

61. $3\sqrt[3]{16} - 4\sqrt[3]{2}$ **62.** $\sqrt[3]{2} - \sqrt[3]{16} + 2\sqrt[3]{54}$ **63.** $2\sqrt[3]{3} + 4\sqrt[3]{24} - \sqrt[3]{81}$

64. $\sqrt[3]{32} - 5\sqrt[3]{4} + 2\sqrt[3]{108}$ **65.** $\dfrac{1}{\sqrt{3}} - \dfrac{2}{\sqrt{12}} + 2\sqrt{3}$ **66.** $\dfrac{1}{\sqrt{2}} + \dfrac{3}{\sqrt{8}} + \dfrac{1}{\sqrt{32}}$

67. $\dfrac{5}{\sqrt[3]{2}} - \dfrac{2}{\sqrt[3]{16}} + \dfrac{1}{\sqrt[3]{54}}$ **68.** $\dfrac{-4}{\sqrt[3]{3}} + \dfrac{1}{\sqrt[3]{24}} - \dfrac{2}{\sqrt[3]{81}}$ **69.** $\sqrt{a^3b^5} - 2\sqrt{a^7b^3} + \sqrt{a^3b^9}$

70. $\sqrt{p^7q^3} - \sqrt{p^5q^9} + \sqrt{p^9q}$ **71.** $(\sqrt{2} + 3)(\sqrt{2} - 3)$ **72.** $(\sqrt{5} + \sqrt{2})(\sqrt{5} - \sqrt{2})$

73. $(\sqrt[3]{11} - 1)(\sqrt[3]{11^2} + \sqrt[3]{11} + 1)$ **74.** $(\sqrt[3]{7} + 3)(\sqrt[3]{7^2} - 3\sqrt[3]{7} + 9)$

75. $(\sqrt{3} + \sqrt{8})^2$ **76.** $(\sqrt{2} - 1)^2$

77. $(3\sqrt{2} + \sqrt{3})(2\sqrt{3} - \sqrt{2})$ **78.** $(4\sqrt{5} - 1)(3\sqrt{5} + 2)$

79. $(2\sqrt[3]{3} + 1)(\sqrt[3]{3} - 4)$ **80.** $(\sqrt[3]{4} + 3)(5\sqrt[3]{4} + 1)$

Rationalize the denominator of each of the following. Assume that all variables represent nonnegative numbers and that no denominators are zero. See Example 8.

81. $\dfrac{3}{1 - \sqrt{2}}$

82. $\dfrac{2}{1 + \sqrt{5}}$

83. $\dfrac{\sqrt{3}}{4 + \sqrt{3}}$

84. $\dfrac{2\sqrt{7}}{3 - \sqrt{7}}$

85. $\dfrac{4}{2 - \sqrt{y}}$

86. $\dfrac{-1}{\sqrt{k} - 2}$

87. $\dfrac{p}{\sqrt{p} + 2}$

88. $\dfrac{\sqrt{r}}{3 - \sqrt{r}}$

89. $\dfrac{a}{\sqrt{a + b} - 1}$

90. $\dfrac{3m}{2 + \sqrt{m + n}}$

91. $\dfrac{\sqrt{x} + \sqrt{x + 1}}{\sqrt{x} - \sqrt{x + 1}}$

92. $\dfrac{\sqrt{p} + \sqrt{p^2 - 1}}{\sqrt{p} - \sqrt{p^2 - 1}}$

In advanced mathematics it is sometimes useful to write a radical expression with a rational numerator. The procedure is similar to rationalizing the denominator. Rationalize the numerator of each of the following expressions.

93. $\dfrac{\sqrt{3}}{2}$

94. $\dfrac{\sqrt{6}}{5}$

95. $\dfrac{1 + \sqrt{2}}{2}$

96. $\dfrac{1 - \sqrt{3}}{3}$

97. $\dfrac{\sqrt{x}}{1 + \sqrt{x}}$

98. $\dfrac{\sqrt{p}}{1 - \sqrt{p}}$

99. $\dfrac{\sqrt{x} + \sqrt{x + 1}}{\sqrt{x} - \sqrt{x + 1}}$

100. $\dfrac{\sqrt{p} + \sqrt{p^2 - 1}}{\sqrt{p} - \sqrt{p^2 - 1}}$

Evaluate each of the following. Write all answers in scientific notation with three significant digits. For fourth roots, use the square root key twice.

101. $\sqrt{2.876 \times 10^7}$

102. $\sqrt{5.432 \times 10^9}$

103. $\sqrt[4]{3.87 \times 10^{-4}}$

104. $\sqrt[4]{5.913 \times 10^{-8}}$

105. $\dfrac{2.04 \times 10^{-3}}{\sqrt{5.97 \times 10^{-5}}}$

106. $\dfrac{3.86 \times 10^{-5}}{\sqrt{4.82 \times 10^{-5}}}$

107. $\sqrt{(4.721)^2 + (8.963)^2 - 2(4.721)(8.963)(.0468)}$

108. $\sqrt{(157.3)^2 + (184.7)^2 - 2(157.3)(184.7)(.9082)}$

Write the following without radicals.

109. $\sqrt{(m + n)^2}$

110. $\sqrt[3]{(a + 2b)^3}$

111. $\sqrt{z^2 - 6zx + 9x^2}$

112. $\sqrt[3]{(r + 2s)(r^2 + 4rs + 4s^2)}$

113. Find an approximate value for $\sqrt{5 + 2\sqrt{6}}$.

114. By squaring both sides, show that $\sqrt{5 + 2\sqrt{6}} = \sqrt{2} + \sqrt{3}$.

Chapter 2   Summary

Key Words	algebraic expression term

Key Words

algebraic expression
coefficient
degree
binomial
division algorithm
factoring
factored form
irreducible polynomial
complex fraction
radical sign
index

term
polynomial
monomial
trinomial
square of a binomial
algorithm
factors
prime polynomial
rational expression
radicals
radicand

Rules for Exponents

Let r and s be rational numbers. The results below are valid for all real numbers a and b for which all indicated powers exist, and as long as no denominators are 0.

$$a^r \cdot a^s = a^{r+s} \qquad (ab)^r = a^r \cdot b^r \qquad (a^r)^s = a^{rs}$$

$$\frac{a^r}{a^s} = a^{r-s} \qquad \left(\frac{a}{b}\right)^r = \frac{a^r}{b^r} \qquad a^{-r} = \frac{1}{a^r}$$

Difference of Two Squares

$$x^2 - y^2 = (x + y)(x - y)$$

Perfect Square Trinomial

$$x^2 + 2xy + y^2 = (x + y)^2$$
$$x^2 - 2xy + y^2 = (x - y)^2$$

Difference and Sum of Cubes

$$x^3 - y^3 = (x - y)(x^2 + xy + y^2)$$
$$x^3 + y^3 = (x + y)(x^2 - xy + y^2)$$

Radical Notation

If a is a real number, if n is a positive integer, and if $a^{1/n}$ exists, then
$$\sqrt[n]{a} = a^{1/n}.$$

Chapter 2   Review Exercises

Simplify each of the following. Write results without negative exponents. Assume that all variables represent nonzero real numbers.

1. 2^{-6}

2. -3^{-2}

3. $\left(\frac{-5}{4}\right)^{-2}$

4. $3^{-1} - 4^{-1}$

5. $(5z^3)(-2z^5)$

6. $(-3a^4b^7)(6a^{-2}b^2)$

7. $(8p^2q^3)(-2p^5q^{-4})$

8. $(-6p^5w^4m^{12})^0$

9. $(-6x^2y^{-3}z^2)^{-2}$

10. $\dfrac{-8y^7p^{-2}}{y^{-4}p^{-3}}$

11. $\dfrac{a^{-6}(a^{-8})}{a^{-2}(a^{11})}$

12. $\dfrac{5^{-1}r^4m^{-2}}{5r^{-2}m^{-4}}$

13. $\dfrac{6r^3s^{-2}}{6^{-1}r^4s^{-3}}$

14. $\dfrac{(p+q)^4(p+q)^{-3}}{(p+q)^6}$

15. $\dfrac{[p^2(m+n)^3]^{-2}}{p^{-2}(m+n)^{-5}}$

16. $[(ab^{-2})^2]^{-1}$

17. $\left(\dfrac{r^{-1}s^2}{t^{-2}}\right)^{-2}$

18. $\dfrac{(2x^{-3})^2(3x^2)^{-2}}{6(x^2y^3)}$

19. $\dfrac{(-5m^3n^{-4})^{-1}}{(10m^{-1}n^{-5})^{-2}}$

20. $\dfrac{(8a^2b)^2(3a^{-1}b^{-2})}{24(a^{-2}b^2)^{-1}(a^{-2})}$

Simplify each of the following. Assume that all variables represent nonzero real numbers and that variables appearing as exponents represent integers. Write results without denominators.

21. $(a^p)(a^{5-2p})$

22. $\dfrac{7k^{r-4}}{3k^{r+6}}$

23. $\dfrac{(p^4)^m}{(3p^m)^2}$

24. $\dfrac{(8-r)^2(8-r)^{a+5}}{(8-r)^{3-a}}$

Perform each of the following operations. Assume that all variables appearing as exponents represent integers.

25. $(-9m^2 + 11m - 7) + (2m^2 - 12m + 4)$

26. $(3q^3 - 9q^2 + 6) + (4q^3 - 8q + 3)$

27. $(5r^4 - 6r^2 + 2r) - (-3r^4 + 2r^2 - 9r)$

28. $(-7z^3 + 8z^2 - z) - (z^3 + 4z^2 + 10z)$

29. $-(r^5 + 2r^4 + 8r^2) + 3(r^4 + 9r^2)$

30. $2(3y^6 - 9y^2 + 2y) - (5y^6 - 10y^2 - 4y)$

31. $(8y - 7)(2y + 7)$

32. $(9k + 2)(4k - 3)$

33. $(7z + 10y)(3z - 5y)$

34. $(2r + 11s)(4r - 9s)$

35. $(3k - 5m)^2$

36. $(4a - 3b)^2$

37. $(5x - 2)^3$

38. $(3m + 2)^4$

39. $(3w - 2)(5w^2 - 4w + 1)$

40. $(2k + 5)(3k^3 - 4k^2 + 8k - 2)$

41. $(a - b + c)(a + b - 2c)$

42. $(x + 2y - z)^2$

43. $(p^q + 1)(p^q - 3)$

44. $(a^y + 6)^2$

Perform each division.

45. $\dfrac{12y^4 - 8y^3 + 2y^2}{2y^2}$

46. $\dfrac{-6p^5 + 12p^3 - 9p}{3p}$

47. $\dfrac{15m^4n^5 - 10m^3n^7 + 20m^7n^3}{10mn^4}$

48. $\dfrac{10r^6m^3 + 8r^4m - 3r^{10}m^2}{12r^3m^4}$

49. $\dfrac{6p^2 - 5p - 56}{2p - 7}$

50. $\dfrac{72r^2 + 59r + 12}{8r + 3}$

51. $\dfrac{28z^2 - 27z - 3}{7z - 5}$

52. $\dfrac{15k^2 + 40k - 47}{3k + 11}$

53. $\dfrac{2y^3 - 11y^2 + 28y - 20}{2y - 3}$

54. $\dfrac{30r^3 + 37r^2 + 34r + 17}{6r + 5}$

55. $\dfrac{5m^3 - 7m^2 - 10m + 14}{m^2 - 2}$

56. $\dfrac{3b^3 - 8b^2 + 12b - 30}{b^2 + 4}$

57. $\dfrac{k^5 - 6k^4 + 9k^3 + k^2 - 12k + 3}{k^2 - 4k + 1}$

58. $\dfrac{z^5 + 2z^4 + 4z^3 - 8z^2 - 15z - 32}{z^2 + 2z + 4}$

Factor as completely as possible.

59. $7z^2 - 9z^3 + z$

60. $3(m - n) + 4k(m - n)$

61. $12p^5 - 8p^4 + 20p^3$

62. $3(z - 4)^2 + 9(z - 4)^3$

63. $r^2 + rp - 42p^2$

64. $z^2 - 6zk - 16k^2$

65. $6m^2 - 13m - 5$

66. $4k^2 + 11k - 3$

67. $15a^2 + 7ab - 2b^2$

68. $6z^2 - 23zx - 4x^2$

69. $30m^5 - 35m^4n - 25m^3n^2$

70. $48a^8 - 12a^7b - 90a^6b^2$

71. $169y^4 - 1$

72. $49m^8 - 9n^2$

73. $4p^2 - 20pr^3 + 25r^6$

74. $49m^6 - 126m^3q^4 + 81q^8$

75. $8y^3 - 1000z^6$

76. $125a^6 + 216b^9$

77. $6(3r - 1)^2 + (3r - 1) - 35$

78. $6(2q - 5)^2 + 13(2q - 5) - 15$

79. $ar - 3as + 5rb - 15sb$

80. $15mp + 9mq - 10np - 6nq$

81. $(16m^2 - 56m + 49) - 25a^2$

82. $r^9 - 8(r^3 - 1)^3$

Perform each of the following operations.

83. $\dfrac{5x^2y}{x + y} \cdot \dfrac{3x + 3y}{30xy^2}$

84. $\dfrac{3m - 9}{8m} \cdot \dfrac{16m + 24}{15}$

85. $\dfrac{k^2 + k}{8k^3} \cdot \dfrac{4}{k^2 - 1}$

86. $\dfrac{3r^3 - 9r^2}{r^2 - 9} \div \dfrac{8r^3}{r + 3}$

87. $\dfrac{x^2 + x - 2}{x^2 + 5x + 6} \div \dfrac{x^2 + 3x - 4}{x^2 + 4x + 3}$

88. $\left(1 - \dfrac{3}{p}\right)\left(1 + \dfrac{3}{p}\right)$

89. $\dfrac{27m^3 - n^3}{3m - n} \div \dfrac{9m^2 + 3mn + n^2}{9m^2 - n^2}$

90. $\dfrac{p^2 - 36q^2}{(p - 6q)^2} \cdot \dfrac{p^2 - 5pq - 6q^2}{p^2 - 6pq + 36q^2} \div \dfrac{5p}{p^3 + 216q^3}$

91. $\dfrac{1}{4y} + \dfrac{8}{5y}$

92. $\dfrac{m}{4 - m} + \dfrac{3m}{m - 4}$

93. $\dfrac{3}{x^2 - 4x + 3} - \dfrac{2}{x^2 - 1}$

94. $\left[\dfrac{1}{(x + h)^2 + 16} - \dfrac{1}{x^2 + 16}\right] \div h$

95. $\dfrac{\dfrac{1}{p} + \dfrac{1}{q}}{1 - \dfrac{1}{pq}}$

96. $\dfrac{3 + \dfrac{2m}{m^2 - 4}}{\dfrac{5}{m - 2}}$

Simplify. Write with only positive exponents. Assume that all variables represent positive real numbers.

97. $\dfrac{p^4(p^{-2})}{p^{5/3}}$

98. $(2 + p)^{5/4}(2 + p)^{-9/4}$

99. $(7r^{1/2})(2r^{3/4})(-r^{1/6})$

100. $(a^{3/4}b^{2/3})(a^{5/8}b^{-5/6})$

101. $\dfrac{y^{5/3} \cdot y^{-2}}{y^{-5/6}}$

102. $\left(\dfrac{25m^3n^5}{m^{-2}n^6}\right)^{-1/2}$

Simplify. Assume that all variables represent positive real numbers. Write without denominators.

103. $\dfrac{k^{2+p} \cdot k^{-4p}}{k^{6p}}$ **104.** $\dfrac{(r^{3+2z})(r^{-5-z})}{(r^{-z/2})^{-4}}$

Find each product. Assume that all variables represent positive real numbers.

105. $2z^{1/3}(5z^2 - 2)$ **106.** $-m^{3/4}(8m^{1/2} + 4m^{-3/2})$

107. $(p + p^{1/2})(3p - 5)$ **108.** $(m^{1/2} - 4m^{-1/2})^2$

Simplify each of the following. Assume that all variables represent positive real numbers.

109. $\dfrac{2}{p^{1/2}} + 5p^{1/2}$ **110.** $\dfrac{3}{r^{2/3}} + r^{1/3}$

111. $\dfrac{(p+1)^{1/2} - p(1/2)(p+1)^{-1/2}}{p+1}$ **112.** $\dfrac{(r-2)^{2/3} - r(2/3)(r-2)^{-1/3}}{(r-2)^{4/3}}$

113. $\dfrac{3(2x^2 + 5)^{1/3} - x(2x^2 + 5)^{-2/3}(4x)}{(2x^2 + 5)^{2/3}}$ **114.** $\dfrac{-(m^3 + m)^{2/3} + m(2/3)(m^3 + m)^{-1/3}(3m^2 + 1)}{(m^3 + m)^{4/3}}$

Simplify. Assume that all variables represent positive numbers.

115. $\sqrt{200}$ **116.** $\sqrt[3]{16}$ **117.** $\sqrt[4]{1250}$ **118.** $-\sqrt{\dfrac{16}{3}}$

119. $\sqrt{\dfrac{7}{3r}}$ **120.** $-\sqrt[3]{\dfrac{2}{5p^2}}$ **121.** $\sqrt{\dfrac{2^7 y^8}{m^3}}$ **122.** $-\sqrt[3]{\dfrac{r^6 h^5}{z^2}}$

123. $\sqrt[4]{\sqrt[3]{m}}$ **124.** $\dfrac{\sqrt[4]{8p^2 q^5} \cdot \sqrt[4]{2p^3 q}}{\sqrt[4]{p^5 q^2}}$

125. $(\sqrt[3]{2} + 4)(\sqrt[3]{2^2} - 4\sqrt[3]{2} + 16)$ **126.** $\dfrac{3}{\sqrt{5}} - \dfrac{2}{\sqrt{45}} + \dfrac{6}{\sqrt{80}}$

127. $\sqrt{18m^3} - 3m\sqrt{32m} + 5\sqrt{m^3}$ **128.** $\dfrac{2}{7 - \sqrt{3}}$

129. $\dfrac{6}{3 - \sqrt{2}}$ **130.** $\dfrac{z}{\sqrt{z+1}}$

131. $\dfrac{k}{\sqrt{k} - 3}$ **132.** $\dfrac{\sqrt{x} - \sqrt{x-2}}{\sqrt{x} + \sqrt{x-2}}$

Equations and Inequalities

For many people, the study of algebra is really the study of equations. Many applications of mathematics require the solution of one or more equations. The study of inequalities has also become important as more and more applications, especially in fields such as business, utilize inequalities.

An **equation** is a statement that two expressions are equal. Examples of equations include

$$x + 2 = 9, \qquad 11y = 5y + 6y, \qquad x^2 - 2x - 1 = 0,$$

and so on. In this chapter we discuss the solution of several different kinds of equations and inequalities.

3.1 Linear Equations

To **solve** an equation means to find all numbers that make the equation a true statement. A number that is a solution of an equation is said to **satisfy** the equation. The set of all solutions for an equation makes up its **solution set.** An equation which is satisfied by every number which is a meaningful replacement for the variable is called an **identity.** Examples of identities are

$$3x + 4x = 7x \qquad \text{and} \qquad x^2 - 3x + 2 = (x - 2)(x - 1).$$

Equations that are satisfied by some numbers, but not satisfied by others, are called **conditional equations.** Examples of conditional equations are

$$2m + 3 = 7 \quad \text{and} \quad \frac{5r}{r - 1} = 7.$$

EXAMPLE 1 Decide whether the following equations are identities or conditional equations.

(a) $9p^2 - 25 = (3p + 5)(3p - 5)$

Since the product of $3p + 5$ and $3p - 5$ is $9p^2 - 25$, the given equation is true for *every* value of p and is an identity.

(b) $5y - 4 = 11$

Replacing y with 3 gives

$$5 \cdot 3 - 4 = 11$$
$$11 = 11,$$

a true statement. On the other hand, $y = 4$ leads to

$$5 \cdot 4 - 4 = 11$$
$$16 = 11,$$

a false statement. The equation $5y - 4 = 11$ is true for some values of y, but not all, and thus is a conditional equation. (By the way, the word *some* in mathematics means "at least one." We can therefore say that the statement $5y - 4 = 11$ is true for *some* replacements of y, even though it turns out to be true only for $y = 3$.) ●

Any two equations with the same solution set are called **equivalent equations.** For example, $x + 1 = 5$ and $6x + 3 = 27$ are equivalent equations since they both have the same solution set, {4}.

EXAMPLE 2 Are the following equations equivalent?

(a) $2x - 1 = 3 \quad \text{and} \quad 12x + 7 = 31$

Each of these equations has solution set {2}. Since the solution sets are equal, the equations are equivalent.

(b) $x = 3 \quad \text{and} \quad x^2 = 9$

The solution set for $x = 3$ is {3}, while the solution set for the equation $x^2 = 9$ is {3, −3}. Since the solution sets are not equal, the equations are not equivalent. ●

One way to solve an equation is to rewrite it as a series of simpler equivalent equations. These simpler equations often can be obtained with the **addition and multiplication properties of equations.**

Addition and Multiplication Properties of Equations

> If P, Q, and R are algebraic expressions, then
>
> **(a)** $P = Q$ and $P + R = Q + R$ are equivalent. (*The same expression may be added to both sides of an equation.*)
>
> **(b)** If $R \neq 0$, then $P = Q$ and $PR = QR$ are equivalent. (*The same nonzero expression may be multiplied on both sides of an equation.*).

EXAMPLE 3   Solve $3(2x - 4) = 7 - (x + 5)$.

Use the distributive property and then collect like terms to get the following sequence of simpler equivalent equations.

$$3(2x - 4) = 7 - (x + 5)$$
$$6x - 12 = 7 - x - 5$$
$$6x - 12 = 2 - x$$

Now add the same expressions to each side of the equation.

$$x + 6x - 12 = x + 2 - x$$
$$7x - 12 = 2$$
$$12 + 7x - 12 = 12 + 2$$
$$7x = 14$$

Finally, multiplying each side by the same number, 1/7, produces

$$\frac{1}{7} \cdot 7x = \frac{1}{7} \cdot 14$$

$$x = 2.$$

To check, replace x with 2 in the original equation, getting

$$3(2x - 4) = 7 - (x + 5) \qquad \text{Original equation}$$
$$3(2 \cdot 2 - 4) = 7 - (2 + 5) \qquad \text{Let } x = 2$$
$$3(4 - 4) = 7 - (7)$$
$$0 = 0. \qquad \text{True}$$

Since replacing x with 2 results in a true statement, 2 is the solution of the given equation. The solution set is therefore {2}.   ●

The equation in Example 3 is a *linear equation,* as are all the equations in this section.

Linear Equation

> A **linear equation** in one variable is an equation that can be written in the form
> $$ax + b = 0,$$
> where a and b are real numbers, with $a \neq 0$.

EXAMPLE 4 Solve $\dfrac{3p - 1}{3} - \dfrac{2p}{p - 1} = p$.

At first glance, this equation does not satisfy the definition of a linear equation given above. However, the equation does appear in proper form after algebraic simplification. To obtain a simpler equivalent equation, first multiply both sides by $3(p - 1)$, where we must assume $p \neq 1$. Doing this gives

$$3(p - 1)\left(\dfrac{3p - 1}{3}\right) - 3(p - 1)\left(\dfrac{2p}{p - 1}\right) = 3(p - 1)p$$
$$(p - 1)(3p - 1) - 3(2p) = 3p(p - 1)$$
$$3p^2 - 4p + 1 - 6p = 3p^2 - 3p.$$

An even simpler equivalent equation comes from combining terms and adding $-3p^2$ to both sides, producing

$$-10p + 1 = -3p.$$

Now add $10p$ to both sides to get $1 = 7p$.

Finally, multiplying both sides by 1/7 gives

$$\dfrac{1}{7} = p.$$

Check 1/7 in the given equation to verify that the solution set is $\{1/7\}$. Our restriction $p \neq 1$ does not affect the solution set here, since $1/7 \neq 1$. ●

EXAMPLE 5 Solve $\dfrac{x}{x - 2} = \dfrac{2}{x - 2} + 2$.

Multiply both sides of the equation by $x - 2$, assuming that $x - 2 \neq 0$.

$$x = 2 + 2(x - 2)$$
$$x = 2 + 2x - 4$$
$$x = 2$$

We had to assume $x - 2 \neq 0$ in order to be able to multiply both sides of the equation by $x - 2$. Since $x = 2$, we have $x - 2 = 0$, and the multiplication property of equations does not apply. The solution set is $\emptyset$. (Substituting 2 for x in the original equation would result in a denominator of 0.) ●

Sometimes an equation with more than one letter must be solved for a specified variable. This process is shown in the next example. (As a general rule letters from the beginning of the alphabet, such as a, b, c, and so on, are used to represent constants, while letters such as x, y, and z represent variables.)

EXAMPLE 6 Solve the equation $3(2x - 5a) + 4b = 4x - 2$ for x.

Using the distributive property gives

$$6x - 15a + 4b = 4x - 2.$$

Treat x as the variable and the other letters as constants. Get all terms with x on one side and all terms without x on the other side.

$$6x - 4x = 15a - 4b - 2$$
$$2x = 15a - 4b - 2$$
$$x = \frac{15a - 4b - 2}{2} \quad \bullet$$

3.1 EXERCISES

Decide whether each of the following equations is an identity or a conditional equation. See Example 1.

1. $x^2 + 5x = x(x + 5)$

2. $3y + 4 = 5(y - 2)$

3. $2(x - 7) = 5x + 3 - x$

4. $2x - 4 = 2(x - 2)$

5. $\dfrac{m + 3}{m} = 1 + \dfrac{3}{m}$

6. $\dfrac{p}{2 - p} = \dfrac{2}{p} - 1$

7. $4q^2 - 25 = (2q + 5)(2q - 5)$

8. $3(k + 2) - 5(k + 2) = -2k - 4$

Decide which of the following pairs of equations are equivalent. See Example 2.

9. $3x - 5 = 7$
$-6x + 10 = 14$

10. $-x = 2x + 3$
$-3x = 3$

11. $\dfrac{3x}{x - 1} = \dfrac{2}{x - 1}$
$3x = 2$

12. $\dfrac{x + 1}{12} = \dfrac{5}{12}$
$x + 1 = 5$

13. $\dfrac{x}{x - 2} = \dfrac{2}{x - 2}$
$x = 2$

14. $\dfrac{x + 3}{x + 1} = \dfrac{2}{x + 1}$
$x = -1$

15. $x = 4$
$x^2 = 16$

16. $z^2 = 9$
$z = 3$

Solve each of the following equations. See Examples 3–5.

17. $4x - 1 = 15$

18. $-3y + 2 = 5$

19. $.2m - .5 = .1m + .7$

20. $.01p + 3.1 = 2.03p - 2.96$

21. $\dfrac{5}{6}k - 2k + \dfrac{1}{3} = \dfrac{2}{3}$

22. $\dfrac{3}{4} + \dfrac{1}{5}r - \dfrac{1}{2} = \dfrac{4}{5}r$

23. $3r + 2 - 5(r + 1) = 6r + 4$

24. $5(a + 3) + 4a - 5 = -(2a - 4)$

25. $2[m - (4 + 2m) + 3] = 2m + 2$

26. $4[2p - (3 - p) + 5] = -7p - 2$

27. $\dfrac{3x - 2}{7} = \dfrac{x + 2}{5}$

28. $\dfrac{2p + 5}{5} = \dfrac{p + 2}{3}$

29. $\dfrac{3k - 1}{4} = \dfrac{5k + 2}{8}$

30. $\dfrac{9x - 1}{6} = \dfrac{2x + 7}{3}$

31. $\dfrac{x}{3} - 7 = 6 - \dfrac{3x}{4}$

32. $\dfrac{y}{3} + 1 = \dfrac{2y}{5} - 4$

33. $\dfrac{1}{4p} + \dfrac{2}{p} = 3$

34. $\dfrac{2}{t} + 6 = \dfrac{5}{2t}$

35. $\dfrac{m}{2} - \dfrac{1}{m} = \dfrac{6m + 5}{12}$

36. $\dfrac{-3k}{2} + \dfrac{9k - 5}{6} = \dfrac{11k + 8}{k}$

37. $\dfrac{2r}{r-1} = 5 + \dfrac{2}{r-1}$

38. $\dfrac{3x}{x+2} = \dfrac{1}{x+2} - 4$

39. $\dfrac{5}{2a+3} + \dfrac{1}{a-6} = 0$

40. $\dfrac{2}{x+1} = \dfrac{3}{2x-5}$

41. $\dfrac{4}{x-3} - \dfrac{8}{2x+5} + \dfrac{3}{x-3} = 0$

42. $\dfrac{5}{2p+3} - \dfrac{3}{p-2} = \dfrac{4}{2p+3}$

43. $\dfrac{3}{2m+4} = \dfrac{1}{m+2} - 2$

44. $\dfrac{8}{3k-9} - \dfrac{5}{k-3} = 4$

45. $\dfrac{2p}{p-2} = 3 + \dfrac{4}{p-2}$

46. $\dfrac{5k}{k+4} = 3 - \dfrac{20}{k+4}$

47. $2(m+1)(m-1) = (2m+3)(m-2)$

48. $(2y-1)(3y+2) = 6(y+2)^2$

49. $(3x-4)^2 - 5 = 3(x+5)(3x+2)$

50. $(2x+5)^2 = 3x^2 + (x+3)^2$

Solve each of the following equations for x. See Example 6.

51. $2(x-a) + b = 3x + a$

52. $5x - (2a+c) = a(x+1)$

53. $ax + b = 3(x-a)$

54. $4a - ax = 3b + bx$

55. $\dfrac{x}{a-1} = ax + 3$

56. $\dfrac{2a}{x-1} = a - b$

57. $a^2x + 3x = 2a^2$

58. $ax + b^2 = bx - a^2$

59. $3x = (2x-1)(m+4)$

60. $-x = (5x+3)(3k+1)$

In the metric system of weights and measures, temperature is measured in degrees Celsius (°C) instead of degrees Fahrenheit (°F). To convert back and forth between the two systems, use the equations

$$C = \dfrac{5(F-32)}{9} \quad and \quad F = \dfrac{9}{5}C + 32.$$

In each of the following exercises, convert to the other system. Round answers to the nearest tenth of a degree if necessary.

61. 20°C

62. 100°C

63. 59°F

64. 86°F

65. 100°F

66. 350°F

67. 40°C

68. 85°C

When a consumer borrows money, the lender must tell the consumer the true annual interest rate of the loan. The method of finding the exact true annual interest rate requires special tables available from the government. However, a quick approximate rate can be found by using the equation

$$A = \dfrac{2pf}{b(q+1)},$$

where p is the number of payments made in one year, f is the finance charge, b is the balance owed on the loan, and q is the total number of payments. Find the value of the variables not given in Exercises 69–74 on the next page. Round A to the nearest percent and round other variables to the nearest whole number. (This formula is not accurate enough for the requirements of federal law.)

69. $p = 12$, $f = \$800$, $b = \$4000$, $q = 36$;   find A

70. $p = 12$, $f = \$60$, $b = \$740$, $q = 12$;   find A

71. $A = 14\%$ (or $.14$), $p = 12$, $b = \$2000$, $q = 36$;   find f

72. $A = 11\%$, $p = 12$, $b = \$1500$, $q = 24$;   find f

73. $A = 16\%$, $p = 12$, $f = \$370$, $q = 36$;   find b

74. $A = 10\%$, $p = 12$, $f = \$490$, $q = 48$;   find b

When a loan is paid off early, a portion of the finance charge must be returned to the borrower. By one method of calculating finance charge (called the rule of 78*), the amount of unearned interest (finance charge to be returned) is given by*

$$u = f \cdot \frac{n(n + 1)}{q(q + 1)},$$

where u represents unearned interest, f is the original finance charge, n is the number of payments remaining when the loan is paid off, and q is the original number of payments. Find the amount of the unearned interest in each of the following.

75. Original finance charge $= \$800$, loan scheduled to run 36 months, paid off with 18 payments remaining

76. Original finance charge $= \$1400$, loan scheduled to run 48 months, paid off with 12 payments remaining

77. Original finance charge $= \$950$, loan scheduled to run 24 months, paid off with 6 payments remaining

78. Original finance charge $= \$175$, loan scheduled to run 12 months, paid off with 3 payments remaining

79. Find the error in the following.

$$x^2 + 2x - 15 = x^2 - 3x$$
$$(x + 5)(x - 3) = x(x - 3)$$
$$x + 5 = x$$
$$5 = 0$$

Find the value of k that will make each equation equivalent to $x = 2$.

80. $9x - 7 = k$

81. $-5x + 11x - 2 = k + 4$

82. $\dfrac{8}{k + x} = 4$

83. $\sqrt{x + k} = 0$

84. $\sqrt{3x - 2k} = 4$

Solve each of the following equations. Round to the nearest hundredth.

85. $9.06x + 3.59(8x - 5) = 12.07x + .5612$

86. $-5.74(3.1 - 2.7p) = 1.09p + 5.2588$

87. $\dfrac{2.5x - 7.8}{3.2} + \dfrac{1.2x + 11.5}{5.8} = 6$

88. $\dfrac{4.19x + 2.42}{.05} - \dfrac{5.03x - 9.74}{.02} = 1$

89. $\dfrac{2.63r - 8.99}{1.25} - \dfrac{3.90r - 1.77}{2.45} = r$

90. $\dfrac{8.19m + 2.55}{4.34} - \dfrac{8.17m - 9.94}{1.04} = 4m$

3.2   Applications

Mathematics is an important problem-solving tool. Many times the solution of a problem depends on the use of a formula which expresses a relationship among several variables. For example, the formula

$$A = \frac{1}{2}(b_1 + b_2)h$$

gives the area of a trapezoid having bases of lengths b_1 and b_2 and height h. To solve the equation for b_1, get b_1 alone, with all other variables or numbers on the other side of the equals sign. To solve for b_1, first multiply both sides of the equation by 2.

$$2 \cdot A = 2 \cdot \frac{1}{2}(b_1 + b_2)h$$

$$2A = (b_1 + b_2)h$$

$$\frac{2A}{h} = b_1 + b_2 \qquad\qquad \text{Multiply by } \frac{1}{h}$$

$$\frac{2A}{h} - b_2 = b_1$$

Thus $$b_1 = \frac{2A}{h} - b_2.$$

This process is called **solving for a specified variable.**

EXAMPLE 1   Solve $J\left(\dfrac{x}{k} + a\right) = x$ for x.

We want to get all terms with x on one side of the equation and all terms without x on the other. To do this, first use the distributive property.

$$J\left(\frac{x}{k}\right) + Ja = x$$

Eliminate the denominator, k, by assuming $k \neq 0$ and multiplying both sides by k.

$$kJ\left(\frac{x}{k}\right) + kJa = kx$$

$$Jx + kJa = kx$$

Then add $-Jx$ to both sides to get the two terms with x together.

$$kJa = kx - Jx$$

$$kJa = x(k - J) \qquad \text{Factor the right side}$$

If we assume $k \neq J$, we can multiply both sides by $1/(k - J)$ to find the solution.

$$x = \frac{kJa}{k - J} \quad \bullet$$

One of the main reasons for learning mathematics is to be able to use it in solving practical problems. However, for most students, learning how to apply mathematical skills to real situations is the most difficult task they face. In the rest of this section we give a few hints that may help you with applications.

A common difficulty with "word problems" is trying to do everything at once. It is usually best to attack the problem in stages.

Solving Word Problems

1. Decide on an unknown, and name it with some variable that you *write down*. Most students try to skip this step. They are eager to get on with writing an equation. But this is an important step. If you don't know what "x" represents, how can you write a meaningful equation or interpret a result?

2. Draw a sketch, if appropriate, showing the information given in the problem.

3. Decide on a variable expression to represent any other unknowns in the problem. For example, if W represents the width of a rectangle, L represents the length, and you know that the length is one more than twice the width, *write down $L = 1 + 2W$*.

4. Finally, use the results of Steps 1 and 3 to write an equation.

Notice how each of the steps listed above is carried out in the following examples.

EXAMPLE 2   If the length of a side of a square is increased by 3 cm, the perimeter of the new square is 40 cm more than twice the length of a side of the original square. Find the dimensions of the original square.

First, what should the variable represent? Since we want to find the length of a side of the original square, let the variable represent that.

$$x = \text{length of side of the original square}$$

Now draw a figure using the given information, as in Figure 3.1.

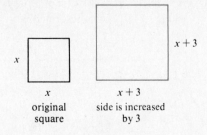

Figure 3.1

The length of a side of the new square is 3 cm more than the length of a side of the old square. Write a variable expression for that.

$$x + 3 = \text{length of side of the new square}$$

Now write a variable expression for the perimeter of the new square. Since the perimeter of a square is 4 times the length of a side,

$$4(x + 3) = \text{perimeter of the new square.}$$

We are now ready to use the given information of the problem to write an equation. The perimeter of the new square is 40 cm more than twice the length of a side of the original square, so the equation is

the new perimeter	is	40	more than	twice the side of the original square
$4(x + 3)$	$=$	40	$+$	$2x.$

Solve the equation as follows:

$$4(x + 3) = 40 + 2x$$
$$4x + 12 = 40 + 2x$$
$$2x = 28$$
$$x = 14.$$

This solution should be checked by using the words of the original problem. The length of a side of the new square would be $14 + 3 = 17$ cm; its perimeter would be $4(17) = 68$ cm. Twice the length of a side of the original square is $2(14) = 28$ cm. Since $40 + 28 = 68$, the solution satisfies the problem. ●

EXAMPLE 3   Chuck travels 80 km in the same time that Mary travels 180 km. Mary travels 50 kph faster than Chuck. Find the rate of each person.

Let x represent Chuck's rate. Since Mary traveled 50 kph faster,

$$x + 50 = \text{rate for Mary.}$$

Constant velocity problems of this kind are solved with the formula $d = rt$, where d is distance traveled in t hours at a constant rate r. We can use a chart to organize the information given in the problem and the necessary variable expressions.

For Chuck, $d = 80$ and $r = x$. From the formula $d = rt$, we have $t = d/r$, so for Chuck, $t = 80/x$. For Mary, $d = 180$, $r = x + 50$, and therefore $t = 180/(x + 50)$. This information is shown in the following chart.

	d	r	t
Chuck	80	x	$\dfrac{80}{x}$
Mary	180	$x + 50$	$\dfrac{180}{x + 50}$

Since they both traveled for the same time,

$$\frac{80}{x} = \frac{180}{x + 50}.$$

Multiply both sides of this equation by $x(x + 50)$, getting

$$x(x + 50) \cdot \frac{80}{x} = x(x + 50) \cdot \frac{180}{x + 50}$$

$$80(x + 50) = 180x$$

$$80x + 4000 = 180x$$

$$4000 = 100x$$

$$40 = x.$$

Since x represents Chuck's rate, Chuck went 40 kph. Mary's rate is $x + 50$, or $40 + 50 = 90$ kph.   ●

EXAMPLE 4   Elizabeth Thornton is a chemist. She needs a 20% solution of potassium permanganate. She has a 15% solution on hand, as well as a 30% solution. How many liters of the 15% solution should she add to 3 liters of the 30% solution to get her 20% solution?

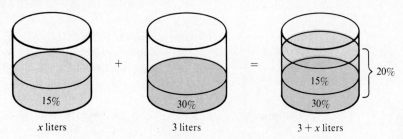

Figure 3.2

Let x be the number of liters of the 15% solution to be added. See Figure 3.2. Arrange the information of the problem in a table.

Strength	Liters of solution	Liters of pure potassium permanganate
15%	x	$.15x$
30%	3	$.30(3)$
20%	$3 + x$	$.20(3 + x)$

Since the number of liters of pure potassium permanganate in the 15% solution plus the number of liters in the 30% solution must equal the number of liters in the final 20% solution,

liters in 15%   liters in 30%   liters in 20%

$$.15x \quad + \quad .30(3) \quad = \quad .20(3 + x).$$

Solve this equation as follows.

$$.15x + .90 = .60 + .20x$$
$$.30 = .05x$$
$$6 = x$$

By this result, 6 liters of the 15% solution should be mixed with 3 liters of the 30% solution, giving $6 + 3 = 9$ liters of 20% solution. •

The next example shows a business application of equations.

EXAMPLE 5

One computer can do a job twice as fast as another. Working together, both computers can do the job in 8/3 hr. How long would it take the faster computer, working alone, to do the job?

Let x be the number of hours it would take the faster computer, working alone, to do the job. Then the slower computer will do the job alone in $2x$ hr.

In 1 hr the faster computer will do $1/x$ of the job, while in 1 hr the slower computer will do $1/(2x)$ of the job. Since the total job can be done by the two machines working together in 8/3 hr, then in 1 hr the fraction $1/(8/3)$ of the job will be done. This information leads to the equation below.

portion done by faster computer in 1 hr		portion done by slower computer in 1 hr		portion done by the two computers in 1 hr
$\dfrac{1}{x}$	$+$	$\dfrac{1}{2x}$	$=$	$\dfrac{1}{\frac{8}{3}}$

Rewrite $1/(8/3)$ as $1 \cdot (3/8)$, or 3/8, getting

$$\frac{1}{x} + \frac{1}{2x} = \frac{3}{8}.$$

Multiply both sides of the equation by $8x$.

$$8x\left(\frac{1}{x}\right) + 8x\left(\frac{1}{2x}\right) = 8x\left(\frac{3}{8}\right)$$
$$8 + 4 = 3x$$
$$12 = 3x$$
$$4 = x$$

The faster computer could do the entire job, working alone, in 4 hr. The slower computer would need $2(4) = 8$ hr. •

EXAMPLE 6

John Miller receives a $14,000 bonus from his company. He invests part of the money in 6% tax-free bonds and the remainder at 15%. He earns $1335 per year in interest from the investments. Find the amount he has invested at each rate.

Let x represent the amount Miller invests at 6%, so that $14,000 - x$ is the amount invested at 15%. Since interest is given by the product of principal, rate, and time ($i = prt$), for one year

$$\text{interest at 6\%} = x \cdot 6\% \cdot 1 = .06x$$
$$\text{interest at 15\%} = (14,000 - x) \cdot 15\% \cdot 1 = .15(14,000 - x).$$

Since the total interest is \$1335,

$$.06x + .15(14,000 - x) = 1335.$$

Solve this equation, getting

$$.06x + .15(14,000) - .15x = 1335$$
$$-.09x + 2100 = 1335$$
$$-.09x = -765$$
$$x = 8500.$$

Miller invested \$8500 at 6%, and $14,000 - \$8500 = \$5500$ at 15%.   ●

3.2   EXERCISES

Solve each equation for the indicated variable. Assume that all denominators are nonzero. See Example 1.

1. $PV = k$ for V

2. $F = whA$ for h

3. $V = lwh$ for l

4. $i = prt$ for p

5. $V = V_0 + gt$ for g

6. $s = s_0 + gt^2 + k$ for g

7. $s = \dfrac{1}{2} gt^2$ for g

8. $A = \dfrac{1}{2} (B + b)h$ for h

9. $A = \dfrac{1}{2} (B + b)h$ for B

10. $C = \dfrac{5}{9} (F - 32)$ for F

11. $S = 2\pi(r_1 + r_2)h$ for r_1

12. $A = P \left(1 + \dfrac{i}{m} \right)$ for m

13. $g = \dfrac{4\pi^2 l}{t^2}$ for l

14. $P = \dfrac{E^2 R}{r + R}$ for R

15. $S = 2\pi rh + 2\pi r^2$ for h

16. $u = f \cdot \dfrac{k(k + 1)}{n(n + 1)}$ for f

17. $A = \dfrac{24f}{B(p + 1)}$ for f

18. $A = \dfrac{24f}{B(p + 1)}$ for B

19. $\dfrac{1}{R} = \dfrac{1}{r_1} + \dfrac{1}{r_2}$ for R

20. $m = \dfrac{Ft}{v_1 - v_2}$ for v_2

Solve each of the following problems that have solutions. See Examples 2–4.

21. A triangle has a perimeter of 30 cm. Two sides of the triangle are both twice as long as the shortest side. Find the length of the short side.

22. The length of a rectangle is 3 cm less than twice the width. The perimeter is 54 cm. Find the width.

23. A box contains 11 nickels. How many quarters must be added so that the box will contain \$2.30?

24. A cash drawer contains 15 twenty-dollar bills. How many five-dollar bills must be added to bring the value of the money up to \$390?

25. A pharmacist wishes to strengthen a mixture which is 10% alcohol to one which is 30% alcohol. How much pure alcohol should he add to 7 liters of the 10% mixture?

26. A student needs 10% hydrochloric acid for a chemistry experiment. How much 5% acid should be mixed with 60 ml of 20% acid to get a 10% solution?

27. The Old Time Goodies Store sells mixed nuts. Cashews sell for $4 per quarter kg, hazelnuts for $3 per quarter kg, and peanuts for $1 per quarter kg. How many kg of peanuts should be added to 10 kg of cashews and 8 kg of hazelnuts to make a mixture which will sell for $2.50 per quarter kg?

28. The Old Time Goodies Store also sells candy. The manager wants to prepare 200 kg of a mixture for a special Halloween promotion to sell at $4.84 per kg. How much $4 per kg candy should be combined with $5.20 per kg candy for the required mix?

Exercises 29 and 30 depend on the idea of the octane rating of gasoline, a measure of its antiknock qualities. In one measure of octane, a standard fuel is made with only two ingredients: heptane and isooctane. For this fuel, the octane rating is the percent of isooctane. An actual gasoline blend is then compared to a standard fuel. For example, a gasoline with an octane rating of 98 has the same antiknock properties as a standard fuel that is 98% isooctane.

29. How many liters of 94-octane gasoline should be mixed with 200 liters of 99-octane gasoline to get a mixture that is 97-octane?

30. A service station has 92-octane and 98-octane gasoline. How many liters of each should be mixed to provide 12 liters of 96-octane gasoline needed for chemical research?

31. On a vacation trip, Jose averaged 50 mph traveling from Amarillo to Flagstaff. Returning by a different route which covered the same number of miles, he averaged 55 mph. What is the distance between the two cities if his total traveling time was 32 hr?

32. Cindy left by plane to visit her mother in Hartford, 420 km away. Fifteen minutes later, her mother left to meet her at the airport. She drove the 20 km to the airport at 40 kph, arriving just as the plane taxied in. What was the speed of the plane?

33. Russ and Janet are running in the Apple Hill Fun Run. Russ runs at 7 mph, Janet at 5 mph. If they start at the same time, how long will it be before they are 1/2 mi apart?

34. If the run in Exercise 33 has a staggered start, and Janet starts first, with Russ starting 10 min later, how long will it be before he catches up with her?

35. Joann took 20 min to drive her boat upstream to water-ski at her favorite spot. Coming back later in the day, at the same boat speed, took her 15 min. If the current in that part of the river is 5 kph, what was her boat speed?

36. Joe traveled against the wind in a small plane for 3 hr. The return trip with the wind took 2.8 hr. Find the speed of the wind if the speed of the plane in still air is 180 mph.

37. Mark can clean the house in 9 hr, while Wendy needs 6 hr. How long will it take them to clean the house if they work together?

38. Helen can paint a room in 5 hr. Jay can paint the same room in 4 hr. (He does a sloppier job.) How long will it take them to paint the room together?

39. Two chemical plants are polluting a river. If plant A produces a predetermined maximum amount of pollution twice as fast as plant B, and together they produce the maximum pollution in 26 hr, how long will it take plant B alone?

40. A sewage treatment plant has two inlet pipes to its settling pond. One can fill the pond in 10 hr, the other in 12 hr. If the first pipe is open for 5 hr and then the second pipe is opened, how long will it take to fill the pond?

41. An inlet pipe can fill Dominic's pool in 5 hr, while an outlet pipe can empty it in 8 hr. In his haste to watch television, Dominic left both pipes open. How long did it take to fill the pool?

42. Suppose Dominic discovered his error (see Exercise 41) after an hour-long program. If he then closed the outlet pipe, how much longer would be needed to fill the pool?

Work the following business oriented problems. See Example 6.

43. A clock-radio is on sale for $49. If the sale price is 15% less than the regular price, what was the regular price?

44. A shopkeeper prices his items 20% over their wholesale price. If a lamp is marked $74, what was its wholesale price?

45. Jim Marshall invests $20,000 received from an insurance settlement in two ways, some at 13% and some at 16%. Altogether, he makes $2840 per year interest. How much is invested at each rate?

46. Candy Bowen received $52,000 profit from the sale of some land. She invested part at 7 1/2% interest and the rest at 9 1/2% interest. She earned a total of $4520 interest per year. How much did she invest at each rate?

47. Bill Cornett won $100,000 in a state lottery. He first paid income tax of 40% on the winnings. Of the rest, he invested some at 8 1/2% and some at 16%, making $5550 interest per year. How much is invested at each rate?

48. Marjorie Williams earned $48,000 from royalties on her cookbook. She paid a 40% income tax on these royalties. The balance was invested in two ways, at 7 1/2% and at 10 1/2%. The investments produced $2550 interest income per year. Find the amount invested at each rate.

49. Karen Harr bought two plots of land for a total of $120,000. On the first plot, she made a profit of 15%. On the second, she lost 10%. Her total profit was $5500. How much did she pay for each piece of land?

50. Suppose $10,000 is invested at 6%. How much additional money must be invested at 8% to produce a yield on the entire amount invested of 7.2%?

51. Kathryn Johnson earns take-home pay of $198 a week. If her deductions for retirement, union dues, medical plan, and so on amount to 26% of her wages, what is her weekly pay before deductions?

52. Barbara Burnett gives 10% of her net income to the church. This amounts to $80 a month. In addition, her paycheck deductions are 24% of her gross monthly income. What is her gross monthly income?

53. A bank pays 7% interest on passbook accounts and 10% interest on long-term deposits. Suppose a depositor divides $20,000 among the two types of deposits. Find the amount deposited at each rate if the total annual income from interest is $2500.

54. Miriam Cross wishes to sell a piece of property for $125,000. She wishes the money to be paid off in two ways—a short-term note at 12% and a long-term note at 10%. Find the amount of each note if the total annual interest income is $2000.

55. In planning his retirement, John Young deposits some money at 12% with twice as much deposited at 10%. The total deposit is $60,000. Find the amount deposited at each rate if the total annual interest income is $6400.

56. A church building fund has invested $75,000 in two ways: part of the money at 9% and four times as much at 12%. Find the amount invested at each rate if the total annual income from interest is $8550.

3.3 Complex Numbers

In the next section we shall see how to solve *quadratic* equations, which are equations with an x^2 term, such as $x^2 - 4x + 3 = 0$. We shall see that the solutions of quadratic equations need not be real numbers. For example, there are no real number solutions to the quadratic equation

$$x^2 + 1 = 0.$$

To get the larger set of numbers that will permit the solution of all quadratic equations, a new number i is defined as

Definition of i

$$i = \sqrt{-1} \quad \text{or} \quad i^2 = -1.$$

Numbers of the form $a + bi$, where a and b are real numbers, are called **complex numbers.** Each real number is a complex number, since a real number a may be thought of as the complex number $a + 0i$. A complex number of the form $0 + bi$, where b is nonzero, is called an **imaginary number** (sometimes a *pure* imaginary number). Both the set of real numbers and the set of imaginary numbers are subsets of the set of complex numbers. (See Figure 3.3, which is an extension of Figure 1.10 in Section 1.2.) A complex number that is written in the form $a + bi$ or $a + ib$ is in **standard form.** (The form $a + ib$ is used to simplify certain symbols such as $i\sqrt{5}$, since $\sqrt{5}i$ could be too easily mistaken for $\sqrt{5i}$.)

EXAMPLE 1   The following statements identify different kinds of complex numbers.

(a) -8 and $\sqrt{7}$ and π are real numbers and complex numbers.

(b) $3i$ and $-11i$ and $i\sqrt{14}$ are imaginary numbers and complex numbers.

(c) $1 - 2i$ and $8 - 8i\sqrt{3}$ are complex numbers.   ●

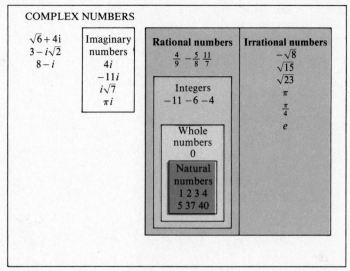

Real numbers are shaded.

Figure 3.3

EXAMPLE 2   The list below shows several numbers, along with the standard form of each number.

Number	Standard form
$6i$	$0 + 6i$
-9	$-9 + 0i$
0	$0 + 0i$
$2 - i$	$2 - i$
$8 + i\sqrt{3}$	$8 + i\sqrt{3}$ ●

Many of the solutions to quadratic equations in the next section will involve expressions such as $\sqrt{-a}$, for a positive real number a. The symbol $\sqrt{-a}$ is defined as follows.

$\sqrt{-a}$

> If $a \geq 0$, then $\sqrt{-a} = i\sqrt{a}$.

EXAMPLE 3   **(a)** $\sqrt{-16} = i\sqrt{16} = 4i$

(b) $\sqrt{-70} = i\sqrt{70}$ ●

Products or quotients with square roots of negative numbers may be simplified using the fact that $\sqrt{-a} = i\sqrt{a}$ for nonnegative numbers a. The next example shows how to do this.

EXAMPLE 4 (a) $\sqrt{-7} \cdot \sqrt{-7} = i\sqrt{7} \cdot i\sqrt{7}$

$$= i^2 \cdot (\sqrt{7})^2$$

$$= (-1) \cdot 7$$

$$= -7$$

(b) $\sqrt{-6} \cdot \sqrt{-10} = i\sqrt{6} \cdot i\sqrt{10}$

$$= i^2 \cdot \sqrt{60}$$

$$= -1 \cdot 2\sqrt{15}$$

$$= -2\sqrt{15}$$

(c) $\dfrac{\sqrt{-20}}{\sqrt{-2}} = \dfrac{i\sqrt{20}}{i\sqrt{2}} = \sqrt{10}$

(d) $\dfrac{\sqrt{-48}}{\sqrt{24}} = \dfrac{i\sqrt{48}}{\sqrt{24}} = i\sqrt{2}$ ●

When working with negative radicands, use the definition $\sqrt{-a} = i\sqrt{a}$ before using any of the other rules for radicals. In particular, the rule $\sqrt{c} \cdot \sqrt{d} = \sqrt{cd}$ is valid only when c and d are *not* both negative. For example,

$$\sqrt{(-4)(-9)} = \sqrt{36} = 6,$$

while

$$\sqrt{-4} \cdot \sqrt{-9} = 2i(3i) = 6i^2 = -6,$$

so

$$\sqrt{(-4)(-9)} \neq \sqrt{-4} \cdot \sqrt{-9}.$$

Operations on Complex Numbers Complex numbers may be added, subtracted, multiplied, and divided, as shown by the following definitions and examples.

The *sum* of two complex numbers $a + bi$ and $c + di$ is defined as follows.

Sum

$$(a + bi) + (c + di) = (a + c) + (b + d)i$$

EXAMPLE 5 (a) $(3 - 4i) + (-2 + 6i) = [3 + (-2)] + [-4 + 6]i$

$$= 1 + 2i$$

(b) $(-9 + 7i) + (3 - 15i) = -6 - 8i$ ●

Since $(a + bi) + (0 + 0i) = a + bi$ for all complex numbers $a + bi$, the number $0 + 0i$ is called the **additive identity** for complex numbers. The sum of $a + bi$ and $-a - bi$ is $0 + 0i$, so the number $-a - bi$ is called the **negative** or **additive inverse** of $a + bi$.

Using this definition of additive inverse, *subtraction* of the complex numbers $a + bi$ and $c + di$ is defined as follows:

$$(a + bi) - (c + di) = (a + bi) + (-c - di).$$

This definition is often written as

Subtraction

$$(a + bi) - (c + di) = (a - c) + (b - d)i.$$

EXAMPLE 6   Subtract as indicated.

(a) $(-4 + 3i) - (6 - 7i) = (-4 + 3i) + (-6 + 7i)$
$$= -10 + 10i$$

(b) $(12 - 5i) - (8 - 3i) = 4 - 2i$ ●

The product of two complex numbers can be found by multiplying as if the numbers were binomials and using the fact that $i^2 = -1$, as follows.

$$(a + bi)(c + di) = ac + adi + bic + bidi$$
$$= ac + adi + bci + bdi^2$$
$$= ac + (ad + bc)i + bd(-1)$$
$$(a + bi)(c + di) = (ac - bd) + (ad + bc)i$$

Based on this, the *product* of the complex numbers $a + bi$ and $c + di$ is defined in the following way.

Product

$$(a + bi)(c + di) = (ac - bd) + (ad + bc)i$$

This definition is hard to remember. To find a given product, it is better to just multiply as with binomials. The next example shows this.

EXAMPLE 7   Find each of the following products.

(a) $(2 - 3i)(3 + 4i) = 2(3) - 3i(3) + 2(4i) - 3i(4i)$
$$= 6 - 9i + 8i - 12i^2$$
$$= 6 - i - 12(-1)$$
$$= 18 - i$$

(b) $(5 - 4i)(7 - 2i) = 5(7) - 4i(7) + 5(-2i) - 4i(-2i)$
$$= 35 - 28i - 10i + 8i^2$$
$$= 35 - 38i + 8(-1)$$
$$= 27 - 38i$$

(c) $(6 + 5i)(6 - 5i) = 6 \cdot 6 - 6 \cdot 5i + 6 \cdot 5i - (5i)^2$
$$= 36 - 25i^2$$
$$= 36 - 25(-1)$$
$$= 36 + 25$$
$$= 61$$

(d) i^{15}

Since $i^2 = -1$, the value of a power of i is found by writing the given power as a product involving i^2. For example, $i^3 = i^2 \cdot i = (-1) \cdot i = -i$. Also, $i^4 = i^2 \cdot i^2 = (-1)(-1) = 1$. Using i^4 to rewrite i^{15} gives

$$i^{15} = i^{12} \cdot i^3 = (i^4)^3 \cdot i^3 = (1)^3 (-i) = -i. \quad \bullet$$

Using methods similar to those of part (d) of this last example, we can construct a table of **powers of i:**

Powers of i

$i^1 = i$	$i^5 = i$	$i^9 = i$
$i^2 = -1$	$i^6 = -1$	$i^{10} = -1$
$i^3 = -i$	$i^7 = -i$	$i^{11} = -i$
$i^4 = 1$	$i^8 = 1$	$i^{12} = 1,$

and so on.

In example 7(c) above, we found that $(6 + 5i)(6 - 5i) = 61$. The numbers $6 + 5i$ and $6 - 5i$ differ only in their middle signs; for this reason the numbers are called **conjugates** of each other.

EXAMPLE 8

The following list shows several pairs of conjugates, along with the products of the conjugates.

Number	Conjugate	Product
$3 - i$	$3 + i$	$(3 - i)(3 + i) = 10$
$2 + 7i$	$2 - 7i$	$(2 + 7i)(2 - 7i) = 53$
$-6i$	$6i$	$(-6i)(6i) = 36 \quad \bullet$

As this example suggests, the product of a complex number and its conjugate is always a real number. (See Exercise 87.)

The conjugate of the divisor is used to find the *quotient* of two complex numbers, as shown in the next example.

EXAMPLE 9

(a) Find $\dfrac{3 + 2i}{5 - i}$.

Multiply numerator and denominator by the conjugate of $5 - i$.

$$\frac{3 + 2i}{5 - i} = \frac{(3 + 2i)(5 + i)}{(5 - i)(5 + i)}$$

$$= \frac{15 + 3i + 10i + 2i^2}{25 - i^2}$$

$$= \frac{13 + 13i}{26} = \frac{1}{2} + \frac{1}{2} i$$

To check this answer, show that

$$(5 - i)\left(\frac{1}{2} + \frac{1}{2}i\right) = 3 + 2i.$$

(b) $\dfrac{3}{i} = \dfrac{3(-i)}{i(-i)}$ $-i$ is the conjugate of i

$$= \frac{-3i}{-i^2}$$

$$= \frac{-3i}{1} \qquad -i^2 = -(-1) = 1$$

$$= -3i \qquad 0 - 3i \text{ in standard form} \quad \bullet$$

3.3 EXERCISES

Identify each number as real, imaginary, or complex. See Example 1.

1. $-9i$ **2.** 6 **3.** π **4.** $-\sqrt{7}$

5. $i\sqrt{6}$ **6.** $-3i$ **7.** $2 + 5i$ **8.** $-7 - 6i$

Write each of the following in standard form. See Examples 2–4.

9. $\sqrt{-100}$ **10.** $\sqrt{-169}$ **11.** $-\sqrt{-400}$ **12.** $-\sqrt{-225}$

13. $-\sqrt{-39}$ **14.** $-\sqrt{-95}$ **15.** $5 + \sqrt{-4}$ **16.** $-7 + \sqrt{-100}$

17. $-6 - \sqrt{-196}$ **18.** $13 + \sqrt{-16}$ **19.** $9 - \sqrt{-50}$ **20.** $-11 - \sqrt{-24}$

21. $\sqrt{-5} \cdot \sqrt{-5}$ **22.** $\sqrt{-20} \cdot \sqrt{-20}$ **23.** $\sqrt{-8} \cdot \sqrt{-2}$ **24.** $\sqrt{-27} \cdot \sqrt{-3}$

25. $\dfrac{\sqrt{-40}}{\sqrt{-10}}$ **26.** $\dfrac{\sqrt{-190}}{\sqrt{-19}}$ **27.** $\dfrac{\sqrt{-6} \cdot \sqrt{-2}}{\sqrt{3}}$ **28.** $\dfrac{\sqrt{-12} \cdot \sqrt{-6}}{\sqrt{8}}$

Add or subtract. Write each result in standard form. See Examples 5 and 6.

29. $(3 + 2i) + (4 - 3i)$ **30.** $(4 - i) + (2 + 5i)$

31. $(-2 + 3i) - (-4 + 3i)$ **32.** $(-3 + 5i) - (-4 + 3i)$

33. $(2 - 5i) - (3 + 4i) - (-2 + i)$ **34.** $(-4 - i) - (2 + 3i) + (-4 + 5i)$

35. $-i - 2 - (3 - 4i) - (5 - 2i)$ **36.** $3 - (4 - i) - 4i + (-2 + 5i)$

Multiply. Write each result in standard form. See Example 7.

37. $(2 + i)(3 - 2i)$ **38.** $(-2 + 3i)(4 - 2i)$ **39.** $(2 + 4i)(-1 + 3i)$

40. $(1 + 3i)(2 - 5i)$ **41.** $(-3 + 2i)^2$ **42.** $(2 + i)^2$

43. $(3 + i)(-3 - i)$ **44.** $(-5 - i)(5 + i)$ **45.** $(2 + 3i)(2 - 3i)$

46. $(6 - 4i)(6 + 4i)$ **47.** $(\sqrt{6} + i)(\sqrt{6} - i)$ **48.** $(\sqrt{2} - 4i)(\sqrt{2} - 4i)$

49. $i(3 - 4i)(3 + 4i)$ **50.** $i(2 + 7i)(2 - 7i)$ **51.** $3i(2 - i)^2$

52. $-5i(4 - 3i)^2$

Divide. Write each result in standard form. See Example 9.

53. $\dfrac{1 + i}{1 - i}$ **54.** $\dfrac{2 - i}{2 + i}$ **55.** $\dfrac{4 - 3i}{4 + 3i}$ **56.** $\dfrac{5 - 2i}{6 - i}$

57. $\dfrac{3 - 4i}{2 - 5i}$ **58.** $\dfrac{1 - 3i}{1 + i}$ **59.** $\dfrac{-3 + 4i}{2 - i}$ **60.** $\dfrac{5 + 6i}{5 - 6i}$

61. $\dfrac{2}{i}$ **62.** $\dfrac{-7}{3i}$ **63.** $\dfrac{1 - \sqrt{-5}}{3 + \sqrt{-4}}$ **64.** $\dfrac{2 + \sqrt{-3}}{1 - \sqrt{-9}}$

Perform the indicated operations and write your answers in standard form.

65. $\dfrac{2 + i}{3 - i} \cdot \dfrac{5 + 2i}{1 + i}$ **66.** $\dfrac{1 - i}{2 + i} \cdot \dfrac{4 + 3i}{1 + i}$ **67.** $\dfrac{6 + 2i}{5 - i} \cdot \dfrac{1 - 3i}{2 + 6i}$

68. $\dfrac{5 - 3i}{1 + 2i} \cdot \dfrac{2 - 4i}{1 + i}$ **69.** $\dfrac{5 - i}{3 + i} + \dfrac{2 + 7i}{3 + i}$ **70.** $\dfrac{4 - 3i}{2 + 5i} + \dfrac{8 - i}{2 + 5i}$

Find each of the following powers of i. See Example 7(d).

71. i^5 **72.** i^8 **73.** i^9 **74.** i^{11}

75. i^{12} **76.** i^{25} **77.** i^{43} **78.** $1/i^9$

79. $1/i^{12}$ **80.** i^{-6} **81.** i^{-15} **82.** i^{-49}

Find all complex numbers $a + bi$ such that the square $(a + bi)^2$ is

83. real; **84.** imaginary.

85. Show that $\dfrac{\sqrt{2}}{2} + \dfrac{\sqrt{2}}{2}i$ is a square root of i.

86. Show that $\dfrac{\sqrt{3}}{2} + \dfrac{1}{2}i$ is a cube root of i.

87. Prove that the product of a complex number and its conjugate is a real number.

3.4 Quadratic Equations

An equation of the form $ax + b = 0$ is a linear equation, while an equation with an x^2 term is called *quadratic*. That is,

Quadratic Equation

> an equation that can be written in the form
> $$ax^2 + bx + c = 0,$$
> where a, b, and c are real numbers with $a \neq 0$, is a **quadratic equation.**

(Why is the restriction $a \neq 0$ necessary?) A quadratic equation written in the form $ax^2 + bx + c = 0$ is in **standard form.**

The simplest method of solving a quadratic equation, but one that is not always easily applied, is by factoring. This method depends on the following **zero-factor property:**

Zero-Factor Property

> If a and b are complex numbers, with $ab = 0$, then $a = 0$ or $b = 0$ or both.

The next example shows how the zero-factor property is used to solve a quadratic equation. ~~Stud $\hat{r}$ = $x^2 + x + 0 = 0$~~

EXAMPLE 1

Solve $6r^2 + 7r = 3$.

First write the equation in standard form, yielding

$$6r^2 + 7r - 3 = 0.$$

Now factor $6r^2 + 7r - 3$ to get

$$(3r - 1)(2r + 3) = 0.$$

By the zero-factor property, the product $(3r - 1)(2r + 3)$ can equal 0 only if

$$3r - 1 = 0 \quad \text{or} \quad 2r + 3 = 0.$$

Solve each of these linear equations separately to find that the solutions of the original equation are $1/3$ and $-3/2$. Check these solutions by substituting back in the original equation. The solution set is $\{1/3, -3/2\}$.   ●

A quadratic equation of the form $x^2 = k$ can be solved by factoring with the following sequence of equivalent equations.

$$x^2 = k$$
$$x^2 - k = 0$$
$$(x - \sqrt{k})(x + \sqrt{k}) = 0$$
$$x - \sqrt{k} = 0 \quad \text{or} \quad x + \sqrt{k} = 0$$
$$x = \sqrt{k} \quad \text{or} \quad x = -\sqrt{k}$$

This proves the following statement, which is sometimes called the **square root property.**

Square Root Property

> The solution set of $x^2 = k$ is $\{\sqrt{k}, -\sqrt{k}\}$.

This solution set is often abbreviated as $\{\pm\sqrt{k}\}$. Both solutions are real if $k > 0$ and imaginary if $k < 0$. (If $k = 0$, there is only one solution.)

EXAMPLE 2   Solve each equation.

(a) $z^2 = 17$

The solution set is $\{\pm\sqrt{17}\}$.

(b) $m^2 = -25$

Since $\sqrt{-25} = 5i$, the solution set of $m^2 = -25$ is $\{\pm 5i\}$.

(c) $(y - 4)^2 = 12$

Use a generalization of the square root property, working as follows.

$$(y - 4)^2 = 12$$
$$y - 4 = \pm\sqrt{12}$$
$$y = 4 \pm \sqrt{12}$$
$$y = 4 \pm 2\sqrt{3}$$

The solution set is $\{4 \pm 2\sqrt{3}\}$.   ●

Completing the Square   As suggested by Example 2(c), any quadratic equation can be solved using the square root property if it is first written in the form $(x + n)^2 = k$ for suitable numbers n and k. The next example shows how to write a quadratic equation in this form.

EXAMPLE 3   Solve $9z^2 - 12z - 1 = 0$.

We need to find constants n and k so that the given equation can be written in the form $(z + n)^2 = k$. Expanding $(z + n)^2$ gives $z^2 + 2zn + n^2$, with 1 the coefficient of z^2. To get a leading coefficient of 1 in the given equation, multiply both sides by 1/9:

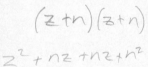

$$z^2 - \frac{4}{3}z - \frac{1}{9} = 0.$$

We need to replace $-1/9$ by a number that will make the left side a perfect square. To find this number, first add 1/9 on both sides, getting

$$z^2 - \frac{4}{3}z = \frac{1}{9}.$$

The term $-4z/3$ is twice the product of z and n, the number needed to make the left side a perfect square. That is, n satisfies the equation

$$2zn = -\frac{4}{3}z$$

$$n = -\frac{2}{3}.$$

If $n = -2/3$, then $n^2 = 4/9$, which is added to both sides, getting

$$z^2 - \frac{4}{3}z + \frac{4}{9} = \frac{1}{9} + \frac{4}{9}.$$

Factoring on the left yields

$$\left(z - \frac{2}{3}\right)^2 = \frac{5}{9}.$$

Now use the square root property and rationalize denominators to get

$$z - \frac{2}{3} = \pm\sqrt{\frac{5}{9}}$$

$$z - \frac{2}{3} = \pm\frac{\sqrt{5}}{3}$$

$$z = \frac{2}{3} \pm \frac{\sqrt{5}}{3}.$$

These two solutions can be written as

$$\frac{2 \pm \sqrt{5}}{3},$$

with the solution set abbreviated $\{(2 \pm \sqrt{5})/3\}$. ●

The process of changing $9z^2 - 12z - 1 = 0$ into the equivalent equation $(z - 2/3)^2 = 5/9$, as in the example above, is called **completing the square.**

Completing the Square

> To solve $ax^2 + bx + c = 0$, $a \neq 0$, by completing the square:
>
> **1.** If $a \neq 1$, multiply both sides by $1/a$. Then rewrite the equation so that the constant is alone on one side of the equals sign.
> **2.** Square half the coefficient of x, and add the square to both sides.
> **3.** Factor, and use the square root property.

Quadratic Formula   The method of completing the square can be used to solve any quadratic equation. However, in the long run it is better to start with the general quadratic equation,

$$ax^2 + bx + c = 0, \quad a \neq 0,$$

and use the method of completing the square to solve this equation for x in terms of the constants a, b, and c. The result will be a general formula for solving any quadratic equation. First, use the fact that $a \neq 0$ and multiply both sides by $1/a$ to get

$$x^2 + \frac{b}{a}x + \frac{c}{a} = 0.$$

Then, adding $-c/a$ to both sides gives

$$x^2 + \frac{b}{a}x = -\frac{c}{a}.$$

Now take half of b/a, and square the result:

$$\frac{1}{2} \cdot \frac{b}{a} = \frac{b}{2a} \quad \text{and} \quad \left(\frac{b}{2a}\right)^2 = \frac{b^2}{4a^2}.$$

Add the square to both sides, producing

$$x^2 + \frac{b}{a}x + \frac{b^2}{4a^2} = \frac{b^2}{4a^2} - \frac{c}{a}.$$

The expression on the left side of the equals sign can be written as the square of a binomial, while the expression on the right can be simplified. Doing all this yields

$$\left(x + \frac{b}{2a}\right)^2 = \frac{b^2 - 4ac}{4a^2}.$$

By the square root property, this last statement leads to

$$x + \frac{b}{2a} = \sqrt{\frac{b^2 - 4ac}{4a^2}} \quad \text{or} \quad x + \frac{b}{2a} = -\sqrt{\frac{b^2 - 4ac}{4a^2}}.$$

Since $4a^2 = (2a)^2$, or $4a^2 = (-2a)^2$,

$$x + \frac{b}{2a} = \frac{\sqrt{b^2 - 4ac}}{2a} \quad \text{or} \quad x + \frac{b}{2a} = \frac{-\sqrt{b^2 - 4ac}}{2a}.$$

Adding $-b/(2a)$ to both sides of each result gives

$$x = \frac{-b + \sqrt{b^2 - 4ac}}{2a} \quad \text{or} \quad x = \frac{-b - \sqrt{b^2 - 4ac}}{2a}.$$

A more compact form of this result, called the **quadratic formula,** is given below.

Quadratic Formula

The solutions of the quadratic equation $ax^2 + bx + c = 0$, where $a \neq 0$, are

$$\frac{-b \pm \sqrt{b^2 - 4ac}}{2a}.$$

EXAMPLE 4

Solve $x^2 - 4x + 1 = 0$.

Here $a = 1$, $b = -4$, and $c = 1$. Substitute these values into the quadratic formula, producing the equation on the next page.

$$x = \frac{-b \pm \sqrt{b^2 - 4ac}}{2a}$$

$$= \frac{-(-4) \pm \sqrt{(-4)^2 - 4(1)(1)}}{2(1)}$$

$$= \frac{4 \pm \sqrt{16 - 4}}{2}$$

$$= \frac{4 \pm 2\sqrt{3}}{2}$$

$$= \frac{2(2 \pm \sqrt{3})}{2}$$

$$= 2 \pm \sqrt{3}$$

The solution set is $\{2 + \sqrt{3}, 2 - \sqrt{3}\}$, abbreviated $\{2 \pm \sqrt{3}\}$.   •

EXAMPLE 5   Solve $2y^2 = y - 4$.

To find the values of a, b, and c, first rewrite the equation as $2y^2 - y + 4 = 0$. Then $a = 2$, $b = -1$, and $c = 4$. By the quadratic formula,

$$y = \frac{-(-1) \pm \sqrt{(-1)^2 - 4(2)(4)}}{2(2)}$$

$$= \frac{1 \pm \sqrt{1 - 32}}{4}$$

$$= \frac{1 \pm \sqrt{-31}}{4}$$

$$= \frac{1 \pm i\sqrt{31}}{4}.$$

The solutions are complex numbers; the solution set is $\{(1 \pm i\sqrt{31})/4\}$.   •

EXAMPLE 6   Solve $im^2 + 5m - 3i = 0$.

Here, $a = i$, $b = 5$, and $c = -3i$, with

$$m = \frac{-5 \pm \sqrt{5^2 - 4(i)(-3i)}}{2i}$$

$$= \frac{-5 \pm \sqrt{25 + 12i^2}}{2i}$$

$$= \frac{-5 \pm \sqrt{25 - 12}}{2i}$$

$$= \frac{-5 \pm \sqrt{13}}{2i}.$$

Simplify by multiplying numerator and denominator by $-i$ (the same result would be obtained if we used i). This gives

$$m = \frac{(-5 \pm \sqrt{13})(-i)}{2i(-i)}.$$

In the denominator, $2i(-i) = -2i^2 = -2(-1) = 2$, so the final result is

$$m = \frac{5i \pm i\sqrt{13}}{2}.$$

The solution set is $\{(5i \pm i\sqrt{13})/2\}$. ●

The Discriminant   The quantity under the radical in the quadratic formula, $b^2 - 4ac$, is called the **discriminant.** When the numbers a, b, and c are *integers* (but not necessarily otherwise), the value of the discriminant can be used to determine whether the solutions will be rational, irrational, or complex numbers. If the discriminant is 0, there will be only one solution. (Why?)

EXAMPLE 7   Use the discriminant to determine whether the solutions of $5x^2 + 2x - 4 = 0$ are rational, irrational, or complex.
    The discriminant is

$$b^2 - 4ac = (2)^2 - (4)(5)(-4) = 84.$$

Since the discriminant is positive, there are two real number solutions. Since 84 is not a perfect square, the solutions will be irrational numbers. ●

    The discriminant of a quadratic equation gives the following information about the solutions of the equation.

Discriminant

Disciminant	*Number of solutions*	*Kind of solutions*
Positive, perfect square	two	rational
Positive, but not a perfect square	two	irrational
Zero	one	rational
Negative	two	complex

EXAMPLE 8   Find a value of k so that the equation

$$16p^2 + kp + 25 = 0$$

has exactly one solution.
    A quadratic equation with real coefficients will have exactly one solution if the discriminant is zero. Here, $a = 16$, $b = k$, and $c = 25$, giving the discriminant

$$b^2 - 4ac = k^2 - 4(16)(25) = k^2 - 1600.$$

The discriminant is 0 if

$$k^2 - 1600 = 0$$

or if $\qquad\qquad\qquad\qquad k^2 = 1600,$

from which $k = \pm 40.$ ●

Word problems often lead to quadratic equations, as in the next example.

EXAMPLE 9   Michael wants to make an exposed gravel border of uniform width around a rectangular pool in his garden. The pool is 10 ft by 6 ft. He has enough material to cover 36 sq ft. How wide will the border be?

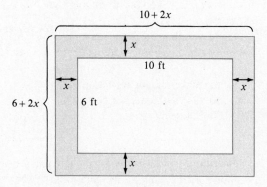

Figure 3.4

A sketch of the pool with border is shown in Figure 3.4. Let x represent the width of the border. Then the width of the large rectangle is $6 + 2x$ and its length is $10 + 2x$. The area of the large rectangle is $(6 + 2x)(10 + 2x)$. The area of the pool is $6 \cdot 10 = 60$. The area of the border is found by subtracting the area of the pool from the area of the larger rectangle. This difference should be 36 sq ft.

$$(6 + 2x)(10 + 2x) - 60 = 36$$

Now solve this equation.

$$60 + 32x + 4x^2 - 60 = 36$$
$$4x^2 + 32x - 36 = 0$$
$$x^2 + 8x - 9 = 0$$
$$(x + 9)(x - 1) = 0$$

The solutions are -9 and 1. We cannot use -9 as the width of the border, so the border should be 1 ft wide.   ●

3.4 EXERCISES

Solve the following equations by factoring or by using the square root property. See Examples 1 and 2.

1. $p^2 = 16$

2. $k^2 = 25$

3. $x^2 = 27$

4. $y^2 = 24$

5. $(m - 3)^2 = 5$

6. $(p + 2)^2 = 7$

7. $(3k - 1)^2 = 12$

8. $(4t + 1)^2 = 20$

9. $p^2 - 5p + 6 = 0$

10. $q^2 + 2q - 8 = 0$

11. $6z^2 - 5z - 50 = 0$

12. $21p^2 = 10 - 29p$

13. $8k^2 + 14k + 3 = 0$

14. $18r^2 - 9r - 2 = 0$

Solve the following equations by completing the square. See Example 3.

15. $p^2 - 8p + 15 = 0$

16. $m^2 + 5m = 6$

17. $x^2 - 2x - 4 = 0$

18. $r^2 + 8r + 13 = 0$

19. $2p^2 + 2p + 1 = 0$

20. $9z^2 - 12z + 8 = 0$

Solve the following equations by using the quadratic formula. See Examples 4 and 5.

21. $m^2 - m - 1 = 0$

22. $y^2 - 3y - 2 = 0$

23. $2s^2 + 2s = 3$

24. $t^2 - t = 3$

25. $x^2 - 6x + 7 = 0$

26. $11p^2 - 7p + 1 = 0$

27. $n^2 + 4 = 3n$

28. $9p^2 = 25 + 30p$

29. $2m^2 = m - 1$

30. $4z^2 - 12z + 11 = 0$

31. $x^2 = 2x - 5$

32. $3k^2 + 2 = k$

33. $4 - \dfrac{11}{x} - \dfrac{3}{x^2} = 0$

34. $3 - \dfrac{4}{p} = \dfrac{2}{p^2}$

35. $2 - \dfrac{5}{r} + \dfrac{3}{r^2} = 0$

36. $2 - \dfrac{5}{k} + \dfrac{2}{k^2} = 0$

Use a calculator to give the solutions of the following equations to the nearest thousandth.

37. $5n^2 - 4 = 3n$

38. $9p^2 - 30p = 25$

39. $2m^2 = m + 2$

40. $4n^2 + 12n - 9 = 0$

41. $x^2 - 2 = 3x$

42. $3k^2 - 2 = 4k$

Use the quadratic formula to solve the following equations. See Example 6.

43. $m^2 - \sqrt{2}m - 1 = 0$

44. $z^2 - \sqrt{3}z - 2 = 0$

45. $\sqrt{2}p^2 - 3p + \sqrt{2} = 0$

46. $-\sqrt{6}k^2 - 2k + \sqrt{6} = 0$

47. $x^2 + ix + 1 = 0$

48. $3im^2 - 2m + i = 0$

49. $ip^2 - 2p + i = 0$

50. $2ix^2 - 3x + 3i = 0$

51. $(1 + i)x^2 - x + (1 - i) = 0$

52. $(2 + i)r^2 - 3r + (2 - i) = 0$

Solve each of the following equations by factoring first and then using the quadratic formula.

53. $x^3 - 1 = 0$

54. $x^3 + 64 = 0$

55. $x^3 + 27 = 0$

56. $x^3 + 1 = 0$

57. $x^3 - 64 = 0$

58. $x^3 - 27 = 0$

59. $8p^3 + 125 = 0$

60. $64r^3 - 343 = 0$

Identify the values of a, b, and c for each of the following; then evaluate the discriminant $b^2 - 4ac$ and use it to predict the type of solutions. Do not solve the equations. See Example 7.

61. $x^2 + 8x + 16 = 0$ **62.** $x^2 - 5x + 4 = 0$ **63.** $3m^2 - 5m + 2 = 0$

64. $8y^2 = 14y - 3$ **65.** $4p^2 = 6p + 3$ **66.** $2r^2 - 4r + 1 = 0$

67. $9k^2 + 11k + 4 = 0$ **68.** $3z^2 = 4z - 5$

Solve the following problems. See Example 9.

69. Find two consecutive even integers whose product is 288.

70. The sum of the squares of two consecutive integers is 481. Find the integers.

71. Two integers have a sum of 10. The sum of the squares of the integers is 148. Find the integers.

72. A shopping center has a rectangular area of 40,000 sq yd enclosed on three sides for a parking lot. The length is 200 yd more than twice the width. What are the dimensions of the lot?

73. An ecology center wants to set up an experimental garden. It has 300 m of fencing to enclose a rectangular area of 5000 sq m. Find the dimensions of the rectangle.

74. Pat Murphy went into a frame-it-yourself shop. He wanted a frame 3 in longer than it was wide. The frame he chose extended 1.5 in beyond the picture on each side. Find the outside dimensions of the frame if the area of the unframed picture is 70 sq in.

75. Joan wants to buy a rug for a room that is 12 ft by 15 ft. She wants to leave a uniform strip of floor around the rug. She can afford 108 sq ft of carpeting. What dimensions should the rug have?

76. Bob Johnson can clean the garage in 9 hr less time than his brother Paul. Working together, they can do the job in 20 hr. How long would it take each one to do the job alone?

77. Dolores drives 10 mph faster than Steve. Both start at the same time for Atlanta from Chattanooga, a distance of about 100 mi. It takes Steve 1/3 hr longer than Dolores to make the trip. What is Steve's average speed?

78. Marjorie Williams walks 1 mph faster than her friend Lisa. In a walk for charity, both walked the full distance of 24 mi. Lisa took 2 hr longer than Marjorie. What was Lisa's average speed?.

Find the solution for each of the following problems. Round to the nearest hundredth.

79. One leg of a right triangle is 4 cm longer than the other leg. The hypotenuse (longest side) is 10 cm longer than the shorter leg. Find each of the three sides of the triangle.

80. A rectangle has a diagonal of 16 cm. The width of the rectangle is 3.2 cm less than the length. Find the length and width of the rectangle.

Find all values of k for which the following equations have exactly one solution. See Example 8.

81. $9x^2 + kx + 4 = 0$ **82.** $25m^2 - 10m + k = 0$ **83.** $y^2 + 11y + k = 0$

84. $z^2 - 5z + k = 0$ **85.** $kr^2 + (2k + 6)r + 16 = 0$ **86.** $ky^2 + 2(k + 4)y + 25 = 0$

Use the methods shown earlier, along with the quadratic formula as necessary, to solve each equation for the indicated variable. Assume that no denominators are zero.

87. $s = \dfrac{1}{2} gt^2$ for t

88. $A = \pi r^2$ for r

89. $L = \dfrac{d^4 k}{h^2}$ for h

90. $F = \dfrac{kMv^2}{r}$ for v

91. $s = s_0 + gt^2 + k$ for t

92. $g = \dfrac{4\pi^2}{t^2}$ for t

93. $P = \dfrac{E^2 R}{(r + R)^2}$ for R

94. $S = 2\pi rh + 2\pi r^2$ for r

For each of the following equations, (a) solve for x in terms of y, (b) solve for y in terms of x.

95. $4x^2 - 2xy + 3y^2 = 2$

96. $3y^2 + 4xy - 9x^2 = -1$

Let r_1 and r_2 be the solutions of the quadratic equation $ax^2 + bx + c = 0$. Show that the following statements are correct.

97. $r_1 + r_2 = -b/a$

98. $r_1 \cdot r_2 = c/a$

3.5 Equations Quadratic in Form

The equation $12m^4 - 11m^2 + 2 = 0$ is not a quadratic equation because of the m^4 term. However, if we make the substitution

$$x = m^2 \quad \text{or} \quad x^2 = m^4,$$

then the given equation becomes

$$12x^2 - 11x + 2 = 0,$$

which is a quadratic equation. This quadratic equation can be solved to find x, and then from $x = m^2$, we can find the values of m, the solutions to our original equation.

An equation such as $12m^4 - 11m^2 + 2 = 0$ is said to be **quadratic in form** if it can be written as

$$au^2 + bu + c = 0,$$

where $a \neq 0$ and u is some algebraic expression.

EXAMPLE 1 Solve $12m^4 - 11m^2 + 2 = 0$.

As mentioned above, this equation is quadratic in form. By making the substitution $x = m^2$, the equation becomes

$$12x^2 - 11x + 2 = 0,$$

which can be solved by factoring in the following way.

$$12x^2 - 11x + 2 = 0$$
$$(3x - 2)(4x - 1) = 0$$
$$x = \frac{2}{3} \quad \text{or} \quad x = \frac{1}{4}$$

Now we know that $x = 2/3$ or $x = 1/4$. To find m, use the fact that $x = m^2$ and replace x with m^2, getting

$$m^2 = \frac{2}{3} \quad \text{or} \quad m^2 = \frac{1}{4}$$

$$m = \pm\sqrt{\frac{2}{3}} \qquad m = \pm\sqrt{\frac{1}{4}}$$

$$m = \pm\frac{\sqrt{2}}{\sqrt{3}}$$

$$m = \frac{\pm\sqrt{6}}{3} \quad \text{or} \quad m = \pm\frac{1}{2}.$$

These four solutions of the given equation $12m^4 - 11m^2 + 2 = 0$ make up the solution set $\{\sqrt{6}/3, -\sqrt{6}/3, 1/2, -1/2\}$, abbreviated $\{\pm\sqrt{6}/3, \pm 1/2\}$ •

EXAMPLE 2

Solve $6p^{-2} + p^{-1} = 2$.

Let $u = p^{-1}$ so that $u^2 = p^{-2}$. Then rearrange terms to get

$$6u^2 + u - 2 = 0.$$

Factor on the left, and then place each factor equal to 0, giving

$$(3u + 2)(2u - 1) = 0$$
$$3u + 2 = 0 \quad \text{or} \quad 2u - 1 = 0$$
$$u = -\frac{2}{3} \qquad u = \frac{1}{2}.$$

Since $u = p^{-1}$,

$$p^{-1} = -\frac{2}{3} \quad \text{or} \quad p^{-1} = \frac{1}{2},$$

from which

$$p = -\frac{3}{2} \quad \text{or} \quad p = 2.$$

The solution set of $6p^{-2} + p^{-1} = 2$ is $\{-3/2, 2\}$. •

EXAMPLE 3   Solve $4p^4 - 16p^2 + 13 = 0$.

Let $u = p^2$ to get

$$4u^2 - 16u + 13 = 0.$$

To solve this equation, use the quadratic formula with $a = 4$, $b = -16$, and $c = 13$. Substituting these values into the quadratic formula gives

$$u = \frac{-(-16) \pm \sqrt{(-16)^2 - 4(4)(13)}}{2(4)}$$

$$= \frac{16 \pm \sqrt{48}}{8}$$

$$= \frac{16 \pm 4\sqrt{3}}{8}$$

$$= \frac{4 \pm \sqrt{3}}{2}.$$

Since $p^2 = u$,

$$p^2 = \frac{4 + \sqrt{3}}{2} \quad \text{or} \quad p^2 = \frac{4 - \sqrt{3}}{2}.$$

Finally,

$$p = \pm\sqrt{\frac{4 + \sqrt{3}}{2}} \quad \text{or} \quad p = \pm\sqrt{\frac{4 - \sqrt{3}}{2}}.$$

Rationalizing the denominators gives the solution set

$$\left\{ \frac{\pm\sqrt{8 + 2\sqrt{3}}}{2}, \frac{\pm\sqrt{8 - 2\sqrt{3}}}{2} \right\}. \quad \bullet$$

To solve equations containing radicals or rational exponents, such as $x = \sqrt{15 - 2x}$, or $(x + 1)^{1/2} = x$, use the following result.

If P and Q are algebraic expressions, then every solution of the equation $P = Q$ is also a solution of the equation $(P)^n = (Q)^n$, for any positive integer n.

Be very careful when using this result. It does *not* say that the equations $P = Q$ and $(P)^n = (Q)^n$ are equivalent; it says only that each solution of the original equation $P = Q$ is also a solution of the new equation $(P)^n = (Q)^n$. However, the new equation may have *more* solutions than the original equation. For example, the solution set of the equation $x = -2$ is $\{-2\}$. If we square both sides of the equation $x = -2$, we get the new equation $x^2 = 4$, which has solution set $\{-2, 2\}$. Since the solution sets are not equal, the equations are not equivalent. Because of this, it is essential to check all proposed solutions back in the original equation.

EXAMPLE 4   Solve $x = \sqrt{15 - 2x}$.

The equation $x = \sqrt{15 - 2x}$ can be solved by squaring both sides as follows.

$$x^2 = (\sqrt{15 - 2x})^2$$
$$x^2 = 15 - 2x$$
$$x^2 + 2x - 15 = 0$$
$$(x + 5)(x - 3) = 0$$
$$x = -5 \quad \text{or} \quad x = 3$$

Now the proposed solutions must be checked in the original equation,

$$x = \sqrt{15 - 2x}.$$

If $x = -5$, does $x = \sqrt{15 - 2x}$?

$$-5 = \sqrt{15 + 10}$$
$$-5 = 5 \qquad \text{False}$$

If $x = 3$, does $x = \sqrt{15 - 2x}$?

$$3 = \sqrt{15 - 6}$$
$$3 = 3 \qquad \text{True}$$

As this check shows, only 3 is a solution, giving the solution set $\{3\}$.   ●

EXAMPLE 5   Solve $\sqrt{2x + 3} - \sqrt{x + 1} = 1$.

Separate the radicals by writing the equation as

$$\sqrt{2x + 3} = 1 + \sqrt{x + 1}.$$

Now we need to square both sides. Be very careful when squaring on the right side of this equation. Recall that $(a + b)^2 = a^2 + 2ab + b^2$; replace a with 1 and b with $\sqrt{x + 1}$ to get the next equation, the result of squaring both sides of $\sqrt{2x + 3} = 1 + \sqrt{x + 1}$.

$$2x + 3 = 1 + 2\sqrt{x + 1} + x + 1$$
$$x + 1 = 2\sqrt{x + 1}$$

One side of the equation still contains a radical; to eliminate it, square both sides again. This gives

$$x^2 + 2x + 1 = 4(x + 1)$$
$$x^2 - 2x - 3 = 0$$
$$(x - 3)(x + 1) = 0$$
$$x = 3 \quad \text{or} \quad x = -1.$$

Check these proposed solutions in the original equation.

$$\text{If } x = 3, \text{ does } \sqrt{2x + 3} - \sqrt{x + 1} = 1?$$
$$\sqrt{9} - \sqrt{4} = 1$$
$$3 - 2 = 1 \qquad \text{True}$$

$$\text{If } x = -1, \text{ does } \sqrt{2x + 3} - \sqrt{x + 1} = 1?$$
$$\sqrt{1} - \sqrt{0} = 1$$
$$1 - 0 = 1 \qquad \text{True}$$

Both proposed solutions 3 and -1 are solutions of the original equation, giving $\{3, -1\}$ as the solution set. ●

EXAMPLE 6   Solve $(5x^2 - 6)^{1/4} = x$.

Since the equation involves a fourth root, begin by raising both sides to the fourth power.

$$[(5x^2 - 6)^{1/4}]^4 = x^4$$
$$5x^2 - 6 = x^4$$
$$x^4 - 5x^2 + 6 = 0$$

Now substitute y for x^2, yielding

$$y^2 - 5y + 6 = 0$$
$$(y - 3)(y - 2) = 0$$
$$y = 3 \quad \text{or} \quad y = 2.$$

Since $y = x^2$,

$$x^2 = 3 \qquad \text{or} \qquad x^2 = 2$$
$$x = \pm\sqrt{3} \qquad \text{or} \qquad x = \pm\sqrt{2}.$$

By checking the four proposed solutions, $\sqrt{3}$, $-\sqrt{3}$, $\sqrt{2}$, and $-\sqrt{2}$, in the original equation, we find that only $\sqrt{3}$ and $\sqrt{2}$ are solutions, so the solution set is $\{\sqrt{3}, \sqrt{2}\}$. ●

3.5 EXERCISES

Solve the following equations by first rewriting in quadratic form. See Examples 1–3.

1. $m^4 + 2m^2 - 15 = 0$

2. $3k^4 + 10k^2 - 25 = 0$

3. $2r^4 - 7r^2 + 5 = 0$

4. $4x^4 - 8x^2 + 3 = 0$

5. $(g - 2)^2 - 6(g - 2) + 8 = 0$

6. $(p + 2)^2 - 2(p + 2) - 15 = 0$

7. $-(r + 1)^2 - 3(r + 1) + 3 = 0$

8. $-2(z - 4)^2 + 2(z - 4) + 3 = 0$

9. $6(k + 2)^4 - 11(k + 2)^2 + 4 = 0$

10. $8(m - 4)^4 - 10(m - 4)^2 + 3 = 0$

11. $7p^{-2} + 19p^{-1} = 6$

12. $5k^{-2} - 43k^{-1} = 18$

13. $(r - 1)^{2/3} + (r - 1)^{1/3} = 12$

14. $(y + 3)^{2/3} - 2(y + 3)^{1/3} - 3 = 0$

15. $5r = r^{-1}$

16. $a^{-1} = a^{-2}$

17. $3 + \dfrac{5}{p^2 + 1} = \dfrac{2}{(p^2 + 1)^2}$

18. $6 = \dfrac{7}{2y - 3} + \dfrac{3}{(2y - 3)^2}$

Solve the following equations. See Examples 4–6.

19. $\sqrt{r} + 1 = 2$

20. $\sqrt{z} - 3 = 5$

21. $\sqrt{2m + 1} = 2\sqrt{m}$

22. $3\sqrt{p} = \sqrt{8p + 16}$

23. $\sqrt{3z + 7} = 3z + 5$

24. $\sqrt{4r + 13} = 2r - 1$

25. $\sqrt{4k + 5} - 2 = 2k - 7$

26. $\sqrt{6m + 7} - 1 = m + 1$

27. $\sqrt{4x} - x + 3 = 0$

28. $\sqrt{2t} - t + 4 = 0$

29. $\sqrt{y} = \sqrt{y - 5} + 1$

30. $\sqrt{2m} = \sqrt{m + 7} - 1$

31. $\sqrt{r + 5} - 2 = \sqrt{r - 1}$

32. $\sqrt{m + 7} + 3 = \sqrt{m - 4}$

33. $\sqrt{y + 2} = \sqrt{2y + 5} - 1$

34. $\sqrt{4x + 1} = \sqrt{x - 1} + 2$

35. $\sqrt{2\sqrt{7x + 2}} = \sqrt{3x + 2}$

36. $\sqrt{3\sqrt{2m + 3}} = \sqrt{5m - 6}$

37. $\sqrt{x + 2} = \sqrt{4 + 7\sqrt{x}}$

38. $3 - \sqrt{x} = \sqrt{2\sqrt{x} - 3}$

39. $\sqrt[3]{4n + 3} = \sqrt[3]{2n - 1}$

40. $\sqrt[3]{2z} = \sqrt[3]{5z + 2}$

41. $\sqrt[3]{t^2 + 2t - 1} = \sqrt[3]{t^2 + 3}$

42. $\sqrt[3]{2x^2 - 5x + 4} = \sqrt[3]{2x^2}$

43. $(2r + 5)^{1/3} = (6r - 1)^{1/3}$

44. $(3m + 7)^{1/3} = (4m + 2)^{1/3}$

45. $\sqrt[4]{q - 15} = 2$

46. $\sqrt[4]{3x + 1} = 1$

47. $\sqrt[4]{y^2 + 2y} = \sqrt[4]{3}$

48. $\sqrt[4]{k^2 + 6k} = 2$

49. $(z^2 + 24z)^{1/4} = 3$

50. $(3t^2 + 52t)^{1/4} = 4$

51. $(2r - 1)^{2/3} = r^{1/3}$

52. $(z - 3)^{2/5} = (4z)^{1/5}$

53. $k^{2/3} = 2k^{1/3}$

54. $3m^{3/4} = m^{1/2}$

Find all solutions for each of the following, rounded to the nearest thousandth.

55. $6p^4 - 41p^2 + 63 = 0$

56. $20z^4 - 67z^2 + 56 = 0$

57. $3k^2 - \sqrt{4k^2 + 3} = 0$

58. $2r^2 - \sqrt{r^2 + 1} = 0$

Solve each equation for the indicated variable. Assume that all denominators are nonzero.

59. $d = k\sqrt{h}$ for h

60. $v = \dfrac{k}{\sqrt{d}}$ for d

61. $P = 2\sqrt{\dfrac{L}{g}}$ for L

62. $c = \sqrt{a^2 + b^2}$ for a

63. $x^{2/3} + y^{2/3} = a^{2/3}$ for y

64. $m^{3/4} + n^{3/4} = 1$ for m

3.6 Inequalities

An equation says that two expressions are equal, while an *inequality* says that one expression is greater than, greater than or equal to, less than, or less than or equal to, another. As with equations, a value of the variable for which the inequality is true is a **solution** of the inequality, and the set of all such solutions is the **solution set** of the inequality. Two inequalities with the same solution set are **equivalent.**

Inequalities are solved with the following properties of inequalities.

Properties of Inequalities

> If P, Q, and R are algebraic expressions, then
> **(a)** $P < Q$ and $P + R < Q + R$ are equivalent.
> *(The same expression may be added to both sides of an inequality.)*
> **(b)** If $R > 0$, then $P < Q$ and $PR < QR$ are equivalent.
> *(The same positive expression may be multiplied on both sides of an inequality.)*
> **(c)** If $R < 0$, then $P < Q$ and $PR > QR$ are equivalent.
> *(The same negative expression may be multiplied on both sides of an inequality, as long as the direction of the inequality symbol is changed.)*

Similar properties are valid if $<$ is replaced with $>$, $\le$, or $\ge$. Pay careful attention to part (c): if both sides of an inequality are multiplied by a negative number, the direction of the inequality symbol must be reversed. For example, if we start with the true statement $-3 < 5$, and multiply both sides by the *positive* number 2, we get

$$-3 \cdot 2 < 5 \cdot 2$$
$$-6 < 10,$$

still a true statement. On the other hand, if we start with $-3 < 5$ and multiply both sides by the *negative* number -2, we get a true result only if we reverse the direction of the inequality symbol.

$$-3(-2) > 5(-2)$$
$$6 > -10$$

EXAMPLE 1   Solve the inequality $-3x + 5 > -7$.

Use the properties of inequalities. Adding -5 on both sides gives

$$-3x + 5 + (-5) > -7 + (-5)$$
$$-3x > -12.$$

Now multiply both sides by $-1/3$. Since $-1/3 < 0$, reverse the direction of the inequality symbol to get the inequality on the next page.

$$-\frac{1}{3}(-3x) < -\frac{1}{3}(-12)$$

$$x < 4$$

The original inequality is satisfied by any real number less than 4. The solution set can be written $\{x | x < 4\}$. A graph of the solution set is shown in Figure 3.5, where the parenthesis is used to show that 4 itself does not belong to the solution set. ●

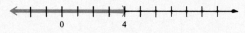

Figure 3.5

The set $\{x | x < 4\}$, the solution set for the inequality in Example 1, is an example of an **interval.** A simplified notation, called **interval notation,** is used for writing intervals. With this notation, the interval in Example 1 can be written as just $(-\infty, 4)$. The symbol $-\infty$ is not a real number; the symbol is used to help show that the interval includes all real numbers less than 4. The interval $(-\infty, 4)$ is an example of an **open interval,** since the endpoint, 4, is not part of the interval. Examples of other sets written in interval notation are shown below. Square brackets are used to show that the given number *is* part of the graph. Whenever two real numbers a and b are used to write an interval, it is assumed that $a < b$.

Type of interval	Set	Interval notation	Graph
Open interval	$\{x | a < x\}$	$(a, +\infty)$	
	$\{x | a < x < b\}$	(a, b)	
	$\{x | x < b\}$	$(-\infty, b)$	
Half-open interval	$\{x | a \leq x\}$	$[a, +\infty)$	
	$\{x | a < x \leq b\}$	$(a, b]$	
	$\{x | a \leq x < b\}$	$[a, b)$	
	$\{x | x \leq b\}$	$(-\infty, b]$	
Closed interval	$\{x | a \leq x \leq b\}$	$[a, b]$	

EXAMPLE 2   Solve $4 - 3y \leq 7 + 2y$. Write the solution in interval notation and graph the solution on a number line.

Write the following series of equivalent inequalities.

$$4 - 3y \leq 7 + 2y$$
$$-4 - 2y + 4 - 3y \leq -4 - 2y + 7 + 2y$$
$$-5y \leq 3$$
$$(-1/5)(-5y) \geq (-1/5)(3)$$
$$y \geq -3/5$$

In set-builder notation, the solution set is $\{y|y \geq -3/5\}$, while in interval notation the solution set is $[-3/5, +\infty)$. See Figure 3.6 for the graph of the solution set.   •

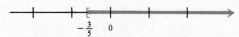

Figure 3.6

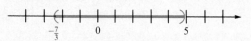

Figure 3.7

From now on, we shall write the solutions of all inequalities with interval notation.

The inequality $-2 < 5 + 3m < 20$ says that $5 + 3m$ is between -2 and 20. Solve this inequality using an extension of the properties of inequalities given above.

EXAMPLE 3   Solve $-2 < 5 + 3m < 20$.

Write equivalent inequalities as follows.

$$-2 < 5 + 3m < 20$$
$$-7 < 3m < 15$$
$$-7/3 < m < 5$$

The solution, graphed in Figure 3.7, is the interval $(-7/3, \qquad$ •

Quadratic Inequalities   Section 3.4 introduced quadratic equations. Now we will look at *quadratic inequalities*.

Quadratic
Inequality

> A **quadratic inequality** is an inequality that can be written in the form
>
> $$ax^2 + bx + c < 0,$$
>
> for real numbers $a \neq 0$, b, and c. (The symbol $<$ can be replaced with $>$, $\leq$, or $\geq$.)

EXAMPLE 4   Solve the quadratic inequality $x^2 - x - 12 < 0$.

Begin by finding the values of x which satisfy $x^2 - x - 12 = 0$.

$$x^2 - x - 12 = 0$$
$$(x + 3)(x - 4) = 0$$
$$x = -3 \quad \text{or} \quad x = 4$$

These two points, -3 and 4, divide a number line into the three regions shown in Figure 3.8. If a point in region A, for example, makes the polynomial $x^2 - x - 12$ negative, then all points in region A will make that polynomial negative.

To find the regions that make $x^2 - x - 12$ negative (< 0), make the **sign graph** shown in Figure 3.9. A sign graph shows where factors are positive or negative. First decide on the sign of the factor $x + 3$ in each of the three regions. Then do the same thing for the factor $x - 4$.

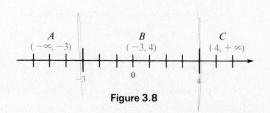

Figure 3.8

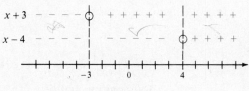

Figure 3.9

Now consider the sign of the product of the two factors in each region. As the sign graph shows, both factors are negative in the interval $(-\infty, -3)$; therefore their product is positive in that interval. For the interval $(-3, 4)$, one factor is positive, while the other is negative, giving a negative product. In the last region, $(4, +\infty)$, both factors are positive so their product is positive. The polynomial $x^2 - x - 12$ is negative (what the original inequality calls for) when the product of its factors is negative, that is, for the interval $(-3, 4)$. The graph of this solution set is shown in Figure 3.10. ●

Figure 3.10

EXAMPLE 5   Solve the inequality $2x^2 + 5x - 12 \geq 0$.

Begin by finding the values of x which satisfy $2x^2 + 5x - 12 = 0$.

$$2x^2 + 5x - 12 = 0$$
$$(2x - 3)(x + 4) = 0$$
$$x = 3/2 \quad \text{or} \quad x = -4$$

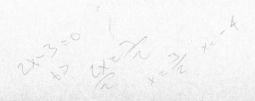

These two points divide the number line into the three regions shown in the sign graph of Figure 3.11. Since both factors are negative in the first interval, their product, $2x^2 + 5x - 12$, is positive there. In the second interval, the factors have opposite signs, and therefore their product is negative. Both factors are positive in the third interval with their product also positive there. Thus, the polynomial $2x^2 + 5x - 12$ is positive or zero in the interval $(-\infty, -4]$ and also in the interval $[3/2, +\infty)$. Since both of these intervals belong to the solution set, the result can be written as the *union** of the two intervals, or

$$(-\infty, -4] \cup [3/2, +\infty).$$

The graph of the solution set is shown in Figure 3.12. ●

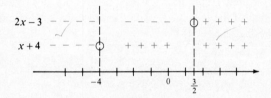

Figure 3.11

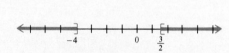

Figure 3.12

Rational Inequalities The inequalities discussed in the remainder of this section involve quotients of algebraic expressions, and for this reason they are called **rational inequalities.** These inequalities can be solved with a sign graph in much the same way that we solved quadratic inequalities.

EXAMPLE 6 Solve the inequality $\dfrac{5}{x + 4} \geq 1$.

We could begin by multiplying both sides of the inequality by $x + 4$, but we would then have to consider whether $x + 4$ is positive or negative. Instead, subtract 1 from both sides of the inequality, getting

$$\frac{5}{x + 4} - 1 \geq 0.$$

Writing the left side as a single fraction yields

$$\frac{5 - (x + 4)}{x + 4} \geq 0$$

or

$$\frac{1 - x}{x + 4} \geq 0.$$

---

*Recall from Chapter 1: The **union** of sets A and B, written $A \cup B$, is defined as $A \cup B = \{x | x$ is an element of A or x is an element of $B\}$.

The quotient can change sign only when the denominator is 0 or when the numerator is 0. This means

$$1 - x = 0 \quad \text{or} \quad x + 4 = 0$$
$$x = 1 \qquad\qquad x = -4.$$

Make a sign graph as before. This time consider the sign of the quotient of the
two quantities rather than their product. See Figure 3.13.

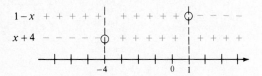

Figure 3.13

The quotient of two numbers is positive if both numbers are positive or if
both numbers are negative. On the other hand, the quotient is negative if the two
numbers have opposite signs. The sign graph in Figure 3.13 shows that values in
the interval $(-4, 1)$ give a positive quotient and are part of the solution. With a
quotient, the endpoints must be considered separately to make sure that no denominator is 0. With this inequality, -4 gives a 0 denominator while 1 satisfies the
given inequality. In interval notation, the solution set is $(-4, 1]$.   •

As suggested by Example 6, be very careful with the endpoints of the intervals in the solution of rational inequalities.

EXAMPLE 7 Solve $\dfrac{2x - 1}{3x + 4} < 5$.

Begin by subtracting 5 on both sides and combining the terms on the left into
a single fraction. This process gives

$$\frac{2x - 1}{3x + 4} < 5$$

$$\frac{2x - 1}{3x + 4} - 5 < 0$$

$$\frac{2x - 1 - 5(3x + 4)}{3x + 4} < 0$$

$$\frac{-13x - 21}{3x + 4} < 0.$$

To draw a sign graph, first solve the equations

$$-13x - 21 = 0 \quad \text{and} \quad 3x + 4 = 0,$$

getting the solutions

$$x = -\frac{21}{13} \quad \text{and} \quad x = -\frac{4}{3}.$$

Use the values $-21/13$ and $-4/3$ to divide the number line into three intervals. Now complete a sign graph and find the intervals where the quotient is negative. See Figure 3.14.

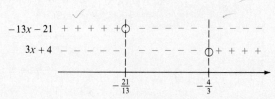

Figure 3.14

From the sign graph, values of x in the two intervals $(-\infty, -21/13)$ and $(-4/3, +\infty)$ make the quotient negative, as required. Neither endpoint satisfies the given inequality, so the solution set should be written $(-\infty, -21/13) \cup (-4/3, +\infty)$. ●

3.6 EXERCISES

Write each of the following in interval notation. Graph each interval.

1. $-1 < x < 4$
2. $x \geq -3$
3. $x < 0$
4. $8 > x > 3$
5. $2 > x \geq 1$
6. $-4 \geq x > -5$
7. $-9 > x$
8. $6 \leq x$

Using the variable x, write each of the following intervals as an inequality.

9. $(-4, 3)$
10. $[2, 7)$
11. $(-\infty, -1]$
12. $(3, +\infty)$

13.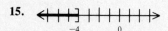
    −2   0   6

14.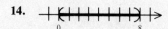
    0   8

15. ![number line graph]
    −4   0

16. ![number line graph]
    0   3

Solve the following inequalities. Write the solutions in interval notation. Graph each solution. See Examples 1–3.

17. $2x + 1 \leq 9$
18. $3y - 2 \leq 10$
19. $-3p - 2 \leq 1$
20. $-5r + 3 \geq -2$

21. $2(m + 5) - 3m + 1 \geq 5$

22. $6m - (2m + 3) \geq 4m - 5$

23. $8k - 3k + 2 < 2(k + 7)$

24. $2 - 4x + 5(x - 1) < -6(x - 2)$

25. $\dfrac{4x + 7}{-3} \leq 2x + 5$

26. $\dfrac{2z - 5}{-8} \geq 1 - z$

27. $2 \leq y + 1 \leq 5$

28. $-3 \leq 2t \leq 6$

29. $-10 > 3r + 2 > -16$

30. $4 > 6a + 5 > -1$

31. $-3 \leq \dfrac{x - 4}{-5} < 4$

32. $1 < \dfrac{4m - 5}{-2} < 9$

Solve the following quadratic inequalities. Write the solutions in interval notation. Graph each solution. (Hint: In Exercises 43–46, use the quadratic formula.) See Examples 4 and 5.

33. $x^2 \leq 9$

34. $p^2 > 16$

35. $r^2 + 4r + 6 \geq 3$

36. $z^2 + 6z + 16 < 8$

37. $x^2 - x \leq 6$

38. $r^2 + r < 12$

39. $2k^2 - 9k > -4$

40. $3n^2 \leq -10 - 13n$

41. $x^2 > 0$

42. $p^2 < -1$

43. $x^2 + 5x - 2 < 0$

44. $4x^2 + 3x + 1 \leq 0$

45. $m^2 - 2m \leq 1$

46. $p^2 + 4p > -1$

The following inequalities are not quadratic inequalities, but they still may be solved in much the same way (with four regions). Solve each inequality. Write solutions in interval notation.

47. $(2r - 3)(r + 2)(r - 3) \geq 0$

48. $(y + 5)(3y - 4)(y + 2) > 0$

49. $x^3 - 4x \leq 0$

50. $r^3 - 9r \geq 0$

51. $4m^3 + 7m^2 - 2m > 0$

52. $6p^3 - 11p^2 + 3p > 0$

Work each of the following problems.

53. Rick has 420 points so far in his math class. He must have at least 80% of a total of 700 points to get a B in the class. If the final is worth 200 points, what is the lowest score he can get on the final to get a B in the class?

54. On a trip to the market, Kevin bought twice as many apples as oranges. If he bought at least a dozen pieces of fruit, what is the smallest number of oranges he bought?

*A product will **break even,** or produce a profit, only if the revenue from selling the product at least equals the cost of producing it. Find all intervals where the following products will at least break even.*

55. The cost to produce x units of wire is $C = 50x + 5000$, while the revenue is $R = 60x$.

56. The cost to produce x units of squash is $C = 100x + 6000$, while the revenue is $R = 500x$.

57. $C = 85x + 900; R = 105x$

58. $C = 70x + 500; R = 60x$

Use the ideas of this section to solve each of the following.

59. The commodity market is very unstable; money can be made or lost quickly when investing in soybeans, wheat, pork bellies, and the like. Suppose that an investor kept track of her total profit, P, at time t, measured in months, after she began investing, and found that

$$P = 4t^2 - 29t + 30.$$

Find the time intervals where she has been ahead. (*Hint: $t > 0$ in this case.*)

60. Suppose the velocity of an object is given by

$$v = 2t^2 - 5t - 12,$$

where t is time in seconds. (Here t can be positive or negative.) Find the intervals where the velocity is negative.

61. An analyst has found that his company's profits, in hundreds of thousands of dollars, are given by

$$P = 3x^2 - 35x + 50,$$

where x is the amount, in hundreds of dollars, spent on advertising. For what values of x does the company make a profit?

62. The manager of a large apartment complex has found that the profit is given by

$$P = -x^2 + 250x - 15{,}000,$$

where x is the number of units rented. For what values of x does the complex produce a profit?

63. The formula for converting from Celsius to Fahrenheit temperature is

$$F = \frac{9}{5}C + 32.$$

What temperature range in °F corresponds to 0°C to 30°C?

64. A projectile is fired from ground level. After t sec its height above the ground is $220t - 16t^2$ ft. For what time period is the projectile at least 624 ft above the ground?

Solve the following inequalities. Give the answer in interval notation. See Examples 6 and 7.

65. $\dfrac{m - 3}{m + 5} \le 0$

66. $\dfrac{r + 1}{r - 4} > 0$

67. $\dfrac{k - 1}{k + 2} > 1$

68. $\dfrac{a - 6}{a + 2} < -1$

69. $\dfrac{3}{x - 6} \le 2$

70. $\dfrac{1}{k - 2} < \dfrac{1}{3}$

71. $\dfrac{1}{m - 1} < \dfrac{5}{4}$

72. $\dfrac{6}{5 - 3x} \le 2$

73. $\dfrac{10}{3 + 2x} \le 5$

74. $\dfrac{1}{x + 2} \ge 3$

75. $\dfrac{7}{k + 2} \ge \dfrac{1}{k + 2}$

76. $\dfrac{5}{p + 1} > \dfrac{12}{p + 1}$

77. $\dfrac{3}{2r - 1} > \dfrac{-4}{r}$ **78.** $\dfrac{-5}{3h + 2} \geq \dfrac{5}{h}$ **79.** $\dfrac{4}{y - 2} \leq \dfrac{3}{y - 1}$

80. $\dfrac{4}{n + 1} < \dfrac{2}{n + 3}$ **81.** $\dfrac{y + 3}{y - 5} \leq 1$ **82.** $\dfrac{a + 2}{3 + 2a} \leq 5$

Solve the following rational inequalities using methods similar to those used above.

83. $\dfrac{2x - 3}{x^2 + 1} \geq 0$ **84.** $\dfrac{9x - 8}{4x^2 + 25} < 0$ **85.** $\dfrac{(3x - 5)^2}{(2x - 5)^3} > 0$

86. $\dfrac{(5x - 3)^3}{(8x - 25)^2} \leq 0$ **87.** $\dfrac{(2x - 3)(3x + 8)}{(x - 6)^3} \geq 0$ **88.** $\dfrac{(9x - 11)(2x + 7)}{(3x - 8)^3} > 0$

Use the discriminant to find the values of k where the following equations have real solutions.

89. $x^2 - kx + 8 = 0$ **90.** $x^2 + kx - 5 = 0$

91. $x^2 + kx + 2k = 0$ **92.** $kx^2 + 4x + k = 0$

93. If $a > b > 0$, show that $1/a < 1/b$.

94. If $a > b$, is it always true that $1/a < 1/b$?

95. Suppose $a > b > 0$. Show that $a^2 > b^2$.

96. Let $b > 0$. When is $b^2 > b$?

3.7   Absolute Value Equations and Inequalities

In this section we study methods of solving equations and inequalities involving absolute value. Recall from Chapter 1 that the absolute value of a number a, written $|a|$, gives the distance from a to 0 on a number line. By this definition, we can solve the absolute value equation $|x| = 3$ by finding all real numbers at a distance of 3 units from 0. As shown in the graph of Figure 3.15, there are two numbers satisfying this condition, 3 and -3, making the solution set of the equation $|x| = 3$ the set $\{3, -3\}$.

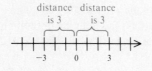

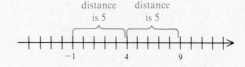

Figure 3.15   **Figure 3.16**

EXAMPLE 1   Solve $|p - 4| = 5$.

      The expression $|p - 4|$ represents the distance between p and 4. The equation $|p - 4| = 5$ can be solved by finding all real numbers that are 5 units from 4. As shown in Figure 3.16, these numbers are -1 and 9. The solution set is $\{-1, 9\}$. ●

The definition of absolute value leads to the following property of absolute value.

> If b is positive, then
> $$|a| = b \quad \text{if and only if} \quad a = b \quad \text{or} \quad a = -b.$$

EXAMPLE 2

Solve $|4m - 3| = |m + 6|$.

By a generalization of the property just given, this equation will be true if either

$$4m - 3 = m + 6 \quad \text{or} \quad 4m - 3 = -(m + 6).$$

Solve each of these equations separately. Starting with $4m - 3 = m + 6$ gives

$$4m - 3 = m + 6$$
$$3m = 9$$
$$m = 3.$$

If $4m - 3 = -(m + 6)$, we get

$$4m - 3 = -(m + 6)$$
$$4m - 3 = -m - 6$$
$$5m = -3$$
$$m = -\frac{3}{5}.$$

The solution set of $|4m - 3| = |m + 6|$ is thus $\{3, -3/5\}$. ●

Absolute Value Inequalities The method used to solve absolute value equations can be generalized to solve inequalities with absolute value.

EXAMPLE 3

Solve $|x| < 5$.

Since absolute value gives the distance from a number to 0, the inequality $|x| < 5$ will be satisfied by all real numbers whose distance from 0 is less than 5. As shown in Figure 3.17, the solution includes all numbers from -5 to 5, or $-5 < x < 5$. In interval notation, the solution is written as the open interval $(-5, 5)$. A graph of the solution set is shown in Figure 3.17. ●

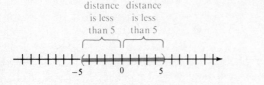

Figure 3.17

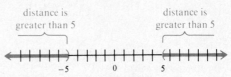

Figure 3.18

$-\infty, -5) (5, \infty)$

In a similar way, the solution of $|x| > 5$ is made up of all real numbers whose distance from 0 is greater than 5. This includes those numbers greater than 5 or those less than -5, or

$$x < -5 \qquad \text{or} \qquad x > 5.$$

In interval notation, the solution is written $(-\infty, -5) \cup (5, +\infty)$. A graph of the solution set is shown in Figure 3.18 on the preceding page.

The following properties of absolute value can be obtained from the definition of absolute value.

lower

left

If b is a positive number,

    (a) $|a| < b$ **if and only if** $-b < a < b;$

    (b) $|a| > b$ **if and only if** $a < -b \quad \text{or} \quad a > b.$

EXAMPLE 4

infinity (

Solve $|x - 2| < 5$.

To solve this inequality, find all real numbers whose distance from 2 is less than 5. As shown in Figure 3.19, the solution set is the interval $(-3, 7)$. We can also find the solution using property (a) above. Let $a = x - 2$ and $b = 5$, so that $|x - 2| < 5$ if and only if

$$-5 < x - 2 < 5.$$

Adding 2 to each portion of this inequality produces

$$-3 < x < 7,$$

again giving the interval solution $(-3, 7)$. ●

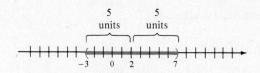

Figure 3.19

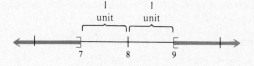

Figure 3.20

EXAMPLE 5

if

all solving

> or

Solve $|x - 8| \geq 1$.

We need to find all numbers whose distance from 8 is greater than or equal to 1. As shown in Figure 3.20 the solution set is $(-\infty, 7] \cup [9, +\infty)$. To find the solution using property (b) above, let $a = x - 8$ and $b = 1$ so that $|x - 8| \geq 1$ if and only if

$$x - 8 \leq -1 \qquad \text{or} \qquad x - 8 \geq 1.$$

Solve each inequality separately to get the same solution set, $(-\infty, 7] \cup [9, +\infty)$, as mentioned above. ●

EXAMPLE 6   Solve $|2 - 7m| - 1 > 4$.

In order to use the properties of absolute value given above, first add 1 to both sides; this gives

$$|2 - 7m| > 5.$$

Now use property (b) above. By this property, $|2 - 7m| > 5$ if and only if

$$2 - 7m < -5 \quad \text{or} \quad 2 - 7m > 5.$$

Solve each of these inequalities separately to get the solution set $(-\infty, -3/7) \cup (1, +\infty)$. ●

EXAMPLE 7   Solve $|2 - 5x| \geq -4$.

Since the absolute value of a number is always nonnegative, $|2 - 5x| \geq -4$ is always true. The solution set includes all real numbers. In interval notation, the solution set is $(-\infty, +\infty)$. ●

3.7 EXERCISES

Solve each of the following equations. See Examples 1 and 2.

1. $|a - 2| = 1$ **2.** $|x - 3| = 2$ **3.** $|3m - 1| = 2$ **4.** $|4p + 2| = 5$

5. $|5 - 3x| = 3$ **6.** $|-3a + 7| = 3$ **7.** $\left|\dfrac{z - 4}{2}\right| = 5$ **8.** $\left|\dfrac{m + 2}{2}\right| = 7$

9. $\left|\dfrac{5}{r - 3}\right| = 10$ **10.** $\left|\dfrac{3}{2h - 1}\right| = 4$ **11.** $\left|\dfrac{6y + 1}{y - 1}\right| = 3$ **12.** $\left|\dfrac{3a - 4}{2a + 3}\right| = 1$

13. $|2k - 3| = |5k + 4|$ **14.** $|p + 1| = |3p - 1|$ **15.** $|4 - 3y| = |7 + 2y|$

16. $|2 + 5a| = |4 - 6a|$ **17.** $|x + 2| = |x - 1|$ **18.** $|y - 5| = |y + 3|$

Solve each of the following inequalities. Give the solution in interval notation. See Examples 3–7.

19. $|x| \leq 3$ **20.** $|y| \leq 10$ **21.** $|m| > 1$ **22.** $|z| > 5$

23. $|a| < -2$ **24.** $|b| > -5$ **25.** $|x| - 3 \leq 7$ **26.** $|r| + 3 \leq 10$

27. $|2x + 5| < 3$ **28.** $\left|x - \dfrac{1}{2}\right| < 2$ **29.** $|3m - 2| > 4$ **30.** $|4x - 6| > 10$

31. $|3z + 1| \geq 7$ **32.** $|8b + 5| \geq 7$ **33.** $\left|5x + \dfrac{1}{2}\right| - 2 < 5$ **34.** $\left|x + \dfrac{2}{3}\right| + 1 < 4$

Write each of the following using absolute value statements. For example, write "k is at least 4 units from 1" as $|k - 1| \geq 4$.

35. x is within 4 units of 2

36. m is no more than 8 units from 9

37. z is no less than 2 units from 12

38. p is at least 5 units from 9

39. k is 6 units from 1

40. r is 5 units from 3

41. If x is within .0004 units of 2, then y is within .00001 units of 7.

42. y is within 10^{-6} units of 10 whenever x is within 2×10^{-4} units of 5.

43. If $|x - 2| < 3$, find the values of m and n such that $m < 3x + 5 < n$.

44. If $|x + 8| < 16$, find values of p and q so that $p < 2x - 1 < q$.

Chapter 3   Summary

Key Words	equation	satisfy an equation
	solution set	equivalent equation
	linear equation	solving for a specified variable
	complex number	equality of complex numbers
	standard form	imaginary number
	additive inverse	additive identity
	quadratic equation	conjugates
	discriminant	zero-factor property
	square root property	quadratic formula
	equivalent inequalities	interval
	interval notation	open interval
	half-open interval	closed interval
	sign graph	

Definition of i $i = \sqrt{-1}$,   with $i^2 = -1$.

Zero-Factor Property   If a and b are complex numbers, with $ab = 0$, then $a = 0$ or $b = 0$ or both.

Square Root Property   The solution set of $x^2 = k$ is $\{\sqrt{k}, -\sqrt{k}\}$.

Quadratic Formula   The solutions of the quadratic equation $ax^2 + bx + c = 0$, $a \neq 0$, are

$$x = \frac{-b \pm \sqrt{b^2 - 4ac}}{2a}.$$

Properties of Absolute Value   If b is a positive number, then

(a) $|a| = b$ if and only if $a = b$ or $a = -b$;

(b) $|a| < b$ if and only if $-b < a < b$;

(c) $|a| > b$ if and only if $a < -b$ or $a > b$.

Chapter 3   Review Exercises

Solve each of the following equations.

1. $2m + 7 = 3m + 1$

2. $4k - 2(k - 1) = 12$

3. $5y - 2(y + 4) = 3(2y + 1)$

4. $\dfrac{x - 3}{2} = \dfrac{2x + 1}{3}$

5. $\dfrac{p}{2} - \dfrac{3p}{4} = 8 + \dfrac{p}{3}$

6. $\dfrac{2r}{5} - \dfrac{r - 3}{10} = \dfrac{3r}{5}$

7. $\dfrac{2z}{5} - \dfrac{4z - 3}{10} = \dfrac{-z + 1}{10}$

8. $\dfrac{p}{p + 2} - \dfrac{3}{4} = \dfrac{2}{p + 2}$

9. $5(x - 3)(2x + 1) = 10(x + 2)(x - 4)$

10. $(3k + 1)^2 = 6k^2 + 3(k - 1)^2$

Solve for x.

11. $3(x + 2b) + a = 2x - 6$

12. $9x - 11(k + p) = x(a - 1)$

13. $\dfrac{x}{m - 2} = kx - 3$

14. $r^2 x - 5x = 3r^2$

Solve each of the following for the indicated variable.

15. $2a + ay = 4y - 4a$ for y

16. $\dfrac{3m}{m - x} = 2m + x$ for x

17. $F = \dfrac{9}{5}C + 32$ for C

18. $A = P + Pi$ for P

19. $A = I\left(1 - \dfrac{j}{n}\right)$ for j

20. $A = \dfrac{24f}{b(p + 1)}$ for f

21. $\dfrac{1}{k} = \dfrac{1}{r_1} + \dfrac{1}{r_2}$ for r_1

22. $m = \dfrac{Ft}{\sqrt{I} - \sqrt{2}}$ for t

23. $V = \pi r^2 L$ for L

24. $P(r + R)^2 = E^2 R$ for P

25. $\dfrac{xy^2 - 5xy + 4}{3x} = 2p$ for x

26. $\dfrac{zx^2 - 5x + z}{z + 1} = 9$ for z

Solve each of the following problems.

27. A stereo is on sale for 15% off. The sale price is $425. What was the original price?

28. To make a special mix for Valentine's Day, the owner of a candy store wants to combine chocolate hearts that sell for $5 per lb with candy kisses that sell for $3.50 per lb. How many lbs of each kind should be used to get 30 lbs of a mix which can be sold for $4.50 per lb?

29. Ed and Susan are stuffing envelopes for a political campaign. Working together, they can stuff 5000 envelopes in 4 hr. When Susan worked at the job alone, it took her 6 hr to stuff the 5000 envelopes. How long would it take Ed working alone to stuff 5000 envelopes?

30. Alison can ride her bike to the university library in 20 min. The trip home, which is all uphill, takes her 30 min. If her rate is 8 mph slower on the return trip, how far does she live from the library?

Perform each operation. Write each result in standard form.

31. $(6 - i) + (4 - 2i)$

32. $(-11 + 2i) - (5 - 7i)$

33. $15i - (3 + 2i) - 5$

34. $-6 + 4i - (5i - 2)$

35. $(5 - i)(3 + 4i)$

36. $(8 - 2i)(1 - i)$

37. $(5 - 11i)(5 + 11i)$

38. $(7 + 6i)(7 - 6i)$

39. $(4 - 3i)^2$

40. $(1 + 7i)^2$

41. $-5i(3 - i)^2$

42. $4i(2 + 5i)(2 - i)$

43. $\dfrac{6 + i}{1 - i}$

44. $\dfrac{5 + 2i}{5 - 2i}$

45. $\dfrac{4 + 3i}{2 + i}$

46. $\dfrac{7 - i}{1 + i}$

Find each of the following powers of i.

47. i^7

48. i^{15}

49. i^{123}

50. i^{507}

51. i^{-18}

52. i^{-23}

Solve each equation.

53. $(b + 7)^2 = 5$

54. $(3y - 2)^2 = 8$

55. $2a^2 + a - 15 = 0$

56. $12x^2 = 8x - 1$

57. $2q^2 - 11q = 21$

58. $3x^2 + 2x = 16$

59. $2 - \dfrac{5}{p} = \dfrac{3}{p^2}$

60. $\dfrac{4}{m^2} = 2 + \dfrac{7}{m}$

61. $ix^2 - 4x + i = 0$

62. $4p^2 - ip + 2 = 0$

Evaluate the discriminant for each of the following, and then use it to predict the type of solutions for the equation.

63. $8y^2 = 2y - 6$

64. $6k^2 - 2k = 3$

65. $16r^2 + 3 = 26r$

66. $8p^2 + 10p = 7$

67. $25z^2 - 110z + 121 = 0$

68. $4y^2 - 8y + 17 = 0$

Work each word problem.

69. Calvin wants to fence off a rectangular playground beside an apartment building. The building forms one boundary, so he needs to fence only the other three sides. The area of the playground is to be 11,250 sq m. He has enough material to build 325 m of fence. Find the length and width of the playground.

70. Steve and Paula sell pies. It takes Paula 1 hr longer than Steve to bake a day's supply of pies. Working together, it takes them 1 1/5 hr to bake the pies. How long would it take Steve working alone?

Solve each equation.

71. $4a^4 + 3a^2 - 1 = 0$

72. $2x^4 - x^2 = 0$

73. $(r + 1)^2 - 3(r + 1) = 4$

74. $4(y - 2)^2 - 9(y - 2) + 2 = 0$

75. $(2z + 3)^{2/3} + (2z + 3)^{1/3} = 6$

76. $\sqrt{x} - 7 = 10$

77. $\sqrt{2p + 1} = 8$

78. $5\sqrt{m} = \sqrt{3m + 2}$

79. $\sqrt{4y - 2} = \sqrt{3y + 1}$

80. $\sqrt{2x + 3} = x + 2$

81. $\sqrt{p + 2} = 2 + p$ **82.** $\sqrt{k} = \sqrt{k + 3} - 1$

83. $\sqrt{x^2 + 3x} - 2 = 0$ **84.** $\sqrt[3]{2r} = \sqrt[3]{3r + 2}$

85. $\sqrt[3]{6y + 2} = \sqrt[3]{4y}$ **86.** $(x - 2)^{2/3} = x^{1/3}$

Solve each of the following inequalities. Write the answers in interval notation.

87. $-9x < 4x + 7$ **88.** $11y \geq 2y - 8$ **89.** $-5z - 4 \geq 3(2z - 5)$

90. $-(4a + 6) < 3a - 2$ **91.** $3r - 4 + r > 2(r - 1)$ **92.** $7p - 2(p - 3) \leq 5(2 - p)$

93. $5 \leq 2x - 3 \leq 7$ **94.** $-8 > 3a - 5 > -12$ **95.** $x^2 + 3x - 4 \leq 0$

96. $p^2 + 4p > 21$ **97.** $6m^2 - 11m - 10 < 0$ **98.** $k^2 - 3k - 5 \geq 0$

99. $z^3 - 16z \leq 0$ **100.** $2r^3 - 3r^2 - 5r < 0$ **101.** $\dfrac{3a - 2}{a} > 4$

102. $\dfrac{5p + 2}{p} < -1$ **103.** $\dfrac{3}{r - 1} \leq \dfrac{5}{r + 3}$ **104.** $\dfrac{3}{x + 2} > \dfrac{2}{x - 4}$

Work the following word problem.

105. Steve and Paula (Exercise 70) have found that the profit from their pie shop is given by

$$P = -x^2 + 28x + 60,$$

where x is the number of units of pies sold daily. For what values of x is the profit positive?

Solve each equation.

106. $|a + 4| = 7$ **107.** $|-y + 2| = 3$ **108.** $\left|\dfrac{r - 5}{3}\right| = 6$ **109.** $\left|\dfrac{7}{2 - 3a}\right| = 9$

110. $\left|\dfrac{8r - 1}{3r + 2}\right| = 7$ **111.** $|5r - 1| = |2r + 3|$ **112.** $|k + 7| = |k - 8|$

Solve each inequality. Write solutions with interval notation.

113. $|m| \leq 7$ **114.** $|r| < 2$ **115.** $|p| > 3$ **116.** $|z| > -1$

117. $|2z + 9| \leq 3$ **118.** $|5m - 8| \leq 2$ **119.** $|7k - 3| < 5$ **120.** $|2p - 1| > 2$

121. $|3r + 7| > 5$ **122.** $|1 - y| \geq 6$

4

Graphing

The graph on the left in Figure 4.1 is from *Road and Track* magazine. The graph shows the velocity in miles per hour at time t in seconds for a Porsche 928 as it accelerates from rest. For example, the graph shows that 15 sec after starting, the car is going 90 mph, with a speed of 100 mph after 19 sec. The graph on the right in Figure 4.1 shows the variation in blood pressure for a typical person.

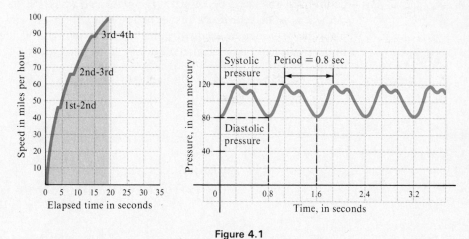

Figure 4.1

Graph on right: from *Calculus for the Life Sciences* by Rodolfo De Sapio. Copyright © 1976, 1978 by W. H. Freeman and Company. Reprinted by permission. Graph on left: from *Road and Track,* April 1978. Reprinted with permission of Road & Track.

(Systolic and diastolic pressures are the upper and lower limits in the periodic changes in pressure that produce the pulse. The length of time between peaks is called the period of the pulse.)

Each of the graphs in Figure 4.1 is an example of a **function:** a rule or procedure giving just one value of one variable from a given value of the other variable. In each of these examples, a given value of time can be used to find just one value of the other variable, speed or blood pressure, respectively. Before we get to a complete discussion of functions in the next chapter, we need to look at the topic of graphing, since functions are often studied by looking at their graphs.

4.1 A Two-Dimensional Coordinate System

In Chapter 1 we saw how each real number corresponds to a point on a number line. This correspondence was set up by establishing a coordinate system for the line. This idea can be extended to the two dimensions of a plane by drawing two perpendicular lines, one horizontal and one vertical. These lines intersect at a point O called the **origin.** The horizontal line is called the **x-axis,** and the vertical line is called the **y-axis.**

Starting at the origin, we can make the x-axis into a number line by placing positive numbers to the right and negative numbers to the left. The y-axis can be made into a number line with positive numbers going up and negative numbers going down.

The x-axis and y-axis set up a **rectangular coordinate system,** or **Cartesian coordinate system** (named for one of its co-inventors, René Descartes; the other co-inventor was Pierre de Fermat). The plane into which the coordinate system is introduced is the **coordinate plane,** or **xy-plane.** The x-axis and y-axis divide the plane into four regions, or **quadrants,** labeled as shown in Figure 4.2. The points of the x-axis and y-axis themselves belong to no quadrant.

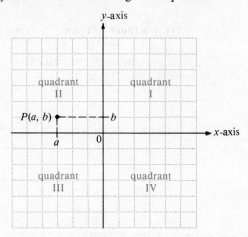

Figure 4.2

Label a given point P as shown in Figure 4.2. Start at P and draw a vertical line cutting the x-axis at a. Draw a horizontal line cutting the y-axis at b. Then point P has **coordinates** (a, b), where (a, b) is an *ordered pair* of numbers. An **ordered pair** of numbers consists of two numbers, written in parentheses, in which the sequence of the numbers is important. For example, $(4, 2)$ and $(2, 4)$ are not the same ordered pair since the sequence of the numbers is different.

A symbol such as $(3, 4)$ has now been used for two different purposes—as an interval on the number line and as an ordered pair of numbers. In virtually every case the intended use can be told from the context of the discussion.

To locate the point on the xy-plane corresponding to the ordered pair $(3, 4)$, draw a vertical line through 3 on the x-axis and a horizontal line through 4 on the y-axis. These two lines would cross at point A of Figure 4.3. This point A corresponds to the ordered pair $(3, 4)$. Also in Figure 4.3, B corresponds to the ordered pair $(-5, 6)$, C to $(-2, -4)$, D to $(4, -3)$, and E to $(-3, 0)$.

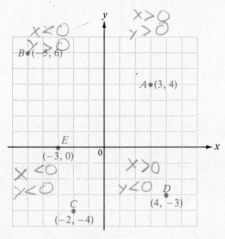

Figure 4.3

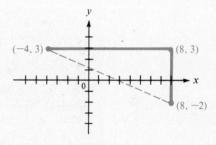

Figure 4.4

The Distance Formula   Figure 4.4 shows the points $(-4, 3)$ and $(8, -2)$. To find the distance between these two points, complete a right triangle as shown in the figure. This right triangle has its $90°$ angle at $(8, 3)$. The horizontal side of this right triangle has length

$$|8 - (-4)| = 12,$$

where absolute value is used to make sure that the distance is not negative. The vertical side of the triangle has length

$$|3 - (-2)| = 5.$$

By the Pythagorean theorem, the length of the remaining side of the triangle is

$$\sqrt{12^2 + 5^2} = \sqrt{144 + 25} = \sqrt{169} = 13.$$

The distance between $(-4, 3)$ and $(8, -2)$ is 13.

To obtain a general formula for the distance between two points on a coordinate plane, let $P(x_1, y_1)$ and $R(x_2, y_2)$ be any two distinct points in a plane. Complete a right triangle by locating point Q with coordinates (x_2, y_1). Again using the Pythagorean theorem, the distance between P and R, written $d(P, R)$ is

$$d(P, R) = \sqrt{(x_2 - x_1)^2 + (y_2 - y_1)^2}.$$

See Figure 4.5.

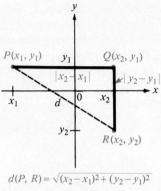

$$d(P, R) = \sqrt{(x_2 - x_1)^2 + (y_2 - y_1)^2}$$

Figure 4.5

The distance formula can be summarized as follows.

Distance Formula

> Suppose that $P(x_1, y_1)$ and $R(x_2, y_2)$ are two points in a coordinate plane. Then the distance between P and R, written $d(P, R)$, is given by the **distance formula,**
>
> $$d(P, R) = \sqrt{(x_2 - x_1)^2 + (y_2 - y_1)^2}.$$

Although the proof of the distance formula assumes that P and R are not on a horizontal or vertical line, the result is true for any two points.

EXAMPLE 1   Find the distance between $P(-8, 4)$ and $Q(3, -2)$.
According to the distance formula,

$$
\begin{aligned}
d(P, Q) &= \sqrt{[3 - (-8)]^2 + (-2 - 4)^2} \\
&= \sqrt{11^2 + (-6)^2} \\
&= \sqrt{121 + 36} \\
&= \sqrt{157}.
\end{aligned}
$$

Using a calculator with a square root key gives 12.530 as an approximate value for $\sqrt{157}$. ●

EXAMPLE 2   Are the points $M(-2, 5)$, $N(12, 3)$, and $Q(10, -11)$ the vertices of a right triangle?

To decide whether or not this triangle is a right triangle, use the converse of the Pythagorean theorem: if the sides a, b, and c of a triangle satisfy $a^2 + b^2 = c^2$, then the triangle is a right triangle. A triangle with the three given points as vertices is shown in Figure 4.6. This triangle is a right triangle if the square of the length of the longest side equals the sum of the squares of the lengths of the other two sides. Use the distance formula to find the length of each side of the triangle.

$$d(M, N) = \sqrt{[12 - (-2)]^2 + (3 - 5)^2} = \sqrt{196 + 4} = \sqrt{200}$$
$$d(M, Q) = \sqrt{[10 - (-2)]^2 + (-11 - 5)^2} = \sqrt{144 + 256} = \sqrt{400} = 20$$
$$d(N, Q) = \sqrt{(10 - 12)^2 + (-11 - 3)^2} = \sqrt{4 + 196} = \sqrt{200}$$

By these results,

$$[d(M, Q)]^2 = [d(M, N)]^2 + [d(N, Q)]^2,$$

proving that the triangle is a right triangle with hypotenuse connecting M and Q.   ●

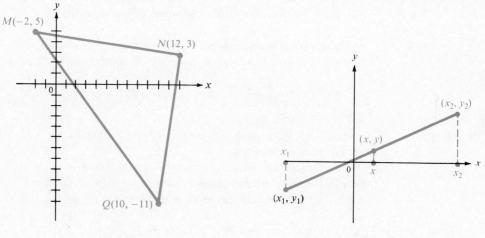

Figure 4.6   **Figure 4.7**

The distance formula is used to find the distance between any two points in a plane. The **midpoint formula** is used to find the coordinates of the midpoint of a line segment.

To develop the midpoint formula, let (x_1, y_1) and (x_2, y_2) be any two distinct points in a plane. (See Figure 4.7.) Assume that the two points are not on a horizontal or vertical line. Let (x, y) be the midpoint of the segment connecting (x_1, y_1) and (x_2, y_2). Draw vertical lines from each of the three points to the x-axis.

As the figure suggests, x is located halfway between x_1 and x_2, so the distance between x and x_1 equals the distance between x and x_2. Subtracting in such a way as to make sure that the results are not negative gives

$$x_2 - x = x - x_1,$$

or

$$x_2 + x_1 = 2x.$$

Finally,

$$x = \frac{x_1 + x_2}{2}.$$

By this result, the x-coordinate of the midpoint is the average of the x-coordinates of the endpoints of the segment. In a similar manner, the y-coordinate of the midpoint is $(y_1 + y_2)/2$, proving the following statement.

Midpoint Formula

> The midpoint of the line segment with endpoints (x_1, y_1) and (x_2, y_2) is
>
> $$\left(\frac{x_1 + x_2}{2}, \frac{y_1 + y_2}{2} \right).$$

In other words, the midpoint formula says that the coordinates of the midpoint of a segment are found by calculating the *average* of the x-coordinates and the average of the y-coordinates of the endpoints of the segment.

EXAMPLE 3   Find the midpoint M of the segment with endpoints $(8, -4)$ and $(-9, 6)$.
Use the midpoint formula to find that the coordinates of M are

$$\left(\frac{8 + (-9)}{2}, \frac{-4 + 6}{2} \right) = \left(-\frac{1}{2}, 1 \right). \quad \bullet$$

EXAMPLE 4   A line segment has an endpoint at $(2, -8)$ and a midpoint at $(-1, -3)$. Find the other endpoint of the segment.
The formula for the x-coordinate of the midpoint is $(x_1 + x_2)/2$. Here the x-coordinate of the midpoint is -1. We can let $x_1 = 2$, getting

$$-1 = \frac{2 + x_2}{2}$$

$$-2 = 2 + x_2$$

$$-4 = x_2.$$

In the same way, $y_2 = 2$ and the endpoint is $(-4, 2)$. $\quad \bullet$

4.1 EXERCISES

Plot the following points in the xy-plane. Identify the quadrant for each.

1. $A(6, -5)$ **2.** $B(8, 3)$ **3.** $C(-4, 7)$ **4.** $D(-9, -8)$

5. $E(0, -5)$ **6.** $F(-8, 0)$

Graph the set of all points satisfying the following conditions for ordered pairs (x, y).

7. $x = 0$

8. $x > 0$

9. $y \leq 0$

10. $y = 0$

11. $xy < 0$

12. $\dfrac{x}{y} > 0$

13. $|x| = 4, y \geq 2$

14. $|y| = 3, x \geq 4$

15. $|y| < 2, x > 1$

16. $|x| < 3, y < -2$

17. $2 \leq |x| \leq 3, y \geq 2$

18. $1 \leq |y| \leq 4, x < 3$

Find the distance d(P, Q) and the midpoint of segment PQ. See Examples 1 and 3.

19. $P(5, 7), Q(13, -1)$

20. $P(-2, 5), Q(4, -3)$

21. $P(-8, -2), Q(-3, -5)$

22. $P(-6, -10), Q(6, 5)$

23. $P(\sqrt{2}, -\sqrt{5}), Q(3\sqrt{2}, 4\sqrt{5})$

24. $P(5\sqrt{7}, -\sqrt{3}), Q(-\sqrt{7}, 8\sqrt{3})$

Give the distance between the following points rounded to the nearest thousandth. See Example 1.

25. $(5, 7), (2, 14)$

26. $(-4, 6), (8, -5)$

27. $(3, -7), (-5, 19)$

28. $(-9, -2), (-1, -15)$

Find the other endpoint of each segment having endpoint and midpoint as given. See Example 4.

29. Endpoint $(-3, 6)$, midpoint $(5, 8)$

30. Endpoint $(2, -8)$, midpoint $(3, -5)$

31. Endpoint $(6, -1)$, midpoint $(-2, 5)$

32. Endpoint $(-5, 3)$, midpoint $(-7, 6)$

Decide whether or not the following points are the vertices of a right triangle. See Example 2.

33. $(-2, 5), (1, 5), (1, 9)$

34. $(-9, -2), (-1, -2), (-9, 11)$

35. $(-4, 0), (1, 3), (-6, -2)$

36. $(-8, 2), (5, -7), (3, -9)$

37. $(\sqrt{3}, 2\sqrt{3} + 3), (\sqrt{3} + 4, -\sqrt{3} + 3), (2\sqrt{3}, 2\sqrt{3} + 4)$

38. $(4 - \sqrt{3}, -2\sqrt{3}), (2 - \sqrt{3}, -\sqrt{3}), (3 - \sqrt{3}, -2\sqrt{3})$

Use the distance formula to decide whether or not the following points lie on a straight line.

39. $(0, 7), (3, -5), (-2, 15)$

40. $(1, -4), (2, 1), (-1, -14)$

41. $(0, -9), (3, 7), (-2, -19)$

42. $(1, 3), (5, -12), (-1, 11)$

43. Show that the points $(-2, 2), (13, 10), (21, -5)$, and $(6, -13)$ are the vertices of a square.

44. Are the points $A(1, 1), B(5, 2), C(3, 4), D(-1, 3)$ the vertices of a parallelogram? Of a rhombus (all sides equal in length)?

45. Use the distance formula to write an equation for all points that are 5 units from $(0, 0)$. Sketch a graph showing these points.

46. Write an equation for all points 3 units from $(-5, 6)$. Sketch a graph showing these points.

47. Find all points (x, y) with $x = y$ that are 4 units from $(1, 3)$.

48. Find all points satisfying $x + y = 0$ that are 8 units from $(-2, 3)$.

4.2 Methods of Graphing

For any set of ordered pairs of real numbers, there is a corresponding set of points in a coordinate plane. If (x, y) is one of the ordered pairs in the set, then the corresponding point (x, y) can be located on the coordinate plane. This set of points is the **graph** of the set of ordered pairs. For now, we shall find graphs of equations by first identifying a reasonable number of ordered pairs. We locate the points corresponding to these ordered pairs on a coordinate plane and then try to decide on the shape of the entire graph. Later we will develop more useful methods of identifying particular graphs.

EXAMPLE 1   Graph $y = -2x + 6$.

Choose several values of x and find the corresponding values of y. For example, if $x = -3$,

$$y = -2(-3) + 6 = 6 + 6 = 12.$$

Also, if $x = 0$, then $y = 6$. These and other ordered pairs are shown in the following table.

x	-3	-2	-1	0	1	2	3	4
y	12	10	8	6	4	2	0	-2

The ordered pairs produced from this table were plotted in Figure 4.8. Studying these points suggests that a straight line will fit through them. This straight line is graphed in Figure 4.8.   ●

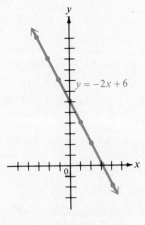

Figure 4.8

There is a danger in this method of graphing an equation. We might choose a few values for x, find the corresponding values of y, graph the resulting points, and then make a wrong guess as to the actual shape of the graph. This danger is remote in the basic graphs of this section, and when we work with more complicated graphs later, we shall develop more accurate methods of drawing them.

EXAMPLE 2   Graph $y = x^2 + 2$.

Again, choose several values of x, and find the corresponding values of y.

x	-4	-3	-2	-1	0	1	2	3	4
y	18	11	6	3	2	3	6	11	18

The ordered pairs obtained from this table were used to plot points in Figure 4.9, and a smooth curve was then drawn through the points. This graph, called a *parabola*, will be studied in Section 4.5. The lowest point on this parabola, the point (0, 2), is called the *vertex* of the parabola. In $y = x^2 + 2$, any value may be used for x. We know that $x^2 \geq 0$ for all x. Since $y = x^2 + 2$, the value of y will always be at least equal to 2, or $y \geq 2$.   ●

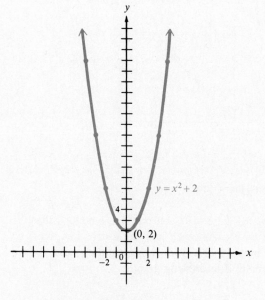

Figure 4.9

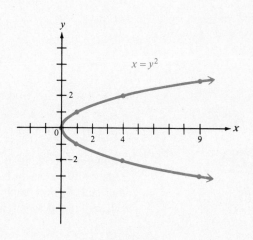

Figure 4.10

EXAMPLE 3   Graph $x = y^2$.

Since y is squared, it is probably easier to choose values of y and then to find the corresponding values of x. Choosing the value 2 for y gives $x = 2^2 = 4$. Choosing -2 for y gives $x = (-2)^2 = 4$, the same result. The following table shows the values of x corresponding to various values of y.

y	0	1	-1	2	-2	3	-3
x	0	1	1	4	4	9	9

The ordered pairs from this table were used to get the points plotted in Figure 4.10. (Don't forget that x always goes first in the ordered pair.) A smooth curve was then drawn through them. This curve is a parabola with vertex $(0, 0)$, opening to the right.

Here, y can take any value. Since $x = y^2$, we have $x \geq 0$.   ●

EXAMPLE 4   Graph $y = |x|$.

Start with a table.

x	-4	-3	-2	-1	0	1	2	3	4
y	4	3	2	1	0	1	2	3	4

Use this table to get the points of Figure 4.11. The graph drawn through these points is made up of portions of two straight lines. Here x may represent any real number, while $y \geq 0$.   ●

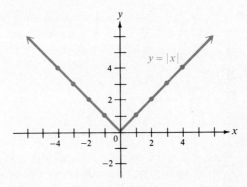

Figure 4.11

Symmetry   Looking at the graph in Figure 4.9, you may have noticed that the graph could be cut in half by the y-axis with each half the mirror image of the other half. Such a graph is said to be *symmetric with respect to the y-axis*. As we shall see, the idea of symmetry is helpful in drawing graphs.

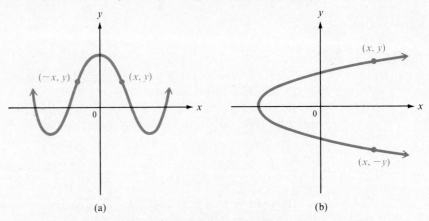

(a)   (b)

Figure 4.12

Figure 4.12(a) shows a graph which is *symmetric with respect to the y-axis.* For a graph to be symmetric with respect to the y-axis, the point $(-x, y)$ must be on the graph whenever (x, y) is on the graph.

If the graph in Figure 4.12(b) were folded in half along the x-axis, the portion at the top would exactly match the portion at the bottom. Such a graph is *symmetric with respect to the x-axis.* As the graph suggests, symmetry with respect to the x-axis means that the point $(x, -y)$ must be on the graph whenever the point (x, y) is on the graph.

The following test tells when a graph is symmetric to the x-axis or y-axis.

Symmetry with Respect to an Axis

> The graph of an equation is **symmetric with respect to the y-axis** if the replacement of x with $-x$ results in an equivalent equation.
>
> The graph of an equation is **symmetric with respect to the x-axis** if the replacement of y with $-y$ results in an equivalent equation.

EXAMPLE 5   Test for symmetry with respect to the x-axis or y-axis.

(a) $y = x^2 + 4$

Replace x with $-x$:

$$y = x^2 + 4 \text{ becomes } y = (-x)^2 + 4 = x^2 + 4.$$

The result is the same as the original equation, so the graph (shown in Figure 4.13) is symmetric with respect to the y-axis. The graph is *not* symmetric with respect to the x-axis.

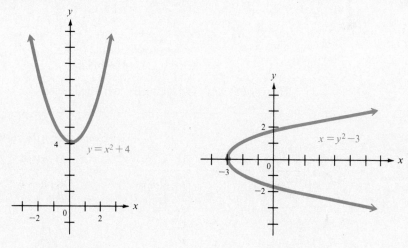

Figure 4.13   Figure 4.14

(b) $x = y^2 - 3$

Replace y with $-y$ to get $x = (-y)^2 - 3 = y^2 - 3$, the same as the original equation. The graph is symmetric with respect to the x-axis, as shown in Figure 4.14. Is the graph symmetric with respect to the y-axis?

(c) $2x + y = 4$

Replace x with $-x$ and then replace y with $-y$; in neither case does an equivalent equation result. This graph is symmetric neither with respect to the x-axis nor the y-axis. ●

Another kind of symmetry is found when it is possible to rotate a graph 180° about the origin and have the result coincide exactly with the original graph. Symmetry of this type is called *symmetry with respect to the origin*. It turns out that rotating a graph 180° is equivalent to saying that the point $(-x, -y)$ is on the graph whenever (x, y) is on the graph. Figure 4.15 shows two graphs that are symmetric with respect to the origin. This type of symmetry is defined below.

Symmetry with Respect to the Origin

> The graph of an equation is **symmetric with respect to the origin** if the replacement of both x with $-x$ and y with $-y$ results in an equivalent equation.

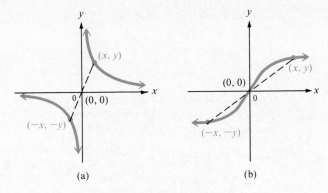

Figure 4.15

The various tests for symmetry are listed below.

Tests for Symmetry

	Symmetric with respect to		
	x-axis	*y-axis*	*origin*
Test	Replace y with $-y$	Replace x with $-x$	Replace x with $-x$ and replace y with $-y$
Example			

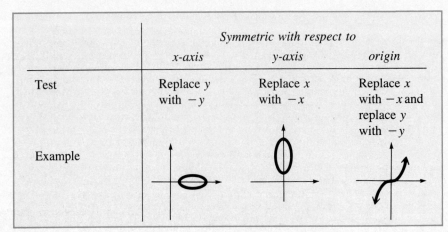

Circles   A good example of symmetry with respect to the origin is given by the graph of a circle having its center at the origin. (See Figure 4.16.)

A **circle** is the set of all points in a plane which lie a given distance from a given point. The given distance is the **radius** of the circle and the given point is the **center.** The equation of a circle can be found by using the distance formula discussed in Section 4.1.

For example, Figure 4.16 shows a circle of radius 3 with center at the origin. To find the equation of this circle, let (x, y) be any point on the circle. The distance between (x, y) and the center of the circle, $(0, 0)$, is given by

$$\sqrt{(x - 0)^2 + (y - 0)^2}.$$

Since this distance equals the radius, 3,

$$\sqrt{(x - 0)^2 + (y - 0)^2} = 3$$
$$\sqrt{x^2 + y^2} = 3$$
$$x^2 + y^2 = 9.$$

As suggested by the graph, the possible values of x are $-3 \leq x \leq 3$, while the possible values of y are $-3 \leq y \leq 3$.

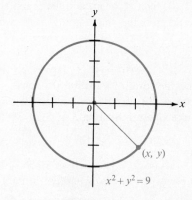

Figure 4.16

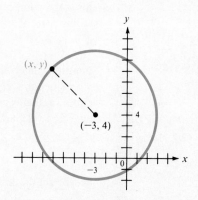

Figure 4.17

EXAMPLE 6   Find an equation for the circle having radius 6 and center at $(-3, 4)$.

This circle is shown in Figure 4.17. Its equation can be found by using the distance formula. Let (x, y) be any point on the circle. The distance from (x, y) to $(-3, 4)$ is given by

$$\sqrt{[x - (-3)]^2 + (y - 4)^2} = \sqrt{(x + 3)^2 + (y - 4)^2}.$$

This same distance is given by the radius, 6. Therefore,

$$\sqrt{(x + 3)^2 + (y - 4)^2} = 6$$

or
$$(x + 3)^2 + (y - 4)^2 = 36.$$

The graph in Figure 4.17 suggests that the possible values of x are $-9 \leq x \leq 3$ while the possible values of y are $-2 \leq y \leq 10$.   ●

Generalizing from the work of Example 6 gives the following result.

Center-Radius Form of the Equation of a Circle

> The circle with center (h, k) and radius r has equation
> $$(x - h)^2 + (y - k)^2 = r^2,$$
> the **center-radius form** of the equation of a circle.

Starting with the center-radius form of the equation of a circle, $(x - h)^2 + (y - k)^2 = r^2$, and squaring $x - h$ and $y - k$ gives a result of the form

$$x^2 + y^2 + cx + dy + e = 0, \qquad (*)$$

where c, d, and e are real numbers. (This result is the **general form of the equation of a circle.**) Also, starting with an equation similar to (*), the process of *completing the square* from Section 3.4 can be used to get an equation of the form

$$(x - h)^2 + (y - k)^2 = m$$

for some number m. If $m > 0$, then $r^2 = m$, and the graph is that of a circle with radius $\sqrt{m}$. If $m = 0$, the graph is the single point (h, k), while there is no graph if $m < 0$.

EXAMPLE 7 Find the center and radius for $x^2 - 6x + y^2 + 10y + 25 = 0$.

Since this equation has the form of equation (*) above, it either represents a circle, a single point, or no points at all. To decide which, complete the square on x and y separately, as explained in Section 3.4. Start with

$$(x^2 - 6x \qquad) + (y^2 + 10y \qquad) = -25.$$

Half of -6 is -3, and $(-3)^2 = 9$. Also, half of 10 is 5, and $5^2 = 25$. Add 9 and 25 on the left, and to compensate, add 9 and 25 on the right:

$$(x^2 - 6x + 9) + (y^2 + 10y + 25) = -25 + 9 + 25$$
$$(x - 3)^2 + (y + 5)^2 = 9.$$

Since $9 > 0$, the equation represents a circle with center at $(3, -5)$ and radius 3. ●

EXAMPLE 8 Graph $x^2 + 10x + y^2 - 4y + 33 = 0$.

Completing the square as above gives

$$(x^2 + 10x + 25) + (y^2 - 4y + 4) = -33 + 25 + 4$$
$$(x + 5)^2 + (y - 2)^2 = -4.$$

Since $-4 < 0$, there are no ordered pairs (x, y), with x and y real numbers, satisfying the equation. Thus, the graph has no points on it. ●

4.2 EXERCISES

Graph each of the following equations. See Examples 1–3.

1. $y = 8x - 3$
2. $y = 2x + 7$
3. $y = 3x$
4. $y = -2x$
5. $3y + 4x = 12$
6. $5y - 3x = 15$
7. $y = 3x^2$
8. $y = -2x^2$
9. $y = x^2 - 8$
10. $y = x^2 + 6$
11. $y = 4 - x^2$
12. $y = -2 - x^2$
13. $4x = y^2$
14. $9x = y^2$
15. $16y^2 = -x$
16. $4y^2 = -x$

Graph each equation involving absolute value. See Example 4.

17. $y = |x| + 4$
18. $y = |x| - 3$
19. $y = |x| - 2$
20. $y = -|x| + 1$
21. $y = 3 - |x|$
22. $y = -4 + |x|$
23. $y = |x + 3|$
24. $y = -|x - 2|$

Graph each of the following circles.

25. $x^2 + y^2 = 36$
26. $x^2 + y^2 = 81$
27. $(x - 2)^2 + y^2 = 36$
28. $x^2 + (y + 3)^2 = 49$
29. $(x - 4)^2 + (y + 3)^2 = 4$
30. $(x + 3)^2 + (y - 2)^2 = 16$
31. $(x + 2)^2 + (y - 5)^2 = 12$
32. $(x - 4)^2 + (y - 3)^2 = 8$

Find equations for each of the following circles. See Example 6.

33. Center $(1, 4)$, radius 3
34. Center $(-2, 5)$, radius 4
35. Center $(-8, 6)$, radius 5
36. Center $(3, -2)$, radius 2
37. Center $(-1, 2)$, passing through $(2, 6)$
38. Center $(2, -7)$, passing through $(-2, -4)$
39. Center $(-3, -2)$, tangent to the x-axis
40. Center $(5, -1)$, tangent to the y-axis

Find the center and the radius of each of the following that are circles. See Examples 7 and 8.

41. $x^2 + 6x + y^2 + 8y = -9$
42. $x^2 - 4x + y^2 + 12y + 4 = 0$
43. $x^2 - 12x + y^2 + 10y + 25 = 0$
44. $x^2 + 8x + y^2 - 6y = -16$
45. $x^2 + 8x + y^2 - 14y + 65 = 0$
46. $x^2 - 2x + y^2 + 1 = 0$
47. $x^2 + y^2 = 2y + 48$
48. $x^2 + 4x + y^2 = 21$
49. $x^2 - 2.84x + y^2 + 1.4y + 1.8664 = 0$
50. $x^2 + 7.4x + y^2 - 3.8y + 16.09 = 0$

51. Find the equation of the circle having the points $(3, -5)$ and $(-7, 2)$ as endpoints of a diameter.

52. Suppose a circle is tangent to both axes, is in the third quadrant, and has a radius of $\sqrt{2}$. Find its equation.

53. One circle has center at $(3, 4)$ and radius 5. A second circle has center at $(-1, -3)$ and radius 4. Do the circles cross?

54. Does the circle with radius 6 and center at $(0, 5)$ cross the circle with center at $(-5, -4)$ with radius 4?

Plot the following points, and then plot the points that are symmetric to the given point with respect to the (a) x-axis, (b) y-axis, (c) origin.

55. $(5, -3)$
56. $(-6, 1)$
57. $(-4, -2)$
58. $(-8, 3)$
59. $(-8, 0)$
60. $(0, -3)$

Use symmetry and point plotting to graph each equation. See Example 5.

61. $x^2 + y^2 = 5$ **62.** $y^2 = 4 - x^2$ **63.** $y = 4 - x^2$ **64.** $y = x^2 - 2$

65. $y = x^2 - 8x$ **66.** $y = 4x - x^2$ **67.** $y = x^3$ **68.** $y = -x^3$

69. $y = \dfrac{1}{1 + x^2}$ **70.** $y = \dfrac{-1}{x^2 + 9}$ **71.** $xy = 2$ **72.** $xy = -6$

73. Let F be some algebraic expression involving x as the only variable. Suppose the equation $y = F$ is changed to $y = -F$. What is the effect on the graph of $y = F$?

Sketch examples of graphs that satisfy the following conditions.

74. Symmetric with respect to the x-axis but not to the y-axis

75. Symmetric with respect to the y-axis but not to the x-axis

76. Symmetric with respect to the origin but to neither the x-axis nor the y-axis

4.3 Graphing Linear Equations

In the last section, Example 1 showed the graph of $y = -2x + 6$. The graph of this equation was a straight line. It turns out that all such equations have straight line graphs. In fact, the name *linear* is given to such equations.

Linear Equation

> A **linear equation** in two variables is one that can be written in the form
> $$Ax + By = C,$$
> where A, B, and C are real numbers, with not both A and B equal to 0.

A linear equation written in the form $Ax + By = C$ is in **standard form.**

EXAMPLE 1   The following table gives examples of linear and nonlinear equations. The standard form is given for linear equations.

Equation	Linear?	Standard form
$4x = 3y + 8$	Yes	$4x - 3y = 8$
$x + y = 2$	Yes	$x + y = 2$
$x + 1 = 0$	Yes	$x = -1$
$y = 2$	Yes	$y = 2$
$y = x^2$	No	——
$xy = 4$	No	——

     ●

   Every linear equation in two variables has a graph which is a straight line. A straight line is determined by two different points on the line. Two points that are useful for sketching the graph of a line are the *x-intercept* and the *y-intercept*. An **x-intercept** is an *x*-value (if any) at which the graph crosses the *x*-axis. A **y-intercept** is a *y*-value (if any) at which the graph crosses the *y*-axis. The graph in Figure 4.18 has an *x*-intercept of x_1 and a *y*-intercept of y_1. While only two different points are needed to graph a line, it is a good idea to find three points, using the third point as a check.

   As suggested by the graph in Figure 4.18, *x*-intercepts can be found by letting $y = 0$, with *y*-intercepts found by letting $x = 0$. The next example shows how the intercepts can be helpful in graphing a line.

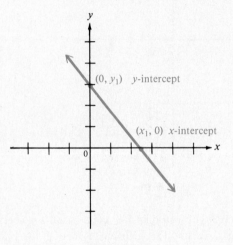

Figure 4.18

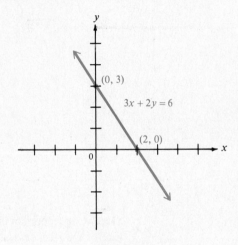

Figure 4.19

EXAMPLE 2   Graph $3x + 2y = 6$. Use the intercepts.
      To find the *y*-intercept, let $x = 0$.

$$3 \cdot 0 + 2y = 6$$
$$2y = 6$$
$$y = 3$$

For the *x*-intercept, let $y = 0$, getting

$$3x + 2 \cdot 0 = 6$$
$$3x = 6$$
$$x = 2.$$

Plotting $(0, 3)$ and $(2, 0)$ gives the graph in Figure 4.19. A third point could be found as a check if desired.   ●

EXAMPLE 3   Graph $y = -3$.

Think of $y = -3$ as the equation $y = -3 + 0 \cdot x$. This alternate form of the equation shows that for any value of x that might be chosen, y equals $-3 + 0 \cdot x = -3 + 0 = -3$. Since y always equals -3, the value of y can never be 0. This means that the graph has no x-intercept. The only way a straight line can have no x-intercept is for it to be parallel to the x-axis, as shown in Figure 4.20.   ●

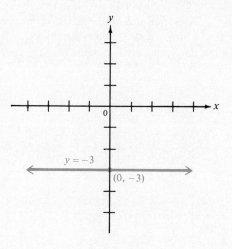

Figure 4.20

Slope   An important characteristic of a straight line is its *slope*, a numerical measure of the steepness of the line. To find this measure, start with the line through the two distinct points (x_1, y_1) and (x_2, y_2), as shown in Figure 4.21. (Since the

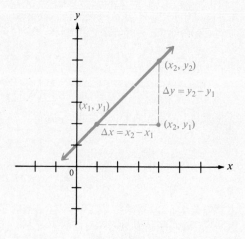

Figure 4.21

points are distinct, $x_1 \neq x_2$.) The difference

$$x_2 - x_1$$

is called the **change in x** and denoted by Δx (read "delta x"), where Δ is the Greek letter *delta*. In the same way, the **change in y** can be written

$$\Delta y = y_2 - y_1.$$

The slope of a nonvertical line (usually denoted m) is defined to be the quotient of the change in y and the change in x, as follows.

Slope

> The **slope m of the line** through (x_1, y_1) and (x_2, y_2) is
>
> $$m = \frac{\Delta y}{\Delta x} = \frac{y_2 - y_1}{x_2 - x_1}, \quad \Delta x \neq 0.$$

The slope of a line can be found only if the line is nonvertical. This guarantees that $x_2 \neq x_1$ so that the denominator $x_2 - x_1 \neq 0$. It is not possible to define the slope of a vertical line.

> **The slope of a vertical line is undefined.**

It can be shown, using theorems for similar triangles, that the slope is independent of the choice of points on the line. That is, the slope of a line is the same no matter which pair of distinct points on the line is used to find it.

EXAMPLE 4 Find the slope of the line through each of the following pairs of points.

(a) $(-4, 8)$, $(2, -3)$

Let $x_1 = -4$, $y_1 = 8$, $x_2 = 2$ and $y_2 = -3$. Then $\Delta y = -3 - 8 = -11$ and $\Delta x = 2 - (-4) = 6$. The slope $m = \Delta y / (\Delta x) = -11/6$.

(b) $(2, 7)$, $(2, -4)$

A sketch shows that the line through $(2, 7)$ and $(2, -4)$ is vertical. As mentioned above, the slope of a vertical line is not defined. (An attempt to use the definition of slope here would produce a zero denominator.)

(c) $(5, -3)$ and $(-2, -3)$

By the definition of slope,

$$m = \frac{-3 - (-3)}{-2 - 5} = \frac{0}{-7} = 0. \quad \bullet$$

Drawing a graph through the points in Example 4(c) produces a line that is horizontal.

Example 4(c) suggests the following generalization.

The slope of a horizontal line is 0.

Figure 4.22 shows lines of various slopes. As the figure shows, a line with a positive slope goes up from left to right, but a line with a negative slope goes down.

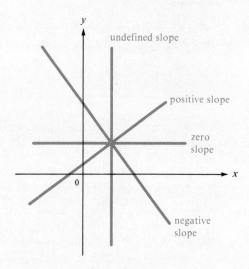

Figure 4.22   Figure 4.23

EXAMPLE 5   Graph the line passing through $(-1, 5)$ and having slope $-5/3$.

First locate the point $(-1, 5)$ as shown in Figure 4.23. Since the slope of this line is $-5/3$, a change of 3 units horizontally produces a change of -5 units vertically. This gives a second point, $(2, 0)$, which can then be used to complete the graph.   ●

4.3 EXERCISES

Which of the following are linear equations? See Example 1.

1. $y = 2x + 3$ **2.** $y = 5x - 1$ **3.** $y = -5$ **4.** $x = 10$

5. $y = x^2 + 1$ **6.** $y = 2x^2 - 3$ **7.** $y = x/5$ **8.** $y = 2/x$

9. $y = \dfrac{1}{x + 1}$ **10.** $y = \dfrac{3x}{2}$ **11.** $y = |x| + 1$ **12.** $y = -|x|$

13. $y^2 = 2x$ **14.** $y^2 = x - 1$ **15.** $x^2 + y^2 = 4$ **16.** $2x^2 - y^2 = 1$

17. $y = \sqrt{x + 1}$ **18.** $y = \sqrt{x} + 1$

Graph each of the following. See Examples 2 and 3.

19. $x - y = 4$ **20.** $x + y = 4$ **21.** $y = 3x + 1$ **22.** $2x + y = 4$

23. $3x - y = 6$ **24.** $2x - 3y = 6$ **25.** $2x + 5y = 10$ **26.** $3x + 4y = 12$

27. $4x - 3y = 9$ **28.** $3y - 5x = 15$ **29.** $x = 2$ **30.** $y = -3$

31. $y = -5$ **32.** $x = 6$ **33.** $y = 3x$ **34.** $x = -2y$

35. $x = -5y$ **36.** $y = x$

Graph the line passing through the given point and having the indicated slope. See Example 5.

37. Through $(-1, 3)$, $m = 3/2$ **38.** Through $(-2, 8)$, $m = -1$

39. Through $(3, -4)$, $m = -1/3$ **40.** Through $(-2, -3)$, $m = -3/4$

41. Through $(-1, 4)$, $m = 0$ **42.** Through $(2, -5)$, $m = 0$

43. Through $(3, 2/3)$, undefined slope   **44.** Through $(9/4, 2)$, undefined slope

Find the slope of each of the following lines. Hint: in Exercises 53–62, find two points on the line. See Example 4.

45. Through $(-2, 1)$ and $(3, 2)$ **46.** Through $(-2, 3)$ and $(-1, 2)$ **47.** Through $(8, 4)$ and $(-1, -3)$

48. Through $(-4, -3)$ and $(5, 0)$ **49.** Through $(4, 5)$ and $(-1, 2)$ **50.** Through $(1, 5)$ and $(-2, 5)$

51. Through $(2, 3)$ and $(2, 7)$ **52.** Through $(10, -2)$ and $(2, -1)$ **53.** $y = 2x$

54. $y = 3x - 2$ **55.** $y = 4x - 5$ **56.** $y = -2x + 1$

57. $3x + 4y = 6$ **58.** $2x + y = 8$ **59.** $y = 4$ 0 slope

60. $x = -6$ **61.** The x-axis   **62.** The y-axis

63. Through $(-1.978, 4.806)$ and $(3.759, 8.125)$

64. Through $(11.72, 9.811)$ and $(-12.67, -5.009)$

Use slopes to decide whether or not the following points lie on a straight line. (Hint: in Exercise 65, find the slopes of the lines through M and N and through M and P. If these slopes are the same, then the points lie on a straight line.)

65. $M(1, -2)$, $N(3, -18)$, $P(-2, 22)$ **66.** $A(0, -2)$, $B(3, 7)$, $C(-4, -14)$

67. $X(1, 3)$, $Y(-4, 73)$, $Z(5, -50)$ **68.** $R(0, -7)$, $S(8, -11)$, $T(-9, -25)$

69. Straight line graphs are often good choices for *supply-demand curves*. In most supply-demand curves, the supply of an item goes up as the price goes up, while the demand goes down as the price goes up. Suppose that the demand and price for a certain model of electric can opener are related by

$$p = 16 - \frac{5}{4}x,$$

where p is price and x is demand, in appropriate units. Find the price when the demand is at the following levels.
   (a) 0 units   **(b)** 4 units   **(c)** 8 units
   Find the demand for the electric can opener at the following prices.
   (d) \$6   **(e)** \$11   **(f)** \$16

(g) Graph $p = 16 - (5/4)x$.

Suppose that the price and supply of the item above are related by

$$p = \frac{3}{4}x,$$

where x represents the supply and p the price. Find the supply at the following prices.

(h) $0   **(i)** $10   **(j)** $20

(k) Graph $p = (3/4)x$ on the same axes used for part (g).

(l) Find the equilibrium supply (the supply at the point where the supply and demand curves cross).

(m) Find the equilibrium price (the price at the equilibrium supply).

70. Let the supply and demand equations for strawberry-flavored licorice be

$$\text{supply: } p = \frac{3}{2}x \quad \text{and} \quad \text{demand: } p = 81 - \frac{3}{4}x.$$

(a) Graph these on the same axes.

(b) Find the equilibrium demand. (See Exercise 69.)

(c) Find the equilibrium price.

71. Let the supply and demand equations for butter pecan ice cream be given by

$$\text{supply: } p = \frac{2}{5}x \quad \text{and} \quad \text{demand: } p = 100 - \frac{2}{5}x.$$

(a) Graph these on the same axes.

(b) Find the equilibrium demand. (See Exercise 69.)

(c) Find the equilibrium price.

72. Let the supply and demand equations for sugar be given by

$$\text{supply: } p = 1.4x - .6 \quad \text{and} \quad \text{demand: } p = -2x + 3.2.$$

(a) Graph these on the same axes.

(b) Find the equilibrium demand. (See Exercise 69.)

(c) Find the equilibrium price.

4.4 Equations of a Line

In the previous section we saw how to draw the graph of a line, given its equation. This section shows how to find the equation of a line if we know two distinct points on the line, or if we know one point on the line as well as the slope of the line.

Figure 4.24 shows the line passing through the fixed point (x_1, y_1) and having slope m. (Assuming that the line has a slope guarantees that it is not vertical.) Let (x, y) be any other point on the line. By the definition of slope, the slope of the line is

$$\frac{y - y_1}{x - x_1}.$$

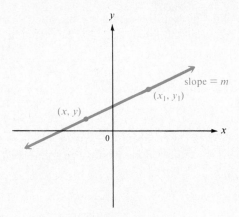

Figure 4.24

Since the slope of the line is m,

$$\frac{y - y_1}{x - x_1} = m.$$

Multiplying both sides by $x - x_1$ gives

$$y - y_1 = m(x - x_1).$$

This result, called the *point-slope form* of the equation of a line, identifies points on a given line: a point (x, y) lies on the line through (x_1, y_1) with slope m if and only if

$$y - y_1 = m(x - x_1).$$

In summary,

Point-Slope Form

> the line with slope m passing through the point (x_1, y_1) has an equation
>
> $$y - y_1 = m(x - x_1),$$
>
> the **point-slope form** of the equation of a line.

EXAMPLE 1   Write an equation of the line through $(-4, 1)$ with slope -3.

Here $x_1 = -4$, $y_1 = 1$, and $m = -3$. Use the point-slope form of the equation of a line to get

$$y - 1 = -3[x - (-4)]$$
$$y - 1 = -3(x + 4)$$
$$y - 1 = -3x - 12$$
$$3x + y = -11. \quad \bullet$$

EXAMPLE 2   Find an equation of the line through $(-3, 2)$ and $(2, -4)$.
Find the slope first. By the definition of slope,

$$m = \frac{-4 - 2}{2 - (-3)}$$

$$= \frac{-6}{5}.$$

We can use either $(-3, 2)$ or $(2, -4)$ for (x_1, y_1). If we let $x_1 = -3$ and $y_1 = 2$ and use the point-slope form, we get

$$y - 2 = \frac{-6}{5}[x - (-3)]$$

$$5(y - 2) = -6(x + 3)$$

$$5y - 10 = -6x - 18$$

$$6x + 5y = -8.$$

Verify that we get the same equation if we use $(2, -4)$ instead of $(-3, 2)$ in the point-slope form. ●

Slope-Intercept Form   As a special case of the point-slope form of the equation of a line, suppose that a line passes through the point $(0, b)$, so the line has y-intercept b. If the line has slope m, then using the point-slope form with $x_1 = 0$ and $y_1 = b$ gives

$$y - y_1 = m(x - x_1)$$

$$y - b = m(x - 0)$$

$$y = mx + b$$

as an equation of the line. Since this result shows the slope of the line and the y-intercept, it is called the *slope-intercept form* of the equation of a line.

Slope-Intercept Form

> The line with slope m and y-intercept b has an equation
>
> $$y = mx + b,$$
>
> the **slope-intercept** form of the equation of a line.

EXAMPLE 3   Find the slope and y-intercept of $3x - y = 2$. Graph the line.
First write $3x - y = 2$ in the slope-intercept form, $y = mx + b$. Do this by solving for y, getting $y = 3x - 2$. This result shows that the slope is $m = 3$ and the y-intercept is $b = -2$. To draw the graph, first locate the y-intercept. See Figure 4.25. Then, as in the previous section, use the slope of 3, or 3/1, to get a second point on the graph. The line through these two points is the graph of $3x - y = 2$. ●

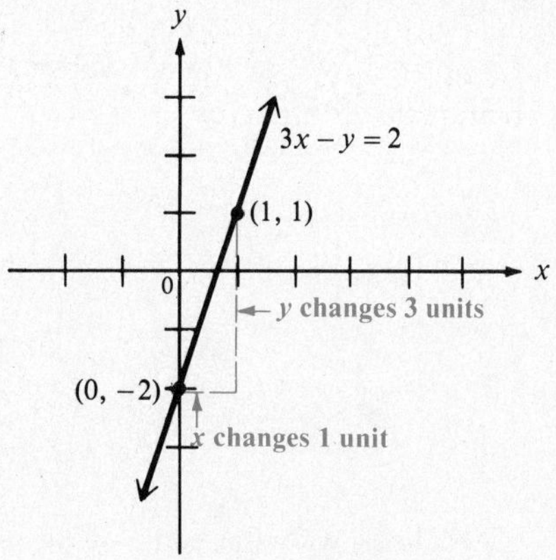

Figure 4.25

In the discussion above, we assumed that the given line had a slope. The only lines having undefined slope are vertical lines, so we can complete our discussion of the equations of lines by giving the equation of a vertical line. Since the vertical line through the point (a, b) passes through all points of the form (a, y), for any value of y,

Equation of a Vertical Line

> the equation of the vertical line through the point (a, b) is $x = a$.

For example, the vertical line through $(-4, 9)$ has equation $x = -4$, while the vertical line through $(0, 1/4)$ has equation $x = 0$.

Parallel and Perpendicular Lines   One application of slopes involves deciding whether or not two lines are parallel. Since two parallel lines are equally "steep," they should have the same slope. Also, two lines with the same "steepness" are parallel. We therefore have the following result.

Parallel Lines

> Two nonvertical lines are parallel if and only if they have the same slope.

EXAMPLE 4

Find the equation of the line through the point $(3, 5)$ and parallel to the line $2x + 5y = 4$.

We know that the point $(3, 5)$ is on the line, and we need to know the slope to use the point-slope form. Find the slope by writing the equation of the given line in slope-intercept form. (That is, solve for y.)

$$2x + 5y = 4$$

$$y = -\frac{2}{5}x + \frac{4}{5}$$

The slope is $-2/5$. Since the lines are parallel, $-2/5$ is also the slope of the line whose equation we are to find. Substituting $m = -2/5$, $x_1 = 3$, and $y_1 = 5$ into the point-slope form gives

$$y - y_1 = m(x - x_1)$$

$$y - 5 = -\frac{2}{5}(x - 3)$$

$$5(y - 5) = -2(x - 3)$$

$$5y - 25 = -2x + 6$$

$$2x + 5y = 31. \quad \bullet$$

Two lines with the same slope are parallel. As is stated below, two lines having slopes with a product of -1 are perpendicular.

Perpendicular Lines

Two lines, neither of which is vertical, are perpendicular if and only if their slopes have a product of -1.

Exercises 43–46 outline a proof of this statement.

EXAMPLE 5

Find the equation of the line L through $(3, 0)$ and perpendicular to $5x - y = 4$.

Use the point-slope form to get the equation of line L. To find the slope of line L, first find the slope of $5x - y = 4$ by solving for y.

$$5x - y = 4$$

$$y = 5x - 4$$

This line has a slope of 5. Since the lines are perpendicular, if line L has slope m, then

$$5m = -1$$

$$m = \frac{-1}{5}.$$

Thus, the required equation is

$$y - 0 = \frac{-1}{5}(x - 3)$$

$$5y = -x + 3$$

$$x + 5y = 3. \quad \bullet$$

All the lines above have equations that could be placed in the form $Ax + By = C$ for real numbers A, B, and C. The equation $Ax + By = C$ is the standard form of the equation of a line. The various forms of linear equations are listed at the top of the next page.

Linear Equations

General equation	Type of equation
$Ax + By = C$	*Standard form*, if $A \neq 0$ and $B \neq 0$, a line with *x*-intercept C/A and *y*-intercept C/B, slope $-A/B$
$x = k$	*Vertical line*, *x*-intercept k, no *y*-intercept, line has undefined slope
$y = k$	*Horizontal line*, *y*-intercept k, no *x*-intercept, line has slope 0
$y = mx + b$	*Slope-intercept form*, slope is m, *y*-intercept is b
$y - y_1 = m(x - x_1)$	*Point-slope form*, slope is m, line passes through (x_1, y_1)

4.4 EXERCISES

For each of the following lines, write the equation in standard form. See Examples 1–2.

1. Through $(1, 3)$, $m = -2$

2. Through $(2, 4)$, $m = -1$

3. Through $(-5, 4)$, $m = -3/2$

4. Through $(-4, 3)$, $m = 3/4$

5. Through $(2, 0)$, $m = -3/4$

6. Through $(0, -1)$, $m = 3/5$

7. Through $(-3, 2)$, $m = 0$

8. Through $(6, 1)$, $m = 0$

9. Through $(-8, 1)$, undefined slope

10. Through $(0, 4)$, undefined slope

11. Through $(4, 2)$ and $(1, 3)$

12. Through $(8, -1)$ and $(4, 3)$

13. Through $(-1, 3)$ and $(3, 4)$

14. Through $(6, 0)$ and $(3, 2)$

15. Through $(0, 3)$ and $(4, 0)$

16. Through $(-3, 0)$ and $(0, -5)$

17. *x*-intercept 3, *y*-intercept -2

18. *x*-intercept -2, *y*-intercept 4

19. *x*-intercept -5, no *y*-intercept

20. *y*-intercept 3, no *x*-intercept

21. Vertical, through $(-6, 5)$

22. Horizontal, through $(8, 7)$

23. Through $(-1.76, 4.25)$, slope -5.081

24. Through $(5.469, 11.08)$, slope 4.723

Write equations in standard form for each of the following lines. See Examples 4 and 5.

25. Through $(-1, 4)$, parallel to $x + 3y = 5$

26. Through $(3, -2)$, parallel to $2x - y = 5$

27. Through $(2, -5)$, parallel to $y - 4 = 2x$

28. Through $(0, 5)$, parallel to $3x - 7y = 8$

29. Through $(3, -4)$, perpendicular to $x + y = 4$

30. Through $(-2, 6)$, perpendicular to $2x - 3y = 5$

31. Through $(1, 6)$, perpendicular to $3x + 5y = 1$

32. Through $(-2, 0)$, perpendicular to $8x - 3y = 7$

33. *x*-intercept -2, parallel to $y = 2x$

34. *y*-intercept 3, parallel to $x + y = 4$

35. Through $(-5, 7)$, perpendicular to $y = -2$

36. Through $(1, -4)$, perpendicular to $x = 4$

37. Do the points $(4, 3)$, $(2, 0)$, and $(-18, -12)$ lie on the same line? (*Hint:* find the equation of the line through two of the points.)

38. Do the points $(4, -5)$, $(3, -5/2)$, and $(-6, 18)$ lie on a line?

39. Find k so that the line through $(4, -1)$ and $(k, 2)$ is
 (a) parallel to $3y + 2x = 6$;
 (b) perpendicular to $2y - 5x = 1$.

40. Find r so that the line through $(2, 6)$ and $(-4, r)$ is
 (a) parallel to $2x - 3y = 4$;
 (b) perpendicular to $x + 2y = 1$.

41. Use slopes to show that the quadrilateral with vertices at $(1, 3)$, $(-5/2, 2)$, $(-7/2, 4)$, and $(2, 1)$ is a parallelogram (has opposite sides parallel).

42. Use slopes to show that the square with vertices at $(-2, 5)$, $(4, 5)$, $(4, -1)$, and $(-2, -1)$ has diagonals that are perpendicular.

To prove that two perpendicular lines, neither of which is vertical, have slopes with a product of -1, go through the following steps. Let line L_1 have equation $y = m_1x + b_1$, and let line L_2 have equation $y = m_2x + b_2$. Assume that L_1 and L_2 are perpendicular, and complete right triangle MPN as shown in the figure.

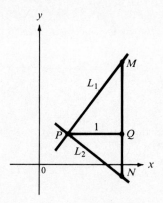

43. Show that MQ has length m_1.

44. Show that QN has length $-m_2$.

45. Show that triangles MPQ and PQN are similar.

46. Show that $m_1/1 = 1/-m_2$ and that $m_1m_2 = -1$.

Many real situations can be described approximately by a straight-line graph. One way to find the equation of such a straight line is to use two typical data points from the graph and the point-slope form of the equation of a line. In each of the following problems, assume that the data can be approximated fairly closely by a straight line. Use the given information to find the equation of the line. Find the slope of each line.

47. A company finds that it can make a total of 20 solar heaters for $13,900, while 10 heaters cost $7500. Let y be the total cost to produce x solar heaters.

48. The sales of a small company were $27,000 in its second year of operation and $63,000 in its fifth year. Let y represent sales in year x.

49. When a certain industrial pollutant is introduced into a river, the reproduction of catfish declines. In a given period of time, three tons of the pollutant results in a fish population of 37,000. Also, 12 tons of pollutant produce a fish population of 28,000. Let y be the fish population when x tons of pollutant are introduced into the river.

50. In the snake *Lampropelbis polyzona,* total length y is related to tail length x in the domain 30 mm $\leq x \leq$ 200 mm by a linear function. Find such a linear function, if a snake 455 mm long has a 60 mm tail, and a 1050 mm snake has a 140 mm tail.

51. According to research done by the political scientist James March, if the Democrats win 45% of the two-party vote for the House of Representatives, they win 42.5% of the seats. If the Democrats win 55% of the vote, they win 67.5% of the seats. Let y be the percent of seats won, and x the percent of the two-party vote.

52. If the Republicans win 45% of the two-party vote, they win 32.5% of the seats (see Exercise 51). If they win 60% of the vote, they get 70% of the seats. Let y represent the percent of the seats, and x the percent of the vote.

4.5 Parabolas

As we saw earlier, the graph of $y = x^2$ is a parabola. It turns out that all expressions of the form $y = ax^2 + bx + c$, where a, b, and c are real numbers, with $a \neq 0$, have parabolas as graphs. For review, the graph of $y = x^2$ is shown in Figure 4.26.

Parabolas are examples of graphs having symmetry about a line (the y-axis in Figure 4.26). The line of symmetry for a parabola is the **axis** of the parabola. The point where the axis intersects the parabola is the **vertex** of the parabola.

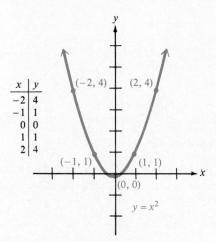

Figure 4.26

Starting with $y = x^2$, there are several possible ways to get a more general expression:

$y = ax^2$ Multiply by a positive or negative coefficient

$y = x^2 + k$ Add a positive or negative constant

$y = (x - h)^2$ Replace x with $x - h$, where h is a constant

$y = a(x - h)^2 + k$ Do all of the above.

The graph of each of these is still a parabola, but modified from that of $y = x^2$. The next few examples show how these changes modify the parabola. The first example shows the result of changing $y = x^2$ to $y = ax^2$.

EXAMPLE 1   Graph each parabola.

(a) $y = -x^2$

For a given value of x, the corresponding value of y will be the negative of what it was for $y = x^2$. Because of this, the graph of $y = -x^2$ is the same shape as that of $y = x^2$ except that it opens downward. See Figure 4.27.

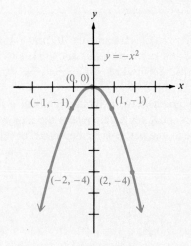

Figure 4.27

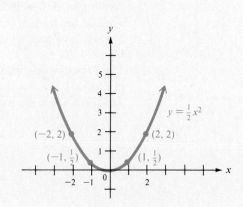

Figure 4.28

(b) $y = \dfrac{1}{2} x^2$

Choose a value of x, and then find x^2. The coefficient of 1/2 will cause the resulting value of y to be smaller than for $y = x^2$, causing the parabola to be "broader" than $y = x^2$. See Figure 4.28. In both parts of this example, the axis is the vertical line $x = 0$. ●

The next two examples show the results of changing $y = x^2$ to $y = x^2 + k$ or to $y = (x - h)^2$, respectively.

EXAMPLE 2   Graph $y = x^2 - 4$.

Each value of y will be 4 less than the corresponding value of y in $y = x^2$. This means that $y = x^2 - 4$ has the same shape as $y = x^2$ but is shifted 4 units down. See Figure 4.29. The vertex of the parabola (on this parabola, the lowest point) is at $(0, -4)$. The axis of the parabola is the vertical line $x = 0$.   •

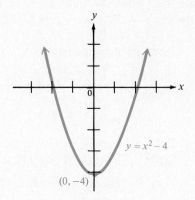

Figure 4.29

The vertical shift of the graph in Example 2 is called a **translation.** Example 3 below shows a horizontal translation, which is a shift to the right or left.

EXAMPLE 3   Graph $y = (x - 4)^2$.

By choosing values of x and finding the corresponding values of y, this parabola can be seen to be translated 4 units to the right when compared to $y = x^2$. The vertex is at $(4, 0)$. The axis of this parabola is the vertical line $x = 4$. See Figure 4.30.   •

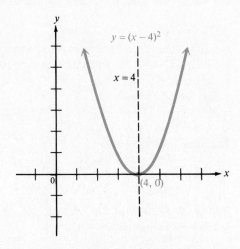

Figure 4.30

A combination of all the changes mentioned above produces the results shown in Example 4.

EXAMPLE 4 Graph $y = -(x + 3)^2 + 1$.

This parabola is translated 3 units to the left and 1 unit up. It opens downward. The vertex, the point $(-3, 1)$, is the *highest* point on the graph. The axis is the line $x = -3$. See Figure 4.31 on the facing page. ●

Generalizing from Example 4,

Graph of a Parabola

> the graph of
>
> $$y = a(x - h)^2 + k, \quad a \neq 0,$$
>
> 1. is a parabola with vertex (h, k), and the vertical line $x = h$ as axis;
> 2. opens upward if $a > 0$ and downward if $a < 0$;
> 3. is broader than $y = x^2$ if $0 < |a| < 1$ and narrower than $y = x^2$ if $|a| > 1$.

Given the expression $y = ax^2 + bx + c$, where a, b, and c are real numbers and $a \neq 0$, the process of *completing the square* (discussed in Chapter 3 and used with circles earlier in this chapter) can be used to change $ax^2 + bx + c$ to the form $a(x - h)^2 + k$, so that the vertex and axis may be identified. Follow the steps given in the next two examples.

EXAMPLE 5 Graph $y = x^2 - 6x + 7$.

To graph this parabola, we need to rewrite $x^2 - 6x + 7$ in the form $(x - h)^2 + k$. Start by writing the equation as follows.

$$y = (x^2 - 6x \qquad) + 7$$

As shown earlier, a number must be added inside the parentheses to get a perfect square trinomial. To find this number, take half the coefficient of x and then square the result. Half of -6 is -3, and $(-3)^2$ is 9. Now add and subtract 9 inside the parentheses.

$$y = (x^2 - 6x + 9 - 9) + 7$$

Factor as follows.

$$y = (x^2 - 6x + 9) - 9 + 7$$
$$y = (x - 3)^2 - 2$$

This result shows that the vertex of the parabola is $(3, -2)$, with axis $x = 3$. A graph is shown in Figure 4.32. ●

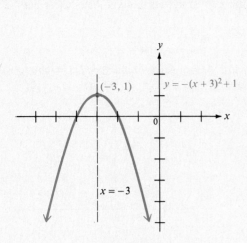

Figure 4.31

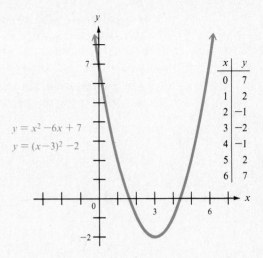

Figure 4.32

EXAMPLE 6   Graph $y = -3x^2 - 2x + 1$.

To complete the square, first factor out -3 to get

$$y = -3\left(x^2 + \frac{2}{3}x \qquad\right) + 1.$$

(This is necessary to make the coefficient of x^2 equal 1.) Half the coefficient of x is 1/3, and $(1/3)^2 = 1/9$. Add and subtract 1/9 inside the parentheses as follows:

$$y = -3\left(x^2 + \frac{2}{3}x + \frac{1}{9} - \frac{1}{9}\right) + 1.$$

Using the distributive property and simplifying gives

$$y = -3\left(x^2 + \frac{2}{3}x + \frac{1}{9}\right) - 3\left(-\frac{1}{9}\right) + 1$$

$$y = -3\left(x^2 + \frac{2}{3}x + \frac{1}{9}\right) + \frac{1}{3} + 1$$

$$y = -3\left(x^2 + \frac{2}{3}x + \frac{1}{9}\right) + \frac{4}{3}.$$

Factor to get

$$y = -3\left(x + \frac{1}{3}\right)^2 + \frac{4}{3}.$$

Now the equation of the parabola is written in the form $y = a(x - h)^2 + k$, and this rewritten equation shows that the axis of the parabola is the vertical line

$$x + \frac{1}{3} = 0 \quad \text{or} \quad x = -\frac{1}{3}$$

and that the vertex is $(-1/3, 4/3)$. Use these results and additional ordered pairs as needed to get the graph in Figure 4.33. ●

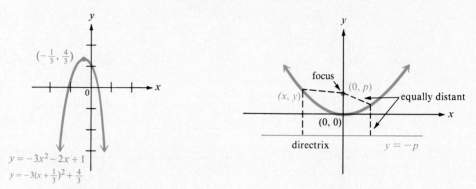

Figure 4.33

Figure 4.34

Geometric Definition of a Parabola   Earlier in this chapter we defined a circle as "the set of all points in the plane a fixed distance from a fixed point." Starting with this definition, we then produced the equation $(x - h)^2 + (y - k)^2 = r^2$. Now we shall do a similar thing with parabolas. Geometrically, a **parabola** is defined as the set of all points in a plane that are equally distant from a fixed point and a fixed line not containing the point. The point is called the **focus** and the line is the **directrix.** The line through the focus and perpendicular to the directrix is the **axis** of the parabola. The point on the axis that is equally distant from the focus and the directrix is the **vertex** of the parabola.

The parabola in Figure 4.34 has focus located at $(0, p)$, with the line $y = -p$ as directrix. The vertex is $(0, 0)$. Let (x, y) be any point on the parabola. The distance from (x, y) to the directrix is $|y - (-p)|$, while the distance from (x, y) to $(0, p)$ is $\sqrt{(x - 0)^2 + (y - p)^2}$. Since (x, y) is equally distant from the directrix and the focus,

$$|y - (-p)| = \sqrt{(x - 0)^2 + (y - p)^2}.$$

Square both sides, getting

$$(y + p)^2 = x^2 + (y - p)^2$$
or
$$y^2 + 2yp + p^2 = x^2 + y^2 - 2yp + p^2,$$
with
$$4yp = x^2,$$

the equation of the parabola with focus $(0, p)$ and directrix $y = -p$. We could extend this result to a parabola with vertex at (h, k), focus p units above (h, k) and directrix p units below (h, k), or to a parabola with focus and directrix exchanged.

Horizontal Parabolas The directrix of a parabola could be the *vertical* line $x = -p$, with focus on the x-axis at $(p, 0)$, producing a parabola opening to the *right*. This parabola would have equation $y^2 = 4px$. The next examples show the graphs of horizontal parabolas.

EXAMPLE 7 Graph $x = y^2$.

The equation $x = y^2$ can be obtained from $y = x^2$ by exchanging x and y. Choosing values of y and finding the corresponding values of x gives the parabola in Figure 4.35. For comparison, the graph of $y = x^2$ is shown as a dashed line. These graphs are symmetric to each other with respect to the line $y = x$. ●

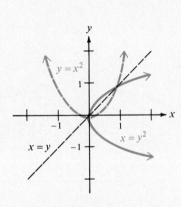

Figure 4.35

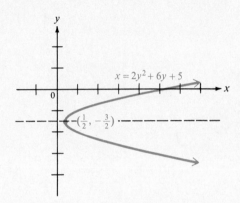

Figure 4.36

EXAMPLE 8 Graph $x = 2y^2 + 6y + 5$.

To write this equation in the form $x = a(y - k)^2 + h$, complete the square on y as follows:

$$x = 2(y^2 + 3y \qquad) + 5$$

$$= 2\left(y^2 + 3y + \frac{9}{4} - \frac{9}{4}\right) + 5$$

$$= 2\left(y^2 + 3y + \frac{9}{4}\right) + 2\left(-\frac{9}{4}\right) + 5$$

$$x = 2\left(y + \frac{3}{2}\right)^2 + \frac{1}{2}.$$

As this result shows, the vertex of the parabola is the point $(1/2, -3/2)$. The axis is the horizontal line

$$y + \frac{3}{2} = 0, \quad \text{or} \quad y = -\frac{3}{2}.$$

Using the vertex and the axis and plotting a few additional points gives the graph in Figure 4.36. ●

Now we can summarize the types of parabolas discussed in this section.

Graphs of Parabolas

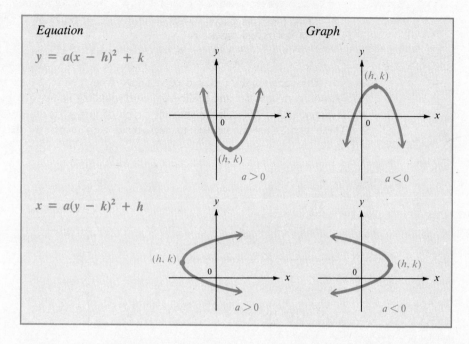

Equation

$y = a(x - h)^2 + k$

$x = a(y - k)^2 + h$

Graph

Applications of Parabolas   Parabolas have many practical applications. For example, if a light source is placed at the focus of a parabolic reflector, as in Figure 4.37, light rays reflect parallel to the axis, making a spotlight or flashlight.

The process also works in reverse. Light rays from a distant source come in parallel to the axis and are reflected to a point at the focus. (If such a reflector is aimed at the sun, a temperature of several thousand degrees may be obtained.)

The vertex of a parabola is often used to help find the maximum or minimum value of a quantity, as in the next example.

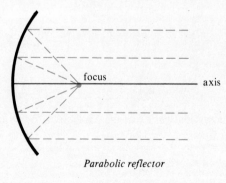

Parabolic reflector

Figure 4.37

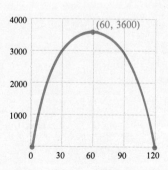

Figure 4.38

EXAMPLE 9   Ms. Whitney owns and operates Aunt Emma's Pie Shop. She has hired a consultant to analyze her business operations. The consultant tells her that her profit P is given by

$$P = 120x - x^2,$$

where x is the number of units of pies that she makes. How many units of pies should be made in order to maximize the profit? What is the maximum possible profit?

　　　Profit can be rewritten as $P = -x^2 + 120x + 0$. Use the methods of this section to find that the vertex of the parabola is (60, 3600). Since $a < 0$, the vertex is the highest point on the graph and produces a *maximum* rather than a minimum. Figure 4.38 shows the portion of the profit graph located in quadrant I. (Why is quadrant I the only one of interest here?) The maximum profit of \$3600 occurs when 60 units of pies are made. In this case, profit increases as more and more pies are made up to 60 units and then decreases as more and more pies are made past this point. ●

4.5 EXERCISES

1. Graph the following on the same coordinate system.

   (a) $y = 2x^2$　　　　　**(b)** $y = 3x^2$　　　　　**(c)** $y = \dfrac{1}{2}x^2$　　　　　**(d)** $y = \dfrac{1}{3}x^2$

   (e) How does the coefficient affect the shape of the graph?

2. Graph the following on the same coordinate system.
   (a) $y = x^2 + 2$　　　　**(b)** $y = x^2 - 1$　　　　**(c)** $y = x^2 + 1$　　　　**(d)** $y = x^2 - 2$
   (e) How do these graphs differ from the graph of $y = x^2$?

3. Graph the following on the same coordinate system.
   (a) $y = (x - 2)^2$　　　**(b)** $y = (x + 1)^2$　　　**(c)** $y = (x + 3)^2$　　　**(d)** $y = (x - 4)^2$
   (e) How do these graphs differ from the graph of $y = x^2$?

4. Use the quadratic formula to find the values of x when $y = 0$. Use the two values you get to locate the x value of the vertex. Find the axis of each parabola.
   (a) $y = x^2 + 8x + 13$　　　　　　　　**(b)** $y = x^2 - 12x + 30$
   (c) $y = 3x^2 - 2x + 6$　　　　　　　　**(d)** $y = 5x^2 + 6x - 3$

Graph each of the following parabolas. Give the vertex and axis of each. See Examples 1–6.

5. $y = (x - 2)^2$　　　　　　　**6.** $y = (x + 4)^2$　　　　　　　**7.** $y = (x + 3)^2 - 4$

8. $y = (x - 5)^2 - 4$　　　　　**9.** $y = -2(x + 3)^2 + 2$　　　**10.** $y = -3(x - 2)^2 + 1$

11. $y = -\dfrac{1}{2}(x + 1)^2 - 3$　　**12.** $y = \dfrac{2}{3}(x - 2)^2 - 1$　　**13.** $y = x^2 - 2x + 3$

14. $y = x^2 + 6x + 5$　　　　　**15.** $y = -x^2 - 4x + 2$　　　　**16.** $y = -x^2 + 6x - 6$

17. $y = 2x^2 - 4x + 5$　　　　　**18.** $y = -3x^2 + 24x - 46$

Find several points satisfying each of the following and then sketch the graph.

19. $y = .14x^2 + .56x - .3$

20. $y = .82x^2 + 3.24x - .4$

21. $y = -.09x^2 - 1.8x + .5$

22. $y = -.35x^2 + 2.8x - .3$

Graph each horizontal parabola. See Examples 7 and 8.

23. $x = y^2 + 2$

24. $x = -y^2$

25. $x = (y + 1)^2$

26. $x = (y - 3)^2$

27. $x = (y + 2)^2 - 1$

28. $x = (y - 4)^2 + 2$

29. $x = -2(y + 3)^2$

30. $x = -3(y - 1)^2 + 2$

31. $x = \frac{1}{2}(y + 1)^2 + 3$

32. $x = \frac{2}{3}(y - 3)^2 + 2$

33. $x = y^2 + 2y - 8$

34. $x = y^2 - 6y + 7$

35. $x = -2y^2 + 2y - 3$

36. $x = -4y^2 - 4y - 3$

Work the following word problems. See Example 9.

37. Glenview Community College wants to construct a rectangular parking lot on land bordered on one side by a highway. It has 320 ft of fencing which it will use to fence off the other three sides. What should be the dimensions of the lot if the enclosed area is to be a maximum? [*Hint:* let x represent the width of the lot and let $320 - 2x$ represent the length. Graph the area parabola, $A = x(320 - 2x)$, and investigate the vertex.]

38. What would be the maximum area that could be enclosed by the college's 320 ft of fencing if it decided to close the entrance by enclosing all four sides of the lot? (See Exercise 37.)

39. George runs a sandwich shop. By studying data concerning his past costs, he has found that the cost of operating his shop is given by

$$C(x) = 2x^2 - 1200x + 180,100,$$

where $C(x)$ is the daily cost to make x sandwiches. Find the number of sandwiches George must sell to minimize the cost. What is the minimum cost?

40. The revenue of a charter bus company depends on the number of unsold seats. If the revenue $R(x)$ is given by

$$R(x) = 5000 + 50x - x^2,$$

where x is the number of unsold seats, find the maximum revenue and the number of unsold seats which produce maximum revenue.

41. The number of mosquitoes $M(x)$, in millions, in a certain area of Kentucky depends on the June rainfall x, in inches, approximately as follows.

$$M(x) = 10x - x^2$$

Find the rainfall that will produce the maximum number of mosquitoes.

42. If an object is thrown upward with an initial velocity of 32 ft per sec, then its height after t sec is given by

$$h = 32t - 16t^2.$$

Find the maximum height attained by the object. Find the number of seconds it takes the object to hit the ground.

43. Find two numbers whose sum is 20 and whose product is a maximum. (*Hint:* let x and $20 - x$ be the two numbers, and write an equation for the product.)

44. A charter flight charges a fare of \$200 per person plus \$4 per person for each unsold seat on the plane. If the plane holds 100 passengers, and if x represents the number of unsold seats, find the following.
  (a) An expression for the total revenue received for the flight (*Hint:* multiply the number of people flying, $100 - x$, by the price per ticket.)
  (b) The graph for the expression of part (a)
  (c) The number of unsold seats that will produce the maximum revenue
  (d) The maximum revenue

45. The demand for a certain type of cosmetic is given by

$$p = 500 - x,$$

where p is the price when x units are demanded.
  (a) Find the revenue $R(x)$ that would be obtained at a price of x. (*Hint:* revenue = demand $\times$ price.)
  (b) Graph the revenue function $R(x)$.
  (c) From the graph of the revenue function, estimate the price that will produce maximum revenue.
  (d) What is the maximum revenue?

46. Between the months of June and October, the percent of maximum possible chlorophyll production in a leaf is approximated by $C(x)$, where

$$C(x) = 10x + 50.$$

Here x is time in months with $x = 1$ representing June. From October through December, $C(x)$ is approximated by

$$C(x) = -20(x - 5)^2 + 100,$$

with x as above. Find the percent of maximum possible chlorophyll production in each of the following months: **(a)** June **(b)** July **(c)** September **(d)** October **(e)** November **(f)** December.

47. Use your results from Exercise 46 to sketch a graph of $y = C(x)$ from June through December. In what month is chlorophyll production a maximum?

48. An arch is shaped like a parabola. It is 30 m wide at the base and 15 m high. How wide is the arch 10 m from the ground?

49. A culvert is shaped like a parabola, 18 cm across the top and 12 cm deep. How wide is the culvert 8 cm from the top?

50. (a) Graph the parabola $y = 2x^2 + 5x - 3$.
  (b) Use the graph to find the solution of the quadratic inequality $2x^2 + 5x - 3 < 0$.
  (c) Use the graph to find the solution of the quadratic inequality $2x^2 + 5x - 3 > 0$.

51. Find a value of c so that $y = x^2 - 10x + c$ has exactly one x-intercept.

52. Find b so that $y = x^2 + bx + 9$ has exactly one x-intercept.

Use the geometric definition of parabola given in the text to find the equation of each parabola having the given point as focus.

53. (0, 3), directrix $y = -3$ **54.** (0, -5), directrix $y = 5$ **55.** (-2, 0), directrix $x = 2$

56. (8, 0), directrix $x = -8$ **57.** (3, 6), vertex (3, 4)   **58.** (-5, 2), vertex (-5, 5)

59. (-4, 1), directrix $x = -7$ **60.** (3, -4), directrix $x = 9$

4.6 Ellipses and Hyperbolas

As the earth travels around the sun over a year's time, it traces out a curve called an *ellipse*.

Ellipse

> An **ellipse** is the set of all points in a plane the sum of whose distances from two fixed points is constant. The two fixed points are called the **foci** of the ellipse.

For example, the ellipse in Figure 4.39 has foci at points F and F'. By the definition, the ellipse is made up of all points P such that the sum $d(P, F) + d(P, F')$ is constant. The ellipse in Figure 4.39 has its **center** at the origin. Points V and V' are the **vertices** of the ellipse, and the line segment connecting V and V' is the **major diameter.**

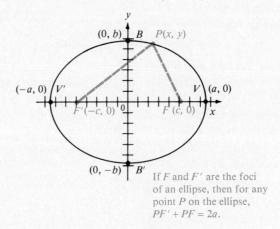

If F and F' are the foci of an ellipse, then for any point P on the ellipse, $PF' + PF = 2a$.

Figure 4.39

If the foci are chosen to be on the *x*-axis (or *y*-axis), with the center of the ellipse at the origin, then the distance formula and the definition of an ellipse can be used to obtain the following result. (See Exercise 56.)

Equation of an Ellipse

The ellipse centered at the origin, crossing the x-axis at a and $-a$ and the y-axis at b and $-b$ has equation

$$\frac{x^2}{a^2} + \frac{y^2}{b^2} = 1.$$

For an equation to represent an ellipse, the coefficients of x^2 and y^2 must be different positive numbers.

EXAMPLE 1   Graph $4x^2 + 9y^2 = 36$.

To get the form of the equation of an ellipse, divide both sides by 36.

$$\frac{x^2}{9} + \frac{y^2}{4} = 1$$

This ellipse is centered at the origin, crosses the x-axis at 3 and -3, and crosses the y-axis at 2 and -2. Additional ordered pairs which satisfy the equation of the ellipse may be found and plotted as needed (a calculator with a square root key will be helpful). The final graph is shown in Figure 4.40.   ●

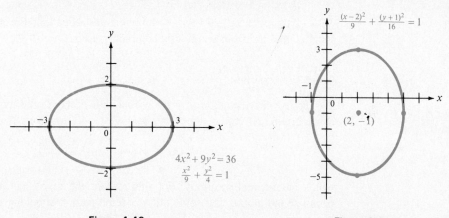

Figure 4.40   **Figure 4.41**

Just as a circle need not have its center at the origin, an ellipse may also be translated away from the origin.

EXAMPLE 2   Graph $\dfrac{(x - 2)^2}{9} + \dfrac{(y + 1)^2}{16} = 1$.

This equation represents an ellipse centered at $(2, -1)$. Graph the ellipse using the fact that $a = 3$ and $b = 4$. Start at $(2, -1)$ and locate two points each 3 units away from $(2, -1)$ on a horizontal line, one to the right of $(2, -1)$ and one to the left. Locate two other points on a vertical line through $(2, -1)$, one 4 units up and one 4 units down. Since $b > a$, the vertices are on the vertical line through the center. The vertices are $(2, 3)$ and $(2, -5)$. Find additional points as necessary. The final graph is shown in Figure 4.41.   ●

Hyperbolas An ellipse was defined as the set of all points in a plane having the sum of the distances from two fixed points as a constant. A **hyperbola** is the set of all points in a plane for which the *difference* of the distances from two fixed points (called **foci**) is constant.

As with ellipses, the equation of a hyperbola can be found from the distance formula and the definition of a hyperbola. (See Exercise 57.) Doing this leads to the following results.

Equations of a Hyperbola

> A hyperbola centered at the origin and crossing the x-axis at a and $-a$ has an equation of the form
>
> $$\frac{x^2}{a^2} - \frac{y^2}{b^2} = 1,$$
>
> while a hyperbola centered at the origin and crossing the y-axis at b and $-b$ has an equation of the form
>
> $$\frac{y^2}{b^2} - \frac{x^2}{a^2} = 1.$$

Some texts use $y^2/a^2 - x^2/b^2 = 1$ for this last equation. For a brief introduction such as this, the form we give is commonly used.

EXAMPLE 3 Graph the hyperbola $\dfrac{x^2}{16} - \dfrac{y^2}{9} = 1$.

By the first equation of a hyperbola shown above, the hyperbola is centered at the origin and crosses the x-axis at 4 and -4. However, if $x = 0$,

$$-\frac{y^2}{9} = 1 \qquad \text{or} \qquad y^2 = -9,$$

which has no real solutions. For this reason, the graph has no y-intercepts. To complete the graph, find some other ordered pairs that belong to it. For example, if $x = 6$,

$$\frac{6^2}{16} - \frac{y^2}{9} = 1$$

$$-\frac{y^2}{9} = 1 - \frac{36}{16}$$

$$\frac{y^2}{9} = \frac{20}{16}$$

$$y^2 = \frac{180}{16} = \frac{45}{4}$$

$$y = \pm \frac{3\sqrt{5}}{2} \approx \pm 3.4.$$

The graph includes the points (6, 3.4) and (6, −3.4). Also, if $x = -6$, we would still have $y \approx \pm 3.4$ with the points (−6, 3.4) and (−6, −3.4) also on the graph. These points, along with other points on the graph, were used to help sketch the final graph shown in Figure 4.42.   •

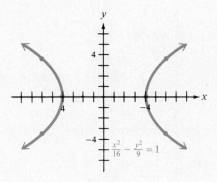

Figure 4.42

Starting with

$$\frac{x^2}{a^2} - \frac{y^2}{b^2} = 1$$

and solving for y gives

$$\frac{x^2}{a^2} - 1 = \frac{y^2}{b^2}$$

$$\frac{x^2 - a^2}{a^2} = \frac{y^2}{b^2}$$

$$y = \pm \frac{b}{a}\sqrt{x^2 - a^2}. \qquad \qquad \textbf{(1)}$$

If x^2 is very large in comparison to a^2, the difference $x^2 - a^2$ would be very close to x^2. If this happens, then the points satisfying equation (1) above would be very close to one of the lines

$$y = \pm \frac{b}{a} x.$$

Thus, as $|x|$ gets larger and larger, the points of the hyperbola $x^2/a^2 - y^2/b^2 = 1$ come closer and closer to the lines $y = (\pm b/a)x$. These lines, called the **asymptotes** of the hyperbola, are very helpful when graphing the hyperbola.

EXAMPLE 4

Graph $\dfrac{x^2}{25} - \dfrac{y^2}{49} = 1$.

For this hyperbola, $a = 5$ and $b = 7$. With these values, $y = (\pm b/a)x$ becomes $y = (\pm 7/5)x$. If $x = 5$, then $y = (\pm 7/5)(5) = \pm 7$, while $x = -5$ also gives $y = \pm 7$. These four points, (5, 7), (5, −7), (−5, 7), and (−5, −7), lead to the rectangle shown in Figure 4.42. The extended diagonals of this rectangle are the asymptotes of the hyperbola. The hyperbola crosses the x-axis at 5 and -5. The final graph is shown in Figure 4.43. ●

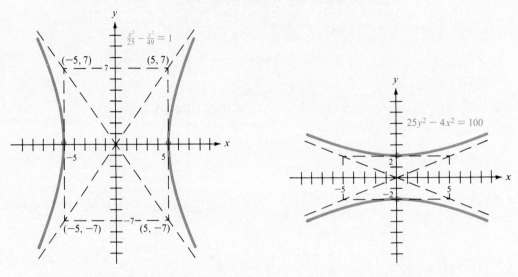

Figure 4.43   Figure 4.44

The rectangle used to graph the hyperbola in Example 4 is called the **fundamental rectangle**. It turns out that for a given hyperbola, the fundamental rectangle and the equations of the asymptotes can be found by replacing the 1 with 0 in either of the general equations of a hyperbola.

EXAMPLE 5   Graph $25y^2 - 4x^2 = 100$.
                 Divide each side by 100 to get

$$\frac{y^2}{4} - \frac{x^2}{25} = 1.$$

This hyperbola is centered at the origin, has foci on the y-axis, and crosses the y-axis at 2 and -2. To find the asymptotes, replace 1 with 0, getting

$$\frac{y^2}{4} - \frac{x^2}{25} = 0$$

$$\frac{y^2}{4} = \frac{x^2}{25}$$

$$y^2 = \frac{4x^2}{25}$$

$$y = \pm\frac{2}{5}x.$$

Use the points $(5, 2)$, $(5, -2)$, $(-5, 2)$, and $(-5, -2)$ to get the fundamental rectangle shown in Figure 4.44. Use the diagonals of this rectangle to determine the asymptotes for the graph, as shown in Figure 4.44.   ●

EXAMPLE 6   Graph $\dfrac{(y + 2)^2}{9} - \dfrac{(x + 3)^2}{4} = 1.$

This equation represents a hyperbola centered at $(-3, -2)$. See Figure 4.45 below.   ●

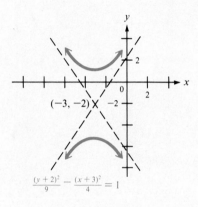

Figure 4.45

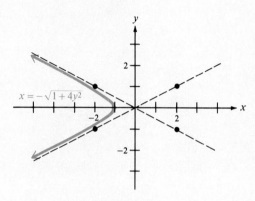

Figure 4.46

EXAMPLE 7   Graph $x = -\sqrt{1 + 4y^2}.$

Squaring both sides gives

$$x^2 = 1 + 4y^2$$

or $$x^2 - 4y^2 = 1.$$

Use the fact that $4 = 1/(1/4)$, and replace 1 with 0 to get the equations of the asymptotes:

$$x^2 - \frac{y^2}{\dfrac{1}{4}} = 0$$

$$\frac{1}{4}x^2 = y^2$$

$$y = \pm\frac{1}{2}x.$$

Since the given equation $x = -\sqrt{1 + 4y^2}$ restricts x to nonpositive values, the graph is the left branch of a hyperbola, as shown in Figure 4.46.   ●

Conic Sections   The graphs of the second-degree equations we have studied, parabolas, hyperbolas, ellipses, and circles, are called **conic sections** since each can be obtained by cutting a cone with a plane, as shown in Figure 4.47.

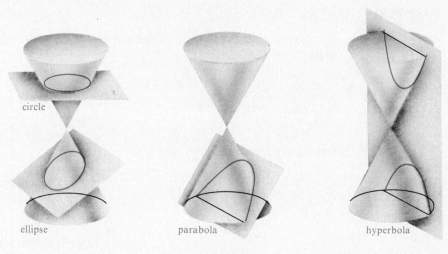

circle

ellipse   parabola   hyperbola

Figure 4.47

It turns out that all conic sections of the types we have studied have equations of the form

$$Ax^2 + Bx + Cy^2 + Dy + E = 0,$$

where either A or C must be nonzero. The special characteristics of each of the conic sections are summarized below.

**Equations of
Conic Sections**

Conic section	Characteristic	Example
Parabola	Either $A = 0$ or $C = 0$, but not both	$x^2 = y + 4$ $(y - 2)^2 = -(x + 3)$
Circle	$A = C \neq 0$	$x^2 + y^2 = 16$
Ellipse	$A \neq C,\, AC > 0$	$\dfrac{x^2}{16} + \dfrac{y^2}{25} = 1$
Hyperbola	$AC < 0$	$x^2 - y^2 = 1$

The chart on the facing page summarizes our work with conic sections.

Equation	Graph	Description	Identification
$y = a(x - h)^2 + k$	parabola	Opens upward if $a > 0$, downward if $a < 0$. Vertex is at (h, k).	x^2 term y is not squared.
$x = a(y - k)^2 + h$	parabola	Opens to right if $a > 0$, to left if $a < 0$. Vertex is at (h, k).	y^2 term x is not squared.
$(x - h)^2 + (y - k)^2 = r^2$	circle	Center is at (h, k), radius is r.	x^2 and y^2 terms have the same positive coefficient.
$\dfrac{x^2}{a^2} + \dfrac{y^2}{b^2} = 1$	ellipse	x-intercepts are a and $-a$. y-intercepts are b and $-b$.	x^2 and y^2 terms have different positive coefficients.
$\dfrac{x^2}{a^2} - \dfrac{y^2}{b^2} = 1$	hyperbola	x-intercepts are a and $-a$. Asymptotes found from (a, b), $(a, -b)$, $(-a, -b)$, and $(-a, b)$.	x^2 has a positive coefficient. y^2 has a negative coefficient.
$\dfrac{y^2}{b^2} - \dfrac{x^2}{a^2} = 1$	hyperbola	y-intercepts are b and $-b$. Asymptotes found from (a, b), $(a, -b)$, $(-a, -b)$, and $(-a, b)$.	y^2 has a positive coefficient. x^2 has a negative coefficient.

4.6 EXERCISES

Sketch the graph of each of the following ellipses or hyperbolas. See Examples 1–6.

1. $\dfrac{x^2}{9} + \dfrac{y^2}{4} = 1$ **2.** $\dfrac{x^2}{16} + \dfrac{y^2}{36} = 1$ **3.** $\dfrac{x^2}{9} + y^2 = 1$ **4.** $\dfrac{y^2}{16} - \dfrac{x^2}{9} = 1$

5. $\dfrac{x^2}{6} + \dfrac{y^2}{9} = 1$ **6.** $\dfrac{x^2}{8} - \dfrac{y^2}{12} = 1$ **7.** $x^2 + 4y^2 = 16$ **8.** $25x^2 + 9y^2 = 225$

9. $x^2 = 9 + y^2$ **10.** $y^2 = 16 + x^2$ **11.** $2x^2 + y^2 = 8$ **12.** $9x^2 - 25y^2 = 225$

13. $25x^2 - 4y^2 = -100$ **14.** $4x^2 - y^2 = -16$ **15.** $\dfrac{x^2}{1/9} + \dfrac{y^2}{1/16} = 1$ **16.** $\dfrac{x^2}{4/25} + \dfrac{y^2}{9/49} = 1$

17. $\dfrac{64x^2}{9} + \dfrac{25y^2}{36} = 1$ **18.** $\dfrac{121x^2}{25} + \dfrac{16y^2}{9} = 1$

19. $\dfrac{(x-1)^2}{9} + \dfrac{(y+3)^2}{25} = 1$ **20.** $\dfrac{(x+3)^2}{16} + \dfrac{(y-2)^2}{36} = 1$

21. $\dfrac{(x-3)^2}{16} - \dfrac{(y+2)^2}{49} = 1$ **22.** $\dfrac{(y-5)^2}{4} - \dfrac{(x+1)^2}{9} = 1$

23. $\dfrac{(y+1)^2}{25} - \dfrac{(x-3)^2}{36} = 1$ **24.** $\dfrac{(x+2)^2}{16} - \dfrac{(y+2)^2}{25} = 1$

Sketch the graph of each of the following equations. See Example 7.

25. $\dfrac{x}{4} = \sqrt{1 - \dfrac{y^2}{9}}$ **26.** $\dfrac{y}{2} = \sqrt{1 - \dfrac{x^2}{25}}$ **27.** $\dfrac{y}{3} = \sqrt{1 + \dfrac{x^2}{16}}$

28. $x = \sqrt{1 + \dfrac{y^2}{36}}$ **29.** $x = -\sqrt{1 - \dfrac{y^2}{64}}$ **30.** $y = \sqrt{1 - \dfrac{x^2}{100}}$

31. $y = -\sqrt{1 + \dfrac{x^2}{25}}$ **32.** $x = -\sqrt{1 + \dfrac{y^2}{9}}$

Identify each of the following conic sections. Draw a graph of each that has a graph.

33. $x^2 = 25 + y^2$ **34.** $x^2 = 25 - y^2$

35. $9x^2 + 36y^2 = 36$ **36.** $x^2 = 4y - 8$

37. $\dfrac{(x+3)^2}{16} + \dfrac{(y-2)^2}{16} = 1$ **38.** $\dfrac{(x-4)}{8} - \dfrac{(y+1)^2}{2} = 0$

39. $y^2 - 4y = x + 4$ **40.** $11 - 3x = 2y^2 - 8y$

41. $(x+7)^2 + (y-5)^2 + 4 = 0$ **42.** $4(x-3)^2 + 3(y+4)^2 = 0$

43. $3x^2 + 6x + 3y^2 - 12y = 12$ **44.** $2x^2 - 8x + 5y^2 + 20y = 12$

45. $x^2 - 6x + y = 0$ **46.** $x - 4y^2 - 8y = 0$

47. $4x^2 - 8x - y^2 - 6y = 6$ **48.** $x^2 + 2x = y^2 - 4y - 2$

49. $4x^2 - 8x + 9y^2 + 54y = -84$ **50.** $3x^2 + 12x + 3y^2 = -11$

51. $6x^2 - 12x + 6y^2 - 18y + 25 = 0$ **52.** $4x^2 - 24x + 5y^2 + 10y + 41 = 0$

Work the following word problems.

53. Draftspeople often use the method shown on the sketch below to draw an ellipse. Explain why the method works.

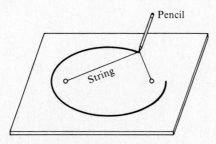

54. Ships and planes often use a location finding system called LORAN. With this system, a radio transmitter at M on the figure sends out a series of pulses. When each pulse is received at transmitter S, it then sends out a pulse. A ship at P receives pulses from both M and S. A receiver on the ship measures the difference in the arrival times of the pulses. The nagivator then consults a special map, showing certain curves according to the differences in arrival times. In this way, the ship can be located as lying on a portion of which curve?

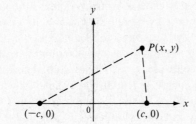

55. Microphones are placed at points $(-c, 0)$ and $(c, 0)$. An explosion occurs at point $P(x, y)$ having positive *x*-coordinate. (See the figure above.) The sound is detected at the closer microphone *t* sec before being detected at the farther microphone. Assume that sound travels at a speed of 330 m per sec, and show that P must be on the hyperbola

$$\frac{x^2}{330^2t^2} - \frac{y^2}{4c^2 - 330^2t^2} = \frac{1}{4}.$$

Ellipses.
Parabola
Hyperbola
Circle

56. Suppose that $(c, 0)$ and $(-c, 0)$ are the foci of an ellipse. Suppose that the sum of the distances from any point (x, y) of the ellipse to the two foci is $2a$. See the accompanying figure.

(a) Use the distance formula to show that the equation of the resulting ellipse is

$$\frac{x^2}{a^2} + \frac{y^2}{a^2 - c^2} = 1.$$

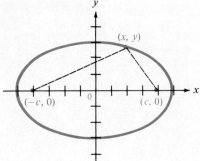

(b) Show that a and $-a$ are the x-intercepts.

(c) Let $b^2 = a^2 - c^2$, and show that b and $-b$ are the y-intercepts.

57. Suppose a hyperbola has center at the origin, foci at $F'(-c, 0)$ and $F(c, 0)$, and the value $d(P, F') - d(P, F) = 2a$. Let $b^2 = c^2 - a^2$, and show that the equation of the hyperbola is

$$\frac{x^2}{a^2} - \frac{y^2}{b^2} = 1.$$

58. The orbit of Mars is an ellipse. The sun is at one focus. An approximate equation for its orbit is

$$\frac{x^2}{5013} + \frac{y^2}{4970} = 1,$$

where x and y are measured in millions of miles. Find the widest possible distance across this ellipse.

4.7 Inequalities

Many mathematical descriptions of real situations are best expressed as inequalities rather than equalities. For example, a firm might be able to use a machine *no more* than 12 hours a day, while a production of *at least* 500 cases of a certain product might be required to meet a contract. Perhaps the simplest way to see the solution of an inequality in two variables is to draw its graph.

A line divides a plane into three sets of points: the points of the line itself and the points belonging to the two regions determined by the line. Each of these two regions is called a **half-plane**. In Figure 4.48, line r divides the plane into three different sets of points: line r, half-plane P and half-plane Q. The points of r belong neither to P nor to Q. Line r is the boundary of each half-plane.

A **linear inequality in two variables** is an inequality of the form

$$Ax + By \leq C,$$

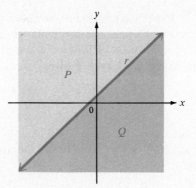

Figure 4.48

where A, B, and C are real numbers, with A and B not both equal to 0. (We could replace $\leq$ with $\geq$, $<$, or $>$.) The graph of a linear inequality turns out to be made up of a half-plane, perhaps with its boundary. For example, to graph the linear inequality $3x - 2y \leq 6$, first graph the boundary, $3x - 2y = 6$, as shown in Figure 4.49.

Since the points of the line $3x - 2y = 6$ satisfy $3x - 2y \leq 6$, this line is part of the solution. To decide which half-plane (the one above the line $3x - 2y = 6$ or the one below the line) is part of the solution, solve the original inequality for y.

$$3x - 2y \leq 6$$
$$-2y \leq -3x + 6$$
$$y \geq \frac{3}{2}x - 3 \qquad \text{Change } \leq \text{ to } \geq$$

For a particular value of x, the inequality will be satisfied by all values of y which are *greater than* or equal to $(3/2)x - 3$. This means that the solution includes the half-plane *above* the line, as well as the line itself. See Figure 4.50.

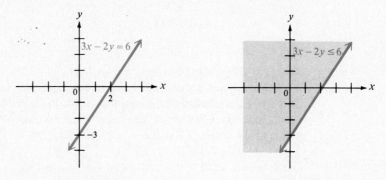

Figure 4.49   **Figure 4.50**

There is an alternative method for deciding which side of the boundary line to shade. Choose as a test point any point not on the graph of the equation. The origin, (0, 0), is often a good test point. Substituting 0 for x and 0 for y in the inequality $3x - 2y \leq 6$ gives

$$3(0) - 2(0) \leq 6$$
$$0 \leq 6,$$

a true statement. Since (0, 0) leads to a true result, it is part of the solution of the inequality. Shade the side of the graph containing (0, 0), as we did in Figure 4.50.

EXAMPLE 1

Graph $x + 4y > 4$.

The boundary here is the straight line $x + 4y = 4$. Since the points on this line do not satisfy $x + 4y > 4$, it is customary to graph the boundary as a dashed line, as in Figure 4.51. To decide which half-plane satisfies the inequality, use a test point. Choosing (0, 0) as a test point gives $0 + 4 \cdot 0 > 4$, or $0 > 4$, a false statement. Since (0, 0) leads to a false statement, shade the side of the graph *not* containing (0, 0), as in Figure 4.51. ●

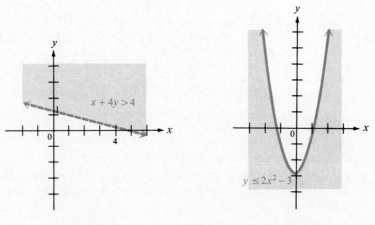

Figure 4.51   **Figure 4.52**

EXAMPLE 2

Graph $y \leq 2x^2 - 3$.

First graph the boundary, the parabola $y = 2x^2 - 3$, as shown in Figure 4.52. Then select any test point not on the parabola, such as (0, 0). Since (0, 0) does not satisfy the original inequality, shade the portion of the graph which does not include (0, 0), as shown in Figure 4.52. Because of the $=$ portion of $\leq$, the points of the parabola itself also belong to the graph. ●

EXAMPLE 3

Graph $y < |x - 2|$.

Begin by graphing the boundary, $y = |x - 2|$. (Graph the boundary as a dashed line.) To determine which region belongs to the graph, select any point not on the boundary line, such as $(5, -3)$. Since $-3 < |5 - 2|$, shade the side of the graph including $(5, -3)$, as shown in Figure 4.53. •

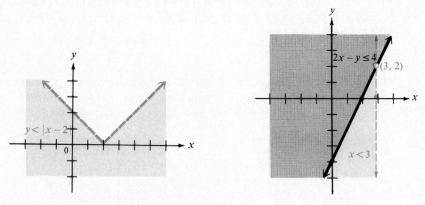

Figure 4.53   Figure 4.54

EXAMPLE 4

Graph the intersection of $2x - y \leq 4$ and $x < 3$.

Figure 4.54 shows the graphs of both inequalities. The heavily shaded area represents the intersection of the two regions. The point of intersection of the boundary lines, $(3, 2)$, belongs to the first inequality, but not the second. For this reason, $(3, 2)$ does not belong to the intersection. This is shown with an open dot on the graph at $(3, 2)$. •

EXAMPLE 5

Graph $y \leq \sqrt{1 + 4x^2}$, $y \geq 0$.

Squaring both sides of $y = \sqrt{1 + 4x^2}$ gives

$$y^2 = 1 + 4x^2 \quad \text{or} \quad y^2 - 4x^2 = 1,$$

a hyperbola with y-intercepts 1 and -1. Since $\sqrt{1 + 4x^2}$ is always nonnegative, the graph is only half a hyperbola, as shown in Figure 4.54.

Select any point not on the hyperbola and try it in the original inequality. Since $(0, 0)$ satisfies this inequality, shade the side of the hyperbola containing $(0, 0)$. Because $y \geq 0$, the shading stops at the x-axis, which is a boundary of the graph, as shown in Figure 4.55. •

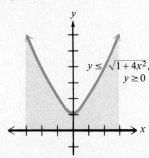

Figure 4.55

EXAMPLE 6 Graph $16x^2 + y^2 \leq 16$.

This inequality has as its boundary the ellipse $16x^2 + y^2 = 16$, graphed in Figure 4.56. To decide which region to shade, choose any point off the ellipse, such as $(0, 0)$. Substituting 0 for x and 0 for y in the inequality $16x^2 + y^2 \leq 16$ gives

$$16 \cdot 0^2 + 0^2 \leq 16 \qquad \text{or} \qquad 0 \leq 16,$$

a true statement. This true statement shows that the solution includes the region containing $(0, 0)$, which is the region inside the ellipse. The final result is made up of this region and the boundary ellipse, as shown in Figure 4.56. ●

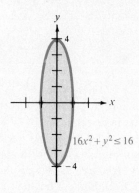

Figure 4.56

4.7 EXERCISES

Graph each of the following inequalities. See Examples 1, 2, 3, and 6.

1. $x \leq 3$
2. $y \leq -2$
3. $x + 2y \leq 6$
4. $x - y \geq 2$

5. $2x + 3y \geq 4$
6. $4y - 3x < 5$
7. $3x - 5y > 6$
8. $x < 3 + 2y$

9. $5x \leq 4y - 2$
10. $2x > 3 - 4y$
11. $y \leq |x|$
12. $y \geq |x + 2|$

13. $x > |y - 3|$
14. $x < |y + 1|$
15. $y < 3x^2 + 2$
16. $y \leq x^2 - 4$

17. $y > (x - 1)^2 + 2$
18. $y > 2(x + 3)^2 - 1$

19. $x^2 + (y + 3)^2 \leq 16$
20. $(x - 4)^2 + (y + 3)^2 \leq 9$

21. $4x^2 \leq 4 - y^2$
22. $x^2 + 9y^2 > 9$

23. $9x^2 - 16y^2 > 144$
24. $4x^2 \leq 36 + 9y^2$

Graph the intersection of each pair of inequalities. See Examples 4 and 5.

25. $x - 3y < 4$ and $x \leq 0$
26. $2x + y > 5$ and $y \geq 0$

27. $3x + 2y \geq 6$ and $y \leq 2$
28. $4x + 3y \leq 6$ and $x \leq -2$

29. $x + 2y < 4$ and $3x - y > 5$
30. $4x + 1 < 2y$ and $-x > y - 3$

31. $2x + 3y < 6$ and $x - 5y \geq 10$
32. $3x - 4y > 6$ and $2x + 3y > 4$

33. $y \leq \sqrt{x + 3}$ and $y \geq 0$
34. $x \leq \sqrt{1 - y}$ and $x \geq 0$

35. $y \leq \sqrt{49 - x^2}$ and $y \geq 0$

36. $x > \sqrt{16 - y^2}$ and $x \geq 0$

37. $\dfrac{x}{6} < \sqrt{1 - \dfrac{y^2}{121}}$ and $x \geq 0$

38. $\dfrac{y}{4} \leq \sqrt{1 + \dfrac{x^2}{25}}$ and $y \geq 0$

39. $x^2 + y^2 \leq 49$ and $x^2 + y^2 \geq 36$

40. $x^2 + 4y^2 \geq 16$ and $4x^2 + y^2 \leq 36$

Chapter 4   Summary

Key Words		
	function	origin
	x-axis	y-axis
	rectangular coordinate system	Cartesian coordinate system
	coordinate plane	xy-plane
	quadrants	coordinates
	ordered pair	distance formula
	midpoint formula	graph
	parabola	vertex, vertices
	axis	translation
	circle	symmetry
	center-radius form	radius
	standard form	linear equation
	y-intercept	x-intercept
	change in x	slope
	slope-intercept form	change in y
	focus, foci	point-slope form
	ellipse	directrix
	center	hyperbola
	asymptotes	major diameter
	conic sections	fundamental rectangle
	linear inequality in two variables	half-plane

Distance Formula

Suppose $P(x_1, y_1)$ and $R(x_2, y_2)$ are two points in a coordinate plane. Then the distance between P and R, written $d(P, R)$, is

$$d(P, R) = \sqrt{(x_2 - x_1)^2 + (y_2 - y_1)^2}.$$

Midpoint Formula

The midpoint of the line segment with endpoints (x_1, y_1) and (x_2, y_2) is

$$\left(\frac{x_1 + x_2}{2}, \frac{y_1 + y_2}{2} \right).$$

Slope

The slope m of the line through (x_1, y_1) and (x_2, y_2) is

$$m = \frac{\Delta y}{\Delta x} = \frac{y_2 - y_1}{x_2 - x_1}.$$

Point-Slope Form

The line with slope m passing through the point (x_1, y_1) has an equation

$$y - y_1 = m(x - x_1),$$

the point-slope form of the equation of a line.

Equations of Conic Sections

Parabola $y = a(x - h)^2 + k$
$x = a(y - k)^2 + h$

Circle $(x - h)^2 + (y - k)^2 = r^2$

Ellipse $\dfrac{x^2}{a^2} + \dfrac{y^2}{b^2} = 1$

Hyperbola $\dfrac{x^2}{a^2} - \dfrac{y^2}{b^2} = 1$ or $\dfrac{y^2}{b^2} - \dfrac{x^2}{a^2} = 1$

Chapter 4   Review Exercises

Find the distance between the following pairs of points.

1. $P(3, -1)$, $Q(-4, 5)$ **2.** $M(-8, 2)$, $N(3, -7)$ **3.** $A(-6, 3)$, $B(-6, 8)$

4. Are the points $(5, 7)$, $(3, 9)$, $(6, 8)$ the vertices of a right triangle?

5. Find all possible values of k so that $(-1, 2)$, $(-10, 5)$, and $(-4, k)$ are the vertices of a right triangle.

6. Find all possible values of x so that the distance between $(x, -9)$ and $(3, -5)$ is 6.

7. Find all points (x, y) with $x = 6$ so that (x, y) is 4 units from $(1, 3)$.

8. Find all points (x, y) with $x + y = 0$ so that (x, y) is 6 units from $(-2, 3)$.

Graph each of the following equations.

9. $x + y = 4$ **10.** $3x - 5y = 20$ **11.** $y = \dfrac{1}{2}x^2$ **12.** $y = 3 - x^2$

13. $y = -\dfrac{8}{x}$ **14.** $y = -2x^3$ **15.** $y = |x| - 3$ **16.** $y = |4 - x|$

Find equations for each of the circles below.

17. Center $(-2, 3)$, radius 5   **18.** Center $(\sqrt{5}, -\sqrt{7})$, radius $\sqrt{3}$

19. Center $(-8, 1)$, passing through $(0, 16)$ **20.** Center $(3, -6)$, tangent to the x-axis

Find the center and radius of each of the following that are circles.

21. $x^2 - 4x + y^2 + 6y + 12 = 0$ **22.** $x^2 - 6x + y^2 - 10y + 30 = 0$

23. $x^2 + 7x + y^2 + 3y + 1 = 0$ **24.** $x^2 + 11x + y^2 - 5y + 46 = 0$

Decide whether the equations below have graphs that are symmetric with respect to the
x-axis, the y-axis, or the origin.

25. $3y^2 - 5x^2 = 15$
26. $x + y^2 = 8$
27. $y^3 = x + 1$
28. $x^2 = y^3$

29. $|y| = -x$
30. $|x + 2| = |y - 3|$
31. $|x| = |y|$
32. $xy = 8$

Find the slope for each of the following lines that has a slope.

33. Through $(8, 7)$ and $(1/2, -2)$
34. Through $(2, -2)$ and $(3, -4)$

35. Through $(5, 6)$ and $(5, -2)$
36. Through $(0, -7)$ and $(3, -7)$

37. $9x - 4y = 2$
38. $11x + 2y = 3$
39. $x - 5y = 0$
40. $x - 2 = 0$

41. $y + 6 = 0$
42. $y = x$

Graph each of the following equations.

43. $3x + 7y = 14$
44. $2x - 5y = 5$
45. $3y = x$
46. $y = 3$

47. $x = -5$
48. $y = x$

For each of the following lines, write the equation in standard form.

49. Through $(-2, 4)$ and $(1, 3)$
50. Through $(-2/3, -1)$ and $(0, 4)$

51. Through $(3, -5)$ with slope -2
52. Through $(-4, 4)$ with slope $3/2$

53. Through $(1/5, 1/3)$ with slope $-1/2$
54. x-intercept -3, y-intercept 5

55. No x-intercept, y-intercept $3/4$
56. Through $(2, -1)$, parallel to $3x - y = 1$

57. Through $(0, 5)$, perpendicular to $8x + 5y = 3$

58. Through $(2, -10)$, perpendicular to a line with undefined slope

59. Through $(3, -5)$, parallel to $y = 4$

60. Through $(-7, 4)$, perpendicular to $y = 8$

Graph each of the following lines.

61. Through $(2, -4)$, $m = 3/4$
62. Through $(0, 5)$, $m = -2/3$

63. Through $(-4, 1)$, $m = 3$
64. Through $(-3, -2)$, $m = -1$

Graph each of the following equations.

65. $y = x^2 - 4$
66. $y = 6 - x^2$
67. $y = 3(x + 1)^2 - 5$

68. $y = -\dfrac{1}{4}(x - 2)^2 + 3$
69. $y = x^2 - 4x + 2$
70. $y = -3x^2 - 12x - 1$

Give the vertex and axis of the graph of each of the following equations.

71. $y = -(x + 3)^2 - 9$
72. $y = (x - 7)^2 + 3$
73. $y = x^2 - 7x + 2$

74. $y = -x^2 - 4x + 1$
75. $y = -3x^2 - 6x + 1$
76. $y = 4x^2 - 4x + 3$

Use parabolas to work each of the following problems.

77. Find two numbers whose sum is 11 and whose product is a maximum.

78. Find two numbers having a sum of 40 such that the sum of the square of one and twice the square of the other is a minimum.

79. Find the rectangular region of maximum area that can be enclosed with 180 m of fencing.

80. Find the rectangular region of maximum area that can be enclosed with 180 m of fencing if no fencing is needed along one side of the region.

Graph each parabola. Give the vertex and axis of each.

81. $x = y^2 - 2$

82. $x = (y + 3)^2$

83. $x = -(y + 1)^2 + 2$

84. $x = y^2 - 4y$

85. $x = -y^2 + 5y + 1$

86. $x = -3y^2 + 6y - 2$

Graph each of the following equations.

87. $y = \sqrt{x + 1}$

88. $y = -\sqrt{3 - x}$

89. $y = -\sqrt{9 - x^2}$

90. $x = \sqrt{36 - y^2}$

91. $\dfrac{x^2}{25} + \dfrac{y^2}{4} = 1$

92. $\dfrac{x^2}{3} + \dfrac{y^2}{16} = 1$

93. $\dfrac{x^2}{4} - \dfrac{y^2}{9} = 1$

94. $\dfrac{y^2}{100} - \dfrac{x^2}{25} = 1$

95. $x^2 = 16 + y^2$

96. $4x^2 + 25y^2 = 100$

97. $\dfrac{25x^2}{9} + \dfrac{4y^2}{25} = 1$

98. $\dfrac{100x^2}{49} + \dfrac{9y^2}{16} = 1$

99. $\dfrac{(x - 2)^2}{9} + \dfrac{(y + 3)^2}{4} = 1$

100. $\dfrac{(x + 1)^2}{16} - \dfrac{(y - 2)^2}{4} = 1$

101. $\dfrac{x}{3} = \sqrt{1 - \dfrac{y^2}{16}}$

102. $x = -\sqrt{1 - \dfrac{y^2}{36}}$

103. $y = -\sqrt{1 - \dfrac{x^2}{25}}$

104. $y = -\sqrt{1 + x^2}$

Graph each inequality.

105. $5x - y \leq 20$

106. $3x - 5y > 10$

107. $y \leq |x - 4|$

108. $y < (x - 1)^2 + 3$

109. $25y^2 - 36x^2 > 900$

110. $(x - 2)^2 + (y + 3)^2 \geq 9$

Graph the intersection of each pair of inequalities.

111. $2x - y \geq 4$ and $y \geq -2$

112. $3x + 7y < 21$ and $x - 4y > 4$

113. $y \leq x^2$ and $x + y \geq 4$

114. $x^2 + y^2 \leq 100$ and $25x^2 + 4y^2 \geq 100$

115. $\dfrac{x^2}{9} + \dfrac{y^2}{4} \geq 1$ and $\dfrac{x^2}{16} + y^2 \leq 1$

116. $y \leq \sqrt{25 - x^2}$ and $x \geq 2$ and $y \geq 0$

5

Functions

Many things in daily life are related. For example, the grade you get in a course is related to the amount of time you spend studying, while the number of miles per gallon you get on a trip depends on the speed you drive. Driving at 55 mph might produce 31 miles per gallon, while speeding at 65 mph might lower this to 28 miles per gallon. In this last example, the speed driven is called the *independent variable,* while the number of miles per gallon that depends on this speed is the *dependent variable*.

In practical applications, the most useful of these relationships are those with a given value of the independent variable leading to exactly one value of the dependent variable. Such relationships, called *functions,* are the subject of this chapter.

5.1 Functions

Suppose a group of students get together each Monday evening to study algebra (and perhaps watch football). To each member of this set of students we could associate a number giving the student's weight to the nearest kilogram. Since each student has only one weight at a given time, the relationship between students and their weights is a function.

Function

> A **function** is a correspondence or rule that assigns to each element of one set (called the **domain** of the function) exactly one element from another set (called the **range** of the function).

In the example of the students, the domain is the set of all the students studying algebra, while the range is the set of all possible weights of the students.

In most mathematical applications of functions, the rule showing the relationship between the elements in the domain and those in the range is given by an equation. For example,

$$y = 5x - 11$$

is a function. To each value of the independent variable x the equation $y = 5x - 11$ produces exactly one value of the dependent variable y. (These equations need not involve only x and y as variables; any appropriate letters may be used. In physics, for example, t is often used to represent the value of the independent variable *time*.)

It is common to use the letters f, g, and h to name functions. If f is a function and x is an element in the domain of f, then $f(x)$, read "f of x," is the corresponding element in the range. For example, if f is the function where a value of x is squared to find the corresponding value in the range, f could be expressed as

$$f(x) = x^2.$$

If $x = -5$ is an element from the domain of f, the corresponding element from the range is found by replacing x with -5:

$$f(-5) = (-5)^2 = 25.$$

The number -5 in the domain corresponds to 25 in the range.

EXAMPLE 1　Let $g(x) = 3\sqrt{x}$ and $h(x) = 1 + 4x$. Find each of the following.

(a) $g(16)$

To find $g(16)$, replace x in $g(x) = 3\sqrt{x}$ with 16, getting

$$g(16) = 3\sqrt{16} = 3 \cdot 4 = 12.$$

(b) $h(-3) = 1 + 4(-3) = -11$

(c) $g(-4)$ does not exist; -4 is not in the domain of g since $\sqrt{-4}$ is not a real number.

(d) $h(\pi) = 1 + 4\pi$

(e) $g(m) = 3\sqrt{m}$, if m represents a nonnegative real number.　●

EXAMPLE 2　Let $f(x) = 2x^2 - 3x$, and find the quotient

$$\frac{f(x + h) - f(x)}{h},$$

which is important in calculus.

To find $f(x + h)$, replace x with $x + h$, to get

$$\frac{f(x + h) - f(x)}{h} = \frac{2(x + h)^2 - 3(x + h) - (2x^2 - 3x)}{h}$$

$$= \frac{2(x^2 + 2xh + h^2) - 3x - 3h - 2x^2 + 3x}{h}$$

$$= \frac{2x^2 + 4xh + 2h^2 - 3x - 3h - 2x^2 + 3x}{h}$$

$$= \frac{4xh + 2h^2 - 3h}{h}$$

$$= 4x + 2h - 3. \quad \bullet$$

Throughout this book, if the domain for a function specified by an algebraic formula is not given, it will be assumed to be the largest possible set of real numbers for which the formula is meaningful. For example, if

$$f(x) = \frac{-4x}{2x - 3},$$

then any real number can be used for x except $x = 3/2$, which makes the denominator equal to 0. Based on our assumption, the domain of this function must be $\{x | x \neq 3/2\}$.

EXAMPLE 3   Give the domain and range of each of the following functions.

(a) $f(x) = 4 - 3x$

Here x can take on any value at all, as can $f(x)$, making both the domain and range the same, the set of all real numbers. In interval notation, both the domain and range are $(-\infty, +\infty)$.

(b) $f(x) = x^2 + 4$

Since any real number can be squared, the domain is the set of all real numbers, $(-\infty, +\infty)$. The square of any number is nonnegative, so that $x^2 \geq 0$ and $x^2 + 4 \geq 4$. The range is the interval $[4, +\infty)$.

(c) $g(x) = \sqrt{x - 2}$

For $\sqrt{x - 2}$ to be a real number, we must have

$$x - 2 \geq 0 \quad \text{or} \quad x \geq 2,$$

with the domain of the function given by the interval $[2, +\infty)$. Since $\sqrt{x - 2}$ must be nonnegative, the range is $[0, +\infty)$.

(d) $h(x) = -\sqrt{100 - x^2}$

For $-\sqrt{100 - x^2}$ to be a real number,

$$100 - x^2 \geq 0.$$

Since $100 - x^2$ factors as $(10 - x)(10 + x)$, use a sign graph to verify that

$$-10 \leq x \leq 10,$$

making the domain of the function $[-10, 10]$. For any value of x in the domain,

$$-10 \leq -\sqrt{100 - x^2} \leq 0,$$

so the range is $[-10, 0]$. ●

For most of the functions that we use, the domain can be found with the methods we have available. The range, however, often must be found by using graphing, complicated algebra, or calculus.

Suppose a function f is given by $f(x) = -3x + 2$. To emphasize that this statement is used to find values in the range of f, it is common to write

$$y = -3x + 2.$$

When a function is written in the form $y = f(x)$, x is the **independent variable,** and y is the **dependent variable.**

For each value of x in the domain of a function f, there is a corresponding value $y = f(x)$ in the range. This correspondence of x and y produces a set of ordered pairs,

$$\{(x, y) \mid y = f(x), \ x \text{ is in the domain of } f\}.$$

The graph of this set of ordered pairs produces the **graph** of the function.

There is a quick way to tell if a given graph is the graph of a function or not. Figure 5.1 shows two graphs. In the graph of part (a), any value of x that we might choose would lead to only one value of y, making this graph the graph of a function. On the other hand, the graph in part (b) is not the graph of a function. For example, if we choose $x = x_1$, the vertical line shows that there are two values of y. We summarize this as the **vertical line test for a function.**

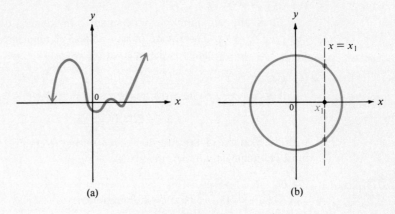

(a)   (b)

Figure 5.1

Vertical Line Test

> If each vertical line cuts a graph in no more than one point, the graph is the graph of a function.

5.1 EXERCISES

For each of the following, find the indicated function values: (a) $f(-2)$, (b) $f(0)$, (c) $f(1)$, and (d) $f(4)$.

1.

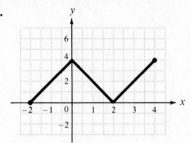

2.

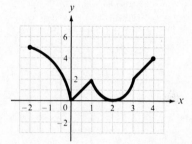

3.

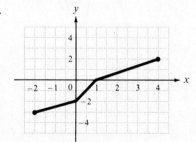

4.

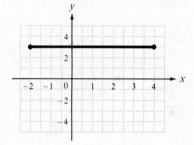

Let $f(x) = 3x - 1$ and $g(x) = x^2$. Find each of the following function values. See Example 1.

5. $f(0)$	**6.** $f(-1)$	**7.** $f(-3)$	**8.** $f(4)$
9. $g(2)$	**10.** $g(0)$	**11.** $f(a)$	**12.** $g(b)$
13. $f(1) + g(1)$	**14.** $f(-2) - g(-2)$	**15.** $f(3) \cdot g(3)$	**16.** $\dfrac{g(-1)}{f(-1)}$
17. $f(-2m)$	**18.** $f(-11p)$	**19.** $f(5a - 2)$	**20.** $f(3 + 2k)$
21. $g(5p - 2)$	**22.** $g(-6k + 1)$	**23.** $f(p) + g(2p)$	**24.** $g(1 - r) - f(1 - r)$
25. $g(m) \cdot f(m)$	**26.** $\dfrac{f(5r)}{g(r)}$		

Let $f(x) = -4.6x^2 - 8.9x + 1.3$. Find each of the following function values.

27. $f(3)$	**28.** $f(-5)$	**29.** $f(-4.2)$	**30.** $f(-1.8)$

Give the domain and range of each of the following functions. Give only the domain in Exercises 43 and 44. See Example 3.

31. $f(x) = 2x - 1$

32. $g(x) = 3x + 5$

33. $g(x) = x^4$

34. $h(x) = (x - 2)^2$

35. $f(x) = \sqrt{8 + x}$

36. $f(x) = -\sqrt{x + 6}$

37. $h(x) = \sqrt{16 - x^2}$

38. $m(x) = \sqrt{x^2 - 25}$

39. $k(x) = (x - 4)^{1/2}$

40. $f(x) = (3x + 2)^{1/2}$

41. $g(x) = \sqrt{\dfrac{1}{x^2 + 25}}$

42. $z(x) = -\sqrt{\dfrac{4}{x^2 + 1}}$

43. $g(x) = \dfrac{2}{x^2 - 3x + 2}$

44. $h(x) = \dfrac{-4}{x^2 + 5x + 4}$

45. $r(x) = \sqrt{x^2 - 4x - 5}$

46. $r(x) = \sqrt{x^2 + 7x + 10}$

47. $f(x) = |x - 4|$

48. $k(x) = -|2x - 7|$

49. $g(x) = -\sqrt{2x + 5}$

50. $h(x) = \sqrt{1 - x}$

51. $f(x) = \sqrt{36 - x^2}$

52. $g(x) = -\sqrt{25 - x^2}$

53. $k(x) = -\sqrt{x^2 - 9}$

54. $z(x) = \sqrt{x^2 - 100}$

55.

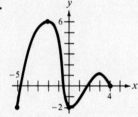

56.

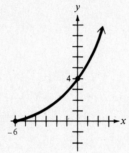

57.

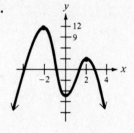

58.

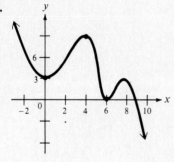

59.

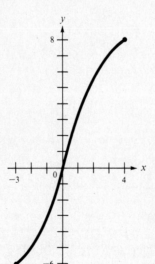

60.

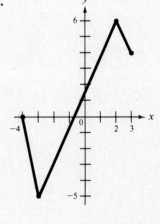

Let $f(x) = 2^x$ for all rational numbers x. Find each of the following function values. Assume that m and r represent rational numbers.

61. $f(1)$ **62.** $f(2)$ **63.** $f(4)$ **64.** $f(-1)$

65. $f(-2)$ **66.** $f(-4)$ **67.** $f(m)$ **68.** $f(-5r)$

69. $f(1/2)$ **70.** $f(1/4)$

*For each of the following functions, find (a) $f(x + h)$,   (b) $f(x + h) - f(x)$,   and
(c) $\dfrac{f(x + h) - f(x)}{h}$. See Example 2.*

71. $f(x) = x^2 - 4$ **72.** $f(x) = 8 - 3x^2$ **73.** $f(x) = 6x + 2$ **74.** $f(x) = 4x + 11$

75. $f(x) = 2x^3 + x^2$ **76.** $f(x) = -4x^3 - 8x$

Functional notation is very common in applications. **Break-even analysis** *provides one example. For example, suppose a firm that produces chicken feed finds that the total cost, $C(x)$, of producing x units is given by*

$$C(x) = 20x + 100,$$

while the revenue from the sale of x units is given by $R(x)$, where

$$R(x) = 24x.$$

The firm will break even (no profit and no loss) as long as revenue just equals cost.

77. Find x so that $R(x) = C(x)$. This value of x gives the *break-even point*.

78. Graph R and C on the same axis. Identify the break-even point. Identify any regions of profit or loss.

You are the manager of a firm. You are considering the manufacture of a new product, so you ask the accounting department to produce cost estimates and the sales department to produce sales estimates. After you receive the data, you must decide whether or not to go ahead with production of the new product. Analyze the following data (find a break-even point) and then decide what you would do. See Exercises 77 and 78.

79. $C(x) = 85x + 900$; $R(x) = 105x$; not more than 38 units can be sold.

80. $C(x) = 105x + 6000$; $R(x) = 250x$; not more that 400 units can be sold.

81. $C(x) = 70x + 500$; $R(x) = 60x$.

82. $C(x) = 950x + 1250$; $R(x) = 950x$.

5.2 Some Useful Functions

Many of the graphs used in Chapter 4 are the graphs of functions. By the vertical line test, any straight line that is not vertical is the graph of a function, as is the graph of any vertical parabola. Because of this, we can define *linear* and *quadratic functions* as follows.

Linear and
Quadratic
Functions

> For real numbers a, b, and c, the function
>
> $$f(x) = ax + b$$
>
> is a **linear function,** while the function
>
> $$f(x) = ax^2 + bx + c, \quad a \neq 0,$$
>
> is a **quadratic function.**

EXAMPLE 1   Graph each function.

(a) $f(x) = -4x + 5$

This linear function has x-intercept 5/4 and y-intercept 5. The graph is shown in Figure 5.2.

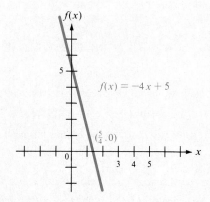

Figure 5.2

(b) $f(x) = -(x - 2)^2 + 5$

The graph of this quadratic function, a parabola with vertex at (2, 5) and opening downward, is shown in Figure 5.3. ●

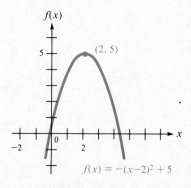

Figure 5.3

In the rest of this section and in the next two sections we shall discuss some useful functions that are not linear and not quadratic.

One common function that is not linear and not quadratic is the **absolute value function,** $f(x) = |x|$, or $y = |x|$. Since we can find $|x|$ for any real number x, the domain is the set of all real numbers. Also, $|x| \geq 0$ for any real number x, so the range is $[0, +\infty)$. When $x \geq 0$, then $y = |x| = x$, so $y = x$ is graphed for nonnegative values of x. On the other hand, if $x < 0$, then $y = |x| = -x$, and $y = -x$ is graphed for negative values of x. The final graph is shown in Figure 5.4. By the vertical line test, the graph is that of a function.

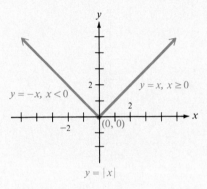

Figure 5.4

EXAMPLE 2

Graph $y = |x - 2|$.

Again, the domain is the set of all real numbers, and the range is $[0, +\infty)$. If $x = 2$, then $y = 0$. This point $(2, 0)$ is the lowest point on the graph since $|x - 2| > 0$ for all other values of x. By graphing the point $(2, 0)$ and a few other selected points, we get the graph in Figure 5.5. This graph has the same shape as that of $y = |x|$, but the "vertex" point is translated 2 units to the right, from $(0, 0)$ to $(2, 0)$. •

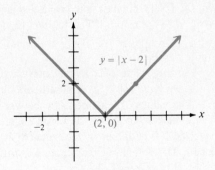

Figure 5.5

EXAMPLE 3   Graph $y = |3x + 4|$.

Plotting a few ordered pairs leads to the graph in Figure 5.6. As the graph shows, the "vertex" is translated to $(-4/3, 0)$. Also, the two parts of the graph form a smaller angle than that of the graph of $y = |x|$.   ●

The graphs of the absolute value functions of Examples 2 and 3 are made up of portions of two different straight lines. Such functions, called **functions defined piecewise,** are often defined with different equations for different parts of the domain.

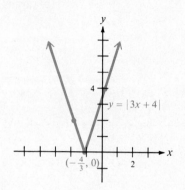

Figure 5.6

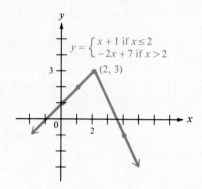

Figure 5.7

EXAMPLE 4   Graph the function

$$y = \begin{cases} x + 1 & \text{if} & x \le 2 \\ -2x + 7 & \text{if} & x > 2. \end{cases}$$

For $x \le 2$, graph $y = x + 1$. For $x > 2$, graph $y = -2x + 7$, as shown in Figure 5.7.   ●

The **greatest integer function,** written $y = [x]$, is defined by saying that $[x]$ is the greatest integer less than or equal to x. For example, $[8] = 8$, $[-5] = -5$, $[\pi] = 3$, $[12\ 1/9] = 12$, $[-2.001] = -3$, and so on.

EXAMPLE 5   Graph $y = [x]$.

For any value of x in the interval $[0, 1)$, we have $[x] = 0$. Also, for x in $[1, 2)$, we have $[x] = 1$. This process continues; for x in $[2, 3)$, the value of $[x]$ is 2. The values of y are constant between integers, but jump at integer values of x. This makes the graph, shown in Figure 5.8, a series of line segments. In each case, the left endpoint of the segment is included, and the right endpoint is excluded. The domain of the function is the set of all real numbers, while the range is the set of integers.   ●

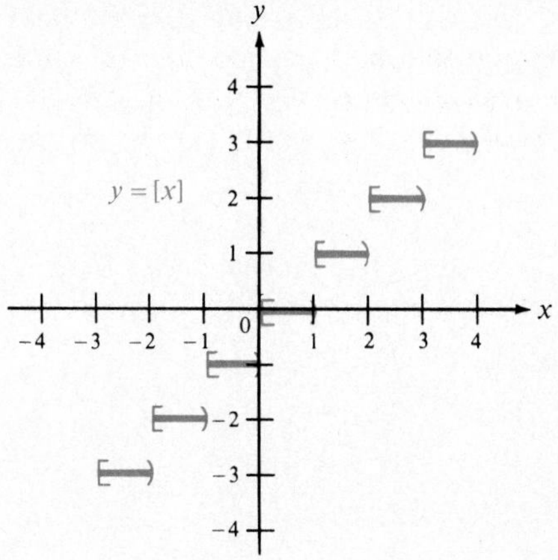

Figure 5.8

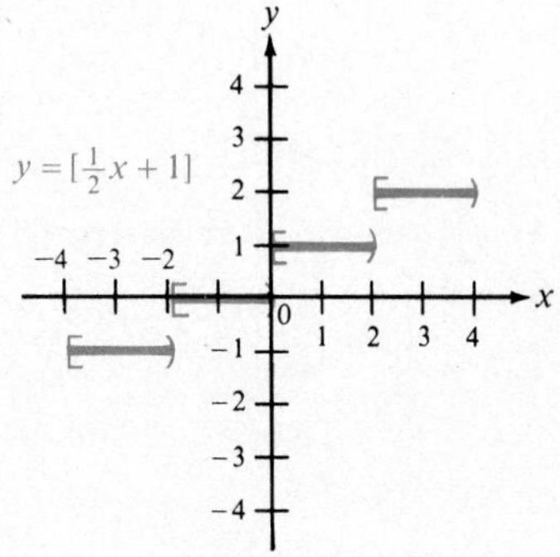

Figure 5.9

Each of the graphs in Examples 5, 6, and 7 is made up of a series of horizontal line segments. (See Figures 5.8, 5.9, and 5.10.) The functions producing these graphs are called **step functions.**

EXAMPLE 6   Graph $y = \left[\dfrac{1}{2}x + 1\right]$.

If x is in the interval $[0, 2)$, then $y = 1$. For x in $[2, 4)$, $y = 2$. The graph is shown in Figure 5.9.   ●

EXAMPLE 7   The Terrific Taxi Company charges \$1 plus 50¢ per mile or part of a mile. Find the cost of a trip of 5 mi; of 4.5 mi; of 9.5 mi. Graph the ordered pairs (miles, cost). Is this graph the graph of a function?

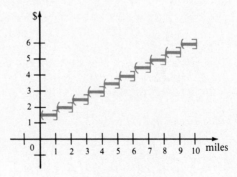

Figure 5.10

The cost for a 5-mile trip is

$$1 + 5(.50) = 3.50 \qquad \text{or} \qquad \$3.50.$$

For a 4.5-mile trip, the cost will be the same as for a 5-mile trip, $3.50. A 9.5-mile trip costs the same as a 10-mile trip, $1 + 10(.50) = 6$, or $6. The graph, a step function, is shown in Figure 5.10. The vertical line test shows that the graph is that of a function. ●

5.2 EXERCISES

Graph each of the following equations. Identify all functions. See Examples 2 and 3.

1. $y = |x + 1|$ **2.** $y = |x - 1|$ **3.** $y = |2 - x|$ **4.** $y = |-3 - x|$

5. $y = |5x + 4|$ **6.** $3y = |x + 2|$ **7.** $y + 1 = |x - 2|$ **8.** $2y - 1 = |x + 3|$

9. $x = |y|$ **10.** $x = |y + 1|$ **11.** $y = |x| + 4$ **12.** $y = 2|x| - 1$

13. $2x = 3|y| - 1$ **14.** $x - 1 = |3y + 2|$ **15.** $y = |x| - x$ **16.** $x = |y| + y$

Graph each of the following functions. See Example 4.

17. $y = \begin{cases} x - 1 & \text{if } x \le 3 \\ 2 & \text{if } x > 3 \end{cases}$ **18.** $y = \begin{cases} 6 - x & \text{if } x \le 3 \\ 3x - 6 & \text{if } x > 3 \end{cases}$

19. $y = \begin{cases} 4 - x & \text{if } x < 2 \\ 1 + 2x & \text{if } x \ge 2 \end{cases}$ **20.** $y = \begin{cases} -2 & \text{if } x \ge 1 \\ 2 & \text{if } x < 1 \end{cases}$

21. $y = \begin{cases} 2x + 1 & \text{if } x \ge 0 \\ x & \text{if } x < 0 \end{cases}$ **22.** $y = \begin{cases} 5x - 4 & \text{if } x \ge 1 \\ x & \text{if } x < 1 \end{cases}$

23. $y = \begin{cases} 2 + x & \text{if } x < -4 \\ -x & \text{if } -4 \le x \le 5 \\ 3x & \text{if } x > 5 \end{cases}$ **24.** $y = \begin{cases} -2x & \text{if } x < -3 \\ 3x + 1 & \text{if } -3 \le x < 2 \\ -4x & \text{if } x \ge 2 \end{cases}$

25. $y = \begin{cases} |x| & \text{if } x > -2 \\ x & \text{if } x \le -2 \end{cases}$ **26.** $y = \begin{cases} |x| - 1 & \text{if } x > -1 \\ x - 1 & \text{if } x \le -1 \end{cases}$

Graph each of the following functions. See Examples 5 and 6.

27. $y = [-x]$ **28.** $y = [2x]$ **29.** $y = [2x - 1]$ **30.** $y = [3x + 1]$

31. $y = [3x]$ **32.** $y = [3x] + 1$ **33.** $y = [3x] - 1$ **34.** $y = x - [x]$

Work the following exercises. See Example 7.

35. Suppose a chain-saw rental firm charges a fixed $4 sharpening fee plus $7 per day or fraction of a day. Let $S(x)$ represent the cost of renting a saw for x days. Find the value of **(a)** $S(1)$, **(b)** $S(1.25)$, **(c)** $S(3.5)$.
(d) Graph $y = S(x)$.
(e) Give the domain and range of S.

36. Assume that it costs 30¢ to mail a letter weighing one ounce or less, and then 27¢ for each additional ounce or fraction of an ounce. Let $L(x)$ be the cost of mailing a letter weighing x oz. Find **(a)** $L(.75)$, **(b)** $L(1.6)$, **(c)** $L(4)$.
(d) Graph $y = L(x)$.
(e) Give the domain and range of L.

37. Use the greatest integer function and write an expression for the number of ounces for which postage will be charged on a letter weighing x oz (See Exercise 36).

38. In a recent year Washington, D. C., taxi rates were 90¢ for the first 1/9 mi and 10¢ for each additional 1/9 mi or fraction of 1/9. Let $C(x)$ be the cost for a taxi ride of x 1/9 mi. Find **(a)** $C(1)$, **(b)** $C(2.3)$, **(c)** $C(8)$.
(d) Graph $y = C(x)$.
(e) Give the domain and range of C.

39. For a lift truck rental of no more than three days, the charge is $300. An additional charge of $75 is made for each day or portion of a day after three. Graph the ordered pairs (number of days, cost).

40. A car rental costs $37 for one day, which includes 50 free miles. Each additional 25 mi or portion costs $10. Graph the ordered pairs (miles, cost).

41. When a diabetic takes long-acting insulin, the insulin reaches its peak effect on the blood sugar level in about 3 hr. This effect remains fairly constant for 5 hr, then declines, and is very low until the next injection. In a typical patient, the level of insulin might be given by the following function.

$$i(t) = \begin{cases} 40t + 100 & \text{if } 0 \le t \le 3 \\ 220 & \text{if } 3 < t \le 8 \\ -80t + 860 & \text{if } 8 < t \le 10 \\ 60 & \text{if } 10 < t \le 24 \end{cases}$$

Here $i(t)$ is the blood sugar level, in appropriate units, at time t measured in hours from the time of the injection. Chuck takes his insulin at 6 a.m. Find the blood sugar level at each of the following times.
(a) 7 A.M.   **(b)** 9 A.M.   **(c)** 10 A.M.   **(d)** noon
(e) 2 P.M.   **(f)** 5 P.M.   **(g)** midnight   **(h)** Graph $y = i(t)$.

42. To rent a midsized car from Avis costs $27 per day or fraction of a day. If you pick up the car in Lansing and drop it in West Lafayette, there is a fixed $25 dropoff charge. Let $C(x)$ represent the cost of renting the car for x days, taking it from Lansing to West Lafayette. Find each of the following.
(a) $C(3/4)$ **(b)** $C(9/10)$ **(c)** $C(1)$
(d) $C(1\ 5/8)$ **(e)** $C(2.4)$ **(f)** Graph $y = C(x)$.
(g) Is C a function?   **(h)** Is C a linear function?

5.3 Polynomial Functions

This section discusses *polynomial functions* such as $f(x) = x^3$ or $f(x) = x^4$. More generally,

Polynomial
Function

> a **polynomial function of degree n** is a function of the form
>
> $$f(x) = a_n x^n + a_{n-1} x^{n-1} + \ldots + a_1 x + a_0,$$
>
> for real numbers a_n, a_{n-1}, . . . , a_1, and a_0, with $a_n \ne 0$.

We have already discussed examples of polynomial functions; a linear function of the form $f(x) = ax + b$, with $a \neq 0$, is a polynomial function of degree 1, while a quadratic function $f(x) = ax^2 + bx + c$ is a polynomial function of degree 2.

In this section we shall graph polynomial functions of degree 3 or more. (This section introduces the idea of graphing these functions, and Chapter 9 provides more detail.) We start by graphing polynomial functions of the form $f(x) = x^n$.

EXAMPLE 1 Graph each function.

(a) $f(x) = x^3$

Choose several values for x, and find the corresponding values of $f(x)$, or y, as shown in the table in Figure 5.11. Plot the resulting ordered pairs and connect the points with a smooth curve. The graph of $f(x) = x^3$ is shown as a solid line in Figure 5.11.

(b) $f(x) = x^5$

Work as in part (a) of this example to get the graph shown as a dashed line in Figure 5.11. Both the graph of $f(x) = x^3$ and the graph of $f(x) = x^5$ are symmetric with respect to the origin. (See the discussion of odd and even functions at the end of this section.)

(c) The graphs of $f(x) = x^4$ and $f(x) = x^6$ are shown in Figure 5.12. These graphs have symmetry about the y-axis, as does the graph of $f(x) = ax^2$ for a nonzero real number a. ●

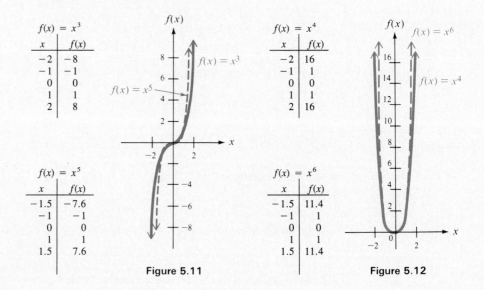

$f(x) = x^3$

x	$f(x)$
-2	-8
-1	-1
0	0
1	1
2	8

$f(x) = x^5$

x	$f(x)$
-1.5	-7.6
-1	-1
0	0
1	1
1.5	7.6

Figure 5.11

$f(x) = x^4$

x	$f(x)$
-2	16
-1	1
0	0
1	1
2	16

$f(x) = x^6$

x	$f(x)$
-1.5	11.4
-1	1
0	0
1	1
1.5	11.4

Figure 5.12

As we saw in Chapter 4 with the graph of $f(x) = ax^2$, the value of a in $f(x) = ax^n$ affects the width of the graph. When $|a| > 1$, the graph is narrower than the graph of $f(x) = x^n$; when $0 < |a| < 1$, the graph is broader. When $a < 0$, the graph of $f(x) = ax^n$ is reflected about the x-axis as compared to the graph of $f(x) = x^n$.

EXAMPLE 2   Graph each of the following functions.

(a) $f(x) = \dfrac{1}{2}x^3$

The graph will be broader than that of $f(x) = x^3$, but will have the same general shape. It includes the points $(-2, -4)$, $(-1, -1/2)$, $(0, 0)$, $(1, 1/2)$, and $(2, 4)$. See Figure 5.13.

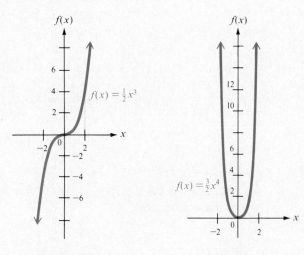

Figure 5.13   Figure 5.14

(b) $f(x) = \dfrac{3}{2}x^4$

The following table gives some ordered pairs.

x	$f(x)$
-2	24
-1	3/2
0	0
1	3/2
2	24

The graph is shown in Figure 5.14. This graph is narrower than that of $f(x) = x^4$. ●

EXAMPLE 3   Graph each of the following.

(a) $f(x) = x^5 - 2$

The graph will be the same as that of $f(x) = x^5$, but translated down 2 units. See Figure 5.15.

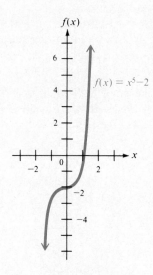

Figure 5.15

(b) $f(x) = (x + 1)^6$

This function has a graph like that of $f(x) = x^6$, but translated 1 unit to the left, as shown in Figure 5.16.

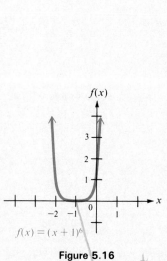

Figure 5.16

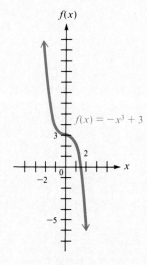

Figure 5.17

(c) $f(x) = -x^3 + 3$

The negative sign causes the graph to be reflected about the x-axis compared to the graph of $f(x) = x^3$. As shown in Figure 5.17, the graph is also translated up 3 units.   ●

Generalizing from the graphs in Example 3, the domain of a polynomial function is the set of all real numbers. The range of a polynomial function of odd degree is also the set of all real numbers. Some typical graphs of polynomial functions of odd degree are shown in Figure 5.18. These graphs suggest that for every polynomial function f of odd degree there is at least one real value of x that makes $f(x) = 0$. Such values of x are called the **real zeros** of f; these values are also the x-intercepts of the graph.

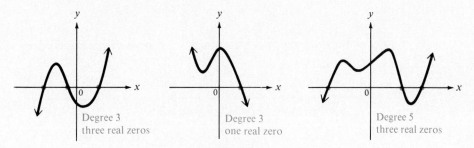

Degree 3
three real zeros

Degree 3
one real zero

Degree 5
three real zeros

Figure 5.18

A polynomial function of even degree will have a range that takes the form $(-\infty, k]$ or else $[k, +\infty)$ for some real number k. Figure 5.19 shows two typical graphs of polynomial functions of even degree.

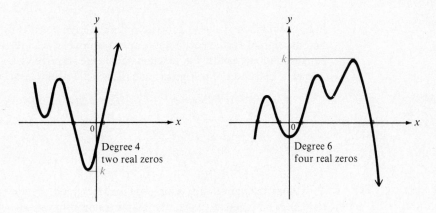

Degree 4
two real zeros

Degree 6
four real zeros

Figure 5.19

It is difficult to graph most polynomial functions without the use of calculus. A large number of points must be plotted to get a reasonably accurate graph. However, if a polynomial function can be factored into linear factors, we can approximate its graph without plotting very many points; this method is shown in the next examples.

EXAMPLE 4   Graph $f(x) = (2x + 3)(x - 1)(x + 2)$.

Multiplying out the expression on the right would show that f is a third degree polynomial, also called a **cubic** polynomial. To sketch the graph of f, first find its zeros by setting each of the three factors equal to 0 and solving the resulting equations.

$$2x + 3 = 0 \qquad \text{or} \qquad x - 1 = 0 \qquad \text{or} \qquad x + 2 = 0$$

$$x = -\frac{3}{2} \qquad\qquad x = 1 \qquad\qquad x = -2$$

The three zeros, $-3/2$, 1, and -2, divide the x-axis into four regions:

$$x < -2, \qquad -2 < x < -\frac{3}{2}, \qquad -\frac{3}{2} < x < 1, \qquad \text{and} \qquad 1 < x.$$

The regions are shown in Figure 5.20.

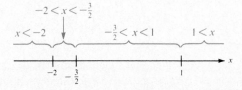

Figure 5.20

In any region, the values of $f(x)$ are either always positive or always negative. To find the sign of $f(x)$ in each region, select a value of x in the region and determine by substitution whether the function values are positive or negative in that region. A typical selection of test points and the results of the tests are shown below.

Region	Test point	Sign of $f(x)$
$x < -2$	-3	Negative
$-2 < x < -3/2$	$-7/4$	Positive
$-3/2 < x < 1$	0	Negative
$1 < x$	2	Positive

When the values of $f(x)$ are negative, the graph is below the x-axis, and when $f(x)$ takes on positive values, the graph is above the x-axis. All these results suggest that the graph looks something like the sketch in Figure 5.21. The sketch could be improved by plotting additional points in each region.   ●

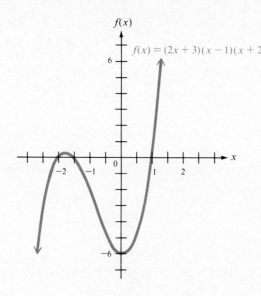

Figure 5.21

Figure 5.22

EXAMPLE 5   Sketch the graph of $f(x) = 3x^4 + x^3 - 2x^2$.

The polynomial can be factored as follows:

$$3x^4 + x^3 - 2x^2 = x^2(3x^2 + x - 2)$$
$$= x^2(3x - 2)(x + 1).$$

The zeros, 0, 2/3, and -1, divide the x-axis into four regions:

$$x < -1, \qquad -1 < x < 0, \qquad 0 < x < 2/3, \qquad \text{and} \qquad 2/3 < x.$$

Determine the sign of $f(x)$ in each region by substituting a value for x from each region to get the following information.

Region	Sign of f(x)	Location relative to axis
$x < -1$	Positive	Above
$-1 < x < 0$	Negative	Below
$0 < x < 2/3$	Negative	Below
$2/3 < x$	Positive	Above

With the values of x used for the test points and the corresponding values of y, sketch the graph as shown in Figure 5.22.   •

Odd and Even Functions   A function $y = f(x)$ with the property that $f(-x) = f(x)$ for every value of x in its domain is called an **even function,** while $y = f(x)$ is an **odd function** if $f(-x) = -f(x)$ for every x in its domain.

EXAMPLE 6 Decide whether the given functions are odd, even, or neither.

(a) $f(x) = 6x^4 - 8x^2 + 1$

Replacing x with $-x$ gives

$$f(-x) = 6(-x)^4 - 8(-x)^2 + 1 = 6x^4 - 8x^2 + 1 = f(x).$$

This function is even.

(b) $f(x) = -25x^5 + 4x^3 + 3x$

Here

$$f(-x) = -25(-x)^5 + 4(-x)^3 + 3(-x)$$
$$= 25x^5 - 4x^3 - 3x$$
$$= -f(x).$$

The function is odd.

(c) $f(x) = 10x - 3$

Since $f(-x) = -10x - 3$, which equals neither $f(x)$ nor $-f(x)$, this function is neither even nor odd. ●

The idea of odd and even functions is useful in graphing because of the following properties, which are based on the tests for symmetry in Chapter 4.

The graph of an even function is symmetric with respect to the y-axis.

The graph of an odd function is symmetric with respect to the origin.

To see that these statements are reasonable, look at the graphs of the even functions $f(x) = x^4$ and $f(x) = x^6$, and the graphs of the odd functions $f(x) = x^3$ and $f(x) = x^5$ at the beginning of this section.

5.3 EXERCISES

Each of the following polynomial functions is symmetric about a line or a point. For each function, (a) sketch the graph and (b) give the line or point of symmetry. See Examples 1–3.

1. $f(x) = \dfrac{1}{4}x^6$ **2.** $f(x) = -\dfrac{2}{3}x^5$ **3.** $f(x) = \dfrac{-5}{4}x^5$ **4.** $f(x) = 2x^4$

5. $f(x) = \dfrac{1}{2}x^3 + 1$ **6.** $f(x) = -x^4 + 2$ **7.** $f(x) = -(x + 1)^3$ **8.** $f(x) = \dfrac{1}{3}(x + 3)^4$

9. $f(x) = (x - 1)^4 + 2$ **10.** $f(x) = (x + 2)^3 - 1$

Graph each of the following polynomial functions. See Examples 4 and 5.

11. $f(x) = 2x(x - 3)(x + 2)$ **12.** $f(x) = x^2(x + 1)(x - 1)$

13. $f(x) = x^2(x - 2)(x + 3)^2$

14. $f(x) = x^2(x - 5)(x + 3)(x - 1)$

15. $f(x) = x^3 - x^2 - 2x$

16. $f(x) = -x^3 - 4x^2 - 3x$

17. $f(x) = 3x^4 + 5x^3 - 2x^2$

18. $f(x) = 4x^3 + 2x^2 - 12x$

19. $f(x) = 2x^3(x^2 - 4)(x - 1)$

20. $f(x) = 5x^2(x^3 - 1)(x + 2)$

21. $f(x) = x^4 - 4x^2$

22. $f(x) = -x^4 + 16x^2$

23. $f(x) = x^2(x - 3)^3(x + 1)$

24. $f(x) = x(x - 4)^2(x + 2)^2$

25. The polynomial function

$$A(x) = -.015x^3 + 1.058x$$

gives the approximate alcohol concentration (in tenths of a percent) in an average person's bloodstream x hr after drinking about 8 oz of 100 proof whiskey. The function is approximately valid for x in the interval [0, 8].

(a) Graph $A(x)$.

(b) Using the graph you drew for part (a), estimate the time of maximum alcohol concentration.

(c) In one state, a person is legally drunk if the blood alcohol concentration exceeds .15%. Use the graph of part (a) to estimate the period in which this average person is legally drunk.

26. A technique for measuring cardiac output depends on the concentration of a dye in the bloodstream after a known amount is injected into a vein near the heart. For a normal heart, the concentration of the dye in the bloodstream at time x (in sec) is given by the function

$$g(x) = -.006x^4 + .140x^3 - .053x^2 + 1.79x.$$

Graph $g(x)$.

27. The pressure of the oil in a reservoir tends to drop with time. By taking sample pressure readings, for a particular oil reservoir, petroleum engineers have found that the change in pressure is given by

$$P(t) = t^3 - 25t^2 + 200t,$$

where t is time in years from the date of the first reading.

(a) Graph $P(t)$.

(b) For what time period is the change in pressure (drop) increasing? decreasing?

28. During the early part of the twentieth century, the deer population of the Kaibab Plateau in Arizona experienced a rapid increase, because hunters had reduced the number of natural predators. The increase in population depleted the food resources and eventually caused the population to decline. For the period from 1905 to 1930, the deer population was approximated by

$$D(x) = -.125x^5 + 3.125x^4 + 4000,$$

where x is time in years from 1905.

(a) Use a calculator to find enough points to graph $D(x)$.

(b) From the graph, over what period of time (from 1905 to 1930) was the population increasing? relatively stable? decreasing?

Decide whether the following polynomials are odd, even, or neither. See Example 6.

29. $f(x) = 2x^3$ **30.** $f(x) = -4x^5$ **31.** $f(x) = 0.2x^4$

32. $f(x) = -x^6$ **33.** $f(x) = -x^5$ **34.** $f(x) = (x - 1)^3$

35. $f(x) = 2x^3 + 3$ **36.** $f(x) = 4x^3 - x$ **37.** $f(x) = x^4 + 3x^2 + 5$

38. $f(x) = 11x + 2$ **39.** $f(x) = (x + 5)^2$ **40.** $f(x) = (3x - 5)^2$

We can find approximate maximum or minimum values of polynomial functions for given intervals by first evaluating the function at the left endpoint of the given interval. Then add 0.1 to the value of x and reevaluate the polynomial. Keep doing this until the right endpoint of the interval is reached. Then identify the approximate maximum and minimum value for the polynomial on the interval.

41. $y = x^3 + 4x^2 - 8x - 8$, $[-3.8, -3]$ **42.** $y = x^3 + 4x^2 - 8x - 8$, $[0.3, 1]$

43. $y = 2x^3 - 5x^2 - x + 1$, $[-1, 0]$ **44.** $y = 2x^3 - 5x^2 - x + 1$, $[1.4, 2]$

45. $y = x^4 - 7x^3 + 13x^2 + 6x - 28$, $[-2, -1]$ **46.** $y = x^4 - 7x^3 + 13x^2 + 6x - 28$, $[2, 3]$

47. The graphs of $f(x) = x$, $f(x) = x^3$, and $f(x) = x^5$ all pass through the points $(-1, -1)$, $(0, 0)$, and $(1, 1)$. Draw a graph showing these three functions for the domain $[-1, 1]$. Use your result to decide how the graph of $f(x) = x^{11}$ would look for the same interval.

48. The graphs of $f(x) = x^2$, $f(x) = x^4$, and $f(x) = x^6$ all pass through the points $(-1, 1)$, $(0, 0)$, and $(1, 1)$. Draw a graph showing these three functions for the domain $[-1, 1]$. Use your result to decide how the graph of $f(x) = x^{12}$ would look for the same interval.

5.4   Rational Functions

A function of the form

$$f(x) = \frac{p(x)}{q(x)},$$

where $p(x)$ and $q(x)$ are polynomial functions, is called a **rational function.** Since any values of x such that $q(x) = 0$ are excluded from the domain, a rational function usually has a graph that has one or more breaks in it.

The simplest rational function with a variable denominator is

$$f(x) = \frac{1}{x}.$$

The domain of this function is the set of all real numbers except 0. To graph the function, first replace x with $-x$, getting $f(-x) = 1/(-x) = -1/x = -f(x)$, showing that f is an odd function. As we saw in the previous section, this means that the graph of f is symmetric with respect to the origin.

The number 0 cannot be used as a value of x, but it is helpful to find the values of $f(x)$ for typical values of x close to 0. The following table shows what happens to $f(x)$ as x gets closer and closer to 0 from either side.

x approaches 0

x	-1	$-.1$	$-.01$	$-.001$	.001	.01	.1	1
$f(x)$	-1	-10	-100	-1000	1000	100	10	1

$|f(x)|$ gets larger and larger

The table suggests that $|f(x)|$ gets larger and larger as x gets closer and closer to 0, written in symbols as

$$|f(x)| \rightarrow \infty \text{ as } x \rightarrow 0.$$

(The symbol $x \rightarrow 0$ means that x approaches as close as desired to 0, without necessarily ever being equal to 0.) Since x cannot equal 0, the graph of $f(x) = 1/x$ will never intersect the vertical line $x = 0$. This line is called a *vertical asymptote*.

On the other hand, as $|x|$ gets larger and larger, the values of $f(x) = 1/x$ get closer and closer to 0. (See the table.)

x	$-10,000$	-1000	-100	-10	10	100	1000	10,000
$f(x)$	$-.0001$	$-.001$	$-.01$	$-.1$	.1	.01	.001	.0001

Letting $|x|$ get larger and larger without bound (written $|x| \rightarrow \infty$) causes the graph of $y = 1/x$ to move closer and closer to the horizontal line $y = 0$. This line is called a *horizontal asymptote*.

As we said, $f(x) = 1/x$ is an odd function, symmetric with respect to the origin. Choosing some positive values of x and finding the corresponding values of $f(x)$ gives the first quadrant part of the graph shown in Figure 5.23. The other part of the graph (in the third quadrant) can be found by symmetry.

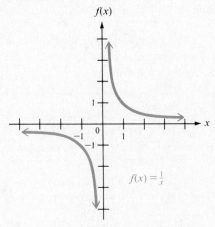

Figure 5.23

EXAMPLE 1   Graph $f(x) = \dfrac{-2}{x}$.

The function can be written as

$$f(x) = -2 \cdot \frac{1}{x}.$$

Compared to $f(x) = 1/x$, the graph will be reflected about the x-axis (because of the negative sign), and each point will be twice as far from the x-axis. See Figure 5.24. ●

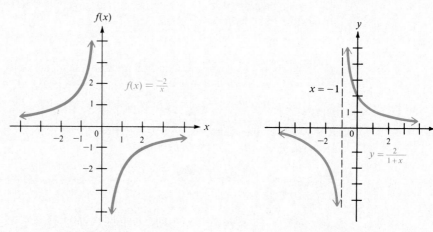

Figure 5.24   Figure 5.25

EXAMPLE 2   Graph $f(x) = \dfrac{2}{1 + x}$.

The domain of this function is the set of all real numbers except -1. As shown in Figure 5.25, the graph is that of $f(x) = 1/x$, translated 1 unit to the left, with each y-value doubled. ●

The examples above suggest the following definition of vertical and horizontal asymptotes. Here we assume that the rational function $f(x) = p(x)/q(x)$ is written in lowest terms.

Asymptotes

> For the rational function $y = f(x)$,
>
> if $|f(x)| \to \infty$ as $x \to a$, then the line $x = a$ is a **vertical asymptote**;
> if $f(x) \to a$ as $|x| \to \infty$, then the line $y = a$ is a **horizontal asymptote**.

EXAMPLE 3   Graph $f(x) = \dfrac{1}{(x - 1)(x + 4)}$.

This graph is not symmetric with respect to the x-axis, the y-axis, or the origin. The denominator is 0 when x is replaced by either 1 or -4, and the lines $x = 1$ and $x = -4$ are vertical asymptotes.

As $|x|$ gets larger and larger, $|(x - 1)(x + 4)|$ also gets larger and larger, bringing y closer and closer to 0. This means that the x-axis is a horizontal asymptote. The vertical asymptotes divide the x-axis into three regions. It is often convenient to consider each region separately. Finding the y-intercept by letting $x = 0$ and plotting a few points gives the result shown in Figure 5.26. •

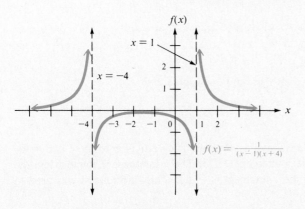

Figure 5.26

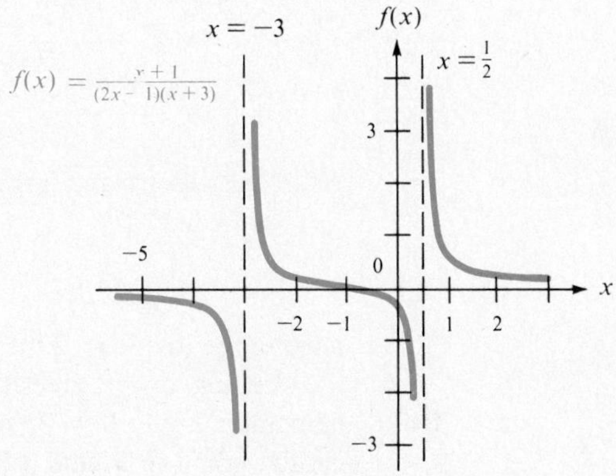

Figure 5.27

EXAMPLE 4   Graph $f(x) = \dfrac{x + 1}{(2x - 1)(x + 3)}$.

The graph has as vertical asymptotes the lines $x = 1/2$ and $x = -3$. Replacing x with 0 gives $-1/3$ as the y-intercept. For the x-intercept, the value of y must be 0. The only way that y can equal 0 is if the numerator, $x + 1$, is 0. This happens when $x = -1$, making -1 the x-intercept.

To find the horizontal asymptote, multiply the factors in the denominator to get

$$f(x) = \frac{x + 1}{(2x - 1)(x + 3)} = \frac{x + 1}{2x^2 + 5x - 3}.$$

Now divide each term in the numerator and denominator by x^2, since 2 is the highest exponent on x. This gives

$$f(x) = \frac{\dfrac{x}{x^2} + \dfrac{1}{x^2}}{\dfrac{2x^2}{x^2} + \dfrac{5x}{x^2} - \dfrac{3}{x^2}} = \frac{\dfrac{1}{x} + \dfrac{1}{x^2}}{2 + \dfrac{5}{x} - \dfrac{3}{x^2}}.$$

As $|x|$ gets larger and larger, the quotients $1/x$, $1/x^2$, $5/x$, and $3/x^2$ all approach 0, with the value of $f(x)$ approaching

$$\frac{0 + 0}{2 + 0 - 0} = \frac{0}{2} = 0.$$

The line $y = 0$ is therefore a horizontal asymptote. Using these asymptotes and intercepts (the graph has no symmetry with respect to either axis or the origin) and plotting a few points gives the graph in Figure 5.27. •

EXAMPLE 5   Graph $f(x) = \dfrac{2x + 1}{x - 3}$.

Work as in the previous example to find any horizontal asymptote. Since the highest exponent on x is 1, divide each term by x.

$$f(x) = \frac{2x + 1}{x - 3} = \frac{\dfrac{2x}{x} + \dfrac{1}{x}}{\dfrac{x}{x} - \dfrac{3}{x}} = \frac{2 + \dfrac{1}{x}}{1 - \dfrac{3}{x}}$$

As $|x|$ gets larger and larger, both $1/x$ and $3/x$ approach 0:

$$\frac{2 + 0}{1 - 0} = \frac{2}{1} = 2,$$

making the line $y = 2$ the horizontal asymptote. The vertical asymptote is $x = 3$.

Letting $x = 0$ gives the y-intercept $-1/3$. The x-intercept is found when the numerator $2x + 1 = 0$, so the x-intercept is $-1/2$. This information was used to get the graph in Figure 5.28.   ●

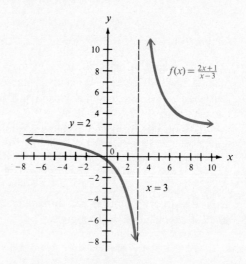

Figure 5.28

EXAMPLE 6   Graph $f(x) = \dfrac{3(x + 1)(x - 2)}{(x + 4)^2}$.

The only vertical asymptote is the line $x = -4$. To find any horizontal asymptotes, multiply the factors in the numerator and denominator.

$$f(x) = \frac{3x^2 - 3x - 6}{x^2 + 8x + 16}$$

Now divide by x^2.

$$f(x) = \frac{\dfrac{3x^2}{x^2} - \dfrac{3x}{x^2} - \dfrac{6}{x^2}}{\dfrac{x^2}{x^2} + \dfrac{8x}{x^2} + \dfrac{16}{x^2}} = \frac{3 - \dfrac{3}{x} - \dfrac{6}{x^2}}{1 + \dfrac{8}{x} + \dfrac{16}{x^2}}$$

Letting $|x|$ get larger and larger shows that $y = 3/1$, or $y = 3$, is the horizontal asymptote. The y-intercept is $-3/8$, and the x-intercepts are -1 and 2. See the final graph in Figure 5.29. ●

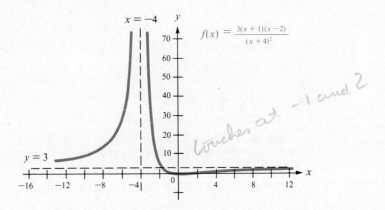

Figure 5.29

The graph of the function in Example 6 actually crosses the horizontal asymptote. While it is possible for the graph of a rational function to cross a horizontal asymptote, such a graph can never cross a vertical asymptote.

The next example shows a rational function having a numerator with higher degree than the denominator. Polynomial division must be used for these functions.

EXAMPLE 7   Graph $f(x) = \dfrac{x^2 + 1}{x - 2}$.

Dividing by x^2 as in the examples above would show that the graph has no horizontal asymptote. The vertical asymptote is the line $x = 2$. Since the numerator has higher degree than the denominator, divide to rewrite the function in another form.

$$
\begin{array}{r}
x + 2 \\
x - 2 \overline{)\,x^2 + 1} \\
\underline{x^2 - 2x } \\
2x + 1 \\
\underline{2x - 4} \\
5
\end{array}
$$

By this result,

$$f(x) = \frac{x^2 + 1}{x - 2} = x + 2 + \frac{5}{x - 2}.$$

For very large values of $|x|$, $5/(x - 2)$ is close to 0, and the graph approaches the line $y = x + 2$. This line is an **oblique asymptote** (neither vertical nor horizontal) for the function. The y-intercept is $-1/2$. There is no x-intercept. Using the asymptotes, the y-intercept, and additional points as needed leads to the graph in Figure 5.30 on the facing page. •

In general, if the degree of the numerator is exactly one more than the degree of the denominator, the rational function may have an oblique asymptote. The equation of this asymptote is found by dividing the numerator by the denominator and dropping the remainder.

To summarize: when graphing a rational function, use the following procedure. (We assume that the rational function is written in lowest terms.)

Graphing Rational Functions

1. Find any vertical asymptotes by setting the denominator equal to 0 and solving for x. If a is a zero of the denominator, then the line $x = a$ is a vertical asymptote.

2. Determine any other asymptotes. There are three possibilities:
   (a) If the numerator has lower degree than the denominator, there is a horizontal asymptote, $y = 0$.
   (b) If the numerator and denominator have the same degree, and the function is of the form

   $$f(x) = \frac{a_n x^n + \ldots + a_0}{b_n x^n + \ldots + b_0}, \quad b_n \neq 0,$$

   dividing by x^n produces the horizontal asymptote

   $$y = \frac{a_n}{b_n}.$$

   (c) If the numerator is of degree exactly one more than the denominator, there may be an oblique asymptote. To find it, divide the numerator by the denominator and drop any remainder. The rest of the quotient gives the equation of the asymptote.

3. Find any intercepts.

4. Check for symmetry.

5. Plot a few selected points—at least one in each region of the domain determined by the vertical asymptotes.

6. Complete the sketch.

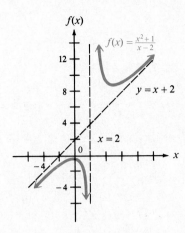

Figure 5.30

5.4 EXERCISES

Find any vertical, horizontal, or oblique asymptotes for the following. See Examples 1–7.

1. $f(x) = \dfrac{2}{x - 5}$

2. $f(x) = \dfrac{-1}{x + 2}$

3. $f(x) = \dfrac{-8}{3x - 7}$

4. $f(x) = \dfrac{5}{4x - 9}$

5. $f(x) = \dfrac{2 - x}{x + 2}$

6. $f(x) = \dfrac{x - 4}{5 - x}$

7. $f(x) = \dfrac{3x - 5}{2x + 9}$

8. $f(x) = \dfrac{4x + 3}{3x - 7}$

9. $f(x) = \dfrac{2}{x^2 - 4x + 3}$

10. $f(x) = \dfrac{-5}{x^2 - 3x - 10}$

11. $f(x) = \dfrac{x^2 - 1}{x + 3}$

12. $f(x) = \dfrac{x^2 + 4}{x - 1}$

13. $f(x) = \dfrac{(x - 3)(x + 1)}{(x + 2)(2x - 5)}$

14. $f(x) = \dfrac{3(x + 2)(x - 4)}{(5x - 1)(x - 5)}$

15. $f(x) = \dfrac{2x^2 - 32}{x^2 + 1}$

16. $f(x) = \dfrac{-5x^2 + 20}{x^2 + 9}$

17. Sketch the following graphs and compare them with the graph of $f(x) = 1/x^2$.

    (a) $f(x) = \dfrac{1}{(x - 3)^2}$

    (b) $f(x) = \dfrac{-2}{x^2}$

    (c) $f(x) = \dfrac{-2}{(x - 3)^2}$

Graph each of the following. See Examples 1–7.

18. $f(x) = \dfrac{4}{5 + 3x}$

19. $f(x) = \dfrac{1}{(x - 2)(x + 4)}$

20. $f(x) = \dfrac{3}{(x + 4)^2}$

21. $f(x) = \dfrac{3x}{(x + 1)(x - 2)}$

22. $f(x) = \dfrac{2x + 1}{(x + 2)(x + 4)}$

23. $f(x) = \dfrac{5x}{x^2 - 1}$

24. $f(x) = \dfrac{-x}{x^2 - 4}$

25. $f(x) = \dfrac{3x}{x - 1}$

26. $f(x) = \dfrac{4x}{1 - 3x}$

27. $f(x) = \dfrac{x + 1}{x - 4}$

28. $f(x) = \dfrac{x - 3}{x + 5}$

29. $f(x) = \dfrac{x - 5}{x + 3}$

30. $f(x) = \dfrac{3x}{x^2 - 16}$

31. $f(x) = \dfrac{x}{x^2 - 9}$

32. $f(x) = \dfrac{x^2 - 5}{x + 2}$

33. $f(x) = \dfrac{x^2 - 3x + 2}{x - 3}$

34. $f(x) = \dfrac{x^2 + 1}{x + 3}$

35. $f(x) = \dfrac{2x^2 + 3}{x - 4}$

36. $f(x) = \dfrac{x^2 - x}{x + 2}$

37. $f(x) = \dfrac{x^2 + 2x}{2x - 1}$

38. $f(x) = \dfrac{x(x - 2)}{(x + 3)^2}$

39. $f(x) = \dfrac{(x - 3)(x + 1)}{(x - 1)^2}$

40. $f(x) = \dfrac{(x + 4)(x - 1)}{x^2 + 1}$

41. $f(x) = \dfrac{(x - 5)(x - 2)}{x^2 + 9}$

42. $f(x) = \dfrac{1}{x^2 + 1}$

43. $f(x) = \dfrac{-9}{x^2 + 9}$

44. Suppose the average cost per unit, $C(x)$, to produce x units of margarine is given by

$$C(x) = \frac{500}{x + 30}.$$

(a) Find $C(10)$, $C(20)$, $C(50)$, $C(75)$, and $C(100)$.
(b) Would a more reasonable domain for C be $(0, +\infty)$ or $[0, +\infty)$? Why?

45. In a recent year, the cost per ton, y, to build an oil tanker of x thousand deadweight tons was approximated by

$$y = \frac{110,000}{x + 225}.$$

(a) Find y for $x = 25$, $x = 50$, $x = 100$, $x = 200$, $x = 300$, and $x = 400$.
(b) Graph the function.

46. Antique-car fans often enter their cars in a *concours d'elegance* in which a maximum of 100 points can be awarded to a particular car. Points are awarded for the general attractiveness of the car. The function

$$C(x) = \frac{10x}{49(101 - x)}$$

expresses the cost, in thousands of dollars, of restoring a car so that it will win x points. Graph the function.

47. In situations involving environmental pollution, a cost-benefit model expresses cost as a function of the percentage of pollutant removed from the environment. Suppose a cost-benefit model is expressed as

$$y = \frac{6.7x}{100 - x},$$

where y is the cost in thousands of dollars of removing x percent of a certain pollutant.
(a) Graph the function.
(b) Is it possible, according to this function, to remove all of the pollutant?

*In recent years the economist Arthur Laffer has been a center of controversy because of his **Laffer curve**, an idealized version of which is shown here.*

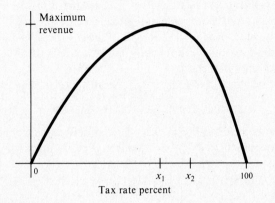

Tax rate percent

According to this curve, increasing a tax rate, say from x_1 percent to x_2 percent on the graph above, can actually lead to a decrease in government revenue. All economists agree on the endpoints, 0 revenue at tax rates of both 0% and 100%, but there is much disagreement on the location of the rate x_1 that produces maximum revenue.

48. Suppose an economist studying the Laffer curve produces the rational function

$$y = \frac{60x - 6000}{x - 120},$$

where y is government revenue in millions of dollars from a tax rate of x percent, with the function valid for $50 \le x \le 100$. Find the revenue from the following tax rates.

(a) 50%

(b) 60%

(c) 80%

(d) 100%

(e) Graph the function.

49. Suppose our economist studies a different tax, this time producing

$$y = \frac{80x - 8000}{x - 110},$$

with y giving government revenue in tens of millions of dollars for a tax rate of x percent, with the function valid for $55 \le x \le 100$. Find the revenue for the following tax rates.

(a) 55%   **(b)** 60%   **(c)** 70%

(d) 90%   **(e)** 100%   **(f)** Graph the function.

50. Let $f(x) = p(x)/q(x)$ be a rational function. If the degree of $p(x)$ is m, and the degree of $q(x)$ is n, and if $m > n$, tell how to find the horizontal or vertical asymptotes of f.

5.5 Algebra of Functions

When a company accountant sits down to estimate the firm's overhead, the first step might be to find functions representing the cost of materials, labor charges, equipment maintenance, and so on. The sum of these various functions could then be used to find the total overhead for the company. Methods of combining functions are discussed in this section.

Given two functions f and g, their *sum*, written $f + g$, is defined as

$$(f + g)(x) = f(x) + g(x),$$

for all x such that both $f(x)$ and $g(x)$ exist. Similar definitions can be given for the difference, $f - g$, product, $f \cdot g$, and quotient, f/g, of functions; however, the quotient,

$$\left(\frac{f}{g}\right)(x) = \frac{f(x)}{g(x)},$$

is defined only for those values of x where both $f(x)$ and $g(x)$ exist, and in addition, $g(x) \neq 0$.

EXAMPLE 1

Let $f(x) = x^2 + 1$, and $g(x) = 3x + 5$. Find each of the following.

(a) $(f + g)(1)$

Since $f(1) = 2$ and $g(1) = 8$, use the definition above to get

$$(f + g)(1) = f(1) + g(1) = 2 + 8 = 10.$$

(b) $(f - g)(-3) = f(-3) - g(-3) = 10 - (-4) = 14$

(c) $(fg)(5) = f(5) \cdot g(5) = 26 \cdot 20 = 520$

(d) $\left(\frac{f}{g}\right)(0) = \frac{f(0)}{g(0)} = \frac{1}{5}$ ●

The various operations on functions are defined as follows.

Operations on Functions

Given two functions f and g, then for all values of x for which both $f(x)$ and $g(x)$ exist,

sum	$(f + g)(x) = f(x) + g(x)$
difference	$(f - g)(x) = f(x) - g(x)$
product	$(fg)(x) = f(x) \cdot g(x)$
quotient	$\left(\dfrac{f}{g}\right)(x) = \dfrac{f(x)}{g(x)}, \quad g(x) \neq 0.$

EXAMPLE 2   Let $f(x) = 8x - 9$ and $g(x) = \sqrt{2x - 1}$.

(a) $(f + g)(x) = f(x) + g(x) = 8x - 9 + \sqrt{2x - 1}$

(b) $(f - g)(x) = f(x) - g(x) = 8x - 9 - \sqrt{2x - 1}$

(c) $(f \cdot g)(x) = f(x) \cdot g(x) = (8x - 9)\sqrt{2x - 1}$

(d) $\left(\dfrac{f}{g}\right)(x) = \dfrac{f(x)}{g(x)} = \dfrac{8x - 9}{\sqrt{2x - 1}}$

The domain of f is the set of all real numbers, while the domain of $g(x) = \sqrt{2x - 1}$ includes just those real numbers that make $2x - 1 \geq 0$; the domain of g is the interval $[1/2, +\infty)$. The domain of $f + g, f - g$, and $f \cdot g$ is thus $[1/2, +\infty)$. With f/g, the denominator cannot be zero, so the value $1/2$ is excluded from the domain. The domain of f/g is $(1/2, +\infty)$.   ●

The domains of $f + g, f - g, f \cdot g$, and f/g are summarized below. (Recall that the intersection of two sets is the set of all elements belonging to *both* sets.)

For functions f and g, the domains of $f + g, f - g$, and $f \cdot g$ include all real numbers in the intersection of the domains of f and g, while the domain of f/g includes those real numbers in the intersection of the domains of f and g for which $g(x) \neq 0$.

Composition of Functions   The sketch in Figure 5.31 shows a function f which assigns to each element x of set X some element y of set Y. Suppose also that a function g takes each element of set Y and assigns a value z of set Z. Using both f and g, then, an element x in X is assigned to an element z in Z. The result of this process is a new function h, which takes an element x in X and assigns an element z in Z.

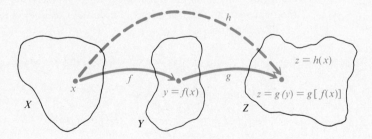

Figure 5.31

This function h is called the *composition* of functions g and f, written $g \circ f$, and defined as follows.

Composition of Functions

> Let f and g be functions. The **composite function,** or **composition,** of g and f is
>
> $$(g \circ f)(x) = g[f(x)],$$
>
> for all x in the domain of f such that $f(x)$ is in the domain of g.

EXAMPLE 3   Let $f(x) = 4x + 1$ and $g(x) = 2x^2 + 5x$. Find each of the following.

(a) $(g \circ f)(x)$

By definition, $(g \circ f)(x) = g[f(x)]$. Using the given functions,

$$(g \circ f)(x) = g[f(x)]$$

$$= g(4x + 1)$$
$$= 2(4x + 1)^2 + 5(4x + 1)$$
$$= 2(16x^2 + 8x + 1) + 20x + 5$$
$$= 32x^2 + 16x + 2 + 20x + 5$$
$$= 32x^2 + 36x + 7.$$

(b) $(f \circ g)(x)$

If we use the definition above with f and g interchanged, $(f \circ g)(x)$ becomes $f[g(x)]$, with

$$(f \circ g)(x) = f[g(x)]$$

$$= f(2x^2 + 5x)$$
$$= 4(2x^2 + 5x) + 1$$
$$= 8x^2 + 20x + 1. \quad \bullet$$

As this example shows, it is not always true that $f \circ g = g \circ f$. In fact, two composite functions are equal only for a special class of functions, discussed in the next section. In Example 3, the domain of both composite functions is the set of all real numbers.

EXAMPLE 4   Let $f(x) = 1/x$ and $g(x) = \sqrt{3 - x}$. Find $f \circ g$ and $g \circ f$. Give the domain of each.

First find $f \circ g$.

$$(f \circ g)(x) = f[g(x)]$$

$$= f(\sqrt{3 - x})$$

$$= \frac{1}{\sqrt{3 - x}}$$

In order for $\sqrt{3 - x}$ to be a nonzero real number, we must have $3 - x > 0$ or $x < 3$, so that the domain of $f \circ g$ is the interval $(-\infty, 3)$.

Use the same functions to find $g \circ f$, as follows.

$$(g \circ f)(x) = g[f(x)]$$

$$= g\left(\frac{1}{x}\right)$$

$$= \sqrt{3 - \frac{1}{x}}$$

$$= \sqrt{\frac{3x - 1}{x}}$$

The domain of $g \circ f$ is the set of all real numbers x such that $(3x - 1)/x \geq 0$. By the methods of Section 3.6, the domain of $g \circ f$ is the set $(-\infty, 0) \cup [1/3, +\infty)$.   ●

5.5   EXERCISES

For each of the following pairs of functions, find $f + g$, $f - g$, $f \cdot g$, and f/g. Give the domain of each. See Example 2.

1. $f(x) = 4x - 1$, $g(x) = 6x + 3$

2. $f(x) = 9 - 2x$, $g(x) = -5x + 2$

3. $f(x) = 3x^2 - 2x$, $g(x) = x^2 - 2x + 1$

4. $f(x) = 6x^2 - 11x$, $g(x) = x^2 - 4x - 5$

5. $f(x) = \sqrt{2x + 5}$, $g(x) = \sqrt{4x - 9}$

6. $f(x) = \sqrt{11x - 3}$, $g(x) = \sqrt{2x - 15}$

7. $f(x) = 4x^2 - 11x + 2$, $g(x) = x^2 + 5$

8. $f(x) = 15x^2 - 2x + 1$, $g(x) = 16 + x^2$

Let $f(x) = 4x^2 - 2x$ and let $g(x) = 8x + 1$. Find each of the following. See Examples 1–3.

9. $(f + g)(3)$

10. $(f + g)(-5)$

11. $(f \cdot g)(4)$

12. $(f \cdot g)(-3)$

13. $\left(\dfrac{f}{g}\right)(-1)$

14. $\left(\dfrac{f}{g}\right)(4)$

15. $(f + g)(m)$

16. $(f - g)(2k)$

17. $(f \circ g)(2)$

18. $(f \circ g)(-5)$

19. $(g \circ f)(2)$

20. $(g \circ f)(-5)$

21. $(f \circ g)(k)$

22. $(g \circ f)(5z)$

Find $f \circ g$ and $g \circ f$ for each of the following pairs of functions. See Examples 3 and 4.

23. $f(x) = 8x + 12$, $g(x) = 3x - 1$

24. $f(x) = -6x + 9$, $g(x) = 5x + 7$

25. $f(x) = 5x + 3$, $g(x) = -x^2 + 4x + 3$

26. $f(x) = 4x^2 + 2x + 8$, $g(x) = x + 5$

27. $f(x) = -x^3 + 2$, $g(x) = 4x$

28. $f(x) = 2x$, $g(x) = 6x^2 - x^3$

29. $f(x) = \dfrac{1}{x}$, $g(x) = x^2$

30. $f(x) = \dfrac{2}{x^4}$, $g(x) = 2 - x$

31. $f(x) = \sqrt{x + 2}$, $g(x) = 8x^2 - 6$

32. $f(x) = 9x^2 - 11x$, $g(x) = 2\sqrt{x + 2}$

33. $f(x) = \dfrac{1}{x - 5}$, $g(x) = \dfrac{2}{x}$

34. $f(x) = \dfrac{8}{x - 6}$, $g(x) = \dfrac{4}{3x}$

35. $f(x) = \sqrt{x + 1}$, $g(x) = \dfrac{-1}{x}$

36. $f(x) = \dfrac{8}{x}$, $g(x) = \sqrt{3 - x}$

For each of the following pairs of functions, show that $(f \circ g)(x) = x$ and $(g \circ f)(x) = x$.

37. $f(x) = 8x, \quad g(x) = \dfrac{1}{8}x$

38. $f(x) = \dfrac{3}{4}x, \quad g(x) = \dfrac{4}{3}x$

39. $f(x) = 8x - 11, \quad g(x) = \dfrac{x + 11}{8}$

40. $f(x) = \dfrac{x - 3}{4}, \quad g(x) = 4x + 3$

41. $f(x) = x^3 + 6, \quad g(x) = \sqrt[3]{x - 6}$

42. $f(x) = \sqrt[5]{x - 9}, \quad g(x) = x^5 + 9$

In each of the following exercises, a function h is given. Find functions f and g such that $h(x) = (f \circ g)(x)$. Many such pairs of functions exist.

43. $h(x) = (6x - 2)^2$

44. $h(x) = (11x^2 + 12x)^2$

45. $h(x) = \sqrt{x^2 - 1}$

46. $h(x) = \dfrac{1}{x^2 + 2}$

47. $h(x) = \dfrac{(x - 2)^2 + 1}{5 - (x - 2)^2}$

48. $h(x) = (x + 2)^3 - 3(x + 2)^2$

49. Suppose the population P of a certain species of fish depends on the number x (in hundreds) of a smaller kind of fish which serves as its food supply, so that

$$P(x) = 2x^2 + 1.$$

Suppose, also, that the number x (in hundreds) of the smaller species of fish depends upon the amount a (in appropriate units) of its food supply, a kind of plankton. Suppose

$$x = f(a) = 3a + 2.$$

Find $(P \circ f)(a)$, the relationship between the population P of the large fish and the amount a of plankton available.

50. Suppose the demand for a certain brand of vacuum cleaner is given by

$$D(p) = \dfrac{-p^2}{100} + 500,$$

where p is the price in dollars. If the price, in terms of the cost, c, is expressed as

$$p(c) = 2c - 10,$$

find the demand in terms of the cost.

51. An oil well off the Gulf Coast is leaking, with the leak spreading oil over the surface as a circle. At any time t, in minutes, after the beginning of the leak, the radius of the circular oil slick on the surface is $r(t) = 4t$ ft. Let $A(r) = \pi r^2$ represent the area of a circle of radius r. Find and interpret $(A \circ r)(t)$.

52. When a thermal inversion layer is over a city (such as happens often in Los Angeles), pollutants cannot rise vertically but are trapped below the layer and must disperse horizontally. Assume that a factory smokestack begins emitting a pollutant at 8 A.M. Assume that the pollutant disperses horizontally, forming a circle. If t represents the time, in hours, since the factory began emitting pollutants ($t = 0$ represents 8 A.M.), assume that the radius of the circle of pollution is $r(t) = 2t$ mi. Let $A(r) = \pi r^2$ represent the area of a circle of radius r. Find and interpret $(A \circ r)(t)$.

5.6   Inverse Functions

Addition and subtraction are inverse operations; starting with a number x, adding 5, and subtracting 5 gives x back as a result. Some functions are inverses of each other also. For example, the functions

$$f(x) = 8x \quad \text{and} \quad g(x) = \frac{1}{8}x$$

are inverses of each other. This means that if we choose a value of x such as $x = 12$, then

$$f(12) = 8 \cdot 12 = 96.$$

Calculating $g(96)$ gives

$$g(96) = \frac{1}{8} \cdot 96 = 12,$$

so that
$$g[f(12)] = 12.$$

Also, $f[g(12)] = 12$. For these functions f and g, it can be shown that

$$f[g(x)] = x \quad \text{and} \quad g[f(x)] = x$$

for any value of x.

This section will show how to start with a function such as $f(x) = 8x$ and obtain the inverse function $g(x) = (1/8)x$. Not all functions have inverse functions. It turns out that the only functions that have inverse functions are *one-to-one functions*.

One-to-One Functions   For the function $y = 5x - 8$, two different values of x produce two different values of y. On the other hand, for the function $y = x^2$, two different values of x can lead to the *same* value of y; for example, both $x = 4$ and $x = -4$ give $y = 4^2 = (-4)^2 = 16$. A function such as $y = 5x - 8$, where different elements from the domain lead to different elements from the range, is called a *one-to-one function*.

One-to-One
Function

> A function f is **one-to-one** if, for elements a and b from the domain of f,
>
> $$a \neq b \text{ implies } f(a) \neq f(b).$$

EXAMPLE 1

Decide whether or not the following functions are one-to-one.

(a) $f(x) = -4x + 12$

Suppose that $a \neq b$. Then $-4a \neq -4b$, and $-4a + 12 \neq -4b + 12$. Thus, the fact that $a \neq b$ implies that $f(a) \neq f(b)$, so f is one-to-one.

(b) $f(x) = \sqrt{25 - x^2}$

If $x = 3$, then $f(3) = \sqrt{25 - 3^2} = \sqrt{16} = 4$. For $x = -3$, we have $f(-3) = \sqrt{25 - (-3)^2} = \sqrt{16} = 4$. Both $x = 3$ and $x = -3$ lead to $y = 4$, so f is not one-to-one. ●

As shown in part (b) of Example 1, a way to show that a function is *not* one-to-one is to produce a pair of unequal numbers that lead to the same function value.

In addition, there is a useful graphical test which tells whether or not a function is one-to-one. Figure 5.32(a) shows the graph of a function cut by a horizontal line. Each point where the horizontal line cuts the graph has the same value of y but a different value of x. Here, three different values of x lead to the same value of y; therefore, the function is not one-to-one. On the other hand, the graph of Figure 5.32(b) can be cut by any horizontal line in no more than one point, so it is the graph of a one-to-one function.

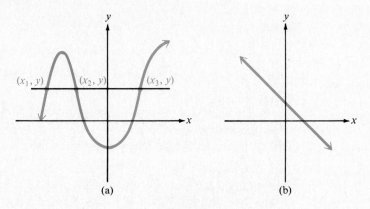

(a)   (b)

Figure 5.32

In summary, we have the **horizontal line test** for one-to-one functions.

Horizontal Line Test

> If every horizontal line cuts the graph of a function in no more than one point, then the function is one-to-one.

Inverse Functions   In the previous section we discussed the composition of two functions. For example, if

$$f(x) = 8 + 3x \qquad \text{and} \qquad g(x) = \frac{x - 8}{3},$$

then

$$(f \circ g)(x) = 8 + 3\left(\frac{x - 8}{3}\right) = 8 + x - 8 = x$$

and

$$(g \circ f)(x) = \frac{(8 + 3x) - 8}{3} = \frac{3x}{3} = x.$$

These two functions f and g have a special property:

$$(f \circ g)(x) = x \quad \text{and} \quad (g \circ f)(x) = x.$$

For this reason, f and g are *inverse functions* of each other.

Inverse Functions

> Let f be a one-to-one function with domain X and range Y. Let g be a function with domain Y and range X. Then g is the **inverse function** of f if
>
> $$(f \circ g)(x) = x \quad \text{for every } x \text{ in } Y,$$
>
> and $\qquad (g \circ f)(x) = x \quad \text{for every } x \text{ in } X.$

EXAMPLE 2

Let $f(x) = x^3 - 1$, and let $g(x) = \sqrt[3]{x + 1}$. Is g the inverse function of f?
Use the definition to get

$$(f \circ g)(x) = f[g(x)] = (\sqrt[3]{x + 1})^3 - 1$$
$$= x + 1 - 1 = x$$
$$(g \circ f)(x) = g[f(x)] = \sqrt[3]{(x^3 - 1) + 1} = \sqrt[3]{x^3} = x.$$

Function g is the inverse function of f. Also, f is the inverse function of g. ●

A special notation is often used for inverse functions: if g is the inverse function of f, then g can be written as f^{-1} (read "f-inverse"). In Example 2,

$$f^{-1}(x) = \sqrt[3]{x + 1}.$$

Do not confuse the -1 in f^{-1} with a negative exponent. The symbol $f^{-1}(x)$ does not represent $1/f(x)$; it represents the inverse function of f. Keep in mind that a function f can have an inverse function f^{-1} if and only if f is one-to-one.

Given a one-to-one function f and a value of x in the domain of f, we can find the corresponding value in the range of f by means of the equation $y = f(x)$. By the inverse function f^{-1}, we can take y and produce x: $x = f^{-1}(y)$. Therefore, to find the equation for f^{-1}, it is only necessary to solve $y = f(x)$ for x.

For example, let $f(x) = 7x - 2$. Then $y = 7x - 2$. This function is one-to-one, so that f^{-1} exists. Solve the equation for x, as follows:

$$y = 7x - 2$$
$$7x = y + 2$$
$$x = \frac{y + 2}{7},$$

or, since $x = f^{-1}(y)$,

$$f^{-1}(y) = \frac{y + 2}{7}.$$

It is common to use x to represent an element from the domain of a function. Replacing y with x gives

$$f^{-1}(x) = \frac{x + 2}{7}.$$

Check that $(f \circ f^{-1})(x) = x$ and $(f^{-1} \circ f)(x) = x$, making f^{-1} the inverse of f.

In summary, finding the equation of an inverse function involves the following steps.

Finding an Equation for f^{-1}

1. Let $y = f(x)$ be a one-to-one function.
2. Solve for x. Let $x = f^{-1}(y)$.
3. Exchange x and y to get $y = f^{-1}(x)$.
4. Check the domains and ranges: the domain of f and the range of f^{-1} are equal, as are the domain of f^{-1} and the range of f.

EXAMPLE 3

For each of the following functions, find its inverse function where possible.

(a) $f(x) = \dfrac{4x + 6}{5}$

This function is one-to-one and thus has an inverse. Let $y = f(x)$, and solve for x, getting

$$y = \frac{4x + 6}{5}$$

$$5y = 4x + 6$$

$$5y - 6 = 4x$$

$$\frac{5y - 6}{4} = x.$$

Finally, exchange x and y, and let $y = f^{-1}(x)$, to get

$$\frac{5x - 6}{4} = y,$$

or

$$f^{-1}(x) = \frac{5x - 6}{4}.$$

(b) $f(x) = x^3 - 1$

Two different values of x will produce two different values of $x^3 - 1$, so the function is one-to-one and has an inverse. To find the inverse, first solve $y = x^3 - 1$ for x, as follows.

$$y = x^3 - 1$$

$$y + 1 = x^3$$

$$\sqrt[3]{y + 1} = x$$

In Example 2, we verified that $\sqrt[3]{x + 1}$ is the inverse of $x^3 - 1$. Exchange x and y, with

$$\sqrt[3]{x + 1} = y,$$

or
$$f^{-1}(x) = \sqrt[3]{x + 1}.$$

(c) $f(x) = x^2$

The two different x-values 4 and -4 give the same value of y, namely 16, showing that the function is not one-to-one and has no inverse function.   ●

Suppose f and f^{-1} are inverse functions of each other. Suppose that $f(a) = b$, for real numbers a and b. Then by the definition of inverse, $f^{-1}(b) = a$. This shows that if a point (a, b) is on the graph of f, then (b, a) will belong to the graph of f^{-1}. As shown in Figure 5.33, the points (a, b) and (b, a) are symmetric with respect to the line $y = x$. Thus, the graph of f^{-1} can be obtained from the graph of f by reflecting the graph of f about the line $y = x$.

For example, Figure 5.34 shows the graph of $f(x) = x^3 - 1$ as a solid line and the graph of $f^{-1}(x) = \sqrt[3]{x + 1}$ as a dashed line. These graphs are symmetric with respect to the line $y = x$.

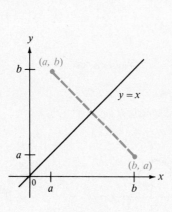

Figure 5.33

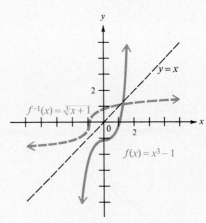

Figure 5.34

EXAMPLE 4   Let $f(x) = \sqrt{x + 5}$ with domain $[-5, +\infty)$. Find $f^{-1}(x)$.

The function f is one-to-one and has an inverse function. To find this inverse function, start with

$$y = \sqrt{x + 5}$$

and solve for x, to get

$$y = \sqrt{x + 5}$$
$$y^2 = x + 5$$
$$y^2 - 5 = x.$$

Exchanging x and y gives

$$x^2 - 5 = y.$$

We cannot just give $x^2 - 5$ as $f^{-1}(x)$. In the definition of f above, the domain was given as $[-5, +\infty)$. The range of f is $[0, +\infty)$. As we have said, the range of f equals the domain of f^{-1}, so f^{-1} must be given as

$$f^{-1}(x) = x^2 - 5, \qquad \text{domain } [0, +\infty).$$

As a check, the range of f^{-1}, $[-5, +\infty)$ equals the domain of f. Graphs of f and f^{-1} are shown in Figure 5.35. The line $y = x$ is included on the graph to show that the graphs of f and f^{-1} are mirror images with respect to this line. ●

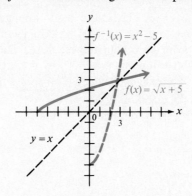

Figure 5.35

5.6 EXERCISES

Which of the following functions are one-to-one? See Example 1.

1.

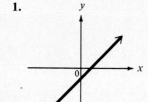

2.

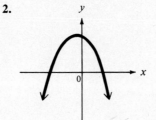

3.

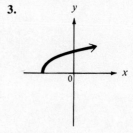

4.

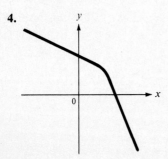

5.

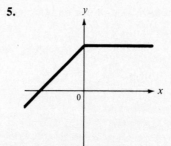

6.

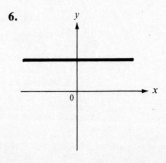

7. $y = 4x - 5$

8. $x + y = 0$

9. $y = 6 - x$

10. $y = -x^2$

11. $y = (x - 2)^2$

12. $y = -(x + 3)^2 - 8$

13. $y = \sqrt{36 - x^2}$

14. $y = -\sqrt{100 - x^2}$

15. $y = |25 - x^2|$

16. $y = -|16 - x^2|$

17. $y = x^3 - 1$

18. $y = -\sqrt[3]{x + 5}$

19. $y = (\sqrt{x} + 1)^2$

20. $y = (3 - 2\sqrt{x})^2$

21. $y = \dfrac{1}{x + 2}$

22. $y = \dfrac{-4}{x - 8}$

23. $y = 9$

24. $y = -4$

Which of the following pairs of functions are inverses of each other? See Example 2.

25.

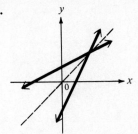

26.

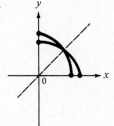

27.

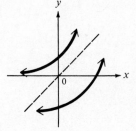

28.

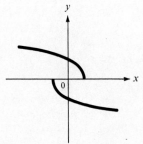

29.

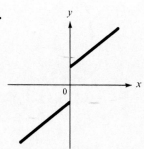

30.

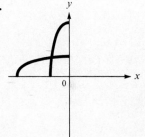

31. $f(x) = -\dfrac{3}{11}x, \quad g(x) = -\dfrac{11}{3}x$

32. $f(x) = 2x + 4, \quad g(x) = \dfrac{1}{2}x - 2$

33. $f(x) = 5x - 5, \quad g(x) = \dfrac{1}{5}x + 1$

34. $f(x) = 8x - 7, \quad g(x) = \dfrac{x + 8}{7}$

35. $f(x) = \dfrac{1}{x + 1}, \quad g(x) = \dfrac{x - 9}{12}$

36. $f(x) = \dfrac{1}{x + 1}, \quad g(x) = \dfrac{1 - x}{x}$

37. $f(x) = \dfrac{2}{x + 6}, \quad g(x) = \dfrac{6x + 2}{x}$

38. $f(x) = \dfrac{1}{x}, \quad g(x) = \dfrac{1}{x}$

39. $f(x) = 4x, \quad g(x) = -\dfrac{1}{4}x$

40. $f(x) = \sqrt{x + 8}$, domain $[-8, +\infty)$, and $g(x) = x^2 - 8$, domain $[0, +\infty)$

41. $f(x) = x^2 + 3$, domain $[0, +\infty)$, and $g(x) = \sqrt{x - 3}$, domain $[3, +\infty)$

42. $f(x) = |x - 1|$, domain $[-1, +\infty)$, and $g(x) = |x + 1|$, domain $[1, +\infty)$

43. $f(x) = -|x + 5|$, domain $[-5, +\infty)$, and $g(x) = |x - 5|$, domain $[5, +\infty)$

Graph the inverse of each one-to-one function.

44.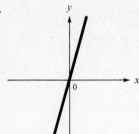

45.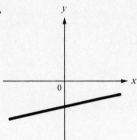

46.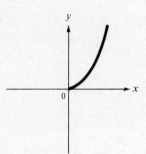

47.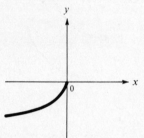

48.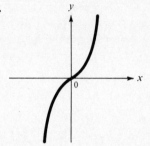

49.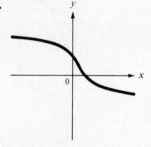

For each of the following functions $y = f(x)$ that is one-to-one, write an equation for the inverse function in the form of $y = f^{-1}(x)$, and then graph f and f^{-1}. See Examples 3 and 4.

50. $y = 3x - 4$

51. $y = 4x - 5$

52. $y = \frac{1}{3}x$

53. $y = -\frac{2}{5}x$

54. $y = x^3 + 1$

55. $y = -x^3 - 2$

56. $x + 4y = 12$

57. $3x + y = 9$

58. $y = x^2$

59. $y = -x^2 + 2$

60. $y = \frac{1}{x}$

61. $xy = 4$

62. $y = \frac{8x + 3}{4x - 1}$

63. $y = \frac{-6x + 5}{3x - 1}$

64. $f(x) = 4 - x^2$, domain $(-\infty, 0]$

65. $f(x) = \sqrt{6 + x}$, domain $[-6, +\infty)$

66. Let $f(x) = ax + b$ and $g(x) = cx + d$. Find all values of $a, b, c,$ and d for which $f^{-1} = g$.

5.7 Variation

In many applications of mathematics, it is necessary to write relationships between variables. For example, in chemistry the ideal gas law shows how temperature, pressure, and volume are related. In physics, various formulas in optics show how the focal length of a lens and the size of an image are related.

1st is expenses

If the quotient of two variables is constant, then one variable *varies directly* or is *directly proportional* to the other. This idea can be stated in a different way as follows.

Directly Proportional

> *y* **varies directly** as *x*, or *y* **is directly proportional** to *x*, means that a nonzero real number *k* (called the **constant of variation**) exists such that
>
> $$y = kx.$$

Here *y* is a function of *x*.

EXAMPLE 1   Suppose the area of a certain rectangle varies directly as the length. If the area is 50 m^2 when the length is 10 m, find the area when the length is 25 m.

Since the area varies directly as the length,

$$A = kl,$$

where *A* represents the area of the rectangle, *l* is the length, and *k* is a nonzero constant that must be found. When $A = 50$ and $l = 10$, the equation $A = kl$ becomes

$$50 = 10k \quad \text{or} \quad k = 5.$$

Using this value of *k*, the relationship between the area and the length can be expressed as

$$A = 5l.$$

To find the area when the length is 25, replace *l* with 25 to get

$$A = 5l = 5(25) = 125.$$

The area of the rectangle is 125 m^2 when the length is 25 m.   ●

Sometimes *y* varies as a power of *x*.

Varies as *n*th Power

> Let *n* be a positive real number. Then *y* **varies directly as the *n*th power** of *x*, or *y* is **directly proportional to the *n*th power** of *x*, if there exists a real number *k* such that
>
> $$y = kx^n.$$

The phrase "directly proportional" is sometimes abbreviated to just "proportional."

For example, the area of a square of side *x* is given by the formula $A = x^2$, so that the area varies directly as the square of the length of a side. Here $k = 1$.

The case where *y* increases as *x* decreases is an example of *inverse variation*.

Inverse Variation

> Let n be a positive real number. Then y **varies inversely as the nth power** of x means that there exists a real number k such that
>
> $$y = \frac{k}{x^n}.$$
>
> If $n = 1$, then $y = k/x$, and y **varies inversely** as x.

EXAMPLE 2   In a certain manufacturing process, the cost of producing a single item varies inversely as the square of the number of items produced. If 100 items are produced, each costs $2. Find the cost per item if 400 items are produced.

We can let x represent the number of items produced and y the cost per item, and write

$$y = \frac{k}{x^2}$$

for some nonzero constant k. Since $y = 2$ when $x = 100$,

$$2 = \frac{k}{100^2} \quad \text{or} \quad k = 20{,}000.$$

Thus, the relationship between x and y is given by

$$y = \frac{20{,}000}{x^2}.$$

When 400 items are produced, the cost per item is given by

$$y = \frac{20{,}000}{400^2} = .125, \text{ or } 12.5\text{¢}. \quad \bullet$$

One variable may depend on more than one other variable. Such variation is called *combined variation*. If a variable depends on the product of two or more other variables, we refer to that as *joint variation*.

Joint Variation

> y **varies jointly** as the nth power of x and the mth power of z if there exists a real number k such that
>
> $$y = kx^n z^m.$$

The next example shows combined variation.

EXAMPLE 3   Suppose that m varies directly as the square of p and inversely as q. Suppose also that $m = 8$ when $p = 2$ and $q = 6$. Find m if $p = 6$ and $q = 10$.

Here m depends on two variables:

$$m = \frac{kp^2}{q}.$$

Since $m = 8$ when $p = 2$ and $q = 6$,

$$8 = \frac{k \cdot 2^2}{6}$$

$$k = 12$$

and it follows that

$$m = \frac{12p^2}{q}.$$

For $p = 6$ and $q = 10$,

$$m = \frac{12 \cdot 6^2}{10}$$

$$m = \frac{216}{5}. \quad \bullet$$

Let us now summarize the steps involved in solving a problem in variation:

**Solving
Variation
Problems**

> 1. Write, in an algebraic form, the general relationship among the variables. Use the constant k.
> 2. Substitute given values of the variables and find the value of k.
> 3. Substitute this value of k into the formula of Step 1, thus obtaining a specific formula.
> 4. Solve for the required unknown.

5.7 EXERCISES

Express each of the following statements as an equation.

1. a varies directly as b.
2. m is proportional to n.
3. x is inversely proportional to y.
4. p varies inversely as y.
5. r varies jointly as s and t.
6. R is proportional to m and p.
7. w is proportional to x^2 and inversely proportional to y.
8. c varies directly as d and inversely as f^2 and g. $\quad c = \dfrac{d}{f^2 g} k$

Solve each of the following problems.

9. If m varies directly as x and y, and $m = 10$ when $x = 4$ and $y = 7$, find m when $x = 11$ and $y = 8$.

10. Suppose m varies directly as z and p. If $m = 10$ when $z = 3$ and $p = 5$, find m when $z = 5$ and $p = 7$.

11. Suppose r varies directly as the square of m, and inversely as s. If $r = 12$ when $m = 6$ and $s = 4$, find r when $m = 4$ and $s = 10$.

12. Suppose p varies directly as the square of z, and inversely as r. If $p = 32/5$ when $z = 4$ and $r = 10$, find p when $z = 2$ and $r = 16$.

13. Let a be proportional to m and n^2, and inversely proportional to y^3. If $a = 9$ when $m = 4$, $n = 9$, and $y = 3$, find a if $m = 6$, $n = 2$, and $y = 5$.

14. If y varies directly as x, and inversely as m^2 and r^2, and $y = 5/3$ when $x = 1$, $m = 2$, and $r = 3$, find y if $x = 3$, $m = 1$, and $r = 8$.

15. Let r vary directly as p^2 and q^3, and inversely as z^2 and x. If $r = 334.6$ when $p = 1.9$, $q = 2.8$, $z = .4$, and $x = 3.7$, find r when $p = 2.1$, $q = 4.8$, $z = .9$, and $x = 1.4$.

16. Let p vary directly as x and z^2, and inversely as m, n, and q. If $p = 9.25$ when $x = 3.2$, $z = 10.1$, $m = 2$, $n = 1.7$, and $q = 8.3$, find p when $x = 8.1$, $z = 4.3$, $m = 4$, $n = 2.1$, and $q = 6.2$.

17. The distance a body falls from rest varies directly as the square of the time it falls (disregarding air resistance). If an object falls 1024 ft in 8 sec, how far will it fall in 12 sec?

18. Hooke's law for an elastic spring states that the distance a spring stretches varies directly as the force applied. If a force of 15 lb stretches a certain spring 8 in, how much will a force of 30 lb stretch the spring?

19. In electric current flow, it is found that the resistance (measured in units called ohms) offered by a fixed length of wire of a given material varies inversely as the square of the diameter of the wire. If a wire .01 in in diameter has a resistance of .4 ohm, what is the resistance of a wire of the same length and material with a diameter of .03 in?

20. The illumination produced by a light source varies inversely as the square of the distance from the source. The illumination of a light source at 5 m is 70 candela. What is the illumination 12 m from the source?

21. The pressure exerted by a certain liquid at a given point is proportional to the depth of the point below the surface of the liquid. If the pressure 20 m below the surface is 70 kg per cm^2, what pressure is exerted 40 m below the surface?

22. The distance that a person can see to the horizon from a point above the surface of the earth varies directly as the square root of the height. A person on a hill 121 m high can see for 15 km to the horizon. How far is the horizon from a hill 900 m high?

23. Simple interest varies jointly as principal and time. If $1000 left at interest for 2 yr earned $110, find the amount of interest earned by $5000 for 5 yr.

24. The volume of a right circular cylinder is jointly proportional to the square of the radius of the circular base and to the height. If the volume is 300 cu cm when the height is 10.62 cm and the radius is 3 cm, find the volume for a cylinder with a radius of 4 cm and a height of 15.92 cm.

25. The Downtown Construction Company is designing a building whose roof rests on round concrete pillars. The company's engineers know that the maximum load a cylindrical column of circular cross section can hold varies directly as the fourth power of the diameter and inversely as the square of the height. If a 9 m column 1 m in diameter will support a load of 8 metric tons, how many metric tons will be supported by a column 12 m high and 2/3 m in diameter?

26. The company's engineers also know that the maximum load of a horizontal beam which is supported at both ends varies directly as the width and square of the height and inversely as the length between supports. If a beam 8 m long, 12 cm wide, and 15 cm high can support a maximum of 400 kg, what will they find to be the maximum load of a beam of the same material 16 m long, 24 cm wide, and 8 cm high?

27. The force needed to keep a car from skidding on a curve varies inversely as the radius of the curve and jointly as the weight of the car and the square of the speed. It takes 3000 lb of force to keep a 2000-lb car from skidding on a curve of radius 500 ft at 30 mph. What force is needed to keep the same car from skidding on a curve of radius 800 ft at 60 mph?

28. The period of a pendulum varies directly as the square root of the length of the pendulum and inversely as the square root of the acceleration due to gravity. Find the period when the length is 121 cm and the acceleration due to gravity is 980 cm per sec squared, if the period is 6π sec when the length is 289 cm and the acceleration due to gravity is 980 cm per sec squared.

29. The pressure on a point in a liquid is directly proportional to the distance from the surface to the point. In a certain liquid the pressure at a depth of 4 m is 60 kg per m^2. Find the pressure at a depth of 10 m.

30. The volume V of a gas varies directly as the temperature T and inversely as the pressure P. If V is 10 when T is 280 and P is 6, find V if T is 300 and P is 10.

31. A sociologist has decided to rank the cities in a nation by population, with population varying inversely as the ranking. If a city with a population of 1,000,000 ranks 8th, find the population of a city that ranks 2nd.

32. Under certain conditions, the length of time that it takes for fruit to ripen during the growing season varies inversely as the average maximum temperature during the season. If it takes 25 days for fruit to ripen with an average maximum temperature of 80°, find the number of days it would take at 75°.

33. The number of long distance phone calls between two cities in a certain time period varies directly as the populations p_1 and p_2 of the cities, and inversely as the distance between them. If 10,000 calls are made between two cities 500 mi apart, having populations of 50,000 and 125,000, find the number of calls between two cities 800 mi apart and having populations of 20,000 and 80,000.

34. The horsepower needed to run a boat through water varies directly as the cube of the speed. If 80 horsepower are needed to go 15 kph in a certain boat, how many horsepower would be needed to go 30 kph?

35. According to Poiseuille's law, the resistance to flow of a blood vessel, R, is directly proportional to the length, l, and inversely proportional to the fourth power of the radius, r. If $R = 25$ when $l = 12$ and $r = 0.2$, find R as r increases to 0.3, while l is unchanged.

36. The Stefan-Boltzmann law says that the radiation of heat from an object is directly proportional to the fourth power of the Kelvin temperature of the object. For a certain object, $R = 213.73$ at room temperature (293° Kelvin). Find R if the temperature increases to 335° Kelvin.

37. Suppose a nuclear bomb is detonated at a certain site. The effects of the bomb will be felt over a distance from the point of detonation that is directly proportional to the cube root of the yield of the bomb. Suppose a 100-kiloton bomb has certain effects to a radius of 3 km from the point of detonation. Find the distance that the effects would be felt for a 1500-kiloton bomb.

38. The maximum speed possible on a length of railroad track is directly proportional to the cube root of the amount of money spent on maintaining the track. Suppose that a maximum speed of 25 kph is possible on a stretch of track for which $450,000 was spent on maintenance. Find the maximum speed if the amount spent on maintenance is increased to $1,750,000.

39. Assume that a person's weight varies directly as the cube of his or her height. Find the weight of a 20-in, 7-lb baby who grows up to be an adult 67 in tall. (How reasonable does our assumption about weight and height seem?)

40. The cost of a pizza varies directly as the square of its radius. If a pizza with a radius of 15 cm costs $7, find the cost of a pizza having a radius of 22.5 cm. (You might want to do some research at a nearby pizza establishment and see if this assumption is reasonable.)

Chapter 5 Summary

Key Words		
independent variable	dependent variable
function	domain
range	graph
vertical line test for a function	linear function
absolute value function	quadratic function
greatest integer function	functions defined piecewise
polynomial function of degree n	step function
rational function	even function
horizontal asymptote	odd function
sum of functions	vertical asymptote
product of functions	oblique asymptote
composite function	difference of functions
horizontal line test	quotient of functions
varies directly	one-to-one function
constant of variation	inverse function
varies jointly	directly proportional
varies inversely	

Function

A function is a correspondence or rule that assigns to each element of one set (called the domain of the function) exactly one element from another set (called the range of the function).

Operations on Functions

Given two functions f and g, for all values of x for which $f(x)$ and $g(x)$ exist,

sum	$(f + g)(x) = f(x) + g(x)$
difference	$(f - g)(x) = f(x) - g(x)$
product	$(fg)(x) = f(x) \cdot g(x)$
quotient	$\left(\dfrac{f}{g}\right)(x) = \dfrac{f(x)}{g(x)}, \quad g(x) \neq 0.$

Composition of Functions

Let f and g be functions. The composite function, or composition, of g and f is

$$(g \circ f)(x) = g[f(x)],$$

for all x in the domain of f such that $f(x)$ is in the domain of g.

Chapter 5   Review Exercises

1-93 ODD

Give the domain of each of the following functions.

1. $y = -x$

2. $x + 6 = -y$

3. $y = |x| - 4$

4. $y = -3|x - 4|$

5. $y = -(x + 3)^2$

6. $y = \sqrt{5 - x}$

7. $y = \sqrt{49 - x^2}$

8. $y = -\sqrt{x^2 - 4}$

9. $y = \sqrt{\dfrac{1}{x^2 + 9}}$

10. $y = \sqrt{x^2 + 3x - 10}$

11. $y = \dfrac{x + 2}{x - 7}$

12. $y = \dfrac{3x - 8}{4x - 7}$

Let $f(x) = -x^2 + 4x + 2$ and $g(x) = 3x + 5$. Find each of the following.

13. $f(-3)$

14. $f(4)$

15. $f(0)$

16. $g(3)$

17. $f[g(-1)]$

18. $g[f(-2)]$

19. $f(3 + z)$

20. $g(5y - 4)$

For each of the following functions, find $\dfrac{f(x + h) - f(x)}{h}$.

21. $f(x) = -2x^2 + 4x - 3$

22. $f(x) = 12x - 5$

23. $f(x) = -x^3 + 2x^2$

24. $f(x) = -3$

Graph each of the following functions.

25. $y = -|x|$

26. $y = |x| - 3$

27. $y = -|x| - 2$

28. $y = -|x + 1| + 3$

29. $y = 2|x - 3| - 4$

30. $y = [x - 3]$

31. $y = \left[\dfrac{1}{2}x - 2\right]$

32. $y = \begin{cases} -4x + 2 & \text{if } x \leq 1 \\ 3x - 5 & \text{if } x > 1 \end{cases}$

33. $y = \begin{cases} 3x + 1 & \text{if } x < 2 \\ -x + 4 & \text{if } x \geq 2 \end{cases}$

34. $y = \begin{cases} |x| & \text{if } x < 3 \\ 6 - x & \text{if } x \geq 3 \end{cases}$

35. A trailer hauling service charges $45, plus $2 per mile or part of a mile. Graph the ordered pairs (miles, cost).

36. Let f be a function which gives the cost to rent a floor polisher for x hr. The cost is a flat \$3 for cleaning the polisher plus \$4 per day or fraction of a day for using the polisher.

(a) Graph f.

(b) Give the domain and range of f.

Graph each of the following functions.

37. $f(x) = x^3 + 5$

38. $f(x) = 1 - x^4$

39. $f(x) = x^2(2x + 1)(x - 2)$

40. $f(x) = (4x - 3)(3x + 2)(x - 1)$

41. $f(x) = 2x^3 + 13x^2 + 15x$

42. $f(x) = x(x - 1)(x + 2)(x - 3)$

Decide whether the following functions are odd, even, or neither.

43. $g(x) = \dfrac{x}{|x|}$ $(x \neq 0)$

44. $h(x) = 9x^6 + 5|x|$

45. $f(x) = x$

46. $f(x) = -x^4 + 9x^2 - 3x^6$

Graph each rational function.

47. $f(x) = \dfrac{8}{x}$

48. $f(x) = \dfrac{2}{3x - 1}$

49. $f(x) = \dfrac{4x - 2}{3x + 1}$

50. $f(x) = \dfrac{6x}{(x - 1)(x + 2)}$

51. $f(x) = \dfrac{2x}{x^2 - 1}$

52. $f(x) = \dfrac{x^2 + 4}{x + 2}$

53. $f(x) = \dfrac{x^2 - 1}{x}$

54. $f(x) = \dfrac{x^2 + 6x + 5}{x - 3}$

Let $f(x) = 3x^2 - 4$ and $g(x) = x^2 - 3x - 4$. Find each of the following values.

55. $(f + g)(x)$

56. $(f \cdot g)(x)$

57. $(f - g)(4)$

58. $(f + g)(-4)$

59. $(f + g)(2k)$

60. $(f \cdot g)(1 + r)$

61. $(f/g)(3)$

62. $(f/g)(-1)$

63. Give the domain of $(f \cdot g)(x)$.

64. Give the domain of $(f/g)(x)$.

Let $f(x) = \sqrt{x - 2}$ and $g(x) = x^2$. Find each of the following.

65. $(f \circ g)(x)$

66. $(g \circ f)(x)$

67. $(f \circ g)(-6)$

68. $(f \circ g)(2)$

69. $(g \circ f)(3)$

70. $(g \circ f)(24)$

Which of the following functions are one-to-one?

71.

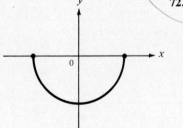

72.

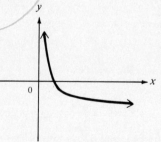

73.

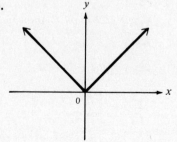

74. $y = \dfrac{8x - 9}{5}$ $\qquad$ **75.** $y = -x^2 + 11$ $\qquad$ **76.** $y = \sqrt{5 - x}$

77. $y = \sqrt{100 - x^2}$ $\qquad$ **78.** $y = -\sqrt{1 - \dfrac{x^2}{100}}$ $(x \geq 0)$

For each of the following functions that is one-to-one, write an equation for the inverse function in the form $y = f^{-1}(x)$ and then graph f and f^{-1}.

79. $f(x) = 12x + 3$ $\qquad$ **80.** $f(x) = \dfrac{2}{x - 9}$

81. $f(x) = x^3 - 3$ $\qquad$ **82.** $f(x) = x^2 - 6$

83. $f(x) = \sqrt{25 - x^2}$, domain $[0, 5]$ $\qquad$ **84.** $f(x) = -\sqrt{x - 3}$

Write each of the following statements as an equation.

85. m varies directly as the square of z.

86. y varies inversely as r and directly as the cube of p.

87. Y varies jointly as M and the square of N and inversely as the cube of X.

88. A varies jointly as the third power of t and the fourth power of s, and inversely as p and the square of h.

Solve each of the following problems.

89. Suppose r varies directly as x and inversely as the square of y. If r is 10 when x is 5 and y is 3, find r when x is 12 and y is 4.

90. Suppose m varies jointly as n and the square of p, and inversely as q. If m is 20 when n is 5, p is 6, and q is 18, find m when n is 7, p is 11, and q is 2.

91. Suppose Z varies jointly as the square of J and the cube of M, and inversely as the fourth power of W. If Z is 125 when J is 3, M is 5 and W is 1, find Z if J is 2, M is 7, and W is 3.

92. The power a windmill obtains from the wind varies directly as the cube of the wind velocity. If a 10 kph wind produces 10,000 units of power, how much power is produced by a wind of 15 kph?

93. Hooke's law for an elastic spring states that the distance a spring stretches varies directly as the force applied. If a force of 32 lb stretches a certain spring 48 in, how much will a force of 24 lb stretch the spring?

94. The weight w of an object varies inversely as the square of the distance d between the object and the center of the earth. If a man weighs 90 kg on the surface of the earth, how much would he weigh 800 km above the surface? (The radius of the earth is about 6400 km.)

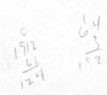

6

Exponential and Logarithmic Functions

In this chapter we will study two kinds of functions that are quite different from those studied before, functions that do not involve just the basic operations of addition, subtraction, multiplication, division, and taking roots. The functions discussed earlier are examples of *algebraic functions*. The functions in this chapter are not algebraic, but are *transcendental functions*. Many applications of mathematics, particularly to growth and decay of populations, involve the closely interrelated exponential and logarithmic functions introduced in this chapter.

6.1 Exponential Functions

Recall from Chapter 2 the definition of a^r, where r is a rational number: if $r = m/n$, then for appropriate values of m and n,

$$a^{m/n} = (\sqrt[n]{a})^m.$$

For example,

$$16^{3/4} = (\sqrt[4]{16})^3 = 2^3 = 8,$$

and

$$27^{-1/3} = \frac{1}{27^{1/3}} = \frac{1}{\sqrt[3]{27}} = \frac{1}{3}.$$

Also,

$$64^{-1/2} = \frac{1}{64^{1/2}} = \frac{1}{\sqrt{64}} = \frac{1}{8}.$$

In this section we want to extend the definition of a^r to include all real (not just rational) values of the exponent r. For example, to give a meaning to the new symbol $2^{\sqrt{3}}$, we might start by approximating the exponent $\sqrt{3}$ by the numbers 1.7, 1.73, 1.732, and so on. Since these decimals approach the value of $\sqrt{3}$ more and more closely, it seems reasonable that $2^{\sqrt{3}}$ should be approximated more and more closely by the numbers $2^{1.7}$, $2^{1.73}$, $2^{1.732}$, and so on. (Recall, for example, that $2^{1.7} = 2^{17/10}$, which means $\sqrt[10]{2^{17}}$.) In fact, this is exactly how $2^{\sqrt{3}}$ is defined (in a more advanced course).

We shall assume that the meaning given to real exponents is such that all rules and theorems for exponents are valid for real number exponents as well as rational ones. In addition to the rules for exponents presented earlier, we will need some additional properties. For example, if $y = 2^x$, then each real value of x leads to exactly one value of 2^x. This shows that $y = 2^x$ defines a function. Furthermore, if

$$3^x = 3^4, \text{ then } x = 4.$$

And, for $p > 0$,

$$\text{if } p^2 = 3^2, \text{ then } p = 3.$$

Also,
$$4^2 < 4^3 \quad \text{but} \quad \left(\frac{1}{2}\right)^2 > \left(\frac{1}{2}\right)^3,$$

so that when $a > 1$, increasing the exponent on a leads to a *larger* number, but if $0 < a < 1$, increasing the exponent on a leads to a *smaller* number.

These properties are generalized below. We have not included a proof of these properties, since the proof requires more advanced mathematics than we have available.

Additional Properties of Exponents

For $a > 0$, $a \neq 1$, and any real number x:

(a) a^x is a unique real number;

(b) $a^b = a^c$ if and only if $b = c$;

(c) if $a > 1$ and $m < n$, then $a^m < a^n$;

(d) if $0 < a < 1$ and $m < n$, then $a^m > a^n$.

Part (a) of the theorem requires $a > 0$ so that a^x is always defined. For example, $(-6)^x$ is not a real number if $x = 1/2$. This means that a^x will always be positive, since a is positive. For part (b) to hold, a must not equal 1 since $1^4 = 1^5$, even though $4 \neq 5$.

As we have said, the expression a^x satisfies all the properties of exponents from Chapter 2. We can now define a function, $f(x) = a^x$, whose domain is the set of all real numbers (and not just the rationals).

**Exponential
Function**

> The function
>
> $$f(x) = a^x, \qquad a > 0 \quad \text{and} \quad a \ne 1,$$
>
> is the **exponential function** with base a.

(If $a = 1$, the function is the constant function $f(x) = 1$.)

EXAMPLE 1   If $f(x) = 2^x$, find each of the following.

(a) $f(-1)$

Replace x with -1.

$$f(-1) = 2^{-1} = \frac{1}{2}$$

(b) $f(3) = 2^3 = 8$

(c) $f(5/2) = 2^{5/2} = (2^5)^{1/2} = 32^{1/2} = \sqrt{32} = 4\sqrt{2}$ •

Figure 6.1(a) shows the graph of the exponential function $f(x) = 2^x$; the base of this exponential function is 2. This graph was found by obtaining a number of ordered pairs satisfying the function and then drawing a smooth curve through them. The domain of the function is the set of all real numbers, and the range is the set of all positive numbers. The x-axis is a horizontal asymptote. The graph of $f(x) = 2^x$ is typical of graphs of $f(x) = a^x$ where $a > 1$. For larger values of a, the graphs rise more steeply, but the general shape is similar to the graph in Figure 6.1(a).

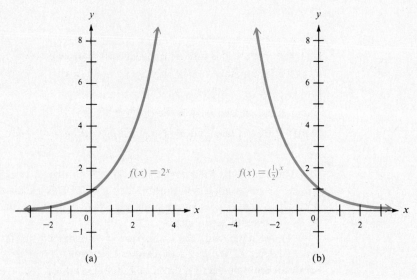

(a)   (b)

Figure 6.1

In Figure 6.1(b), the graph of $f(x) = (1/2)^x$ was obtained in a similar way. The graphs of $f(x) = 2^x$ and $f(x) = (1/2)^x$ are mirror images of each other with respect to the y-axis, because $(1/2)^x = (2^{-1})^x = 2^{-x}$. As shown by the horizontal line test, exponential functions are one-to-one.

EXAMPLE 2   Graph $f(x) = 2^{-x^2}$.

Since $f(x) = 2^{-x^2} = 1/(2^{x^2})$, we have $0 < y \leq 1$ for all values of x. Plotting some typical points, such as $(-2, 1/16)$, $(-1, 1/2)$, $(0, 1)$, $(1, 1/2)$, and $(2, 1/16)$, and drawing a smooth curve through them gives the graph of Figure 6.2. This graph is symmetric with respect to the y-axis and has the x-axis as a horizontal asymptote.   ●

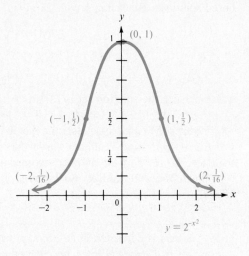

Figure 6.2

We can use the properties given above to help solve equations, as shown by the next example.

EXAMPLE 3   Solve $\left(\dfrac{1}{3}\right)^x = 81$.

First, write $1/3$ as 3^{-1}, so that $(1/3)^x = (3^{-1})^x = 3^{-x}$. Since $81 = 3^4$,

$$\left(\frac{1}{3}\right)^x = 81$$

becomes

$$3^{-x} = 3^4.$$

By the second property above,

$$-x = 4, \quad \text{or} \quad x = -4.$$

The solution set of the given equation is $\{-4\}$. (In Section 6.4 we discuss a method of solving equations with exponents where this approach is not possible.)   ●

EXAMPLE 4   If $81 = b^{4/3}$, find b.

Raise both sides of the equation to the 3/4 power, since $(b^{4/3})^{3/4} = b^1 = b$.

$$81 = b^{4/3}$$
$$81^{3/4} = (b^{4/3})^{3/4}$$
$$(\sqrt[4]{81})^3 = b^1$$
$$3^3 = b$$
$$27 = b$$

Remember that the process of raising both sides of an equation to the same power may result in false "solutions." For this reason, it is necessary to check all proposed solutions. Replacing b with 27 gives

$$27^{4/3} = (\sqrt[3]{27})^4$$
$$= 3^4$$
$$= 81,$$

which checks; the solution set is $\{27\}$.   ●

Perhaps the single most useful exponential function is the function $f(x) = e^x$, where e is an irrational number that occurs often in practical applications, as we shall see. To see how the number e comes up in a practical problem, let us begin with the formula for **compound interest** (interest paid on both principal and interest).

Compound Interest

> If P dollars is deposited in an account paying a rate of interest i compounded (paid) m times per year, then after n years the account will contain
>
> $$A = P\left(1 + \frac{i}{m}\right)^{nm}$$
>
> dollars.

For example, let \$1000 be deposited in an account paying 8% per year compounded quarterly, or four times per year. After 10 years the account will contain

$$P\left(1 + \frac{i}{m}\right)^{nm} = 1000\left(1 + \frac{.08}{4}\right)^{10(4)}$$
$$= 1000(1 + .02)^{40}$$
$$= 1000(1.02)^{40}$$

dollars. The number $(1.02)^{40}$ can be found in financial tables or by using a calculator with an x^y key. To five decimal places, $(1.02)^{40} = 2.20804$. The amount on deposit after 10 years is

$$1000(1.02)^{40} = 1000(2.20804) = 2208.04,$$

or \$2208.04.

Suppose now that a lucky investment produces annual interest of 100%, so that $i = 1.00$, or $i = 1$. Suppose also that only \$1 can be deposited at this rate, and for only one year. Then $P = 1$ and $n = 1$. Substitute into the formula for compound interest:

$$P\left(1 + \frac{i}{m}\right)^{nm} = 1\left(1 + \frac{1}{m}\right)^{1(m)} = \left(1 + \frac{1}{m}\right)^{m}.$$

As interest is compounded more and more often, the value of this expression will increase. If interest is compounded annually, making $m = 1$, the total amount on deposit is

$$\left(1 + \frac{1}{m}\right)^{m} = \left(1 + \frac{1}{1}\right)^{1} = 2^{1} = 2,$$

so an investment of \$1 becomes \$2 in one year.

A calculator with an x^y key gives the results in the following table. These results have been rounded to five decimal places.

m	$\left(1 + \dfrac{1}{m}\right)^{m}$
1	2
2	2.25
5	2.48832
10	2.59374
25	2.66584
50	2.69159
100	2.70481
500	2.71557
1000	2.71692
10,000	2.71815
1,000,000	2.71828

The table suggests that as m increases, the value of $(1 + 1/m)^{m}$ gets closer and closer to some fixed number. It turns out that this is indeed the case. This fixed number is called e. To nine decimal places,

Value of e

$$e = 2.718281828.$$

Table 5 in this book gives various powers of e. Also, some calculators will give values of e^x. In Figure 6.3 the functions $y = 2^x$, $y = e^x$, and $y = 3^x$ are graphed for comparison.

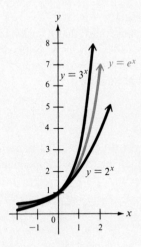

Figure 6.3

It can be shown that in many situations involving growth or decay of a population, the amount or number present at time t can be closely approximated by a function of the form

$$y = y_0 e^{kt},$$

where y_0 is the amount or number present at time $t = 0$ and k is a constant. (In other words, once the numbers y_0 and k have been determined, population is a function of time.) The next example illustrates exponential growth.

EXAMPLE 5   Suppose the population of a midwestern city is

$$P(t) = 10,000e^{0.04t},$$

where t represents time measured in years. The population at time $t = 0$ is

$$P(0) = 10,000e^{(0.04)\,0}$$
$$= 10,000e^0$$
$$= 10,000(1)$$
$$= 10,000.$$

The population of the city is 10,000 at time $t = 0$, written $P_0 = 10,000$. In year $t = 5$ the population is

$$P(5) = 10,000e^{(0.04)\,5}$$
$$= 10,000e^{0.2}.$$

The number $e^{0.2}$ can be found in Table 5 or by using a suitable calculator. By either of these methods, $e^{0.2} = 1.22140$ (to five decimal places), so

$$P(5) = 10,000(1.22140) = 12,214.$$

Thus, in five years the population of the city will be about 12,200. •

6.1 EXERCISES

Evaluate each of the following expressions.

1. 2^3 **2.** 3^{-2} **3.** $25^{1/2}$ **4.** $\left(\dfrac{1}{27}\right)^{1/3}$

Let $f(x) = (2/3)^x$. Find each of the following values. See Example 1.

5. $f(2)$ **6.** $f(-1)$ **7.** $f\left(\dfrac{1}{2}\right)$ **8.** $f(0)$

9. Graph each of the following functions. Compare the graphs to that of $y = 2^x$.
  (a) $y = 2^x + 1$ **(b)** $y = 2^x - 4$ **(c)** $y = 2^{x+1}$ **(d)** $y = 2^{x-4}$

10. Graph each of the following functions. Compare the graphs to that of $y = 3^{-x}$.
  (a) $y = 3^{-x} - 2$ **(b)** $y = 3^{-x} + 4$ **(c)** $y = 3^{-x-2}$ **(d)** $y = 3^{-x+4}$

Graph each of the following functions. See Example 2.

11. $y = 3^x$ **12.** $y = 4^x$ **13.** $y = (3/2)^x$ **14.** $y = 3^{-x}$

15. $y = 10^{-x}$ **16.** $y = 10^x$ **17.** $y = 2^{x+1}$ **18.** $y = 2^{1-x}$

19. $y = 2^{|x|}$ **20.** $y = 2^{-|x|}$ **21.** $y = 2^x + 2^{-x}$ **22.** $y = (1/2)^x + (1/2)^{-x}$

Solve each of the following equations. See Examples 3 and 4.

23. $4^x = 2$ **24.** $125^r = 5$ **25.** $\left(\dfrac{1}{2}\right)^k = 4$ **26.** $\left(\dfrac{2}{3}\right)^x = \dfrac{9}{4}$

27. $2^{3-y} = 8$ **28.** $5^{2p+1} = 25$ **29.** $\dfrac{1}{27} = b^{-3}$ **30.** $\dfrac{1}{81} = k^{-4}$

31. $4 = r^{2/3}$ **32.** $z^{5/2} = 32$ **33.** $27^{4z} = 9^{z+1}$ **34.** $32^t = 16^{1-t}$

35. $125^{-x} = 25^{3x}$ **36.** $216^{3-a} = 36^a$ **37.** $\left(\dfrac{1}{8}\right)^{-2p} = 2^{p+3}$ **38.** $3^{-h} = \left(\dfrac{1}{27}\right)^{1-2h}$

39. $\left(\dfrac{1}{2}\right)^{-x} = \left(\dfrac{1}{4}\right)^{x+1}$ **40.** $\left(\dfrac{2}{3}\right)^{k-1} = \left(\dfrac{81}{16}\right)^{k+1}$

41. For $a > 1$, how does the graph of $y = a^x$ change as a increases? What if $0 < a < 1$?

Use a calculator to help graph each of the following functions.

42. $y = \dfrac{e^x - e^{-x}}{2}$ **43.** $y = \dfrac{e^x + e^{-x}}{2}$ **44.** $y = x \cdot 2^x$ **45.** $y = x^2 \cdot 2^{-x}$

46. $y = (1 - x)e^x$ **47.** $y = x \cdot e^x$

48. Suppose that the population of a city is given by $P(t)$, where

$$P(t) = 1,000,000e^{.02t},$$

and t represents time measured in years from some initial year. Find each of the following.

(a) P_0 (b) $P(2)$ (c) $P(4)$ (d) $P(10)$

(e) Graph $y = P(t)$.

49. Suppose the quantity in grams of a radioactive substance present at time t is

$$Q(t) = 500e^{-.05t}.$$

Let t be time measured in days from some initial day. Find the quantity present at each of the following times.

(a) $t = 0$ (b) $t = 4$ (c) $t = 8$ (d) $t = 20$

(e) Graph $y = Q(t)$.

50. Experiments have shown that the sales of a product, under relatively stable market conditions, but in the absence of promotional activities such as advertising, tend to decline at a constant yearly rate. This rate of sales decline varies considerably from product to product, but seems to remain the same for any particular product. The sales decline can be expressed by a function of the form

$$S(t) = S_0 e^{-at},$$

where $S(t)$ is the rate of sales at time t measured in years, S_0 is the rate of sales at time $t = 0$, and a is the sales decay constant.

(a) Suppose the sales decay constant for a particular product is $a = 0.10$. Let $S_0 = 50,000$ and find $S(1)$ and $S(3)$.

(b) Find $S(2)$ and $S(10)$ if $S_0 = 80,000$ and $a = 0.05$.

51. It can be shown that P dollars compounded continuously (every instant) at an interest rate of i per year would amount to

$$A = Pe^{ni}$$

dollars at the end of n years. How much would $20,000 amount to at 8% compounded continuously for the following numbers of years? (More detail on this result is given in Section 6.3.)

(a) 1 year   (b) 5 years

Use the formula for compound interest and a calculator to find the total amount on deposit for each of the following investments.

52. $4292 at 6% compounded annually for 10 years

53. $965.43 at 9% compounded annually for 15 years

54. $10,765 at 11% compounded semiannually for 7 years

55. $8906.54 at 15% compounded semiannually for 9 years

56. $1567 at 13.5% compounded semiannually for 11 semiannual periods

57. $23,675 at 17.2% compounded semiannually for 17 semiannual periods

58. $1593.24 at 10 1/2% compounded quarterly for 14 years

59. $45,675 at 13.2% compounded quarterly for 9 years

60. $3256 at 14% compounded quarterly for 13 quarters

61. $56,780 at 15.3% compounded quarterly for 23 quarters

62. $13,675 at 14.4% compounded monthly for 36 months

63. $2398.45 at 13.2% compounded monthly for 48 months

64. $68,922 at 12% compounded daily (ignoring leap years) for 3 years of 365 days

65. $2964.58 at 11 1/4% compounded daily (ignoring leap years) for 9 years of 365 days

66. $45,788 at 15% compounded daily (ignoring leap years) for 11 years of 365 days

67. Suppose $10,000 is left at interest for 3 years at 12%. Find the interest earned if the interest is compounded
    (a) annually;   **(b)** quarterly;   **(c)** daily (365 days).

In the formula for compound interest given in this section, sometimes A is called the **future value,** *while P is the* **present value.** *Find the present value for the following future values.*

68. $10,000, if interest is 12% compounded semiannually for 5 years

69. $25,000, if interest is 16% compounded quarterly for 11 quarters

70. $45,678.93, if interest is 12.6% compounded monthly for 11 months

71. $123,788, if interest is 14.7% compounded daily for 195 days

72. $32,558.57, if interest is 18.45% compounded daily for 450 days

Solve each of the following problems. See Example 5.

73. *Escherichia coli* is a strain of bacteria that occurs naturally in many different organisms. Under certain conditions, the number of bacteria present in a colony is

$$E(t) = E_0 \cdot 2^{t/30},$$

where $E(t)$ is the number of bacteria present t min after the beginning of an experiment, and E_0 is the number present when $t = 0$. Let $E_0 = 2,400,000$ and find the number of bacteria at the following times.
    (a) $t = 5$ **(b)** $t = 10$ **(c)** $t = 60$ **(d)** $t = 120$

74. The higher a student's grade-point average, the fewer applications that the student need send to medical schools (other things being equal). Using information given in a guidebook for prospective medical students, we constructed the function $y = 540e^{-1.3x}$ for the number of applications a student should send out. Here y is the number of applications for a student whose grade-point average is x. The domain of x is the interval $[2.0, 4.0]$. Find the number of applications that should be sent out by students having the following grade-point averages.
    (a) 2.0   **(b)** 3.0   **(c)** 3.5   **(d)** 3.9

In calculus, it is shown that

$$e^x = 1 + x + \frac{x^2}{2 \cdot 1} + \frac{x^3}{3 \cdot 2 \cdot 1} + \frac{x^4}{4 \cdot 3 \cdot 2 \cdot 1} + \frac{x^5}{5 \cdot 4 \cdot 3 \cdot 2 \cdot 1} + \cdots .$$

75. Use the terms shown here and replace x with 1 to approximate $e^1 = e$ to three decimal places. Then check your results in Table 5 or with a calculator.

76. Use the terms shown here and replace x with $-.05$ to approximate $e^{-.05}$ to four decimal places. Check your results in Table 5 or with a calculator.

6.2   Logarithmic Functions

In the previous section we discussed exponential functions of the form $y = a^x$ for all positive values of a, where $a \neq 1$. As mentioned there, the horizontal line test shows that exponential functions are one-to-one, and thus have inverse functions. In this section we discuss the inverses of exponential functions. The equation of the inverse comes from exchanging x and y. Doing this with $y = a^x$ gives

$$x = a^y$$

as the inverse of the exponential function $y = a^x$. To solve this equation for y, use the following definition.

Definition of Logarithm

> For all real numbers y, and all positive numbers a and x, where $a \neq 1$,
>
> $$y = \log_a x \quad \text{if and only if} \quad x = a^y.$$

"Log" is an abbreviation for *logarithm*. Read $\log_a x$ as "the logarithm of x to the base a."

EXAMPLE 1   The chart below shows several pairs of equivalent statements. The same statement is written in both exponential and logarithmic forms.

Exponential form	*Logarithmic form*
$2^3 = 8$	$\log_2 8 = 3$
$(1/2)^{-4} = 16$	$\log_{1/2} 16 = -4$
$10^5 = 100,000$	$\log_{10} 100,000 = 5$
$3^{-4} = 1/81$	$\log_3 (1/81) = -4$
$5^1 = 5$	$\log_5 5 = 1$
$(3/4)^0 = 1$	$\log_{3/4} 1 = 0$ ●

Using the definition of logarithm, we can define the logarithmic function with base a.

Logarithmic Function

> If $a > 0$, $a \neq 1$, and $x > 0$, then
>
> $$f(x) = \log_a x$$
>
> is the **logarithmic function with base a**.

Exponential and logarithmic functions are inverses of each other. Since the domain of an exponential function is the set of all real numbers, the range of a logarithmic function will also be the set of all real numbers. In the same way, both the range of an exponential function and the domain of a logarithmic function are the set of all positive real numbers, so logarithms can be found for positive numbers only.

The graph of $y = 2^x$ is shown in Figure 6.4(a). To get the graph of its inverse, reflect $y = 2^x$ about the line $y = x$. The graph of the inverse function, $y = \log_2 x$, is shown in color. As the graph suggests, $y = \log_2 x$ is one-to-one and has the y-axis as a vertical asymptote.

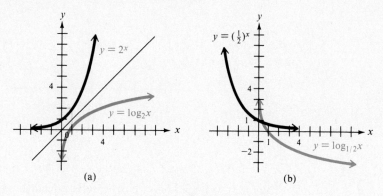

Figure 6.4

The graph of $y = (1/2)^x$ is shown in Figure 6.4(b). Its inverse, $y = \log_{1/2} x$, in color, is found by reflecting $y = (1/2)^x$ about the line $y = x$. The function $y = \log_{1/2} x$ is one-to-one and has the y-axis for a vertical asymptote.

EXAMPLE 2   Graph $y = \log_2 (x - 1)$.

The domain of this function is $(1, +\infty)$, since logarithms can be found only for positive numbers. The graph of $y = \log_2 (x - 1)$ will be the graph of $y = \log_2 x$, translated one unit to the right. See Figure 6.5.   ●

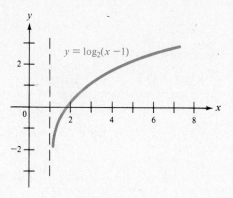

Figure 6.5

EXAMPLE 3   Graph $y = \log_3 |x|$.

There are two ways to graph this function: choose values of x and find the corresponding values of y, as shown in the table below, or first use the definition of logarithm to replace $y = \log_3 |x|$ with $|x| = 3^y$. Either method gives the graph in Figure 6.6.

x	-3	-1	$-1/3$	$1/3$	1	3
y	1	0	-1	-1	0	1

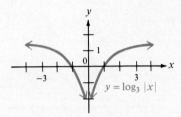

Figure 6.6

The next example shows how to solve equations involving logarithms.

EXAMPLE 4   Solve each of the following equations.

(a) $\log_x \dfrac{8}{27} = 3$

First, write the expression in exponential form.

$$x^3 = \frac{8}{27}$$

$$x^3 = \left(\frac{2}{3}\right)^3$$

$$x = \frac{2}{3}$$

The solution set is $\{2/3\}$.

(b) $\log_4 x = 5/2$

In exponential form, the given statement becomes

$$4^{5/2} = x$$

$$(4^{1/2})^5 = x$$

$$2^5 = x$$

$$32 = x.$$

The solution set is $\{32\}$.

Logarithms were originally important as an aid for numerical calculations, but the availability of inexpensive calculators has greatly reduced the need for this application of logarithms. Yet the principles behind the use of logarithms for calculation are important; these principles are based on the properties listed below.

Properties of Logarithms

If x and y are any positive real numbers, r is any real number, and a is any positive real number, $a \neq 1$, then

(a) $\log_a xy = \log_a x + \log_a y$ (b) $\log_a \dfrac{x}{y} = \log_a x - \log_a y$

(c) $\log_a x^r = r \cdot \log_a x$ (d) $\log_a a = 1$

(e) $\log_a 1 = 0.$

To prove part (a) of the properties of logarithms, let $m = \log_a x$ and $n = \log_a y$. Then, by the definition of logarithm,

$$a^m = x \quad \text{and} \quad a^n = y.$$

Multiplication gives

$$a^m \cdot a^n = xy.$$

By a property of exponents,

$$a^{m+n} = xy.$$

Now use the definition of logarithm to write

$$\log_a xy = m + n.$$

Since $m = \log_a x$ and $n = \log_a y$,

$$\log_a xy = \log_a x + \log_a y.$$

Properties (b) and (c) are proven in a similar way. (See Exercises 83 and 84.) Properties (d) and (e) follow directly from the definition of logarithm since $a^1 = a$ and $a^0 = 1$.

The properties of logarithms are useful for rewriting expressions with logarithms in different forms, as shown in the next examples.

EXAMPLE 5 Assuming that all variables represent positive real numbers, use the properties of logarithms to rewrite each of the following expressions.

(a) $\log_6 7 \cdot 9 = \log_6 7 + \log_6 9$

(b) $\log_9 \dfrac{15}{7} = \log_9 15 - \log_9 7$

(c) $\log_5 \sqrt{8} = \log_5 8^{1/2} = \dfrac{1}{2} \log_5 8$

(d) $\log_a \dfrac{mnq}{p^2} = \log_a m + \log_a n + \log_a q - 2 \log_a p$

(e) $\log_a \sqrt[3]{m^2} = \dfrac{2}{3} \log_a m$

(f) $\log_b \sqrt[n]{\dfrac{x^3 y^5}{z^m}} = \dfrac{1}{n} \log_b \dfrac{x^3 y^5}{z^m}$

$$= \frac{1}{n}(\log_b x^3 + \log_b y^5 - \log_b z^m)$$

$$= \frac{3}{n} \log_b x + \frac{5}{n} \log_b y - \frac{m}{n} \log_b z \quad \bullet$$

EXAMPLE 6   Use the properties of logarithms to write each of the following as a single logarithm with a coefficient of 1. Assume that all variables represent positive real numbers.

(a) $\log_3 (x + 2) + \log_3 x - \log_3 2 = \log_3 \dfrac{(x + 2)x}{2}$

(b) $2 \log_a m - 3 \log_a n = \log_a m^2 - \log_a n^3 = \log_a \dfrac{m^2}{n^3}$

(c) $\dfrac{1}{2} \log_b m + \dfrac{3}{2} \log_b 2n - \log_b m^2 n$

$$= \log_b m^{1/2} + \log_b (2n)^{3/2} - \log_b m^2 n$$

$$= \log_b \frac{m^{1/2}(2n)^{3/2}}{m^2 n}$$

$$= \log_b \frac{2^{3/2} n^{1/2}}{m^{3/2}}$$

$$= \log_b \sqrt{\frac{8n}{m^3}} \quad \bullet$$

EXAMPLE 7   Assume that $\log_{10} 2 = .3010$. Find the base 10 logarithms of 4 and 5.

By the properties of logarithms,

$$\log_{10} 4 = \log_{10} 2^2 = 2 \log_{10} 2 = 2(.3010) = .6020$$

$$\log_{10} 5 = \log_{10} \frac{10}{2} = \log_{10} 10 - \log_{10} 2 = 1 - .3010 = .6990. \quad \bullet$$

We can use compositions of the exponential and logarithmic functions to get two more useful properties. If $f(x) = a^x$ and $g(x) = \log_a x$, then

$$f[g(x)] = a^{\log_a x}$$

and
$$g[f(x)] = \log_a a^x.$$

In Section 5.6, we saw that for functions f and g that are inverses of each other, $f[g(x)] = g[f(x)] = x$. Since exponential and logarithmic functions of the same base are inverses of each other, we can apply this to get the following theorem.

Theorem

$$a^{\log_a x} = x \qquad \text{and} \qquad \log_a a^x = x.$$

By the results of the last theorem,

$$\log_5 5^3 = 3, \qquad 7^{\log_7 10} = 10, \qquad \text{and} \qquad \log_r r^{k+1} = k + 1.$$

6.2 EXERCISES

For each of the following statements, write an equivalent statement in logarithmic form.
See Example 1.

1. $3^4 = 81$

2. $2^5 = 32$

3. $10^4 = 10,000$

4. $8^2 = 64$

5. $(1/2)^{-4} = 16$

6. $(2/3)^{-3} = 27/8$

7. $10^{-4} = .0001$

8. $(1/100)^{-2} = 10,000$

For each of the following statements, write an equivalent statement in exponential form.
See Example 1.

9. $\log_6 36 = 2$

10. $\log_5 5 = 1$

11. $\log_{\sqrt{3}} 81 = 8$

12. $\log_4 (1/64) = -3$

13. $\log_{10} .0001 = -4$

14. $\log_3 \sqrt[3]{9} = 2/3$

15. $\log_m k = n$

16. $\log_2 r = y$

Find the value of each of the following expressions.

17. $\log_5 25$

18. $\log_3 81$

19. $\log_8 8$

20. $\log_7 1$

21. $\log_{10} .001$

22. $\log_6 \dfrac{1}{216}$

23. $\log_{25} 5$

24. $\log_{16} 2$

25. $\log_4 \dfrac{\sqrt[3]{4}}{2}$

26. $\log_9 \dfrac{\sqrt[4]{27}}{3}$

27. $\log_{1/3} \dfrac{9^{-4}}{3}$

28. $\log_{1/4} \dfrac{16^2}{2^{-3}}$

29. $\log_6 36^4$

30. $\log_5 125^2$

31. $2^{\log_2 9}$

32. $8^{\log_8 11}$

Solve each of the following equations. See Example 4.

33. $x = \log_2 32$

34. $x = \log_2 128$

35. $\log_x 25 = -2$

36. $\log_x (1/16) = -2$

Write each of the following expressions as a sum, difference, or product of logarithms.
Simplify the result if possible. Assume that all variables represent positive real numbers.
See Example 5.

37. $\log_3 (2/5)$

38. $\log_4 (6/7)$

39. $\log_2 \dfrac{6x}{y}$

40. $\log_3 \dfrac{4p}{q}$

41. $\log_5 \dfrac{5\sqrt{7}}{3}$

42. $\log_2 \dfrac{2\sqrt{3}}{5}$

43. $\log_4 (2x + 5y)$

44. $\log_6 (7m - 3q)$

45. $\log_k \dfrac{pq^2}{m}$

46. $\log_z \dfrac{x^5 y^3}{3}$

47. $\log_m \sqrt{\dfrac{5r^3}{z^5}}$

48. $\log_p \sqrt[3]{\dfrac{m^5 n^4}{t^2}}$

Write each of the following expressions as a single logarithm. Assume that all variables represent positive real numbers. See Example 6.

49. $\log_a x + \log_a y - \log_a m$

50. $(\log_b k - \log_b m) - \log_b a$

51. $2 \log_m a - 3 \log_m b^2$

52. $\frac{1}{2} \log_y p^3 q^4 - \frac{2}{3} \log_y p^4 q^3$

53. $-\frac{3}{4} \log_x a^6 b^8 + \frac{2}{3} \log_x a^9 b^3$

54. $\log_a (pq^2) + 2 \log_a (p/q)$

55. $\log_b (x + 2) + \log_b 7x - \log_b 8$

56. $\log_h (2m - 3) + \log_h 2m - \log_h 3$

57. $2 \log_a (z - 1) + \log_a (3z + 2)$

58. $\log_b (2y + 5) - \frac{1}{2} \log_b (y + 3)$

59. $-\frac{2}{3} \log_5 5m^2 + \frac{1}{2} \log_5 25m^2$

60. $-\frac{3}{4} \log_3 16p^4 - \frac{2}{3} \log_3 8p^3$

61. Graph each of the following functions. Compare the graphs to that of $y = \log_2 x$.
   (a) $y = (\log_2 x) + 3$
   (b) $y = \log_2 (x + 3)$
   (c) $y = |\log_2 (x + 3)|$

62. Graph each of the following functions. Compare the graphs to that of $y = \log_{1/2} x$.
   (a) $y = (\log_{1/2} x) - 2$
   (b) $y = \log_{1/2} (x - 2)$
   (c) $y = |\log_{1/2} (x - 2)|$

Graph each of the following functions. See Example 2.

63. $y = \log_3 x$

64. $y = \log_{10} x$

65. $y = \log_{1/2} (1 - x)$

66. $y = \log_{1/3} (3 - x)$

67. $y = \log_2 x^2$

68. $y = \log_3 (x - 1)$

69. $y = x \cdot \log_{10} x$

70. $y = x^2 \cdot \log_{10} x$

Given $\log_{10} 2 = 0.3010$ and $\log_{10} 3 = 0.4771$, find each of the following logarithms without using calculators or tables. See Example 7.

71. $\log_{10} 6$

72. $\log_{10} 12$

73. $\log_{10} 9$

74. $\log_{10} 20$

75. $\log_{10} 30$

76. $\log_{10} 36$

77. The population of an animal species that is introduced into a certain area may grow rapidly at first but then grow more slowly as time goes on. A logarithmic function can provide an excellent description of such growth. Suppose that the population of foxes, $F(t)$, in an area t months after the foxes were introduced there is

$$F(t) = 500 \log_{10} (2t + 3).$$

Use a calculator with a log key to find the population of foxes at the following times:
   (a) when they are first released into the area (that is, when $t = 0$);
   (b) after 3 months;
   (c) after 15 months.
   (d) Graph $y = F(t)$.

*The loudness of sounds is measured in a unit called a **decibel**. To measure with this unit, an intensity of I_0 is assigned to a very faint sound, called the **threshold sound**. If a particular sound has intensity I, then the decibel rating of this louder sound is*

$$10 \cdot \log_{10} \frac{I}{I_0}.$$

78. Find the decibel ratings of sounds having the following intensities.
   (a) $100 I_0$
   (b) $1000 I_0$
   (c) $100,000 I_0$
   (d) $1,000,000 I_0$

79. Find the decibel ratings of the following sounds, having intensities as given. (You will need a calculator with a log key.) Round answers to the nearest whole number.
  (a) whisper, $115I_0$
  (b) busy street, $9,500,000I_0$
  (c) heavy truck, 20 m away, $1,200,000,000I_0$
  (d) rock music, $895,000,000,000I_0$
  (e) jetliner at takeoff, $109,000,000,000,000I_0$

80. The intensity of an earthquake, measured on the *Richter scale,* is $\log_{10}(I/I_0)$, where I_0 is the intensity of an earthquake of a certain (small) size. Find the Richter scale ratings of earthquakes having the following intensities.
  (a) $1000I_0$ **(b)** $1,000,000I_0$ **(c)** $100,000,000I_0$

81. The San Francisco earthquake of 1906 had a Richter scale rating of 8.6. Use a calculator with an x^y key to express the intensity of this earthquake as a multiple of I_0. (See Exercise 80.)

82. How much more powerful is an earthquake with a Richter scale rating of 8.6 than one with a rating of 8.2?

Prove the following properties of logarithms.

83. $\log_a(x/y) = \log_a x - \log_a y$ **84.** $\log_a x^r = r \cdot \log_a x$

6.3   Natural Logarithms

Since our number system uses base 10, logarithms to base 10 are most convenient for numerical calculation, historically the main application of logarithms. Base 10 logarithms are called **common logarithms.** The common logarithm of the number x, or $\log_{10} x$, is often abbreviated as just $\log x$. Common logarithms are discussed in Section 6.5.

In most other practical applications of logarithms, however, the number $e \approx 2.718281828$ is used as base. Logarithms to base e are called **natural logarithms,** since they occur in many natural-world applications, such as those involving growth and decay. The abbreviation $\ln x$ is used for the natural logarithm of x, so that $\log_e x = \ln x$.

Natural logarithms can be found with a calculator that has an ln key or with a table of natural logarithms. A table of natural logarithms is given in Table 5. Reading directly from this table,

$$\ln 55 = 4.0073,$$
$$\ln 1.9 = 0.6419,$$
and $$\ln 0.4 = -.9163.$$

EXAMPLE 1   Use a calculator or Table 5 to find the following logarithms.
  (a) $\ln 85$
    If you are using a calculator, enter 85, press the ln key, and read the result, 4.4427.

Table 5 does not give ln 85. However, the value of ln 85 can be found using the properties of logarithms.

$$\ln 85 = \ln (8.5 \times 10)$$
$$= \ln 8.5 + \ln 10$$
$$\approx 2.1401 + 2.3026$$
$$= 4.4427$$

A result found in this way is sometimes slightly different from the answer found using a calculator. Such a difference is due to rounding error.

(b) ln 36

A calculator gives $\ln 36 = 3.5835$. To use the table, first use properties of logarithms, since 36 is not listed in Table 5.

$$\ln 36 = \ln 6^2$$
$$= 2 \ln 6$$
$$\approx 2(1.7918)$$
$$= 3.5836$$

Alternatively, find ln 36 as follows:

$$\ln 36 = \ln 9 \cdot 4 = \ln 9 + \ln 4 = 2.1972 + 1.3863 = 3.5835. \quad \bullet$$

Logarithms to Other Bases   We can use a calculator or a table to find the values of either natural logarithms (base e) or common logarithms (base 10). However, sometimes it is convenient to use logarithms to other bases. The following theorem can be used to convert logarithms from one base to another.

Change of Base Theorem

If x is any positive number and if a and b are positive real numbers, $a \neq 1, b \neq 1$, then

$$\log_a x = \frac{\log_b x}{\log_b a}.$$

To prove this result, use the definition of logarithm to write $y = \log_a x$ as $x = a^y$ or $x = a^{\log_a x}$ (for positive x and positive a, $a \neq 1$). Now take base b logarithms of both sides of this last equation:

$$\log_b x = \log_b a^{\log_a x}$$

or

$$\log_b x = (\log_a x)(\log_b a),$$

from which

$$\log_a x = \frac{\log_b x}{\log_b a}.$$

EXAMPLE 2   Use natural logarithms to find each of the following. Round to the nearest hundredth.

(a) $\log_5 17$

Let $x = 17$, $a = 5$, and $b = e$. Substituting into the change of base theorem gives

$$\log_5 17 = \frac{\log_e 17}{\log_e 5}$$
$$= \frac{\ln 17}{\ln 5}.$$

Now use a calculator or Table 5.

$$\log_5 17 \approx \frac{2.8332}{1.6094}$$
$$\approx 1.76$$

To check, use a calculator with an x^y key, along with the definition of logarithm, to verify that $5^{1.76} \approx 17$.

(b) $\log_2 .1$

$$\log_2 .1 = \frac{\ln .1}{\ln 2} \approx \frac{-2.3026}{.6931} = -3.32 \quad \bullet$$

EXAMPLE 3   One measure of the diversity of the species in an ecological community is given by the formula

$$H = -[P_1 \log_2 P_1 + P_2 \log_2 P_2 + \cdots + P_n \log_2 P_n],$$

where $P_1, P_2 \ldots, P_n$ are the proportions of a sample belonging to each of n species found in the sample. For example, in a community with two species, where there are 90 of one species and 10 of the other, $P_1 = 90/100 = .9$ and $P_2 = 10/100 = .1$. Thus,

$$H = -[.9 \log_2 .9 + .1 \log_2 .1].$$

We found $\log_2 .1$ in Example 2(b) above. Now find $\log_2 .9$.

$$\log_2 .9 = \frac{\ln .9}{\ln 2}$$
$$\approx \frac{-.1054}{.6931}$$
$$\approx -.152$$

Therefore,

$$H \approx -[(.9)(-.152) + (.1)(-3.32)] \approx .469. \quad \bullet$$

Continuous Compounding   Section 6.1 introduced the formula for compound interest,

$$A = P\left(1 + \frac{i}{m}\right)^{nm}.$$

By this formula, the more often interest is compounded within a given time period, the more interest will be earned. Using a calculator with an x^y key, and the formula above, we can get the results shown in the following chart.

		Interest on \$1000 at \$12% per Year for 10 Years		
Compounded	*Number of periods*	*Total on deposit*		*Interest*
Not at all (simple interest)	—	—		\$1200.00
Annually	10	$1000(1 + .12)^{10}$	$= \$3105.85$	\$2105.85
Semiannually	20	$1000\left(1 + \dfrac{.12}{2}\right)^{20}$	$= \$3207.14$	\$2207.14
Quarterly	40	$1000\left(1 + \dfrac{.12}{4}\right)^{40}$	$= \$3262.04$	\$2262.04
Monthly	120	$1000\left(1 + \dfrac{.12}{12}\right)^{120}$	$= \$3300.39$	\$2300.39
Daily	3650	$1000\left(1 + \dfrac{.12}{365}\right)^{3650}$	$= \$3319.46$	\$2319.46
Hourly	87,600	$1000\left(1 + \dfrac{.12}{8760}\right)^{87,600}$	$= \$3320.09$	\$2320.09
Every minute	5,256,000	$1000\left(1 + \dfrac{.12}{525,600}\right)^{5,256,000}$	$= \$3320.11$	\$2320.11

As suggested by the chart, it makes a big difference whether interest is compounded or not. Interest differs by $2105.85 - \$1200 = \$905.85$ when simple interest is compared to interest compounded annually. However, increasing the frequency of compounding makes smaller and smaller differences in the amount of interest earned. In fact, it can be shown that even if interest is compounded at intervals of time as small as one chooses (such as each hour, each minute, or each second), the total amount of interest earned will be only slightly more than for daily compounding. This is true even for a process called **continuous compounding,** which can be described loosely as compounding every instant. It turns out (although we shall not prove it) that the formula for continuous compounding involves the number e.

Continuous Compounding

> If P dollars is deposited at a rate of interest i compounded continuously for n years, the final amount on deposit is
>
> $$A = Pe^{ni}$$
>
> dollars.

EXAMPLE 4

Suppose $5000 is deposited in an account paying 8% compounded continuously for five years. Find the total on deposit.

Let $P = 5000$, $n = 5$, and $i = .08$. Then

$$A = 5000e^{5(.08)} = 5000e^{.4}.$$

From Table 5 or a calculator, $e^{.4} \approx 1.49182$, and

$$A = 5000(1.49182) = 7459.10,$$

or $7459.10. Check that daily compounding would have produced a compound amount about 30¢ less.   ●

EXAMPLE 5

Suppose $24,000 is deposited at 16% for 9 years. Find the interest earned by (a) daily, (b) hourly, and (c) continuous compounding.

(a) In 9 years there are $9 \times 365 = 3285$ days. The total on deposit with daily compounding is

$$24{,}000\left(1 + \frac{.16}{365}\right)^{3285} = 101{,}264.74,$$

or $101,264.74. The interest earned is

$$\$101{,}264.74 - \$24{,}000 = \$77{,}264.74.$$

(b) In one year, there are $365 \times 24 = 8760$ hours, while in 9 years there are $9 \times 8760 = 78{,}840$ hours. The total on deposit is

$$24{,}000\left(1 + \frac{.16}{8760}\right)^{78{,}840} = 101{,}295.37,$$

or $101,295.37. This amount includes interest of

$$\$101{,}295.37 - \$24{,}000 = \$77{,}295.37,$$

only $30.63 more than when interest is compounded daily.

(c) The total amount for continuous compounding is

$$24{,}000e^{9(.16)} = 24{,}000e^{1.44}$$
$$= 24{,}000(4.2206958) \qquad \text{Use a calculator}$$
$$= 101{,}296.70,$$

or $101,296.70. With continuous compounding, the amount of interest earned, $101,296.70 - $24,000 = $77,296.70, is only $1.33 more than when interest is compounded hourly.   ●

6.3 EXERCISES

Find each of the following logarithms to four decimal places. Use a calculator or Table 5. See Example 1.

1. ln 4	**2.** ln 6	**3.** ln 17	**4.** ln 29
5. ln 350	**6.** ln 900	**7.** ln 49	**8.** ln 64
9. ln 42	**10.** ln 72	**11.** ln 81,000	**12.** ln 121,000

Find each of the following logarithms to the nearest hundredth. See Example 2.

13. $\log_5 10$	**14.** $\log_9 12$	**15.** $\log_{15} 5$	**16.** $\log_6 8$
17. $\log_{1/2} 3$	**18.** $\log_{12} 62$	**19.** $\log_{100} 83$	**20.** $\log_{200} 175$
21. $\log_{2.9} 7.5$	**22.** $\log_{5.8} 12.7$	**23.** $\log_{.6} 13.2$	**24.** $\log_{.9} 5.77$

Graph each of the following functions.

25. $y = \ln x$	**26.** $y = |\ln x|$	**27.** $y = x \ln x$	**28.** $y = x^2 \ln x$

Work the following problems. For Exercises 29 and 30, see Example 3.

29. Suppose a sample of a small community shows two species with 50 individuals each. Find the index of diversity H.

30. A virgin forest in northwestern Pennsylvania has 4 species of large trees with the following proportions of each: hemlock, .521; beech, .324; birch, .081; maple, .074. Find the index of diversity H.

31. The number of species in a sample is given by

$$S(n) = a \ln \left(1 + \frac{n}{a} \right).$$

Here n is the number of individuals in the sample and a is a constant that indicates the diversity of species in the community. If $a = .36$, find $S(n)$ for the following values of n.

    (a) 100   **(b)** 200   **(c)** 150   **(d)** 10

32. In Exercise 31, find $S(n)$ if a changes to .88. Use the following values of n.

    (a) 50   **(b)** 100   **(c)** 250

33. If an object is fired vertically upward and is subject only to the force of gravity, g, and to air resistance, then the maximum height, H, attained by the object is

$$H = \frac{1}{K} \left(V_0 - \frac{g}{K} \ln \frac{g + V_0 K}{g} \right),$$

where V_0 is the initial velocity of the object and K is a constant. Find H if $K = 2.5$, $V_0 = 1000$ ft per sec, and $g = 32$ ft per sec per sec.

34. The pull, P, of a tracked vehicle on dry sand under certain conditions is approximated by

$$P = W \left[.2 + .16 \ln \frac{G(bl)^{3/2}}{W} \right],$$

where G is an index of sand strength, W is the load on the vehicle, b is the width of the track, and l is the length of the track.* Find P if $W = 10$, $G = 5$, $b = 30.5$ cm, and $l = 61.0$ cm.

35. The following formula† can be used to estimate the population of the United States, where t is time in years measured from 1914 (times before 1914 are negative):

$$N = \frac{197,273,000}{1 + e^{-.03134t}}.$$

(a) Complete the following chart.

Year	Observed population	Predicted population
1790	3,929,000	3,929,000
1810	7,240,000	
1860	31,443,000	30,412,000
1900	75,995,000	
1970	204,000,000	
1980	224,000,000	

(b) Estimate the number of years after 1914 that it would take for the population to increase to 197,273,000.

36. Many environmental situations place effective limits on the growth of the number of an organism in an area. Many such limited growth situations are described by the *logistic function*,

$$G(t) = \frac{m \cdot G_0}{G_0 + (m - G_0)e^{-kmt}},$$

where G_0 is the initial number present, m is the maximum possible size of the population, k is a positive constant, and t is time in appropriate units. Assume $G_0 = 1000$, $m = 25,000$, and $k = .04$.
(a) Find $G(.005)$.   (b) Find $G(.01)$.   (c) Find a value of t so that $G(t) = m/2$.

To find the maximum permitted levels of certain pollutants in fresh water, the EPA has established the functions of Exercises 37–38, where $M(h)$ is the maximum permitted level of pollutant for a water hardness of h milligrams per liter. Find $M(h)$ in each case. (These results give the maximum permitted average concentration in micrograms per liter for a 24-hour period.)

37. Pollutant: copper, $M(h) = e^r$, where $r = 0.65 \cdot \ln h - 1.94$ and $h = 9.7$

38. Pollutant: lead, $M(h) = e^r$, where $r = 1.51 \cdot \ln h - 3.37$ and $h = 8.4$

---

*Gerald W. Turnage, *Prediction of Track Pull Performance in a Desert Sand*, unpublished MS thesis, The Florida State University, 1971.

†The formula is given in Alfred J. Lotka, *Elements of Mathematical Biology*, reprinted by Dover Press, 1957, p. 67.

Find the total amount on deposit if $20,000 is invested at 8% compounded continuously for the following numbers of years. (Use a calculator with an x^y key for Exercises 43 and 44.) See Examples 4 and 5.

39. 1   **40.** 5   **41.** 10   **42.** 15

43. 3   **44.** 7

Find the total amount on deposit if the following amounts are deposited at the given rate of interest for the given time periods. Assume continuous compounding. See Examples 4 and 5.

45. $16,500 at 11.2% for 9 years   **46.** $23,889 at 17.1% for 13 years

47. $67,312 at 11.9% for 18 months   **48.** $43,789.12 at 15.67% for 30 months

49. Suppose $56,890 is deposited at 15.1% for 3 years. Find the interest earned by
   (a) daily   **(b)** hourly   and   **(c)** continuous compounding.

50. If $34,678.79 is deposited at 15.6% for 5 years, find the interest earned by
   (a) daily   **(b)** hourly   and   **(c)** continuous compounding.

*In the formula $A = Pe^{ni}$, P is the **present value** of the **future value** A. Find the present value of the following future values. Assume continuous compounding.*

51. $18,000 at 10% for 5 years   **52.** $25,000 at 12% for 10 years

In the central Sierra Nevada mountains of California, it turns out that the percent of moisture that falls as snow rather than rain is approximated reasonably well by

$$p = 86.3 \ln h - 680,$$

where p is the percent of snow at an altitude h in feet. (Assume $h \geq 3000$.)

53. Find the percent of moisture that falls as snow at the following altitudes.
   (a) 3000 ft   **(b)** 4000 ft   **(c)** 7000 ft

54. Graph p.

6.4 Exponential and Logarithmic Equations

In Section 6.1 we solved exponential equations such as $(1/3)^x = 81$ by writing each side of the equation as a power of 3. However, that method cannot be used to solve an equation such as $7^x = 12$, since 12 cannot easily be written as a power of 7. However, the equation $7^x = 12$ can be solved by taking the logarithm of each side, a process that depends on the fact that a logarithmic function is one-to-one.

> If $x > 0$, $y > 0$, $b > 0$, $b \neq 1$, then
>
> $x = y$ if and only if $\log_b x = \log_b y$.

EXAMPLE 1

Solve the equation $7^x = 12$.

While the result is valid for any appropriate base b, the best practical base to use is often base e. Taking base e (natural) logarithms of both sides gives

$$\ln 7^x = \ln 12$$
$$x \cdot \ln 7 = \ln 12$$
$$x = \frac{\ln 12}{\ln 7}.$$

To get a decimal approximation for x, use Table 5 or a calculator:

$$x = \frac{\ln 12}{\ln 7} \approx \frac{2.4849}{1.9459}.$$

Using a calculator to divide 2.4849 by 1.9459 gives

$$x \approx 1.277.$$

A calculator with an x^y key can be used to check this answer. Evaluate $7^{1.277}$; the result should be approximately 12. This step verifies that, to the nearest thousandth, the solution set is $\{1.277\}$.   ●

EXAMPLE 2

Solve $3^{2x-1} = 4^{x+2}$.

Taking natural logarithms on both sides gives

$$\ln 3^{2x-1} = \ln 4^{x+2}.$$

Now use properties of logarithms.

$$(2x - 1) \ln 3 = (x + 2) \ln 4$$
$$2x \ln 3 - \ln 3 = x \ln 4 + 2 \ln 4$$
$$2x \ln 3 - x \ln 4 = 2 \ln 4 + \ln 3$$

Factor out x on the left to get

$$x(2 \ln 3 - \ln 4) = 2 \ln 4 + \ln 3$$

or
$$x = \frac{2 \ln 4 + \ln 3}{2 \ln 3 - \ln 4}.$$

Using the properties of logarithms, this can be expressed as

$$x = \frac{\ln 16 + \ln 3}{\ln 9 - \ln 4}$$

or finally,
$$x = \frac{\ln 48}{\ln \dfrac{9}{4}} = \frac{\ln 48}{\ln 2.25}.$$

This quotient could be approximated by a decimal if desired:

$$x = \frac{\ln 48}{\ln 2.25} \approx \frac{3.8712}{.8109} \approx 4.774.$$

To the nearest thousandth, the solution set is $\{4.774\}$. To find $\ln 2.25$ using Table 5, write $\ln 2.25$ as $\ln 1.5^2 = 2 \ln 1.5$.   •

Logarithmic Equations   The properties of logarithms given in Section 6.2 are useful in solving logarithmic equations, as shown in the next examples.

EXAMPLE 3

Solve $\log_a (x + 4) - \log_a (x + 2) = \log_a x$.
   Using a property of logarithms, rewrite the equation as

$$\log_a \frac{x + 4}{x + 2} = \log_a x.$$

Then

$$\frac{x + 4}{x + 2} = x$$

$$x + 4 = x(x + 2)$$

$$x + 4 = x^2 + 2x$$

$$x^2 + x - 4 = 0.$$

By the quadratic formula,

$$x = \frac{-1 \pm \sqrt{1 + 16}}{2},$$

so that

$$x = \frac{-1 + \sqrt{17}}{2} \quad \text{or} \quad x = \frac{-1 - \sqrt{17}}{2}.$$

We cannot evaluate $\log_a x$ for $x = (-1 - \sqrt{17})/2$ since this number is negative and thus not in the domain of $\log_a x$. For this reason, the only valid solution is the positive number

$$x = \frac{-1 + \sqrt{17}}{2},$$

giving the solution set $\{(-1 + \sqrt{17})/2\}$.   •

EXAMPLE 4

Solve $\log (3x + 2) + \log(x - 1) = 1$.
   Since $\log x$ is an abbreviation for $\log_{10} x$, and $1 = \log_{10} 10$, the properties of logarithms give

$$\log (3x + 2)(x - 1) = \log 10$$

$$(3x + 2)(x - 1) = 10$$

$$3x^2 - x - 2 = 10$$

$$3x^2 - x - 12 = 0.$$

Now use the quadratic formula to arrive at

$$x = \frac{1 \pm \sqrt{1 + 144}}{6}.$$

If $x = (1 - \sqrt{145})/6$, then $x - 1 < 0$; therefore, $\log (x - 1)$ does not exist. For this reason, this proposed solution must be discarded, giving the solution set $\{(1 + \sqrt{145})/6\}$. ●

Solving Exponential and Logarithmic Equations

To solve an exponential or logarithmic equation, first use algebra to change the equation into one of the following forms, where a and b are real numbers.

1. $a^{f(x)} = b$
   Solve by taking logarithms of each side. (Natural logarithms are often a good choice.)
2. $\log_a f(x) = \log_a g(x)$
   From the given equation, $f(x) = g(x)$, which is solved algebraically.

The next examples show applications of exponential and logarithmic equations.

EXAMPLE 5 Suppose that the function

$$P(t) = 10,000e^{.4t}$$

gives the population of a city at time t measured in years. In how many years will the population of the city double?

Since the population of the city is 10,000 when $t = 0$, we want to find the value of t for which $P(t) = 20,000$. Replace $P(t)$ with 20,000 to get

$$20,000 = 10,000e^{.4t},$$

or

$$2 = e^{.4t}.$$

To solve for t, take natural logarithms on both sides.

$$\ln 2 = \ln e^{.4t}$$

Since $\ln e^{.4t} = (.4t)\ln e = (.4t)(1) = .4t$,

$$\ln 2 = .4t,$$

with

$$t = \frac{\ln 2}{.4}.$$

Using a calculator or Table 5 to find $\ln 2$ and then using a calculator to find the quotient gives $t = 1.73$ to the nearest hundredth. The population of the city will double in about 1.73 years. ●

EXAMPLE 6   Suppose the amount, y, in grams, of a certain radioactive substance at a time t is given by

$$y = y_0 e^{-.1t},$$

where y_0 is the amount of the substance present initially (when $t = 0$) and t is measured in days. Find the *half-life* of the substance—that is, the time it takes for half a given amount of the substance to decay.

We want to know the time t that must elapse for y to be reduced to a value equal to $y_0/2$. That is, we want to solve the equation

$$\frac{1}{2}y_0 = y_0 e^{-.1t}.$$

Dividing both sides by y_0 yields $1/2 = e^{-.1t}$, and if we now take natural logarithms of both sides,

$$\ln \frac{1}{2} = \ln e^{-.1t}$$

or

$$\ln \frac{1}{2} = -.1t \ln e.$$

Since $\ln e = 1$,

$$\ln \frac{1}{2} = -.1t,$$

$$t = \frac{-\ln 1/2}{.1}.$$

From Table 5 or a calculator, $\ln 1/2 = \ln .5 = -.6931$. Dividing gives

$$t \approx 6.9 \text{ days.} \quad \bullet$$

EXAMPLE 7   Carbon 14 is a radioactive isotope of carbon which has a half-life of about 5600 yr. The atmosphere contains much carbon, mostly in the form of carbon dioxide, with small traces of carbon 14. Most of this atmospheric carbon is in the form of the nonradioactive isotope carbon 12. The ratio of carbon 14 to carbon 12 is virtually constant in the atmosphere. However, as a plant absorbs carbon dioxide from the air in the process of photosynthesis, the carbon 12 stays in the plant while the carbon 14 decays by conversion to nitrogen. Thus, the ratio of carbon 14 to carbon 12 is smaller in the plant than it is in the atmosphere. Even when the plant is eaten by an animal, this ratio will continue to decrease. Based on these facts, a method of dating objects called *carbon 14 dating* has been developed.

Let R be the (nearly constant) ratio of carbon 14 to carbon 12 found in the atmosphere, and let r be the ratio found in a fossil. It can be shown that the relationship between R and r is given by

$$\frac{R}{r} = e^{(t \ln 2)/5600},$$

where t is the age of the fossil in years.

(a) Verify the formula for $t = 0$.
   If $t = 0$,

$$\frac{R}{r} = e^0 = 1.$$

Thus $R = r$, so that the ratio in the fossil is the same as the ratio in the atmosphere. This is true only when $t = 0$.

(b) Verify the formula for $t = 5600$.
   Substitute 5600 for t. Then

$$\frac{R}{r} = e^{(5600 \ln 2)/5600}$$

$$\frac{R}{r} = e^{\ln 2} = 2 \qquad \text{(Recall that } a^{\log_a x} = x\text{)}$$

$$r = \frac{1}{2}R.$$

From this last result, we see that the ratio in the fossil is half the ratio in the atmosphere. Since the half-life of carbon 14 is 5600 yr, we expect only half of it to remain at the end of that time. Thus, the formula gives the correct result for $t = 5600$. ●

6.4 EXERCISES

Solve the following equations. Give answers as decimals rounded to the nearest thousandth. See Examples 1–4.

1. $3^x = 6$

2. $4^x = 12$

3. $7^x = 8$

4. $13^p = 55$

5. $3^{a+2} = 5$

6. $5^{2-x} = 12$

7. $6^{1-2k} = 8$

8. $2^{k-3} = 11$

9. $4^{3m-1} = 12$

10. $3^{2m-5} = 13$

11. $e^{k-1} = 4$

12. $e^{2-y} = 12$

13. $2e^{5a+2} = 8$

14. $10e^{3z-7} = 5$

15. $2^x = -3$

16. $(1/4)^p = -4$

17. $\left(1 + \frac{r}{2}\right)^5 = 9$

18. $\left(1 + \frac{n}{4}\right)^3 = 12$

19. $100(1 + .02)^{3+n} = 150$

20. $500(1 + .05)^{p/4} = 200$

21. $\log (t - 1) = 1$

22. $\log q^2 = 1$

23. $\log (x - 3) = 1 - \log x$

24. $\log (z - 6) = 2 - \log (z + 15)$

25. $\ln (y + 2) = \ln (y - 7) + \ln 4$

26. $\ln p - \ln (p + 1) = \ln 5$

27. $\ln (3x - 1) - \ln (2 + x) = \ln 2$

28. $\ln (8k - 7) - \ln (3 + 4k) = \ln (9/11)$

29. $\ln (5 + 4y) - \ln (3 + y) = \ln 3$

30. $\ln m + \ln (2m + 5) = \ln 7$

31. $\ln x + 1 = \ln (x - 4)$

32. $\ln (4x - 2) = \ln 4 - \ln (x - 2)$

33. $2 \ln (x - 3) = \ln (x + 5) + \ln 4$

34. $\ln (k + 5) + \ln (k + 2) = \ln 14k$

35. $\log_5 (r + 2) + \log_5 (r - 2) = 1$

36. $\log_4 (z + 3) + \log_4 (z - 3) = 1$

37. $\log_3 (a - 3) = 1 + \log_3 (a + 1)$

38. $\log w + \log (3w - 13) = 1$

39. $\log_2 \sqrt{2y^2 - 1} = 1/2$

40. $\log_2 (\log_2 x) = 1$

41. $\log z = \sqrt{\log z}$

42. $\log x^2 = (\log x)^2$

Find the solution of each of the following equations. Round to the nearest thousandth.

43. $\log_x 5.87 = 2$ **44.** $\log_x 11.9 = 3$ **45.** $1.8^{p+4} = 9.31$ **46.** $3.7^{5z-1} = 5.88$

47. The amount of a radioactive substance present at time t measured in seconds is $A(t) = 5000e^{-.02t}$, where $A(t)$ is measured in grams. Find the half-life of the substance, that is, the time it takes until only half the substance remains.

48. Find the half-life of a radioactive specimen if the amount y in grams present at time t in days is $y = 1000e^{-.4t}$.

49. Suppose the number of rabbits in a colony is $y = y_0 e^{.4t}$, where t represents time in months and y_0 is the rabbit population when $t = 0$.
   (a) If $y_0 = 100$, find the number of rabbits present at time $t = 4$.
   (b) How long will it take for the number of rabbits to triple?

50. A midwestern city finds its residents moving to the suburbs. Its population is declining according to the relationship $P = P_0 e^{-.04t}$, where t is time measured in years and P_0 is the population at time $t = 0$. Assume that $P_0 = 1,000,000$.
   (a) Find the population at time $t = 1$.
   (b) Estimate the time it will take for the population to be reduced to 750,000.
   (c) How long will it take for the population to be cut in half?

For Exercises 51–56, refer to Example 7 in this section of the text.

51. Suppose an Egyptian mummy is discovered in which the ratio of carbon 14 to carbon 12 is only about half the ratio found in the atmosphere. About how long ago did the Egyptian die?

52. If the ratio of carbon 14 to carbon 12 in an object is 1/4 the atmospheric ratio, how old is the object? How old if the ratio is 1/8?

53. Paint from the Lascaux caves of France contains 15% of the normal amount of carbon 14. Estimate the age of the caves.

54. Solve the formula for t.

55. Suppose a specimen is found in which $r = (2/3) R$. Estimate the age of the specimen.

56. Estimate the age of a specimen having $r = R/12$.

A large cloud of radioactive debris from a nuclear explosion has floated over the Pacific Northwest, contaminating much of the hay supply. Consequently, farmers in the area are concerned that the cows who eat this hay will give contaminated milk. (The tolerance level for radioactive iodine in milk is 0.) The percent of the initial amount of radioactive iodine still present in the hay after t days is approximated by $P(t) = 100e^{-.1t}$.

57. Some scientists feel that the hay is safe after the percent of radioactive iodine has declined to 10% of the original amount. Find the number of days before the hay can be used.

58. Other scientists believe that the hay is not safe until the level of radioactive iodine has declined to only 1% of the level. Find the number of days this would take.

Solve each of the following equations for the indicated variables. Use logarithms to the appropriate bases.

59. $P = P_0 e^{kt/1000}$, for t

60. $I = \dfrac{E}{R}(1 - e^{-Rt/2})$, for t

61. $T = T_0 + (T_1 - T_0)\,10^{-kt}$, for t

62. $A = \dfrac{Pi}{1 - (1 + i)^{-n}}$, for n

Recall (from the exercises for Section 6.2) the formula for the decibel rating of a sound:

$$d = 10 \log \frac{I}{I_0}.$$

63. Solve this formula for I.

64. A few years ago, there was a controversy about a proposed government limit on factory noise. One group wanted a maximum of 89 decibels, while another group wanted 86. This difference seemed very small to many people. Find the percent by which the 89 decibel intensity exceeds that for 86 decibels.

In Section 6.1 we gave the formula for compound interest:

$$A = P\left(1 + \frac{i}{m}\right)^{nm}.$$

65. Use natural logarithms and solve for i.

66. Use the result of Exercise 65 and find i to the nearest hundredth of a percent if \$1786 becomes \$2063.40 after 1.25 years, with interest compounded monthly.

67. The turnover of legislators is a problem of interest to political scientists. One model of legislative turnover in the U.S. House of Representatives is given by

$$M = 434e^{-0.08t},$$

where M is the number of continuously serving members at time t.* This model is based on the 1965 membership of the House. Find the number of continuously serving members in each of the following years.
(a) 1969 (b) 1973 (c) 1979

68. Solve the formula in Exercise 67 for t.

69. The growth of bacteria in food products makes it necessary to time-date some products (such as milk) so that they will be sold and consumed before the bacteria count is too high. Suppose for a certain product that the number of bacteria present is given by

$$f(t) = 500e^{.1t},$$

under certain storage conditions, where t is time in days after packing of the product and the value of $f(t)$ is in millions. Find the number of bacteria present at each of the following times.
(a) 2 days (b) 1 week (c) 2 weeks

70. (a) Suppose the product of Exercise 69 cannot be safely eaten after the bacteria count reaches 3,000,000,000. How long will this take?
(b) If $t = 0$ corresponds to January 1, what date should be placed on the product?

---

*Thomas W. Casstevens, "Exponential Models of Legislative Turnover," UMAP, Unit 296.

6.5 Common Logarithms (Optional)

Before small calculators became easily available, computation involving multiplication, division, powers, and roots was performed by using the properties of logarithms and tables of base 10 logarithms. Now many of those problems can be worked on a calculator. However, it is still not possible to calculate numbers like $1.4^{17.3}$ and $12.8^{0.004}$ with non-scientific calculators. In this section we will see how to do such calculations with logarithms.

As we have said, logarithms to base 10, called common logarithms, are most convenient for numerical calculations. With $\log_{10} x$ abbreviated as $\log x$,

$$\log 1000 = \log 10^3 \quad = 3$$
$$\log 100 = \log 10^2 \quad = 2$$
$$\log 10 = \log 10^1 \quad = 1$$
$$\log 1 = \log 10^0 \quad = 0$$
$$\log .1 = \log 10^{-1} = -1$$
$$\log .01 = \log 10^{-2} = -2$$
$$\log .001 = \log 10^{-3} = -3,$$

and so on.

Since there is no rational number x such that $10^x = 6$, there is no rational number x such that $x = \log 6$. However, we can find a decimal approximation for $\log 6$. The excerpt of the logarithm table included on the facing page shows that a decimal approximation for $\log 6$ is given by .7782:

$$\log 6 \approx .7782$$

(or, equivalently, $10^{.7782} \approx 6$). Since most logarithms are approximations, it is common to replace $\approx$ with $=$ and write $\log 6 = .7782$.

To find $\log 6240$ with the table, start with

$$1000 < 6240 < 10,000,$$

so that $\log 1000 < \log 6240 < \log 10,000$. Since $\log 1000 = 3$ and $\log 10,000 = 4$,

$$3 < \log 6240 < 4.$$

Therefore, $\log 6240$ equals 3 plus some decimal. This decimal part of the logarithm is found in the table. The decimal part of a logarithm is called the **mantissa** and the integer part, the **characteristic.** The entries in the table of logarithms are all positive, so the mantissa must always be positive. However, the characteristic may be any integer.

One way to find the characteristic is to first write the number in scientific notation. For example, to find $\log 6240$, write 6240 as 6.24×10^3. Then, by the properties of logarithms discussed earlier in this chapter,

$$\log 6240 = \log (6.24 \times 10^3)$$
$$= \log 6.24 + \log 10^3$$
$$= \log 6.24 + 3 \log 10$$
$$= \log 6.24 + 3.$$

To find the decimal part of log 6240 in the table, locate the first two significant figures, 62, in the left column of the table. Then find the third digit, 4, across the top.

	0	1	2	3	4	5	6	7	8	9
5.7	.7559	.7566	.7574	.7582	.7589	.7597	.7604	.7612	.7619	.7627
5.8	.7634	.7642	.7649	.7657	.7664	.7672	.7679	.7686	.7694	.7701
5.9	.7709	.7716	.7723	.7731	.7738	.7745	.7752	.7760	.7767	.7774
6.0	.7782	.7789	.7796	.7803	.7810	.7818	.7825	.7832	.7839	.7846
6.1	.7853	.7860	.7868	.7875	.7882	.7889	.7896	.7903	.7910	.7917
6.2	.7924	.7931	.7938	.7945	.7952	.7959	.7966	.7973	.7980	.7987
6.3	.7993	.8000	.8007	.8014	.8021	.8028	.8035	.8041	.8048	.8055
6.4	.8062	.8069	.8075	.8082	.8089	.8096	.8102	.8109	.8116	.8122
6.5	.8129	.8136	.8142	.8149	.8156	.8162	.8169	.8176	.8182	.8189
6.6	.8195	.8202	.8209	.8215	.8222	.8228	.8235	.8241	.8248	.8254

The intersection of this row and column gives the mantissa, .7952. Since the characteristic is 3,

$$\log 6240 = 3.7952.$$

EXAMPLE 1 Find log .00587.

Use scientific notation and the properties of logarithms to get

$$\log .00587 = \log (5.87 \times 10^{-3})$$
$$= \log 5.87 + \log 10^{-3}$$
$$= .7686 + (-3)$$
$$= .7686 - 3.$$

The logarithm is usually left in this form. A calculator would give the answer as -2.2314, the algebraic sum of .7686 and -3. The decimal portion in the calculator answer is a negative number. This is not the best form for the logarithm when using tables, since it is not clear which number is the mantissa.

It is possible to write the characteristic in other forms. For example, log .00587 could be written as

$$\log .00587 = 7.7686 - 10.$$

The best choice depends on the anticipated use of the logarithm. ●

Let us now use Table 4 and the properties of logarithms to do some numerical calculations.

EXAMPLE 2   Find $(76.9)(.00282)$.

Use the properties of logarithms and Table 4.

$$
\begin{aligned}
\log (76.9)(.00282) &= \log 76.9 + \log .00282 \\
&= 1.8859 + (.4502 - 3) \\
&= 1.8859 + .4502 - 3 \\
&= 2.3361 - 3 \\
&= .3361 - 1
\end{aligned}
$$

(This final logarithm could also be written as $9.3361 - 10$.)

Now look in Table 4 for a mantissa of $.3361$. The closest mantissa, $.3365$, corresponds to 2.17, so

$$(76.9)(.00282) \approx 2.17 \times 10^{-1}.$$

Thus, $(76.9)(.00282) \approx .217$.   ●

In Example 2, we used the logarithm table to find a number, given its logarithm. Such a number is called the **antilogarithm,** abbreviated "antilog." For example, $.217$ is the antilogarithm of $.3361 - 1$.

EXAMPLE 3   Find $\dfrac{93.1}{2.69}$.

By a property of logarithms,

$$
\begin{aligned}
\log \frac{93.1}{2.69} &= \log 93.1 - \log 2.69 \\
&= 1.9689 - .4298 \\
&= 1.5391.
\end{aligned}
$$

Now look in Table 4 for the number whose logarithm has a mantissa closest to $.5391$. The number is 3.46, with

$$\frac{93.1}{2.69} \approx 3.46 \times 10^1 = 34.6.$$ ●

EXAMPLE 4   Find each of the following.

(a) $\log(2.73)^4$

Use a property of logarithms to get

$$
\begin{aligned}
\log(2.73)^4 &= 4 \log 2.73 \\
&= 4(.4362) \\
&= 1.7448.
\end{aligned}
$$

(b) $\log \sqrt[3]{.0762}$

We find that

$$\log\sqrt[3]{.0762} = \log (.0762)^{1/3}$$

$$= \frac{1}{3} \log .0762$$

$$= \frac{1}{3} (.8820 - 2).$$

To preserve the characteristic as an integer, change the characteristic to a multiple of 3 before multiplying by 1/3. One way to do this is to add and subtract 1 (which adds 0 to the characteristic) as follows.

$$.8820 - 2 = 1 - 1 + .8820 - 2$$
$$= 1.8820 - 3$$

Now complete the work above.

$$\log\sqrt[3]{.0762} = \frac{1}{3} (1.8820 - 3)$$

$$= .6273 - 1 \quad \bullet$$

EXAMPLE 5   In chemistry, the pH of a solution is defined as

$$pH = -\log [H_3O^+],$$

where $[H_3O^+]$ is the hydronium ion concentration in moles per liter. The number pH is a measure of the acidity or alkalinity of solutions. Pure water has a pH of 7.0 with values greater than that indicating alkalinity and values less than 7.0 indicating acidity. Find the following.

(a) The pH of a solution with $[H_3O^+] = 2.5 \times 10^{-4}$

$$pH = -\log [H_3O^+]$$
$$pH = -\log (2.5 \times 10^{-4})$$
$$= -(\log 2.5 + \log 10^{-4})$$
$$= -(.3979 - 4)$$
$$= -.3979 + 4$$
$$\approx 3.6$$

It is customary to round pH values to the nearest tenth.

(b) The hydronium ion concentration of a solution with pH = 7.1

$$pH = -\log [H_3O^+]$$
$$7.1 = -\log [H_3O^+]$$
$$-7.1 = \log [H_3O^+]$$

We want to find the antilogarithm of the number -7.1. First write -7.1 as $-7.1 + 8 - 8 = .9 - 8$. Look up the mantissa $.9$ in Table 4. Write the antilogarithm in scientific notation, rounding to the nearest tenth.

$$H_3O^+ = 7.9 \times 10^{-8} \quad \bullet$$

Interpolation The table of logarithms included in this text contains decimal approximations of common logarithms to four places of accuracy. If greater accuracy is necessary, a calculator or more accurate tables can be used. However, if desired, more accuracy can be obtained from the table included in this text by the process of **linear interpolation.** As an example, use linear interpolation to approximate log 75.37. First,

$$\log 75.3 < \log 75.37 < \log 75.4.$$

Figure 6.7 shows the portion of the curve $y = \log x$ between $x = 75.3$ and $x = 75.4$. We shall use the line segment PR to approximate the logarithm curve (this approximation is usually adequate for values of x relatively close to one another). From the figure, log 75.37 is given by the length of segment MQ, which we cannot find directly from our table. We can, however, find MN, which we shall use as our approximation to log 75.37. By properties of similar triangles,

$$\frac{PS}{PT} = \frac{SN}{TR}.$$

In this case, $PS = 75.37 - 75.3 = .07$, $PT = 75.4 - 75.3 = .1$, and $RT = \log 75.4 - \log 75.3 = 1.8774 - 1.8768 = .0006$. Hence,

$$\frac{.07}{.1} = \frac{SN}{.0006}$$

or
$$SN = .7(.0006) \approx .0004.$$

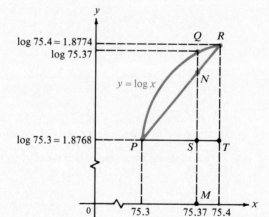

Figure 6.7

(Note that 75.37 is .7 of the distance from 75.3 to 75.4.) Since log 75.37 = $MN = MS + SN$, and since $MS = \log 75.3 = 1.8768$,

$$\log 75.37 = 1.8768 + .0004 = 1.8772.$$

EXAMPLE 6   Find log .008726.

Here log .00872 = .9405 − 3 and log .00873 = .9410 − 3. Since .008726 is .6 of the distance from .00872 to .00873, take .6 of the difference between log .00873 and log .00872.

$$.6[(.9410 − 3) − (.9405 − 3)] = .6(.0005) \approx .0003$$

Hence,

$$\log .008726 = (.9405 − 3) + .0003$$
$$= .9408 − 3. \quad \bullet$$

We can also interpolate when finding an antilogarithm, as shown in the next example.

EXAMPLE 7   Find x such that $\log x = .3275$.

From the logarithm table,

$$\log 2.12 = .3263 < .3275 < .3284 = \log 2.13.$$

The distance between .3263 and .3275 is .0012, while the distance between .3263 and .3284 is .0021. Hence, .3275 is

$$\frac{.0012}{.0021} \approx .6$$

of the way from .3263 to .3284. From this last result, we see that x should be .6 of the way from 2.12 to 2.13, so

$$\log 2.126 = .3275. \quad \bullet$$

6.5   EXERCISES

Find the characteristic of the logarithm of each of the following.

1. 875
2. 9462
3. 2,400,000
4. 875,000
5. .00023
6. .098
7. .000042
8. .000000257

Find the logarithms of the following numbers. Use interpolation if necessary. See Examples 1, 4, and 6.

9. 875
10. 3,750
11. 12,800
12. 653,000
13. 7.63
14. 9.37
15. .000893
16. .00376
17. 3.876
18. 2.975
19. 68,250
20. 103,200
21. .0003824
22. .00008632

Find the antilogarithms of each of the following logarithms.

23. 1.5366   **24.** 2.9253   **25.** .8733 − 2   **26.** .2504 − 1

27. 3.4947   **28.** 4.6863   **29.** .8039 − 3   **30.** .8078 − 2

Use logarithms (and interpolation) to find approximations to three-digit accuracy (four digits in 41, 42, 45, 46) for each of the following. See Examples 2 and 3.

31. $(79.6)(4.83)$ **32.** $(123)(7.89)$ **33.** $\dfrac{9.03}{43.8}$ **34.** $\dfrac{2.81}{.436}$

35. $(8.16)^{1/3}$ **36.** $\sqrt{276}$ **37.** $(.463)^{.4}$ **38.** π^{π} (Use $\pi = 3.14$)

39. $\dfrac{(21.3)^2}{(.86)^3}$ **40.** $\dfrac{(.893)^2}{(.624)^3}$ **41.** $\dfrac{(38.47)(258.6)}{\sqrt{124.8}}$ **42.** $\dfrac{(968.4)\sqrt[3]{25.68}}{29.83}$

43. $\dfrac{(2.51)^2}{(\sqrt{1.52})(3.94)}$ **44.** $\dfrac{(7.06)^3}{(31.7)(\sqrt{1.09})}$

45. $(49.83)^{1/2} + (2.917)^2$ **46.** $(38.42)^{1/3} + (86.13)^{1/4}$

Use common logarithms and Table 4 to find each of the following logarithms. Round to the nearest hundredth. Use the change of base formula from Section 6.3.

47. ln 125   **48.** ln 63.1   **49.** ln 0.98   **50.** ln 1.53

51. ln 275   **52.** ln 39.8   **53.** $\log_2 10$ **54.** $\log_5 8$

55. $\log_4 12$ **56.** $\log_7 4$

Use common logarithms to solve the following equations. Round to the nearest thousandth.

57. $10^x = 2.5$ **58.** $10^{x-1} = 143$ **59.** $100^{2x+1} = 17$ **60.** $1000^{5x} = 2010$

61. $50^{x+2} = 8$ **62.** $200^{x+3} = 12$ **63.** $.01^{3x} = .005$ **64.** $.001^{2x} = .0004$

Use logarithms to solve each of the following applied problems.

65. Philadelphia Chemicals, Inc., has to store a new gas it plans to produce. The pressure p (lb per cu ft) and volume v (cu ft) of the gas are related by the formula $pv^{1.6} = 800$. What is p if $v = 7.6$ cu ft?

66. The approximate period T (in seconds) of a simple pendulum of length L (in feet) is

$$T \approx 2\pi \sqrt{\frac{L}{32}}.$$

Find the period of a pendulum of length $L = 26$ ft.

67. The number of years, n, since two independently evolving languages split off from a common ancestral language is approximated by $n \approx -7600 \log r$, where r is the proportion of words from the ancestral language common to both languages.
   (a) Find n if $r = .9$.   **(b)** Find n if $r = .3$.
   (c) How many years have elapsed since the split if half of the words of the ancestral language are common to both languages?

A common problem in archaeology is to determine estimates of populations. Several methods have been used to calculate the number of people who occupied a site. One method relates the total surface area of a site to the number of occupants. If P represents the population of a site which covers an area of a square units, then

$$\log P = k \log a,$$

where k is an appropriate constant which varies for hilly, coastal, or desert environments, or for sites with single family dwellings or multiple family dwellings.* Find the population of sites with the following areas using k = .8.

68. $a = 230$ m^2 **69.** $a = 95$ m^2 **70.** $a = 20{,}000$ m^2

Find the population of a site with an area of 100,000 square meters for the following values of k.

71. $k = 1.2$ **72.** $k = .5$ **73.** $k = .7$

For each of the following substances, find the pH from the given hydronium ion concentration. See Example 5.

74. Grapefruit, 6.3×10^{-4} **75.** Crackers, 3.9×10^{-9}

76. Limes, 1.6×10^{-2} **77.** Sodium hydroxide (lye), 3.2×10^{-14}

Find the $[H_3O^+]$ for each of the following substances from the given pH.

78. Soda pop, 2.7   **79.** Wine, 3.4   **80.** Beer, 4.8   **81.** Drinking water, 6.5

The area of a triangle having sides of length a, b, and c is given by

$$\sqrt{s(s - a)(s - b)(s - c)},$$

where s = (a + b + c)/2. Use logarithms to find the areas of the following triangles.

82. $a = 114, b = 196, c = 153$ **83.** $a = .0941, b = .0873, c = .0896$

84. $a = 96.5, b = 103, c = 38.9$

Chapter 6   Summary

Key Words

exponential function
logarithm
common logarithm
continuous compounding
characteristic
linear interpolation

compound interest
logarithmic function with base a
natural logarithms
mantissa
antilogarithm

---

*Cook, S. F., 1950, "The Quantitative Investigation of Indian Mounds." Berkeley: University of California Publications in American Archaeology and Ethnology, 40: 231–233.

Additional Properties of Exponents

For $a > 0$, $a \neq 1$, and any real number x:

(a) a^x is a unique real number;

(b) $a^b = a^c$ if and only if $b = c$;

(c) if $a > 1$ and $m < n$, then $a^m < a^n$;

(d) if $0 < a < 1$ and $m < n$, then $a^m > a^n$.

Definition of Logarithm

For all real numbers y, and all positive numbers a, where $a \neq 1$,

$$y = \log_a x \quad \text{if and only if} \quad x = a^y.$$

Properties of Logarithms

If x and y are any positive real numbers, r is any real number, and a is any positive real number, $a \neq 1$, then

(a) $\log_a xy = \log_a x + \log_a y$ (b) $\log_a \dfrac{x}{y} = \log_a x - \log_a y$

(c) $\log_a x^r = r \cdot \log_a x$ (d) $\log_a a = 1$

(e) $\log_a 1 = 0.$

Change of Base Theorem

If x is any positive number and if a and b are positive real numbers, $a \neq 1$, $b \neq 1$, then

$$\log_a x = \frac{\log_b x}{\log_b a}.$$

Chapter 6 Review Exercises

Graph each of the following functions.

1. $y = 2^x$ **2.** $y = 2^{-x}$ **3.** $y = (1/2)^{x+1}$ **4.** $y = \log_3 x$

Solve each of the following equations.

5. $8^p = 32$ **6.** $9^{2y-1} = 27^y$ **7.** $\dfrac{8}{27} = b^{-3}$ **8.** $\dfrac{1}{2} = \left(\dfrac{b}{4}\right)^{1/4}$

The amount of a certain radioactive material, in grams, present after t days is given by

$$A(t) = 800e^{-0.04t}.$$

Use Table 5 or a calculator to find A(t) if t has the following values.

9. $t = 0$ **10.** $t = 5$

Suppose P dollars is compounded continuously at rate of interest i annually; the final amount on deposit would be $A = Pe^{ni}$ dollars after n years. How much would $1200 amount to at 10% compounded continuously for the following numbers of years.

11. 4 yr   **12.** 10 yr

13. Historically, the consumption of electricity has increased at a continuous rate of 6% per year. If it continued to increase at this rate, find the number of years before exactly twice as much electricity would be needed.

14. Suppose a conservation campaign together with higher rates caused demand for electricity to increase at only 2% per year. (See Exercise 13.) Find the number of years before twice as much electricity would be needed.

Write each of the following expressions in logarithmic form.

15. $2^5 = 32$ **16.** $100^{1/2} = 10$ **17.** $(1/16)^{1/4} = 1/2$ **18.** $(3/4)^{-1} = 4/3$

19. $10^{.4771} = 3$ **20.** $e^{2.4849} = 12$

Write each of the following logarithms in exponential form.

21. $\log_{10} .001 = -3$ **22.** $\log_2 \sqrt{32} = 5/2$ **23.** $\log 3.45 = .537819$ **24.** $\ln 45 = 3.806662$

Use properties of logarithms to write each of the following logarithms as a sum, difference, or product of logarithms.

25. $\log_3 \dfrac{mn}{5r}$ **26.** $\log_2 \dfrac{\sqrt{7}}{15}$ **27.** $\log_5 x^2 y^4 \sqrt[5]{m^3 p}$ **28.** $\log_7 (7k + 5r^2)$

Find each of the following logarithms. Round to the nearest thousandth.

29. $\ln 35$ **30.** $\ln 470$ **31.** $\ln 144,000$ **32.** $\ln 98,000$

33. $\log_{3.4} 15.8$ **34.** $\log_{1/2} 9.45$ **35.** $\log_3 769$ **36.** $\log_{2/3} 5/8$

The height in meters of the members of a certain tribe is approximated by

$$h = .5 + \log t,$$

where t is the tribe member's age in years, and $1 \le t \le 20$. Estimate the heights of tribe members of the following ages.

37. 2 yr   **38.** 5 yr   **39.** 10 yr   **40.** 20 yr

Use the formula for compound interest to find the total amount on deposit if the following amounts are placed on deposit at the given interest rates for the given numbers of years.

41. $1000 at 8% compounded annually for 9 years

42. $2800 at 6% compounded annually for 10 years

43. $19,456.11 at 12% compounded semiannually for 7 years

44. $312.45 at 16% compounded semiannually for 16 years

45. $1900 at 16% compounded quarterly for 9 years

46. $57,809.34 at 12% compounded quarterly for 5 years

47. $2500 at 18% compounded monthly for 3 years

48. $11,702.55 at 18% compounded monthly for 4 years

Find the amount of interest earned by each deposit.

49. $3954 at 8% compounded annually for 12 years

50. $12,699.36 at 16% compounded semiannually for 7 years

51. $7801.72 at 12% compounded quarterly for 5 years

52. $48,121.91 at 18% compounded monthly for 2 years

The formula

$$A = P\left(1 + \frac{i}{m}\right)^{nm}$$

from Section 6.1 gives the amount of money in an account after n years if P dollars is deposited at rate i compounded m times per year. If A is known, then P is called the present value of A. Find the present value of the following sums.

53. $4500 at 6% compounded annually for 9 years

54. $11,500 at 4% compounded annually for 12 years

55. $2000 at 4% compounded semiannually for 11 years

56. $2000 at 6% compounded quarterly for 8 years

57. $3456.78 at 15.6% compounded monthly for 30 months

58. How long would it take for $1000 at 5% compounded quarterly to double?

59. In Exercise 58, how long would it take at 10%?

60. If the inflation rate were 10%, use the formula for continuous compounding to find the number of years for a $1 item to cost $2.

61. In Exercise 60, find the number of years if the rate of inflation were 13%.

If R dollars is deposited at the end of each year in an account paying a rate of interest i per year compounded annually, then after n years the account will contain a total of

$$R\left[\frac{(1 + i)^n - 1}{i}\right]$$

dollars. (This sequence of equal payments is called an annuity.) Find the final amount on deposit for each of the following. Use logarithms or a calculator.

62. $800, 12%, 10 years   **63.** $1500, 14%, 7 years   **64.** $375, 10%, 12 years

65. Manuel deposits $10,000 at the end of each year for 12 years in an account paying 12% compounded annually. He then puts this total amount on deposit in another account paying 10% compounded semiannually for another 9 years. Find the total amount on deposit after the entire 21-year period.

66. Scott Hardy deposits $12,000 at the end of each year for 8 years in an account paying 14% compounded annually. He then leaves the money alone with no further deposits for an additional 6 years. Find the total amount on deposit after the entire 14-year period.

Solve each of the following equations. Round to the nearest thousandth.

67. $5^r = 11$

68. $10^{2r-3} = 17$

69. $e^{p+1} = 10$

70. $(1/2)^{3k+1} = 3$

71. $\log_{64} y = 1/3$

72. $\log_2 (y + 3) = 5$

73. $\ln 6x - \ln (x + 1) = \ln 4$

74. $\log_{16} \sqrt{x + 1} = 1/4$

Find each logarithm.

75. log 45.6

76. log 348

77. log 1,230,000

78. log .0411

79. log .000567

80. log .000002

Find the antilogarithms of each of the following logarithms.

81. 3.4983

82. .0043

83. 8.6493 − 10

84. −1.8297

Use common logarithms to approximate each of the following.

85. $\dfrac{(23.4)(.0349)}{.00453}$

86. $2.43^{3.2}$

87. $\dfrac{6^{2.1}}{\sqrt{52}}$

88. $\sqrt[5]{\dfrac{27.1}{4.33}}$

Use interpolation to find each of the following.

89. log 18.99

90. log 4.763

91. log .009814

Interpolate as necessary to find the antilogarithms of each of the following.

92. 1.9255

93. 8.7468 − 10

94. −3.7067

Use interpolation and logarithms to find each of the following to four significant digits of accuracy.

95. $\dfrac{8.937}{.4288}$

96. $(12.11)^2 + (15.09)^2$

97. $\sqrt[4]{4.213}$

98. $(4.067)^{1.24}$

7

Matrices

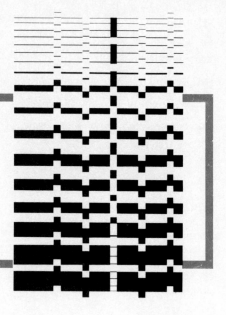

Suppose you are the manager of a health food store and you receive the following shipments of vitamins from two suppliers: from Dexter, 2 cartons of vitamin A pills, 7 cartons of vitamin E pills, and 5 cartons of vitamin K pills; from Sullivan 4 cartons of vitamin A pills, 6 cartons of vitamin E pills, and 9 cartons of vitamin K pills. It might be helpful to rewrite the information in a chart to make it more understandable.

	Cartons of Vitamin		
Manufacturer	*A*	*E*	*K*
Dexter	2	7	5
Sullivan	4	6	9

The information is clearer when presented this way. In fact, as long as you remember what each number represents, you could remove all the labels and write just the numbers, enclosing them in square brackets or parentheses to avoid confusion.

$$\begin{bmatrix} 2 & 7 & 5 \\ 4 & 6 & 9 \end{bmatrix}$$

This array of numbers gives all the information needed.

A rectangular array of numbers, such as the one above, is called a **matrix** (plural: matrices). Each number in the array is an **element** of the array.

Matrices have been of interest to mathematicians for some time. Recently, the use of matrices has gained importance in the life sciences, management, and the social sciences. Matrices, together with the computer, are useful in solving systems of equations such as those discussed in the next chapter.

7.1 Basic Properties of Matrices

Matrices are classified by their **dimension** or **order,** that is, by the number of rows and columns that they contain. For example, the matrix

$$\begin{bmatrix} 2 & 7 & 5 \\ 4 & 6 & 9 \end{bmatrix} \begin{matrix} \leftarrow \\ \leftarrow \end{matrix} \text{rows}$$

$$\uparrow \quad \uparrow \quad \uparrow$$

columns

has two rows and three columns. This matrix has dimension 2×3 (read "2 by 3"), or order 2×3. By definition, a matrix with m rows and n columns is of **dimension $m \times n$** or **order $m \times n$.** The number of rows is always given first.

EXAMPLE 1

(a) The matrix $\begin{bmatrix} 6 & 5 \\ 3 & 4 \\ 5 & -1 \end{bmatrix}$ is of order 3×2. *columns*

# of rows

(b) $\begin{bmatrix} 5 & 8 & 9 \\ 0 & 5 & -3 \\ -4 & 0 & 5 \end{bmatrix}$ is of order 3×3.

(c) $[1 \quad 6 \quad 5 \quad -2 \quad 5]$ is of order 1×5.

(d) $\begin{bmatrix} 3 \\ -5 \\ 0 \\ 2 \end{bmatrix}$ is of order 4×1. ●

A matrix having the same number of rows as columns is called a **square matrix.** The matrix given in Example 1(b) above is a square matrix, as are

$$\begin{bmatrix} -5 & 6 \\ 8 & 3 \end{bmatrix} \quad \text{and} \quad \begin{bmatrix} 0 & 0 & 0 & 0 \\ -2 & 4 & 1 & 3 \\ 0 & 0 & 0 & 0 \\ -5 & -4 & 1 & 8 \end{bmatrix}.$$

Same # rows as columns

A matrix containing only one row is called a **row matrix.** The matrix in Example 1(c) is a row matrix, as are

$$[5 \quad 8], \quad [6 \quad -9 \quad 2], \quad \text{and} \quad [-4 \quad 0 \quad 0 \quad 0].$$

Finally, a matrix of only one column, as in part (d) of Example 1, is a **column matrix.**

It is customary to use capital letters to name matrices. Subscript notation is used to name the elements of a matrix, as follows.

$$A = \begin{bmatrix} a_{11} & a_{12} & a_{13} & \cdots & a_{1n} \\ a_{21} & a_{22} & a_{23} & \cdots & a_{2n} \\ a_{31} & a_{32} & a_{33} & \cdots & a_{3n} \\ \cdot & \cdot & \cdot & & \cdot \\ \cdot & \cdot & \cdot & & \cdot \\ \cdot & \cdot & \cdot & & \cdot \\ a_{m1} & a_{m2} & a_{m3} & & a_{mn} \end{bmatrix}$$

Using this notation, the first row, first column element is denoted a_{11}, the second row, third column element is denoted a_{23}, and the ith row, jth column element is denoted a_{ij}.

Two matrices are **equal** if they are of the same order and if each corresponding element, position by position, is equal. Using this definition, the matrices

$$\begin{bmatrix} 2 & 1 \\ 3 & -5 \end{bmatrix} \quad \text{and} \quad \begin{bmatrix} 1 & 2 \\ -5 & 3 \end{bmatrix}$$

are *not* equal (even though they contain the same elements and are of the same order), since the corresponding elements differ.

EXAMPLE 2   From the definition of equality given above, the only way that the statement

$$\begin{bmatrix} 2 & 1 \\ p & q \end{bmatrix} = \begin{bmatrix} x & y \\ -1 & 0 \end{bmatrix}$$

can be true is if $2 = x$, $1 = y$, $p = -1$, and $q = 0$.   ●

EXAMPLE 3   The statement

$$\begin{bmatrix} x \\ y \end{bmatrix} = \begin{bmatrix} 1 \\ 4 \\ 0 \end{bmatrix}$$

can never be true, since the two matrices are of different order. (One is 2×1 and the other is 3×1.)   ●

In the introduction to this chapter we presented the matrix

$$\begin{bmatrix} 2 & 7 & 5 \\ 4 & 6 & 9 \end{bmatrix},$$

where the columns represent the numbers of cartons of three different types of vitamins (A, E, and K, respectively), and the rows represent two different manufacturers (Dexter and Sullivan, respectively). For example, the element 7 represents 7 cartons of vitamin E pills from Dexter, and so on. Suppose another shipment from these two suppliers is described by the following matrix.

$$\begin{bmatrix} 3 & 12 & 10 \\ 15 & 11 & 8 \end{bmatrix}$$

Here, for example, 8 cartons of vitamin K pills arrived from Sullivan. Let us now find the number of cartons of each kind of pill that were received from these two shipments.

In the first shipment, 2 cartons of vitamin A pills were received from Dexter, and in the second shipment, 3 cartons of vitamin A pills were received. Altogether, $2 + 3$, or 5, cartons of these pills were received. Corresponding elements can be added to find the total number of cartons of each type of pill received.

$$\begin{bmatrix} 2 & 7 & 5 \\ 4 & 6 & 9 \end{bmatrix} + \begin{bmatrix} 3 & 12 & 10 \\ 15 & 11 & 8 \end{bmatrix} = \begin{bmatrix} 2 + 3 & 7 + 12 & 5 + 10 \\ 4 + 15 & 6 + 11 & 9 + 8 \end{bmatrix}$$

$$= \begin{bmatrix} 5 & 19 & 15 \\ 19 & 17 & 17 \end{bmatrix}$$

The last matrix gives the total number of cartons of each type of pill that were received. For example, 15 cartons of vitamin K pills were received from Dexter. Generalizing from this example leads to the following definition.

Addition of Matrices

> The sum of two $m \times n$ matrices A and B is the $m \times n$ matrix $A + B$ in which each element is the sum of the corresponding elements of A and B.

Only matrices of the same order can be added.

EXAMPLE 4   Find each of the following sums.

(a) $\begin{bmatrix} 5 & -6 \\ 8 & 9 \end{bmatrix} + \begin{bmatrix} -4 & 6 \\ 8 & -3 \end{bmatrix} = \begin{bmatrix} 5 + (-4) & -6 + 6 \\ 8 + 8 & 9 + (-3) \end{bmatrix} = \begin{bmatrix} 1 & 0 \\ 16 & 6 \end{bmatrix}$

(b) $\begin{bmatrix} 2 \\ 5 \\ 8 \end{bmatrix} + \begin{bmatrix} -6 \\ 3 \\ 12 \end{bmatrix} = \begin{bmatrix} -4 \\ 8 \\ 20 \end{bmatrix}$

(c) The matrices

$$A = \begin{bmatrix} 5 & 8 \\ 6 & 2 \end{bmatrix} \quad \text{and} \quad B = \begin{bmatrix} 3 & 9 & 1 \\ 4 & 2 & 5 \end{bmatrix}$$

are of different order. Therefore, the sum $A + B$ does not exist. ●

A matrix containing only zero elements is called a **zero matrix.** For example, $[0 \quad 0 \quad 0]$ is the 1×3 zero matrix, while

$$\begin{bmatrix} 0 & 0 & 0 \\ 0 & 0 & 0 \end{bmatrix}$$

is the 2×3 zero matrix. A zero matrix can be written for any order.

In Chapter 1 the additive inverse of a real number a was defined as the real number $-a$ such that $a + (-a) = 0$ and $-a + a = 0$. Given a matrix A, we can find a matrix $-A$ such that $A + (-A) = 0$ and $-A + A = 0$. For example, if

$$A = \begin{bmatrix} -5 & 2 & -1 \\ 3 & 4 & -6 \end{bmatrix},$$

then find matrix $-A$ by finding the additive inverse of each element of A. (Remember that each element of A is a real number and thus has an additive inverse.)

$$-A = \begin{bmatrix} 5 & -2 & 1 \\ -3 & -4 & 6 \end{bmatrix}$$

To check, first test that $A + (-A)$ equals 0.

$$A + (-A) = \begin{bmatrix} -5 & 2 & -1 \\ 3 & 4 & -6 \end{bmatrix} + \begin{bmatrix} 5 & -2 & 1 \\ -3 & -4 & 6 \end{bmatrix} = \begin{bmatrix} 0 & 0 & 0 \\ 0 & 0 & 0 \end{bmatrix}$$

Then test that $-A + A$ is also 0. Matrix $-A$ is the **additive inverse,** or **negative,** of matrix A. Every matrix of any order has an additive inverse.

Subtraction of real numbers was defined in Chapter 1 by saying that $a - b = a + (-b)$. The same definition works for subtraction of matrices.

Subtraction of Matrices

> If A and B are matrices of the same order, then
> $$A - B = A + (-B).$$

EXAMPLE 5 Find each of the following differences.

(a) $\begin{bmatrix} -5 & 6 \\ 2 & 4 \end{bmatrix} - \begin{bmatrix} -3 & 2 \\ 5 & -8 \end{bmatrix} = \begin{bmatrix} -5 & 6 \\ 2 & 4 \end{bmatrix} + \begin{bmatrix} 3 & -2 \\ -5 & 8 \end{bmatrix} = \begin{bmatrix} -2 & 4 \\ -3 & 12 \end{bmatrix}$

(b) $[8 \quad 6 \quad -4] - [3 \quad 5 \quad -8] = [5 \quad 1 \quad 4]$

(c) The matrices

$$\begin{bmatrix} -2 & 5 \\ 0 & 1 \end{bmatrix} \quad \text{and} \quad \begin{bmatrix} 3 \\ 5 \end{bmatrix}$$

are of different order and cannot be subtracted.   ●

In work with matrices, a real number is called a **scalar** to distinguish it from a matrix. Multiply a scalar and a matrix as follows.

Multiplication of a Matrix by a Scalar

> The product of a scalar k and a matrix X is the matrix kX, each of whose elements is k times the corresponding element of X.

EXAMPLE 6   **(a)** $5\begin{bmatrix} 2 & -3 \\ 0 & 4 \end{bmatrix} = \begin{bmatrix} 10 & -15 \\ 0 & 20 \end{bmatrix}$

(b) $\dfrac{3}{4}\begin{bmatrix} 20 & 36 \\ 12 & -16 \end{bmatrix} = \begin{bmatrix} 15 & 27 \\ 9 & -12 \end{bmatrix}$ ●

7.1   EXERCISES

1-43 ODD

Find the order of each of the following matrices. Identify any square, column, or row matrices. See Example 1.

1. $\begin{bmatrix} -4 & 8 \\ 2 & 3 \end{bmatrix}$

2. $\begin{bmatrix} -9 & 6 & 2 \\ 4 & 1 & 8 \end{bmatrix}$

3. $\begin{bmatrix} -6 & 8 & 0 & 0 \\ 4 & 1 & 9 & 2 \\ 3 & -5 & 7 & 1 \end{bmatrix}$

4. $[8 \quad -2 \quad 4 \quad 6 \quad 3]$

5. $\begin{bmatrix} 2 \\ 4 \end{bmatrix}$

6. $\begin{bmatrix} -9 & 6 & 5 & 1 & 2 \\ -4 & 0 & 8 & 7 & 3 \end{bmatrix}$

7. $\begin{bmatrix} -5 \\ -8 \\ -4 \\ -6 \end{bmatrix}$

8. $[-9]$

9. $\begin{bmatrix} -4 & 2 & 3 \\ -8 & 2 & 1 \\ 4 & 6 & 8 \end{bmatrix}$

10. $\begin{bmatrix} -4 & 2 \\ 3 & 5 \end{bmatrix}$

Find the values of the variables in each of the following matrices. See Examples 2 and 3.

11. $\begin{bmatrix} 2 & 1 \\ 4 & 8 \end{bmatrix} = \begin{bmatrix} x & 1 \\ y & z \end{bmatrix}$

12. $\begin{bmatrix} -5 \\ y \end{bmatrix} = \begin{bmatrix} -5 \\ 8 \end{bmatrix}$

13. $\begin{bmatrix} 2 & 5 & 6 \\ 1 & m & n \end{bmatrix} = \begin{bmatrix} z & y & w \\ 1 & 8 & -2 \end{bmatrix}$

14. $\begin{bmatrix} 0 & 0 \\ a & b \end{bmatrix} = \begin{bmatrix} p & q \\ 5 & -7 \end{bmatrix}$

15. $\begin{bmatrix} x+6 & y+2 \\ 8 & 3 \end{bmatrix} = \begin{bmatrix} -9 & 7 \\ 8 & k \end{bmatrix}$

16. $\begin{bmatrix} 9 & 7 \\ r & 0 \end{bmatrix} = \begin{bmatrix} m-3 & n+5 \\ 8 & 0 \end{bmatrix}$

17. $\begin{bmatrix} 3 & 5 \\ 8 & 9 \end{bmatrix} + \begin{bmatrix} m & 3 \\ 5 & n \end{bmatrix} = \begin{bmatrix} 9 & 8 \\ 13 & 0 \end{bmatrix}$

18. $[8 \quad p + 9 \quad q + 5] + [9 \quad -3 \quad 12] = [k - 2 \quad 12 \quad 2q]$

19. $\begin{bmatrix} -7 + z & 4r & 8s \\ 6p & 2 & 5 \end{bmatrix} + \begin{bmatrix} -9 & 8r & 3 \\ 2 & 5 & 4 \end{bmatrix} = \begin{bmatrix} 2 & 36 & 27 \\ 20 & 7 & 12a \end{bmatrix}$

20. $\begin{bmatrix} a + 2 & 3z + 1 & 5m \\ 4k & 0 & 3 \end{bmatrix} + \begin{bmatrix} 3a & 2z & 5m \\ 2k & 5 & 6 \end{bmatrix} = \begin{bmatrix} 10 & -14 & 80 \\ 10 & 5 & 9 \end{bmatrix}$

Perform each of the following operations, whenever possible. See Examples 4 and 5.

21. $\begin{bmatrix} 6 & -9 & 2 \\ 4 & 1 & 3 \end{bmatrix} - \begin{bmatrix} -8 & 2 & 5 \\ 6 & -3 & 4 \end{bmatrix}$

22. $\begin{bmatrix} 9 & 4 \\ -8 & 2 \end{bmatrix} + \begin{bmatrix} -3 & 2 \\ -4 & 7 \end{bmatrix}$

23. $2\begin{bmatrix} -6 & 8 \\ 0 & 0 \end{bmatrix} - \begin{bmatrix} 0 & 0 \\ -4 & -2 \end{bmatrix}$

24. $3\begin{bmatrix} 6 & -1 & 4 \\ 2 & 8 & -3 \\ -4 & 5 & 6 \end{bmatrix} + 5\begin{bmatrix} -2 & -8 & -6 \\ 4 & 1 & 3 \\ 2 & -1 & 5 \end{bmatrix}$

25. $4\begin{bmatrix} 1 & -4 \\ 2 & -3 \\ -8 & 4 \end{bmatrix} - 3\begin{bmatrix} -6 & 9 \\ -2 & 5 \\ -7 & -12 \end{bmatrix}$

26. $\begin{bmatrix} -8 & -4 & -2 & -3 \\ 2 & -1 & 5 & 9 \end{bmatrix} - 2\begin{bmatrix} 4 & 6 & 8 & -3 \\ -2 & -5 & 1 & 6 \end{bmatrix}$

27. $\begin{bmatrix} -8 & 4 & 0 \\ 2 & 5 & 0 \end{bmatrix} + \begin{bmatrix} 6 & 3 \\ 8 & 9 \end{bmatrix}$

28. $\begin{bmatrix} 2 \\ 3 \end{bmatrix} - \begin{bmatrix} 8 & 1 \\ 9 & 4 \end{bmatrix}$

29. $\begin{bmatrix} 9 & 4 & 1 & -2 \\ 5 & -6 & 3 & 4 \\ 2 & -5 & 1 & 2 \end{bmatrix} - \begin{bmatrix} -2 & 5 & 1 & 3 \\ 0 & 1 & 0 & 2 \\ -8 & 3 & 2 & 1 \end{bmatrix} + \begin{bmatrix} 2 & 4 & 0 & 3 \\ 4 & -5 & 1 & 6 \\ 2 & -3 & 0 & 8 \end{bmatrix}$

30. $\begin{bmatrix} 6 & -2 & 4 \\ -2 & 5 & 8 \\ 1 & 0 & 2 \end{bmatrix} + \begin{bmatrix} 3 & 0 & 8 \\ 1 & -2 & 4 \\ 6 & 9 & -2 \end{bmatrix} - \begin{bmatrix} -4 & 2 & 1 \\ 0 & 3 & -2 \\ 4 & 2 & 0 \end{bmatrix}$

31. $\begin{bmatrix} -4x + 2y & -3x + y \\ 6x - 3y & 2x - 5y \end{bmatrix} + \begin{bmatrix} -8x + 6y & 2x \\ 3y - 5x & 6x + 4y \end{bmatrix}$

32. $\begin{bmatrix} 4k - 8y \\ 6z - 3x \\ 2k + 5a \\ -4m + 2n \end{bmatrix} - \begin{bmatrix} 5k + 6y \\ 2z + 5x \\ 4k + 6a \\ 4m - 2n \end{bmatrix}$

33. $\begin{bmatrix} 1.003 & 4.728 \\ 2.341 & 1.572 \\ 4.918 & 2.763 \end{bmatrix} + \begin{bmatrix} 9.871 & 12.283 \\ 10.268 & 14.001 \\ 3.974 & 11.754 \end{bmatrix}$

34. $\begin{bmatrix} .0012 & .9164 & .0915 \\ .3571 & .3752 & .1134 \\ .4263 & .8036 & .2042 \end{bmatrix} + \begin{bmatrix} .6813 & .1119 & .2971 \\ .2471 & .0743 & .7642 \\ .0065 & .3685 & .6603 \end{bmatrix}$

Let $A = \begin{bmatrix} -2 & 4 \\ 0 & 3 \end{bmatrix}$ and $B = \begin{bmatrix} -6 & 2 \\ 4 & 0 \end{bmatrix}$.

Find each of the following matrices. See Example 6.

35. $2A$

36. $-3B$

37. $-4A$

38. $5B$

39. $2A - B$

40. $-4A + 5B$

41. $3A - 11B$

42. $-2A + 4B$

43. $3A - 2B$

44. $-9A + B$

Work each of the following problems.

45. Mary found that her new car averaged 18.2 miles per gallon the first week, 19.0 the second week, 17.6 the third week, and 18.5 the fourth week. Write this information as a row matrix and then as a column matrix.

46. Joan's scores on the first three tests in her math class were 82, 77, and 85. Paula scored 91, 80, and 82 on the same three tests. Janet's scores on the three tests were 90, 82, and 79. Write this information as a 3×3 square matrix in two different ways.

47. Richard Macias bought 7 shares of Sears stock, 9 shares of IBM stock, and 8 shares of Chrysler stock. The following month, he bought 2 shares of Sears stock, no IBM, and 6 shares of Chrysler. Write this information first as a 3×2 matrix and then as a 2×3 matrix.

48. Margie Bezzone works in a computer store. The first week she sold 5 computers, 3 printers, 4 disc drives, and 6 monitors. The next week she sold 4 computers, 2 printers, 6 disc drives, and 5 monitors. Write this information first as a 2×4 matrix and then as a 4×2 matrix.

Addition of real numbers satisfies certain properties, such as the commutative, associative, identity, inverse, and closure properties. In the following exercises, you can check to see which of these properties are satisfied by addition of matrices.

Let $A = \begin{bmatrix} a & b \\ c & d \end{bmatrix}$, $B = \begin{bmatrix} e & f \\ g & h \end{bmatrix}$, and $C = \begin{bmatrix} j & m \\ k & n \end{bmatrix}$.

Decide which of the following statements are true for these square matrices. Then decide if a similar property holds for any matrices of the same order.

49. $A + B = B + A$ (commutative property)

50. $A + (B + C) = (A + B) + C$ (associative property)

51. $A + B$ is a 2×2 matrix. (closure property)

52. There exists a matrix 0 such that $A + 0 = A$ and $0 + A = A$. (identity property)

53. There exists a matrix $-A$ such that $A + (-A) = 0$ and $-A + A = 0$. (inverse property)

54. Are these properties valid for matrices that are not square?

7.2 Matrix Products

In the last section we saw how to multiply a matrix by a scalar. Multiplication of two matrices is more complicated; however, it is important in practical problems. To see the reasoning behind matrix multiplication, we continue with the example about the vitamin pills from Section 7.1.

Below is the matrix from the previous section that shows the number of cartons of each type of vitamin received from each company.

$$\begin{bmatrix} 2 & 7 & 5 \\ 4 & 6 & 9 \end{bmatrix}$$

The two rows of this matrix represent the number of cartons of each kind of vitamin received from Dexter and Sullivan respectively. Now suppose that each carton of viatmin A pills costs the store $12, each carton of vitamin E pills costs $18, and each carton of vitamin K pills costs $9.

To find the total cost of the pills from Dexter, multiply as follows.

Vitamin	Number of Cartons	Cost per Carton	Total Cost
A	2	$12	$ 24
E	7	$18	$126
K	5	$ 9	$ 45
			$195 Total from Dexter

The Dexter pills cost a total of $195.

This result is the sum of three products:

$$2(\$12) + 7(\$18) + 5(\$9) = \$195.$$

In the same way, using the second row of the matrix and the three costs, the total cost of the Sullivan pills is

$$4(\$12) + 6(\$18) + 9(\$9) = \$237.$$

The costs, $12, $18, and $9, can be written as a column matrix.

$$\begin{bmatrix} 12 \\ 18 \\ 9 \end{bmatrix}$$

The total costs for each supplier, $195 and $237, also can be written as a column matrix.

$$\begin{bmatrix} 195 \\ 237 \end{bmatrix}$$

The product of the matrices

$$\begin{bmatrix} 2 & 7 & 5 \\ 4 & 6 & 9 \end{bmatrix} \quad \text{and} \quad \begin{bmatrix} 12 \\ 18 \\ 9 \end{bmatrix}$$

can be written as follows.

$$\begin{bmatrix} 2 & 7 & 5 \\ 4 & 6 & 9 \end{bmatrix} \cdot \begin{bmatrix} 12 \\ 18 \\ 9 \end{bmatrix} = \begin{bmatrix} 2 \cdot 12 + 7 \cdot 18 + 5 \cdot 9 \\ 4 \cdot 12 + 6 \cdot 18 + 9 \cdot 9 \end{bmatrix} = \begin{bmatrix} 195 \\ 237 \end{bmatrix}$$

Each element of the product was found by multiplying the elements of the *rows* of the matrix on the left and the corresponding elements of the *columns* of the matrix on the right, and then finding the sum of these products. Notice that the product of a 2×3 matrix and a 3×1 matrix is a 2×1 matrix.

Generalizing from this example gives the following definition of matrix multiplication.

Multiplication of Matrices

The product AB of an $m \times n$ matrix A and an $n \times k$ matrix B is found as follows. To get the ith row, jth column element of AB, multiply each element in the ith row of A by the corresponding element in the jth column of B. The sum of these products will give the element of row i, column j of AB.

EXAMPLE 1

Find the product AB, where

$$A = \begin{bmatrix} 2 & 4 \\ 5 & 6 \end{bmatrix} \quad \text{and} \quad B = \begin{bmatrix} -3 & 5 \\ 4 & -2 \end{bmatrix}.$$

Step 1   Multiply the elements of the first row of A and the corresponding elements of the first column of B, and add these products.

$$AB = \begin{bmatrix} \mathbf{2} & \mathbf{4} \\ 5 & 6 \end{bmatrix} \begin{bmatrix} -3 & 5 \\ 4 & -2 \end{bmatrix} \qquad \mathbf{2}(-3) + \mathbf{4}(4) = -6 + 16 = 10$$

The first row, first column entry of the product matrix AB is 10.

Step 2   Multiply the elements of the first row of A and the second column of B and then add the products to get the first row, second column entry of the product matrix.

$$AB = \begin{bmatrix} \mathbf{2} & \mathbf{4} \\ 5 & 6 \end{bmatrix} \begin{bmatrix} -3 & 5 \\ 4 & -2 \end{bmatrix} \qquad \mathbf{2}(5) + \mathbf{4}(-2) = 10 + (-8) = 2$$

Step 3

$$AB = \begin{bmatrix} 2 & 4 \\ 5 & 6 \end{bmatrix} \begin{bmatrix} -3 & 5 \\ 4 & -2 \end{bmatrix} \qquad 5(-3) + 6(4) = -15 + 24 = 9$$

The second row, first column entry of the product matrix is 9.

Step 4

$$AB = \begin{bmatrix} 2 & 4 \\ 5 & 6 \end{bmatrix} \begin{bmatrix} -3 & 5 \\ 4 & -2 \end{bmatrix} \qquad 5(5) + 6(-2) = 25 + (-12) = 13.$$

Finally, 13 is the second row, second column entry.

Step 5   Write the product. The four entries in the product matrix come from the four steps above.

$$AB = \begin{bmatrix} 2 & 4 \\ 5 & 6 \end{bmatrix} \begin{bmatrix} -3 & 5 \\ 4 & -2 \end{bmatrix} = \begin{bmatrix} 10 & 2 \\ 9 & 13 \end{bmatrix} \quad \bullet$$

EXAMPLE 2   Find the product

$$\begin{bmatrix} -3 & 4 & 2 \\ 5 & 0 & 4 \end{bmatrix} \begin{bmatrix} -6 & 4 \\ 2 & 3 \\ 3 & -2 \end{bmatrix}.$$

Step 1 $\begin{bmatrix} -3 & 4 & 2 \\ 5 & 0 & 4 \end{bmatrix} \begin{bmatrix} -6 & 4 \\ 2 & 3 \\ 3 & -2 \end{bmatrix} \qquad (-3)(-6) + 4(2) + 2(3) = 32$

Step 2 $\begin{bmatrix} -3 & 4 & 2 \\ 5 & 0 & 4 \end{bmatrix} \begin{bmatrix} -6 & 4 \\ 2 & 3 \\ 3 & -2 \end{bmatrix} \qquad (-3)(4) + 4(3) + 2(-2) = -4$

Step 3 $\begin{bmatrix} -3 & 4 & 2 \\ 5 & 0 & 4 \end{bmatrix} \begin{bmatrix} -6 & 4 \\ 2 & 3 \\ 3 & -2 \end{bmatrix} \qquad 5(-6) + 0(2) + 4(3) = -18$

Step 4 $\begin{bmatrix} -3 & 4 & 2 \\ 5 & 0 & 4 \end{bmatrix} \begin{bmatrix} -6 & 4 \\ 2 & 3 \\ 3 & -2 \end{bmatrix} \qquad 5(4) + 0(3) + 4(-2) = 12$

Step 5   Write the product.

$$\begin{bmatrix} -3 & 4 & 2 \\ 5 & 0 & 4 \end{bmatrix} \begin{bmatrix} -6 & 4 \\ 2 & 3 \\ 3 & -2 \end{bmatrix} = \begin{bmatrix} 32 & -4 \\ -18 & 12 \end{bmatrix}$$

As this example shows, the product of a 2 × 3 matrix and a 3 × 2 matrix is a 2 × 2 matrix.   ●

The examples suggest the following restriction on matrix multiplication.

> Two matrices A and B can be multiplied only if the number of *columns* of A is the same as the number of *rows* of B. The final product will have as many rows as A and as many columns as B.

EXAMPLE 3   Suppose matrix A is 2 × 2, while matrix B is 2 × 4. Can the product AB be calculated? What is the order of the product?

The following diagram helps answer these questions.

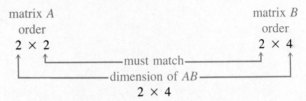

The product AB can be calculated because A has two columns and B has two rows. The order of the product is 2 × 4. (By the way, the product BA could not be found.)   ●

EXAMPLE 4   Find AB and BA if

$$A = \begin{bmatrix} 1 & -3 \\ 7 & 2 \end{bmatrix} \quad \text{and} \quad B = \begin{bmatrix} 1 & 0 & -1 & 2 \\ 3 & 1 & 4 & -1 \end{bmatrix}.$$

Use the definition of matrix multiplication to find AB.

$$AB = \begin{bmatrix} 1 & -3 \\ 7 & 2 \end{bmatrix} \begin{bmatrix} 1 & 0 & -1 & 2 \\ 3 & 1 & 4 & -1 \end{bmatrix}$$

$$= \begin{bmatrix} 1(1) + (-3)(3) & 1(0) + (-3)1 & 1(-1) + (-3)4 & 1(2) + (-3)(-1) \\ 7(1) + 2(3) & 7(0) + 2(1) & 7(-1) + 2(4) & 7(2) + 2(-1) \end{bmatrix}$$

$$= \begin{bmatrix} -8 & -3 & -13 & 5 \\ 13 & 2 & 1 & 12 \end{bmatrix}$$

Since B is a 2 × 4 matrix, and A is a 2 × 2 matrix, the product BA cannot be found.   ●

EXAMPLE 5   Find MN and NM, given

$$M = \begin{bmatrix} 1 & 3 \\ -2 & 4 \end{bmatrix} \quad \text{and} \quad N = \begin{bmatrix} 2 & 5 \\ 10 & -3 \end{bmatrix}.$$

By the definition of matrix multiplication,

$$MN = \begin{bmatrix} 1 & 3 \\ -2 & 4 \end{bmatrix} \begin{bmatrix} 2 & 5 \\ 10 & -3 \end{bmatrix}$$

$$= \begin{bmatrix} 1(2) + 3(10) & 1(5) + 3(-3) \\ -2(2) + 4(10) & -2(5) + 4(-3) \end{bmatrix}$$

$$= \begin{bmatrix} 32 & -4 \\ 36 & -22 \end{bmatrix}.$$

Similarly, $$NM = \begin{bmatrix} 2 & 5 \\ 10 & -3 \end{bmatrix} \begin{bmatrix} 1 & 3 \\ -2 & 4 \end{bmatrix}$$

$$= \begin{bmatrix} 2(1) + 5(-2) & 2(3) + 5(4) \\ 10(1) + (-3)(-2) & 10(3) + (-3)4 \end{bmatrix}$$

$$= \begin{bmatrix} -8 & 26 \\ 16 & 18 \end{bmatrix}. \quad \bullet$$

In Example 4 the product AB could be found, but not BA. In Example 5, although both MN and NM could be found, they were not equal, showing that multiplication of matrices is not commutative. This fact distinguishes matrix arithmetic from the arithmetic of real numbers. Other properties of matrix arithmetic are given as Exercises 39–45 at the end of this section.

EXAMPLE 6   A contractor builds three kinds of houses, models A, B, and C, with a choice of two styles, colonial or ranch. Matrix P below shows the number of each kind of house the contractor is planning to build for a new 100-home subdivision. The amounts for each of the main materials used depend on the style of the house. These amounts are shown in matrix Q below, while matrix R gives the cost for each kind of material. Concrete is measured here in cubic yards, lumber in 1000 board feet, brick in 1000's, and shingles in 100 square feet.

$$\begin{array}{c} \\ \text{model A} \\ \text{model B} \\ \text{model C} \end{array} \overset{\begin{array}{cc} \text{colonial} & \text{ranch} \end{array}}{\begin{bmatrix} 0 & 30 \\ 10 & 20 \\ 20 & 20 \end{bmatrix}} = P$$

$$\begin{array}{c} \\ \text{colonial} \\ \text{ranch} \end{array} \overset{\begin{array}{cccc} \text{concrete} & \text{lumber} & \text{brick} & \text{shingles} \end{array}}{\begin{bmatrix} 10 & 2 & 0 & 2 \\ 50 & 1 & 20 & 2 \end{bmatrix}} = Q$$

$$\begin{array}{r}\text{concrete}\\\text{lumber}\\\text{brick}\\\text{shingles}\end{array}\overset{\substack{\text{cost per}\\\text{unit}}}{\begin{bmatrix}20\\180\\60\\25\end{bmatrix}}=R$$

(a) What is the total cost of materials for all houses of each model?

To find the materials cost for each model, first find matrix PQ, which will show the total amount of each material needed for all houses of each model.

$$PQ=\begin{bmatrix}0&30\\10&20\\20&20\end{bmatrix}\begin{bmatrix}10&2&0&2\\50&1&20&2\end{bmatrix}=\overset{\substack{\text{concrete}\quad\text{lumber}\quad\text{brick}\quad\text{shingles}}}{\begin{bmatrix}1500&30&600&60\\1100&40&400&60\\1200&60&400&80\end{bmatrix}}\begin{array}{l}\text{model A}\\\text{model B}\\\text{model C}\end{array}$$

Multiplying PQ and the cost matrix R gives the total cost of materials for each model.

$$\begin{bmatrix}1500&30&600&60\\1100&40&400&60\\1200&60&400&80\end{bmatrix}\overset{\text{cost}}{\begin{bmatrix}20\\180\\60\\25\end{bmatrix}}=\begin{bmatrix}72{,}900\\54{,}700\\60{,}800\end{bmatrix}\begin{array}{l}\text{model A}\\\text{model B}\\\text{model C}\end{array}$$

(b) How much of each of the four kinds of material must be ordered?

The totals of the columns of matrix PQ will give a matrix whose elements represent the total amounts of each material needed for the subdivision. Let us call this matrix T and write it as a row matrix.

$$T=[3800\quad130\quad1400\quad200]$$

(c) What is the total cost of the materials?

The total cost of all the materials is given by the product of matrix R, the cost matrix, and matrix T, the total amounts matrix. To multiply these and get a 1×1 matrix, representing the total cost, requires multiplying a 1×4 matrix and a 4×1 matrix. This is why in (b) we chose to write a row matrix. The total materials cost is given by TR, so

$$TR=[3800\quad130\quad1400\quad200]\begin{bmatrix}20\\180\\60\\25\end{bmatrix}=[188{,}400].\quad\bullet$$

To help keep track of the quantities a matrix represents, let matrix P, from Example 6, represent models/styles, matrix Q represent styles/materials, and matrix R represent materials/cost. In each case the meaning of the rows is written first and that of the columns second. When the product PQ was found in Example 6, the rows of the matrix represented models and the columns represented materials. Therefore, the matrix product PQ represents models/materials. The common quantity, styles, in both P and Q was eliminated in the product PQ. Do you see that the product $(PQ)R$ represents models/cost?

In practical problems this notation helps to identify the order in which two matrices should be multiplied so that the results are meaningful. In Example 6(c), either product RT or product TR could have been found. However, since T represents subdivisions/materials and R represents materials/cost, only TR gave the required matrix representing subdivisions/cost.

7.2 EXERCISES

In each of the following exercises, the order of two matrices A and B is given. Find the order of the product AB and the product BA, whenever these products exist.

1. A is 2×2, B is 2×2

2. A is 3×3, B is 3×3

3. A is 4×2, B is 2×4

4. A is 3×1, B is 1×3

5. A is 3×5, B is 5×2

6. A is 4×3, B is 3×6

7. A is 4×2, B is 3×4

8. A is 7×3, B is 2×7

9. A is 4×3, B is 2×5

10. A is 1×6, B is 2×4

Find each of the following matrix products, whenever possible. See Examples 1 and 2.

11. $\begin{bmatrix} 1 & 2 \\ 3 & 4 \end{bmatrix} \begin{bmatrix} -1 \\ 7 \end{bmatrix}$

12. $\begin{bmatrix} -1 & 5 \\ 7 & 0 \end{bmatrix} \begin{bmatrix} 6 \\ 2 \end{bmatrix}$

13. $\begin{bmatrix} 3 & -4 & 1 \\ 5 & 0 & 2 \end{bmatrix} \begin{bmatrix} -1 \\ 4 \\ 2 \end{bmatrix}$

14. $\begin{bmatrix} -6 & 3 & 5 \\ 2 & 9 & 1 \end{bmatrix} \begin{bmatrix} -2 \\ 0 \\ 3 \end{bmatrix}$

15. $\begin{bmatrix} 5 & 2 \\ -1 & 4 \end{bmatrix} \begin{bmatrix} 3 & -2 \\ 1 & 0 \end{bmatrix}$

16. $\begin{bmatrix} -4 & 0 \\ 1 & 3 \end{bmatrix} \begin{bmatrix} -2 & 4 \\ 0 & 1 \end{bmatrix}$

17. $\begin{bmatrix} 2 & 2 & -1 \\ 3 & 0 & 1 \end{bmatrix} \begin{bmatrix} 0 & 2 \\ -1 & 4 \\ 0 & 2 \end{bmatrix}$

18. $\begin{bmatrix} -9 & 2 & 1 \\ 3 & 0 & 0 \end{bmatrix} \begin{bmatrix} 2 \\ -1 \\ 4 \end{bmatrix}$

19. $\begin{bmatrix} -1 & 2 & 0 \\ 0 & 3 & 2 \\ 0 & 1 & 4 \end{bmatrix} \begin{bmatrix} 2 & -1 & 2 \\ 0 & 2 & 1 \\ 3 & 0 & -1 \end{bmatrix}$

20. $\begin{bmatrix} -2 & -3 & -4 \\ 2 & -1 & 0 \\ 4 & -2 & 3 \end{bmatrix} \begin{bmatrix} 0 & 1 & 4 \\ 1 & 2 & -1 \\ 3 & 2 & -2 \end{bmatrix}$

21. $\begin{bmatrix} -2 & 4 & 1 \end{bmatrix} \begin{bmatrix} 3 & -2 & 4 \\ 2 & 1 & 0 \\ 0 & -1 & 4 \end{bmatrix}$

22. $\begin{bmatrix} 0 & 3 & -4 \end{bmatrix} \begin{bmatrix} -2 & 6 & 3 \\ 0 & 4 & 2 \\ -1 & 1 & 4 \end{bmatrix}$

23. $\begin{bmatrix} -2 & 1 & 4 \\ 0 & 1 & 2 \end{bmatrix} \begin{bmatrix} -2 & 1 & 0 \\ 0 & -2 & 0 \\ 4 & 1 & 2 \end{bmatrix}$

24. $\begin{bmatrix} -1 & 0 & 0 \\ 2 & 1 & 4 \end{bmatrix} \begin{bmatrix} 4 & -2 & 5 \\ 0 & 1 & 4 \\ 2 & -9 & 0 \end{bmatrix}$

25. $\begin{bmatrix} -3 & 0 & 2 & 1 \\ 4 & 0 & 2 & 6 \end{bmatrix} \begin{bmatrix} -4 & 2 \\ 0 & 1 \end{bmatrix}$

26. $\begin{bmatrix} -1 & 2 & 4 & 1 \\ 0 & 2 & -3 & 5 \end{bmatrix} \begin{bmatrix} 1 & 2 & 4 \\ -2 & 5 & 1 \end{bmatrix}$

27. $\begin{bmatrix} -2 & 4 & 6 \end{bmatrix} \begin{bmatrix} 3 \\ -2 \\ 1 \end{bmatrix}$

28. $\begin{bmatrix} 4 & 0 & 2 \end{bmatrix} \begin{bmatrix} -5 \\ 1 \\ 6 \end{bmatrix}$

29. $\begin{bmatrix} 3 \\ -2 \\ 1 \end{bmatrix} \begin{bmatrix} -2 & 4 & 6 \end{bmatrix}$

30. $\begin{bmatrix} -5 \\ 1 \\ 6 \end{bmatrix} \begin{bmatrix} 4 & 0 & 2 \end{bmatrix}$

31. $\begin{bmatrix} 1.25 & 2.83 \\ 6.94 & 5.07 \end{bmatrix} \begin{bmatrix} 1.04 & 2.93 \\ 3.68 & 6.55 \end{bmatrix}$

32. $\begin{bmatrix} .041 & .002 \\ .610 & .345 \\ .992 & .613 \end{bmatrix} \begin{bmatrix} .116 & .581 \\ .423 & .704 \end{bmatrix}$

Let $A = \begin{bmatrix} -2 & 4 \\ 1 & 3 \end{bmatrix}$, $B = \begin{bmatrix} -2 & 1 \\ 3 & 6 \end{bmatrix}$,   and $C = \begin{bmatrix} 5 & -2 & 1 \\ 0 & 3 & 7 \end{bmatrix}$.

Find each of the following products. See Examples 4 and 5.

33. *AB*   **34.** *BA*   **35.** *AC*   **36.** *CA*

37. Did you get the same answer in Exercises 33 and 34? Do you think that matrix multiplication is commutative?

38. In general, for matrices *A* and *B* such that *AB* and *BA* both exist, does *AB* always equal *BA?*

For the following exercises, let

$$A = \begin{bmatrix} a & b \\ c & d \end{bmatrix}, \quad B = \begin{bmatrix} e & f \\ g & h \end{bmatrix} \quad and \quad C = \begin{bmatrix} j & m \\ k & n \end{bmatrix}.$$

Decide which of the following statements are true for these three matrices. Then decide if a similar property holds for any square matrices of the same order.

39. $(AB)C = A(BC)$ (associative property)

40. $A(B + C) = AB + AC$ (distributive property)

41. *AB* is a 2 × 2 matrix   (closure property)

42. $k(A + B) = kA + kB$ for any real number *k*

43. $(k + p)A = kA + pA$ for any real numbers *k* and *p*

44. $(A + B)(A - B) = A^2 - B^2$ $(A^2 = AA)$

45. $(A + B)^2 = A^2 + 2AB + B^2$

46. The Bread Box, a small neighborhood bakery, sells four main items: sweet rolls, bread, cakes, and pies. The amount of each ingredient required for these items is given by matrix A.

	eggs	flour*	sugar*	shortening*	milk*
rolls (doz)	1	4	1/4	1/4	1
bread (loaves)	0	3	0	1/4	0
cakes	4	3	2	1	1
pies (crust)	0	1	0	1/3	0

$= A$

The cost (in cents) for each ingredient when purchased in large lots or small lots is given by matrix B.

	cost	
	large lot	small lot
eggs	5	5
flour	8	10
sugar	10	12
shortening	12	15
milk	5	6

$= B$

(a) Use matrix multiplication to find a matrix giving the comparative cost per item for the two purchase options.

Suppose a day's orders consist of 20 dozen sweet rolls, 200 loaves of bread, 50 cakes and 60 pies.

(b) Writing the orders as a 1×4 matrix and using matrix multiplication, write as a matrix the amount of each ingredient needed to fill the day's orders.

(c) Use matrix multiplication to find a matrix giving the costs under the two purchase options to fill the day's orders.

47. For an article in the Daily News, reporter Mary Starr compared prices at the three leading supermarkets in town on four items: a quart of milk, a head of lettuce, a loaf of bread, and a pound of hamburger. Her results are summarized in the following matrix.

	market		
	I	II	III
milk	.45	.46	.45
lettuce	.59	.63	.58
bread	.95	.89	.92
hamburger	1.65	1.68	1.60

*In cups.

For her survey, Starr bought 3 quarts of milk, 1 head of lettuce, 2 loaves of bread and 4 pounds of hamburger at each market.

(a) Write the amount of each item purchased as a 1×4 row matrix.
(b) Multiply the two matrices to find Starr's total bill at each market.

7.3   The Multiplicative Inverse of a Matrix

In this section multiplicative identity elements and multiplicative inverses are introduced and used to solve matrix equations. These ideas will be useful in Chapter 8 for solving systems of equations.

The identity property for real numbers says that $a \cdot 1 = a$ and $1 \cdot a = a$ for any real number a. If there is to be a multiplicative identity matrix I, such that

$$AI = A \quad \text{and} \quad IA = A,$$

for any matrix A, then A and I must be square matrices of the same order. Otherwise it would not be possible to find both products. For example, let A be the 2×2 matrix

$$A = \begin{bmatrix} a_{11} & a_{12} \\ a_{21} & a_{22} \end{bmatrix},$$

and let

$$I = \begin{bmatrix} x_{11} & x_{12} \\ x_{21} & x_{22} \end{bmatrix}$$

represent the 2×2 identity matrix. To find I, use the fact that $IA = A$, or

$$\begin{bmatrix} x_{11} & x_{12} \\ x_{21} & x_{22} \end{bmatrix} \begin{bmatrix} a_{11} & a_{12} \\ a_{21} & a_{22} \end{bmatrix} = \begin{bmatrix} a_{11} & a_{12} \\ a_{21} & a_{22} \end{bmatrix}.$$

Multiplying the two matrices on the left side of this equation and setting the elements of the product matrix equal to the corresponding elements of A gives the following four restrictions on x_{11}, x_{12}, x_{21}, and x_{22}.

$$x_{11}a_{11} + x_{12}a_{21} = a_{11}$$
$$x_{21}a_{11} + x_{22}a_{21} = a_{21}$$
$$x_{11}a_{12} + x_{12}a_{22} = a_{12}$$
$$x_{21}a_{12} + x_{22}a_{22} = a_{22}$$

These equations are solved when $x_{11} = 1$, $x_{12} = x_{21} = 0$, and $x_{22} = 1$. (It can be shown that these are the only solutions.) From these solutions, the 2×2 identity matrix is

$$I = \begin{bmatrix} 1 & 0 \\ 0 & 1 \end{bmatrix}.$$

Check that with this definition of I, both $AI = A$ and $IA = A$.

EXAMPLE 1 Let $M = \begin{bmatrix} -2 & 6 \\ 3 & 5 \end{bmatrix}$. Verify that $MI = M$ and $IM = M$.

$$MI = \begin{bmatrix} -2 & 6 \\ 3 & 5 \end{bmatrix}\begin{bmatrix} 1 & 0 \\ 0 & 1 \end{bmatrix} = \begin{bmatrix} -2(1) + 6(0) & -2(0) + 6(1) \\ 3(1) + 5(0) & 3(0) + 5(1) \end{bmatrix}$$

$$= \begin{bmatrix} -2 & 6 \\ 3 & 5 \end{bmatrix} = M$$

$$IM = \begin{bmatrix} 1 & 0 \\ 0 & 1 \end{bmatrix}\begin{bmatrix} -2 & 6 \\ 3 & 5 \end{bmatrix} = \begin{bmatrix} 1(-2) + 0(3) & 1(6) + 0(5) \\ 0(-2) + 1(3) & 0(6) + 1(5) \end{bmatrix}$$

$$= \begin{bmatrix} -2 & 6 \\ 3 & 5 \end{bmatrix} = M \quad \bullet$$

The idea of an identity matrix for square matrices can be generalized.

$n \times n$ Identity Matrix

For any value of n there is an $n \times n$ identity matrix having 1's down the diagonal and 0's elsewhere. The $n \times n$ **identity matrix** is given by I, where

$$I = \begin{bmatrix} 1 & 0 & \cdots & 0 \\ 0 & 1 & \cdots & 0 \\ \cdot & \cdot & & \cdot \\ \cdot & \cdot & a_{ij} & \cdot \\ \cdot & \cdot & & \cdot \\ 0 & 0 & \cdots & 1 \end{bmatrix}.$$

Here $a_{ij} = 1$ when $i = j$ (the diagonal elements) and $a_{ij} = 0$ otherwise.

EXAMPLE 2 Let $K = \begin{bmatrix} -2 & 4 & 0 \\ 3 & 5 & 9 \\ 0 & 8 & -6 \end{bmatrix}$. Give the 3×3 identity matrix I and show that $KI = K$.

The 3×3 identity matrix is

$$I = \begin{bmatrix} 1 & 0 & 0 \\ 0 & 1 & 0 \\ 0 & 0 & 1 \end{bmatrix}.$$

By the definition of matrix multiplication.

$$KI = \begin{bmatrix} -2 & 4 & 0 \\ 3 & 5 & 9 \\ 0 & 8 & -6 \end{bmatrix} \begin{bmatrix} 1 & 0 & 0 \\ 0 & 1 & 0 \\ 0 & 0 & 1 \end{bmatrix} = \begin{bmatrix} -2 & 4 & 0 \\ 3 & 5 & 9 \\ 0 & 8 & -6 \end{bmatrix} = K. \quad \bullet$$

For every nonzero real number a, there is a multiplicative inverse $1/a$ such that

$$a \cdot \frac{1}{a} = 1 \quad \text{and} \quad \frac{1}{a} \cdot a = 1.$$

(Recall that $1/a$ can also be written a^{-1}.) In the rest of this section, a method is developed for finding a multiplicative inverse for square matrices. The **multiplicative inverse** of a matrix A is written A^{-1}. This matrix must satisfy the statements

$$AA^{-1} = I \quad \text{and} \quad A^{-1}A = I.$$

The method we shall use depends on three basic *row operations* that may be performed on any row of any given matrix. These operations are chosen because they are useful in solving systems of equations, as in Section 8.3.

Matrix Row Operations

> The **matrix row operations** are:
>
> **1.** interchanging any two rows of a matrix,
>
> **2.** multiplying the elements of any row of a matrix by the same nonzero scalar k, and
>
> **3.** adding a multiple of the elements of one row to the elements of another row.

EXAMPLE 3

(a) The first row operation is used to change the matrix

$$\begin{bmatrix} 1 & 3 & 5 \\ 0 & 1 & 2 \\ 1 & -1 & -2 \end{bmatrix} \quad \text{to} \quad \begin{bmatrix} 0 & 1 & 2 \\ 1 & 3 & 5 \\ 1 & -1 & -2 \end{bmatrix}$$

by interchanging the first two rows.

(b) Using the second row operation with $k = -2$ changes

$$\begin{bmatrix} 1 & 3 & 5 \\ 0 & 1 & 2 \\ 1 & -1 & -2 \end{bmatrix} \quad \text{to} \quad \begin{bmatrix} -2 & -6 & -10 \\ 0 & 1 & 2 \\ 1 & -1 & -2 \end{bmatrix},$$

where the elements of the first row of the original matrix were multiplied by -2.

(c) The third row operation is used to change

$$\begin{bmatrix} 1 & 3 & 5 \\ 0 & 1 & 2 \\ 1 & -1 & -2 \end{bmatrix} \quad \text{to} \quad \begin{bmatrix} 0 & 4 & 7 \\ 0 & 1 & 2 \\ 1 & -1 & -2 \end{bmatrix},$$

by multiplying each element in the third row of the original matrix by -1 and adding the results to the corresponding elements in the first row of that matrix. That is, the elements in the new first row were found as follows.

$$\begin{bmatrix} 1 + 1(-1) & 3 + (-1)(-1) & 5 + (-2)(-1) \\ 0 & 1 & 2 \\ 1 & -1 & -2 \end{bmatrix} = \begin{bmatrix} 0 & 4 & 7 \\ 0 & 1 & 2 \\ 1 & -1 & -2 \end{bmatrix}$$

Rows two and three were left unchanged. ●

An **augmented matrix** is a combination of two matrices written as one, such as the matrix

$$\begin{bmatrix} 1 & 2 & | & 2 \\ 3 & 4 & | & 5 \end{bmatrix}.$$

The vertical bar separates the two original matrices

$$\begin{bmatrix} 1 & 2 \\ 3 & 4 \end{bmatrix} \quad \text{and} \quad \begin{bmatrix} 2 \\ 5 \end{bmatrix},$$

but does not affect any operation on the rows, so the resulting augmented matrix can be treated as a single matrix.

To get A^{-1} for any $n \times n$ matrix A, first form the augmented matrix $[A|I]$ where A is any $n \times n$ matrix and I is the $n \times n$ multiplicative identity matrix. Perform row operations on $[A|I]$ until an augmented matrix of the form $[I|B]$ is obtained. The matrix B is the desired matrix A^{-1}. To verify that the matrix B from $[I|B]$ is indeed A^{-1}, show that $BA = I$. As an example, let

$$A = \begin{bmatrix} 2 & 3 \\ 1 & -1 \end{bmatrix}.$$

Find A^{-1} by going through the following steps.

Step 1 Form the augmented matrix $[A|I]$.

$$[A|I] = \begin{bmatrix} 2 & 3 & | & 1 & 0 \\ 1 & -1 & | & 0 & 1 \end{bmatrix}$$

Perform row operations on $[A|I]$ until a new matrix of the form $[I|B]$ is obtained. Proceed as follows.

Step 2   Interchanging the first and second rows gives a 1 in the upper left-hand corner:

$$\begin{bmatrix} 1 & -1 & \vline & 0 & 1 \\ 2 & 3 & \vline & 1 & 0 \end{bmatrix}.$$

Step 3   Multiplying the elements of the first row by -2 and adding the results to the second row gives a 0 in the lower left-hand corner:

$$\begin{bmatrix} 1 & -1 & \vline & 0 & 1 \\ 0 & 5 & \vline & 1 & -2 \end{bmatrix}.$$

Step 4   Multiplying the elements of the second row by 1/5 gives a 1 for the second element in that row:

$$\begin{bmatrix} 1 & -1 & \vline & 0 & 1 \\ 0 & 1 & \vline & 1/5 & -2/5 \end{bmatrix}.$$

Step 5   Replacing the first row by the sum of the first row and the second row gives a 0 as the second element in the first row:

$$\begin{bmatrix} 1 & 0 & \vline & 1/5 & 3/5 \\ 0 & 1 & \vline & 1/5 & -2/5 \end{bmatrix}.$$

This last matrix is in the form $[I|B]$, where

$$B = \begin{bmatrix} 1/5 & 3/5 \\ 1/5 & -2/5 \end{bmatrix},$$

which should equal A^{-1}. To check, multiply B and A. The result should be I.

$$BA = \begin{bmatrix} 1/5 & 3/5 \\ 1/5 & -2/5 \end{bmatrix} \begin{bmatrix} 2 & 3 \\ 1 & -1 \end{bmatrix} = \begin{bmatrix} 2/5 + 3/5 & 3/5 - 3/5 \\ 2/5 - 2/5 & 3/5 + 2/5 \end{bmatrix} = \begin{bmatrix} 1 & 0 \\ 0 & 1 \end{bmatrix} = I$$

Verify that AB also equals I, so that

$$A^{-1} = \begin{bmatrix} 1/5 & 3/5 \\ 1/5 & -2/5 \end{bmatrix}.$$

EXAMPLE 4   Find A^{-1} if $A = \begin{bmatrix} 1 & 0 & 1 \\ 2 & -2 & -1 \\ 3 & 0 & 0 \end{bmatrix}$.

Use row transformations as follows.

Step 1   Write the augmented matrix $[A|I]$.

$$\begin{bmatrix} 1 & 0 & 1 & \vline & 1 & 0 & 0 \\ 2 & -2 & -1 & \vline & 0 & 1 & 0 \\ 3 & 0 & 0 & \vline & 0 & 0 & 1 \end{bmatrix}$$

Step 2   Since 1 is already in the upper left-hand corner as desired, begin by selecting the row operation which will result in a 0 for the first element in the second row. Multiply the elements of the first row by -2, and add the results to the second row.

$$\left[\begin{array}{ccc|ccc} 1 & 0 & 1 & 1 & 0 & 0 \\ 0 & -2 & -3 & -2 & 1 & 0 \\ 3 & 0 & 0 & 0 & 0 & 1 \end{array}\right]$$

Step 3   To get 0 for the first element in the third row, multiply the elements of the first row by -3 and add to the third row.

$$\left[\begin{array}{ccc|ccc} 1 & 0 & 1 & 1 & 0 & 0 \\ 0 & -2 & -3 & -2 & 1 & 0 \\ 0 & 0 & -3 & -3 & 0 & 1 \end{array}\right]$$

Step 4   To get 1 for the second element in the second row, multiply the elements of the second row by $-1/2$.

$$\left[\begin{array}{ccc|ccc} 1 & 0 & 1 & 1 & 0 & 0 \\ 0 & 1 & 3/2 & 1 & -1/2 & 0 \\ 0 & 0 & -3 & -3 & 0 & 1 \end{array}\right]$$

Step 5   To get 1 for the third element in the third row, multiply the elements of the third row by $-1/3$.

$$\left[\begin{array}{ccc|ccc} 1 & 0 & 1 & 1 & 0 & 0 \\ 0 & 1 & 3/2 & 1 & -1/2 & 0 \\ 0 & 0 & 1 & 1 & 0 & -1/3 \end{array}\right]$$

Step 6   To get 0 for the third element in the first row, multiply the elements of the third row by -1 and add to the first row.

$$\left[\begin{array}{ccc|ccc} 1 & 0 & 0 & 0 & 0 & 1/3 \\ 0 & 1 & 3/2 & 1 & -1/2 & 0 \\ 0 & 0 & 1 & 1 & 0 & -1/3 \end{array}\right]$$

Step 7   To get 0 for the third element in the second row, multiply the elements of the third row by $-3/2$ and add to the second row.

$$\left[\begin{array}{ccc|ccc} 1 & 0 & 0 & 0 & 0 & 1/3 \\ 0 & 1 & 0 & -1/2 & -1/2 & 1/2 \\ 0 & 0 & 1 & 1 & 0 & -1/3 \end{array}\right]$$

The last transformation shows that the inverse is

$$A^{-1} = \left[\begin{array}{ccc} 0 & 0 & 1/3 \\ -1/2 & -1/2 & 1/2 \\ 1 & 0 & -1/3 \end{array}\right].$$

Confirm this by forming the products $A^{-1}A$ and AA^{-1}, each of which should equal the matrix I. ●

As illustrated by the example, the most efficient order of steps is to make the changes column by column from left to right, so that for each column the required 1 is the result of the first change. Next, perform the steps that obtain the zeros in that column. Then proceed to another column. Since it is tedious to find an inverse with paper and pencil, these same steps can be adapted for a computer program. A computer can produce the inverse of a large matrix, even a 40×40 matrix, in a few seconds.

EXAMPLE 5   Find A^{-1} given $A = \begin{bmatrix} 2 & -4 \\ 1 & -2 \end{bmatrix}$.

Using row operations to transform the first column of the augmented matrix

$$\begin{bmatrix} 2 & -4 & | & 1 & 0 \\ 1 & -2 & | & 0 & 1 \end{bmatrix}$$

results in the following matrices:

$$\begin{bmatrix} 1 & -2 & | & 1/2 & 0 \\ 1 & -2 & | & 0 & 1 \end{bmatrix} \quad \text{and} \quad \begin{bmatrix} 1 & -2 & | & 1/2 & 0 \\ 0 & 0 & | & -1/2 & 1 \end{bmatrix}.$$

At this point, the matrix should be changed so that the second row, second column element will be 1. Since that element is now 0, there is no way to complete the desired transformation.

What is wrong? Just as there is no multiplicative inverse for the real number 0, not every matrix has an inverse. Matrix A is an example of such a matrix. ●

7.3 EXERCISES

Use the third row operation to change each of the following matrices as indicated. See Example 3.

1. $\begin{bmatrix} 2 & 4 \\ 4 & 7 \end{bmatrix}$; -2 times row 1 added to row 2

2. $\begin{bmatrix} -1 & 4 \\ 7 & 0 \end{bmatrix}$;   7 times row 1 added to row 2

3. $\begin{bmatrix} 1 & 1/5 \\ 5 & 2 \end{bmatrix}$; -5 times row 1 added to row 2

4. $\begin{bmatrix} 5 & 7 \\ 2 & -4 \end{bmatrix}$;   4/7 times row 1 added to row 2

5. $\begin{bmatrix} 1 & 5 & 6 \\ -2 & 3 & -1 \\ 4 & 7 & 0 \end{bmatrix}$;   2 times row 1 added to row 2

6. $\begin{bmatrix} 2 & 5 & 6 \\ 4 & -1 & 2 \\ 3 & 7 & 1 \end{bmatrix}$; -6 times row 3 added to row 1

7. $\begin{bmatrix} -3 & 1 & -4 \\ 2 & 1 & 3 \\ -7 & 5 & 2 \end{bmatrix}$; -5 times row 2 added to row 3

8. $\begin{bmatrix} 4 & 10 & -8 \\ 7 & 4 & 3 \\ -1 & 1 & 0 \end{bmatrix}$; -4 times row 3 added to row 2

Decide whether the given matrices are inverses of each other.

9. $\begin{bmatrix} 2 & 3 \\ 1 & 1 \end{bmatrix}$, $\begin{bmatrix} -1 & 3 \\ 1 & -2 \end{bmatrix}$

10. $\begin{bmatrix} 5 & 7 \\ 2 & 3 \end{bmatrix}$, $\begin{bmatrix} 3 & -7 \\ -2 & 5 \end{bmatrix}$

11. $\begin{bmatrix} 2 & 1 \\ 3 & 2 \end{bmatrix}$, $\begin{bmatrix} 2 & 1 \\ -3 & 2 \end{bmatrix}$

12. $\begin{bmatrix} -1 & 2 \\ 3 & -5 \end{bmatrix}$, $\begin{bmatrix} -5 & -2 \\ -3 & -1 \end{bmatrix}$

13. $\begin{bmatrix} 1 & -2 & -3 \\ 2 & -2 & -5 \\ -1 & 1 & 4 \end{bmatrix}$, $\begin{bmatrix} -1 & 5/3 & 4/3 \\ -1 & 1/3 & -1/3 \\ 0 & 1/3 & 2/3 \end{bmatrix}$

14. $\begin{bmatrix} 1 & 2 & -1 \\ 2 & -1 & 3 \\ 3 & -2 & 3 \end{bmatrix}$, $\begin{bmatrix} 3/10 & -2/5 & 1/2 \\ 3/10 & 3/5 & -1/2 \\ -1/10 & 4/5 & -1/2 \end{bmatrix}$

15. $\begin{bmatrix} 1 & 2 & -1 \\ 0 & 1 & 3 \\ 2 & 1 & -2 \end{bmatrix}$, $\begin{bmatrix} 1 & 1 & 2 \\ 1 & 1 & 1 \\ 2 & 3 & 4 \end{bmatrix}$

16. $\begin{bmatrix} 2 & -1 & 4 \\ 0 & 5 & 0 \\ 3 & 2 & -1 \end{bmatrix}$, $\begin{bmatrix} 1 & 0 & 1 \\ 6 & 4 & 2 \\ 1 & 1 & 0 \end{bmatrix}$

17. $\begin{bmatrix} 4 & 3 & 3 \\ -1 & 0 & -1 \\ -4 & -4 & -3 \end{bmatrix}$, $\begin{bmatrix} 4 & 3 & 3 \\ -1 & 0 & -1 \\ -4 & -4 & -3 \end{bmatrix}$

18. $\begin{bmatrix} 1 & 0 & 2 \\ -1 & 0 & -2 \\ 1 & 1 & 1 \end{bmatrix}$, $\begin{bmatrix} 1 & 0 & -2 \\ -1 & 0 & 2 \\ 1 & 1 & 1 \end{bmatrix}$

Use row operations to find any inverses of the following matrices which exist. See Examples 4 and 5.

19. $\begin{bmatrix} 1 & -1 \\ 2 & 0 \end{bmatrix}$

20. $\begin{bmatrix} 3 & -1 \\ -5 & 2 \end{bmatrix}$

21. $\begin{bmatrix} -6 & 4 \\ -3 & 2 \end{bmatrix}$

22. $\begin{bmatrix} -1 & 2 \\ -2 & -1 \end{bmatrix}$

23. $\begin{bmatrix} -1 & -2 \\ 3 & 4 \end{bmatrix}$

24. $\begin{bmatrix} 5 & 10 \\ -3 & -6 \end{bmatrix}$

25. $\begin{bmatrix} .6 & .2 \\ .5 & .1 \end{bmatrix}$

26. $\begin{bmatrix} .8 & -.3 \\ .5 & -.2 \end{bmatrix}$

27. $\begin{bmatrix} .4 & .3 \\ .2 & .1 \end{bmatrix}$

28. $\begin{bmatrix} -.5 & .7 \\ -.1 & .2 \end{bmatrix}$

29. $\begin{bmatrix} 1 & 0 & 0 \\ 0 & -1 & 0 \\ 1 & 0 & 1 \end{bmatrix}$

30. $\begin{bmatrix} 1 & 3 & 3 \\ -1 & 0 & 0 \\ -4 & -4 & -3 \end{bmatrix}$

31. $\begin{bmatrix} 3 & 6 & 3 \\ 6 & 4 & -2 \\ 0 & 1 & -1 \end{bmatrix}$

32. $\begin{bmatrix} 2 & 6 & 0 \\ 1 & 5 & 3 \\ 0 & 0 & 1 \end{bmatrix}$

33. $\begin{bmatrix} -1 & -1 & -1 \\ 4 & 5 & 0 \\ 0 & 1 & -3 \end{bmatrix}$

34. $\begin{bmatrix} 2 & 0 & 4 \\ 3 & 1 & 5 \\ -1 & 1 & -2 \end{bmatrix}$

35. $\begin{bmatrix} 1 & -2 & 3 & 0 \\ 0 & 1 & -1 & 1 \\ -2 & 2 & -2 & 4 \\ 0 & 2 & -3 & 1 \end{bmatrix}$

36. $\begin{bmatrix} 1 & 1 & 0 & 2 \\ 2 & -1 & 1 & -1 \\ 3 & 3 & 2 & -2 \\ 1 & 2 & 1 & 0 \end{bmatrix}$

37. $\begin{bmatrix} -.4 & .1 & .2 \\ 0 & .6 & .8 \\ .3 & 0 & -.2 \end{bmatrix}$

38. $\begin{bmatrix} .8 & .2 & .1 \\ -.2 & 0 & .3 \\ 0 & 0 & .5 \end{bmatrix}$

Let $A = \begin{bmatrix} a & b \\ c & d \end{bmatrix}$. Show that each of the following statements is true.

39. $IA = A$

40. $AI = A$

41. $A \cdot 0 = 0$

42. Find A^{-1}. (Assume $ad - bc \neq 0$.) Show that $AA^{-1} = I$.

43. Show that $A^{-1}A = I$.

7.4 Determinants

Every square matrix A is associated with a real number called the determinant of A, written $\delta(A)$.

Determinant of a 2 × 2 Matrix

> The **determinant of a 2 × 2 matrix** A,
>
> $$A = \begin{bmatrix} a_{11} & a_{12} \\ a_{21} & a_{22} \end{bmatrix},$$
>
> is defined as
>
> $$\delta(A) = |A| = \begin{vmatrix} a_{11} & a_{12} \\ a_{21} & a_{22} \end{vmatrix} = a_{11}a_{22} - a_{21}a_{12}.$$

EXAMPLE 1 If $P = \begin{bmatrix} -3 & 4 \\ 6 & 8 \end{bmatrix}$, then

$$\delta(P) = \begin{vmatrix} -3 & 4 \\ 6 & 8 \end{vmatrix} = -3(8) - 6(4) = -48. \quad \bullet$$

The definition of a determinant can be extended to a 3 × 3 matrix as follows.

Determinant of a 3 × 3 Matrix

> The **determinant of a 3 × 3 matrix** A,
>
> $$A = \begin{bmatrix} a_{11} & a_{12} & a_{13} \\ a_{21} & a_{22} & a_{23} \\ a_{31} & a_{32} & a_{33} \end{bmatrix},$$
>
> is defined as
>
> $$\delta(A) = \begin{vmatrix} a_{11} & a_{12} & a_{13} \\ a_{21} & a_{22} & a_{23} \\ a_{31} & a_{32} & a_{33} \end{vmatrix} = (a_{11}a_{22}a_{33} + a_{12}a_{23}a_{31} + a_{13}a_{21}a_{32}) \\ - (a_{31}a_{22}a_{13} + a_{32}a_{23}a_{11} + a_{33}a_{21}a_{12}).$$

The terms on the right side of the equation above can be rearranged and factored to get

$$\begin{vmatrix} a_{11} & a_{12} & a_{13} \\ a_{21} & a_{22} & a_{23} \\ a_{31} & a_{32} & a_{33} \end{vmatrix} = a_{11}(a_{22}a_{33} - a_{32}a_{23}) - a_{21}(a_{12}a_{33} - a_{32}a_{13}) \\ + a_{31}(a_{12}a_{23} - a_{22}a_{13}). \quad (*)$$

Each of the quantities in parentheses represents a 2×2 determinant that is the determinant of that part of the 3×3 matrix remaining when the row and column of the multiplier are eliminated, as shown below.

$$a_{11}(a_{22}a_{33} - a_{32}a_{23}) \qquad \begin{bmatrix} a_{11} & a_{12} & a_{13} \\ a_{21} & a_{22} & a_{23} \\ a_{31} & a_{32} & a_{33} \end{bmatrix}$$

$$a_{21}(a_{12}a_{33} - a_{32}a_{13}) \qquad \begin{bmatrix} a_{11} & a_{12} & a_{13} \\ a_{21} & a_{22} & a_{23} \\ a_{31} & a_{32} & a_{33} \end{bmatrix}$$

$$a_{31}(a_{12}a_{23} - a_{22}a_{13}) \qquad \begin{bmatrix} a_{11} & a_{12} & a_{13} \\ a_{21} & a_{22} & a_{23} \\ a_{31} & a_{32} & a_{33} \end{bmatrix}$$

The 2×2 determinant is called a **minor** of the element in the 3×3 determinant. Thus, the minors of a_{11}, a_{21}, and a_{31} are as shown below.

Element	a_{11}	a_{21}	a_{31}
Minor	$\begin{vmatrix} a_{22} & a_{23} \\ a_{32} & a_{33} \end{vmatrix}$	$\begin{vmatrix} a_{12} & a_{13} \\ a_{32} & a_{33} \end{vmatrix}$	$\begin{vmatrix} a_{12} & a_{13} \\ a_{22} & a_{23} \end{vmatrix}$

The 3×3 determinant can be evaluated by multiplying each element in the first column by its minor and combining the products as indicated in (*) on the preceding page. This is called the **expansion** of the determinant by minors about the first column.

EXAMPLE 2

Evaluate the determinant

$$\begin{vmatrix} 1 & 3 & -2 \\ -1 & -2 & -3 \\ 1 & 1 & 2 \end{vmatrix}$$

by expanding by minors about the first column.

Using the procedure above,

$$\begin{vmatrix} 1 & 3 & -2 \\ -1 & -2 & -3 \\ 1 & 1 & 2 \end{vmatrix} = 1 \begin{vmatrix} -2 & -3 \\ 1 & 2 \end{vmatrix} - (-1) \begin{vmatrix} 3 & -2 \\ 1 & 2 \end{vmatrix} + 1 \begin{vmatrix} 3 & -2 \\ -2 & -3 \end{vmatrix}$$

$$= 1[(-2)(2) - (1)(-3)] + 1[(3)(2) - (1)(-2)]$$
$$\qquad + 1[(3)(-3) - (-2)(-2)]$$
$$= 1(-1) + 1(8) + 1(-13)$$
$$= -1 + 8 - 13 = -6. \quad \bullet$$

To get equation (*) the terms could have been rearranged and factored differently by factoring out the three elements of the second or third columns or of any of the three rows of the array. Therefore, expanding by minors about any row or any column results in the same value for the determinant. To find the correct signs for the terms of other expansions, the following array of signs is helpful.

Array of Signs

$$
\begin{array}{ccc}
+ & - & + \\
- & + & - \\
+ & - & +
\end{array}
$$

The signs alternate in each row and each column, beginning with $+$ in the first row, first column position. Thus, the array of signs can be reproduced when needed. If we expand about the second column, for example, the first term would have a minus sign, the second a plus sign, and the third a minus sign. These signs are independent of the sign of the corresponding element of the matrix. The sign array can be extended for larger determinants.

EXAMPLE 3   Evaluate the determinant of Example 2 by expansion by minors about the second column.

Using the methods described above,

$$
\begin{vmatrix}
1 & 3 & -2 \\
-1 & -2 & -3 \\
1 & 1 & 2
\end{vmatrix}
= -3 \begin{vmatrix} -1 & -3 \\ 1 & 2 \end{vmatrix}
+ (-2) \begin{vmatrix} 1 & -2 \\ 1 & 2 \end{vmatrix}
- 1 \begin{vmatrix} 1 & -2 \\ -1 & -3 \end{vmatrix}
$$

$$
= -3(1) - 2(4) - 1(-5)
$$

$$
= -3 - 8 + 5
$$

$$
= -6. \quad \bullet
$$

In evaluating the determinant of Examples 2 and 3, we expanded by minors, first about the first column and then about the second column. In evaluating the determinant by expansion about the second column, we needed to evaluate minors as follows.

Element of second column	*Minor*	*Sign associated with minor*
3	$\begin{vmatrix} -1 & -3 \\ 1 & 2 \end{vmatrix}$	$-$
-2	$\begin{vmatrix} 1 & -2 \\ 1 & 2 \end{vmatrix}$	$+$
1	$\begin{vmatrix} 1 & -2 \\ -1 & -3 \end{vmatrix}$	$-$

The minor of an element, together with the sign of the element, is called the **cofactor** of the element. Using the determinant above, we have

Element		Cofactor
3	$-$	$\begin{vmatrix} -1 & -3 \\ 1 & 2 \end{vmatrix} = -1$
-2	$+$	$\begin{vmatrix} 1 & -2 \\ 1 & 2 \end{vmatrix} = 4$
1	$-$	$\begin{vmatrix} 1 & -2 \\ -1 & -3 \end{vmatrix} = 5.$

The method of expansion by minors and cofactors can be extended to find determinants of any $n \times n$ matrix, as shown in the next example.

EXAMPLE 4 Evaluate

$$\begin{vmatrix} -1 & -2 & 3 & 2 \\ 0 & 1 & 4 & -2 \\ 3 & -1 & 4 & 0 \\ 2 & 1 & 0 & 3 \end{vmatrix}.$$

Expanding by minors about the fourth row gives

$$-2 \begin{vmatrix} -2 & 3 & 2 \\ 1 & 4 & -2 \\ -1 & 4 & 0 \end{vmatrix} + 1 \begin{vmatrix} -1 & 3 & 2 \\ 0 & 4 & -2 \\ 3 & 4 & 0 \end{vmatrix} - 0 \begin{vmatrix} -1 & -2 & 2 \\ 0 & 1 & -2 \\ 3 & -1 & 0 \end{vmatrix} + 3 \begin{vmatrix} -1 & -2 & 3 \\ 0 & 1 & 4 \\ 3 & -1 & 4 \end{vmatrix}$$

$$= -2(6) + 1(-50) - 0 + 3(-41)$$

$$= -185. \quad \bullet$$

Each of the four 3×3 determinants in Example 4 must be evaluated by expansion of three 2×2 minors, with much work required to get the final value.

7.4 EXERCISES

1-47 odd

Find the value of each of the following 2×2 determinants. See Example 1.

1. $\begin{vmatrix} 5 & 8 \\ 2 & -4 \end{vmatrix}$ **2.** $\begin{vmatrix} -3 & 0 \\ 0 & 9 \end{vmatrix}$ **3.** $\begin{vmatrix} -1 & -2 \\ 5 & 3 \end{vmatrix}$ **4.** $\begin{vmatrix} 6 & -4 \\ 0 & -1 \end{vmatrix}$

5. $\begin{vmatrix} 9 & 3 \\ -3 & -1 \end{vmatrix}$ **6.** $\begin{vmatrix} 0 & 2 \\ 1 & 5 \end{vmatrix}$ **7.** $\begin{vmatrix} 3 & 4 \\ 5 & -2 \end{vmatrix}$ **8.** $\begin{vmatrix} -9 & 7 \\ 2 & 6 \end{vmatrix}$

9. $\begin{vmatrix} 0 & 4 \\ 4 & 0 \end{vmatrix}$ **10.** $\begin{vmatrix} 1 & 0 \\ 0 & 2 \end{vmatrix}$ **11.** $\begin{vmatrix} 8 & 3 \\ 8 & 3 \end{vmatrix}$ **12.** $\begin{vmatrix} 9 & -4 \\ -4 & 9 \end{vmatrix}$

13. $\begin{vmatrix} x & 4 \\ 8 & 2 \end{vmatrix}$ 　　**14.** $\begin{vmatrix} k & 3 \\ 0 & 4 \end{vmatrix}$ 　　**15.** $\begin{vmatrix} y & 2 \\ 8 & y \end{vmatrix}$ 　　**16.** $\begin{vmatrix} 3 & 8 \\ m & n \end{vmatrix}$

17. $\begin{vmatrix} x & y \\ y & x \end{vmatrix}$ 　　**18.** $\begin{vmatrix} 2m & 8n \\ 8n & 2m \end{vmatrix}$ 　　▣ **19.** $\begin{vmatrix} 1.4 & 2.5 \\ 3.7 & 6.2 \end{vmatrix}$ 　　**20.** $\begin{vmatrix} .123 & .054 \\ .691 & .302 \end{vmatrix}$

Find the value of each of the following 3 × 3 determinants. See Examples 2 and 3.

21. $\begin{vmatrix} 1 & 0 & 0 \\ 0 & -1 & 0 \\ 1 & 0 & 1 \end{vmatrix}$ 　**22.** $\begin{vmatrix} -2 & 0 & 1 \\ 0 & 1 & 0 \\ 0 & 0 & -1 \end{vmatrix}$ 　**23.** $\begin{vmatrix} -2 & 0 & 0 \\ 4 & 0 & 1 \\ 3 & 4 & 2 \end{vmatrix}$ 　**24.** $\begin{vmatrix} 3 & -2 & 0 \\ 0 & -1 & 1 \\ 4 & 0 & 2 \end{vmatrix}$

25. $\begin{vmatrix} 1 & 2 & 0 \\ -1 & 2 & -1 \\ 0 & 1 & 4 \end{vmatrix}$ 　**26.** $\begin{vmatrix} 2 & 1 & -1 \\ 4 & 7 & -2 \\ 2 & 4 & 0 \end{vmatrix}$ 　**27.** $\begin{vmatrix} 10 & 2 & 1 \\ -1 & 4 & 3 \\ -3 & 8 & 10 \end{vmatrix}$ 　**28.** $\begin{vmatrix} 7 & -1 & 1 \\ 1 & -7 & 2 \\ -2 & 1 & 1 \end{vmatrix}$

29. $\begin{vmatrix} 1 & -2 & 3 \\ 0 & 0 & 0 \\ 1 & 10 & -12 \end{vmatrix}$ 　**30.** $\begin{vmatrix} 2 & 3 & 0 \\ 1 & 9 & 0 \\ -1 & -2 & 0 \end{vmatrix}$ 　**31.** $\begin{vmatrix} 3 & 3 & -1 \\ 2 & 6 & 0 \\ -6 & -6 & 2 \end{vmatrix}$ 　**32.** $\begin{vmatrix} 5 & -3 & 2 \\ -5 & 3 & -2 \\ 1 & 0 & 1 \end{vmatrix}$

33. $\begin{vmatrix} 3 & 2 & 0 \\ 0 & 1 & x \\ 2 & 0 & 0 \end{vmatrix}$ 　**34.** $\begin{vmatrix} 0 & 3 & y \\ 0 & 4 & 2 \\ 1 & 0 & 1 \end{vmatrix}$ 　▣ **35.** $\begin{vmatrix} .12 & .43 & .18 \\ .65 & .09 & .71 \\ .42 & .55 & .68 \end{vmatrix}$ 　**36.** $\begin{vmatrix} 1.04 & 2.87 & 3.69 \\ 5.42 & 3.66 & 7.81 \\ 10.3 & 6.41 & 1.17 \end{vmatrix}$

Solve each of the following determinant equations for x.

37. $\begin{vmatrix} -2 & 0 & 1 \\ -1 & 3 & x \\ 5 & -2 & 0 \end{vmatrix} = 3$ 　　　　**38.** $\begin{vmatrix} 4 & 3 & 0 \\ 2 & 0 & 1 \\ -3 & x & -1 \end{vmatrix} = 5$

39. $\begin{vmatrix} 5 & 3x & -3 \\ 0 & 2 & -1 \\ 4 & -1 & x \end{vmatrix} = -7$ 　　　　**40.** $\begin{vmatrix} 2x & 1 & -1 \\ 0 & 4 & x \\ 3 & 0 & 2 \end{vmatrix} = x$

Find the cofactor of each element in the second row for the following determinants.

41. $\begin{vmatrix} -2 & 0 & 1 \\ 3 & 2 & -1 \\ 1 & 0 & 2 \end{vmatrix}$ 　**42.** $\begin{vmatrix} 0 & -1 & 2 \\ 1 & 0 & 2 \\ 0 & -3 & 1 \end{vmatrix}$ 　**43.** $\begin{vmatrix} 1 & 2 & -1 \\ 2 & 3 & -2 \\ -1 & 4 & 1 \end{vmatrix}$ 　**44.** $\begin{vmatrix} 2 & -1 & 4 \\ 3 & 0 & 1 \\ -2 & 1 & 4 \end{vmatrix}$

Find the value of each of the following 4 × 4 determinants. See Example 4.

45. $\begin{vmatrix} 4 & 0 & 0 & 2 \\ -1 & 0 & 3 & 0 \\ 2 & 4 & 0 & 1 \\ 0 & 0 & 1 & 2 \end{vmatrix}$ 　　　　**46.** $\begin{vmatrix} -2 & 0 & 4 & 2 \\ 3 & 6 & 0 & 4 \\ 0 & 0 & 0 & 3 \\ 9 & 0 & 2 & -1 \end{vmatrix}$

47. $\begin{vmatrix} 1 & 1 & 0 & 1 \\ 2 & 1 & 0 & 2 \\ 0 & 1 & -1 & 1 \\ 1 & -1 & 1 & 1 \end{vmatrix}$ 　　　　**48.** $\begin{vmatrix} 2 & 7 & 0 & -1 \\ 1 & 0 & 1 & 3 \\ 2 & 4 & -1 & -1 \\ -1 & 1 & 0 & 8 \end{vmatrix}$

Determinants can be used to find the area of a triangle given the coordinates of its vertices. Given a triangle PQR with vertices (x_1, y_1), (x_3, y_3), and (x_2, y_2), as shown in the following figure, we can introduce segments PM, RN, and QS perpendicular to the x-axis, forming trapezoids PMNR, NSQR, and PMSQ. Recall that the area of a trapezoid is given by half the sum of the parallel bases times the altitude. For example, the area of trapezoid PMSQ equals $(1/2)(y_1 + y_3)(x_3 - x_1)$. The area of triangle PQR can be found by subtracting the area of PMSQ from the sum of the areas of PMNR and RNSQ. Thus, for the area A of triangle PQR,

$$A = (1/2)(x_3y_1 - x_1y_3 + x_2y_3 - x_3y_2 + x_1y_2 - x_2y_1).$$

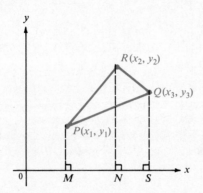

By evaluating the determinant

$$\begin{vmatrix} x_1 & y_1 & 1 \\ x_2 & y_2 & 1 \\ x_3 & y_3 & 1 \end{vmatrix},$$

it can be shown that the area of the triangle is given by the absolute value of A, where

$$A = \frac{1}{2} \begin{vmatrix} x_1 & y_1 & 1 \\ x_2 & y_2 & 1 \\ x_3 & y_3 & 1 \end{vmatrix}.$$

Use the formula given to find the area of the following triangles.

49. $P(0, 1)$, $Q(2, 0)$, $R(1, 3)$

50. $P(2,5)$, $Q(-1, 3)$, $R(4, 0)$

51. $P(2, -2)$, $Q(0, 0)$, $R(-3, -4)$

52. $P(4, 7)$, $Q(5, -2)$, $R(1, 1)$

53. $P(3, 8)$, $Q(-1, 4)$, $R(0, 1)$

54. $P(-3, -1)$, $Q(4, 2)$, $R(3, -3)$

Determinants can be used to find the equation of a line passing through two given points. The next two exercises show how to do this.

55. Expand the determinant below and show that the result is the equation of the line through $(2, 3)$ and $(-1, 4)$.

$$\begin{vmatrix} x & y & 1 \\ 2 & 3 & 1 \\ -1 & 4 & 1 \end{vmatrix} = 0$$

56. **(a)** Write the equation of the line through the points (x_1, y_1) and (x_2, y_2) using the point-slope formula.

(b) Expanding the determinant in the equation below, show that the equation is equivalent to the equation in part (a).

$$\begin{vmatrix} x & y & 1 \\ x_1 & y_1 & 1 \\ x_2 & y_2 & 1 \end{vmatrix} = 0$$

7.5 Properties of Determinants

Some determinants can be evaluated only after much tedious calculation. The more calculation is involved, the greater the chance for error. To evaluate a 3×3 determinant, it is necessary to evaluate three different 2×2 determinants and then combine them correctly. For a 4×4 determinant, twelve 2×2 determinants must be found. This section gives several properties of determinants that make it easier to calculate them. All of these properties of determinants are summarized at the end of this section.

The value of

$$\begin{vmatrix} 3 & 0 \\ 5 & 0 \end{vmatrix}$$

is $3 \cdot 0 - 5 \cdot 0 = 0$. Also, expanded about the second row,

$$\begin{vmatrix} 1 & 2 & 3 \\ 0 & 0 & 0 \\ 4 & 6 & 8 \end{vmatrix} = -0 \begin{vmatrix} 2 & 3 \\ 6 & 8 \end{vmatrix} + 0 \begin{vmatrix} 1 & 3 \\ 4 & 8 \end{vmatrix} - 0 \begin{vmatrix} 1 & 2 \\ 4 & 6 \end{vmatrix} = 0.$$

The fact that each of these determinants had a row or column of zeros led to a 0 value. Generalizing from these examples leads to the following result.

Property 1 If every element in a row or column of a determinant is 0, then the determinant equals 0.

To prove Property 1, expand the determinant about the 0 row or column. Each term of this expansion would have a 0 factor, so the value would be 0. For example,

$$\begin{vmatrix} a & b & c \\ 0 & 0 & 0 \\ d & e & f \end{vmatrix} = -0 \begin{vmatrix} b & c \\ e & f \end{vmatrix} + 0 \begin{vmatrix} a & c \\ d & f \end{vmatrix} - 0 \begin{vmatrix} a & b \\ d & e \end{vmatrix} = 0.$$

EXAMPLE 1

$$\begin{vmatrix} 2 & 4 \\ 0 & 0 \end{vmatrix} = 0 \quad \text{and} \quad \begin{vmatrix} -3 & 7 & 0 \\ 4 & 9 & 0 \\ -6 & 8 & 0 \end{vmatrix} = 0 \quad \bullet$$

Suppose the determinant

$$\begin{vmatrix} 1 & 5 \\ 3 & 7 \end{vmatrix}$$

is rewritten as

$$\begin{vmatrix} 1 & 3 \\ 5 & 7 \end{vmatrix},$$

where the rows of the first determinant are the columns of the second determinant. Evaluating both determinants gives

$$\begin{vmatrix} 1 & 5 \\ 3 & 7 \end{vmatrix} = 1 \cdot 7 - 3 \cdot 5 = -8$$

$$\begin{vmatrix} 1 & 3 \\ 5 & 7 \end{vmatrix} = 1 \cdot 7 - 3 \cdot 5 = -8.$$

The value of the determinant is unchanged. This suggests the next property of determinants.

Property 2 If corresponding rows and columns of a determinant are interchanged, the value is not changed.

The proof of Property 2 is given in Exercise 35.

EXAMPLE 2   By Property 2, since the rows of the first determinant are the columns of the second,

$$\begin{vmatrix} 2 & 1 & 6 \\ 3 & 0 & 5 \\ -4 & 6 & 9 \end{vmatrix} = \begin{vmatrix} 2 & 3 & -4 \\ 1 & 0 & 6 \\ 6 & 5 & 9 \end{vmatrix}. \quad \bullet$$

Exchanging the rows of the determinant

$$\begin{vmatrix} 1 & 6 \\ 5 & 3 \end{vmatrix} \quad \text{gives} \quad \begin{vmatrix} 5 & 3 \\ 1 & 6 \end{vmatrix}.$$

While

$$\begin{vmatrix} 1 & 6 \\ 5 & 3 \end{vmatrix} = 1 \cdot 3 - 5 \cdot 6 = -27,$$

the value of the second determinant is

$$\begin{vmatrix} 5 & 3 \\ 1 & 6 \end{vmatrix} = 5 \cdot 6 - 1 \cdot 3 = 27.$$

The second determinant is the negative of the first. This suggests the third property of determinants.

Property 3   Exchanging two rows (or columns) of a determinant reverses the sign of the determinant.

See Exercise 36 for the proof of Property 3.

EXAMPLE 3   **(a)** Exchange the two columns of

$$\begin{vmatrix} 2 & 5 \\ 3 & 4 \end{vmatrix} \quad \text{to get} \quad \begin{vmatrix} 5 & 2 \\ 4 & 3 \end{vmatrix}.$$

By Property 3, since

$$\begin{vmatrix} 2 & 5 \\ 3 & 4 \end{vmatrix} = -7,$$

the value of

$$\begin{vmatrix} 5 & 2 \\ 4 & 3 \end{vmatrix} = 7.$$

(b)
$$\begin{vmatrix} 2 & 1 & 6 \\ 3 & 0 & 5 \\ -4 & 6 & 9 \end{vmatrix} = - \begin{vmatrix} -4 & 6 & 9 \\ 3 & 0 & 5 \\ 2 & 1 & 6 \end{vmatrix} \quad \bullet$$

Multiplying each element of the second row of the determinant

$$\begin{vmatrix} 2 & -3 \\ 4 & 1 \end{vmatrix}$$

by -5 gives the new determinant

$$\begin{vmatrix} 2 & -3 \\ 4(-5) & 1(-5) \end{vmatrix} = \begin{vmatrix} 2 & -3 \\ -20 & -5 \end{vmatrix}.$$

The values of these determinants are

$$\begin{vmatrix} 2 & -3 \\ 4 & 1 \end{vmatrix} = 2 \cdot 1 - 4(-3) = 14,$$

and

$$\begin{vmatrix} 2 & -3 \\ -20 & -5 \end{vmatrix} = 2(-5) - (-20)(-3) = -70.$$

Since $-70 = -5(14)$, the value of the new determinant is -5 times the value of the original determinant. The next property generalizes this idea.

Property 4   If every element of a row (or column) of a determinant is multiplied by the real number k, then the new determinant is k times the value of the original determinant.

EXAMPLE 4   By Property 4,

$$\begin{vmatrix} -2 & 3 & 7 \\ 0 & 5 & 2 \\ -16 & 6 & 4 \end{vmatrix} = -2 \begin{vmatrix} 1 & 3 & 7 \\ 0 & 5 & 2 \\ 8 & 6 & 4 \end{vmatrix},$$

since the first column of the determinant on the left is -2 times the first column of the determinant on the right.   ●

The determinant

$$\begin{vmatrix} 2 & 1 & 2 \\ 5 & 10 & 5 \\ 3 & 6 & 3 \end{vmatrix}$$

has first and third columns which are identical. Expanding about the third column gives

$$\begin{vmatrix} 2 & 1 & 2 \\ 5 & 10 & 5 \\ 3 & 6 & 3 \end{vmatrix} = -1 \begin{vmatrix} 5 & 5 \\ 3 & 3 \end{vmatrix} + 10 \begin{vmatrix} 2 & 2 \\ 3 & 3 \end{vmatrix} - 6 \begin{vmatrix} 2 & 2 \\ 5 & 5 \end{vmatrix}$$

$$= -1(15 - 15) + 10(6 - 6) - 6(10 - 10)$$

$$= 0.$$

This result suggests the next property.

Property 5   A determinant with two rows (or columns) identical equals 0.

EXAMPLE 5   Since two rows are identical,

$$\begin{vmatrix} -4 & 2 & 3 \\ 0 & 1 & 6 \\ -4 & 2 & 3 \end{vmatrix} = 0.   ●$$

Multiply each element of the second row of the determinant

$$\begin{vmatrix} -3 & 5 \\ 1 & 2 \end{vmatrix}$$

by 3 and add the results to the first row.

$$\begin{vmatrix} -3 + 1(3) & 5 + 2(3) \\ 1 & 2 \end{vmatrix} = \begin{vmatrix} -3 + 3 & 5 + 6 \\ 1 & 2 \end{vmatrix} = \begin{vmatrix} 0 & 11 \\ 1 & 2 \end{vmatrix}$$

Verify that the new determinant has the same value as the original determinant, -11. This idea, which is generalized below, is perhaps the most useful property of determinants presented in this section.

Property 6   The value of a determinant is unchanged if a multiple of a row (or column) is added to the corresponding elements of another row (or column).

EXAMPLE 6   Multiply each element of the first column of the determinant

$$\begin{vmatrix} -2 & 4 & 1 \\ 2 & 1 & 5 \\ 3 & 0 & 2 \end{vmatrix}$$

by 3, and add the results to the third column to get the new determinant

$$\begin{vmatrix} -2 & 4 & 1 + 3(-2) \\ 2 & 1 & 5 + 3(2) \\ 3 & 0 & 2 + 3(3) \end{vmatrix} = \begin{vmatrix} -2 & 4 & -5 \\ 2 & 1 & 11 \\ 3 & 0 & 11 \end{vmatrix}.$$

By Property 6, these two determinants have the same value, 37.   ●

The following examples show how the properties of determinants can be used to simplify the calculation of determinants.

EXAMPLE 7   Without expanding, show that the value of the following determinant is zero.

$$\begin{vmatrix} 2 & 5 & -1 \\ 1 & -15 & 3 \\ -2 & 10 & -2 \end{vmatrix}$$

Each element in the second column is -5 times the corresponding element in the third column. Using Property 6, multiply the elements of the third column by 5 and then add the results to the corresponding elements in the second column to get the equivalent determinant

$$\begin{vmatrix} 2 & 0 & -1 \\ 1 & 0 & 3 \\ -2 & 0 & -2 \end{vmatrix}.$$

By Property 1, the value of this determinant is zero.   ●

EXAMPLE 8   Evaluate

$$\begin{vmatrix} 4 & 2 & 1 & 0 \\ -2 & 4 & -1 & 7 \\ -5 & 2 & 3 & 1 \\ 6 & 4 & -3 & 2 \end{vmatrix}.$$

Use Property 6 to change the first row (any row or column could be used) to a row in which every element but one is zero. Begin by multiplying the elements of the second column by -2 and adding the results to the first column, replacing the first column with this new column. The result is a 0 for the first element in the first row.

$$\begin{vmatrix} 0 & 2 & 1 & 0 \\ -10 & 4 & -1 & 7 \\ -9 & 2 & 3 & 1 \\ -2 & 4 & -3 & 2 \end{vmatrix}$$

The first column replaced by the sum of -2 times the second column and the first column

To get a zero for the second element in the first row, multiply the elements of column three by -2 and add to the second column.

$$\begin{vmatrix} 0 & 0 & 1 & 0 \\ -10 & 6 & -1 & 7 \\ -9 & -4 & 3 & 1 \\ -2 & 10 & -3 & 2 \end{vmatrix}$$

The second column replaced by the sum of -2 times the third column and the second column

Since the first row now has only one nonzero number, expand about the first row to get

$$1 \begin{vmatrix} -10 & 6 & 7 \\ -9 & -4 & 1 \\ -2 & 10 & 2 \end{vmatrix} = \begin{vmatrix} -10 & 6 & 7 \\ -9 & -4 & 1 \\ -2 & 10 & 2 \end{vmatrix}.$$

Repeat the process with this 3×3 determinant. Change the third column to a column with two zeros, as shown below.

$$\begin{vmatrix} 53 & 34 & 0 \\ -9 & -4 & 1 \\ -2 & 10 & 2 \end{vmatrix}$$

The first row replaced by the sum of -7 times the second row and the first row

$$\begin{vmatrix} 53 & 34 & 0 \\ -9 & -4 & 1 \\ 16 & 18 & 0 \end{vmatrix}$$

The third row replaced by the sum of -2 times the second row and the third row

Expand about the third column to find the value of the determinant:

$$-1 \begin{vmatrix} 53 & 34 \\ 16 & 18 \end{vmatrix} = -1(954 - 544) = -410.$$

When applying Property 6, work with sums of *rows* to get a *column* with only one nonzero number or with sums of *columns* to get a *row* with one nonzero number. ●

The following summary lists the properties of determinants presented in this section.

Properties of Determinants

1. If every element in a row or column of a determinant is 0, then the determinant equals 0.

2. If corresponding rows and columns of a determinant are interchanged, the value is not changed.

3. Exchanging two rows (or columns) of a determinant reverses the sign of the determinant.

4. If every element of a row (or column) of a determinant is multiplied by the real number k, then the new determinant is k times the value of the original determinant.

5. A determinant with two rows (or columns) identical equals 0.

6. The value of a determinant is unchanged if a multiple of a row (or column) is added to the corresponding elements of another row (or column).

7.5 EXERCISES

Tell why each of the following determinants has a value of 0.

1. $\begin{vmatrix} 2 & 3 \\ 2 & 3 \end{vmatrix}$

2. $\begin{vmatrix} -5 & -5 \\ 6 & 6 \end{vmatrix}$

3. $\begin{vmatrix} 2 & 0 \\ 3 & 0 \end{vmatrix}$

4. $\begin{vmatrix} -8 & 0 \\ -6 & 0 \end{vmatrix}$

5. $\begin{vmatrix} -1 & 2 & 4 \\ 4 & -8 & -16 \\ 3 & 0 & 5 \end{vmatrix}$

6. $\begin{vmatrix} 2 & -8 & 3 \\ 0 & 2 & -1 \\ -6 & 24 & -9 \end{vmatrix}$

7. $\begin{vmatrix} 3 & 6 & 6 \\ 2 & 0 & 4 \\ 1 & 4 & 2 \end{vmatrix}$

8. $\begin{vmatrix} 1 & 0 & 0 \\ 1 & 0 & 1 \\ 3 & 0 & 0 \end{vmatrix}$

9. $\begin{vmatrix} m & 2 & 2m \\ 3n & 1 & 6n \\ 5p & 6 & 10p \end{vmatrix}$

10. $\begin{vmatrix} 7z & 8x & 2y \\ z & x & y \\ 7z & 7x & 7y \end{vmatrix}$

Use the appropriate properties from this section to tell why each of the following is true. Do not evaluate the determinants.

11. $\begin{vmatrix} 2 & 1 & 6 \\ 3 & 0 & 2 \\ 4 & 1 & 8 \end{vmatrix} = \begin{vmatrix} 2 & 3 & 4 \\ 1 & 0 & 1 \\ 6 & 2 & 8 \end{vmatrix}$

12. $\begin{vmatrix} 4 & -2 \\ 3 & 8 \end{vmatrix} = \begin{vmatrix} 4 & 3 \\ -2 & 8 \end{vmatrix}$

13. $\begin{vmatrix} 2 & 6 \\ 3 & 5 \end{vmatrix} = - \begin{vmatrix} 3 & 5 \\ 2 & 6 \end{vmatrix}$

14. $\begin{vmatrix} -1 & 8 & 9 \\ 0 & 2 & 1 \\ 3 & 2 & 0 \end{vmatrix} = - \begin{vmatrix} 8 & -1 & 9 \\ 2 & 0 & 1 \\ 2 & 3 & 0 \end{vmatrix}$

15. $3 \begin{vmatrix} 6 & 0 & 2 \\ 4 & 1 & 3 \\ 2 & 8 & 6 \end{vmatrix} = \begin{vmatrix} 6 & 0 & 2 \\ 4 & 3 & 3 \\ 2 & 24 & 6 \end{vmatrix}$

16. $-\dfrac{1}{2} \begin{vmatrix} 5 & -8 & 2 \\ 3 & -6 & 9 \\ 2 & 4 & 4 \end{vmatrix} = \begin{vmatrix} 5 & 4 & 2 \\ 3 & 3 & 9 \\ 2 & -2 & 4 \end{vmatrix}$

17. $\begin{vmatrix} 3 & -4 \\ 2 & 5 \end{vmatrix} = \begin{vmatrix} 3 & -4 \\ 5 & 1 \end{vmatrix}$

18. $\begin{vmatrix} -1 & 6 \\ 3 & -5 \end{vmatrix} = \begin{vmatrix} -1 & 6 \\ 2 & 1 \end{vmatrix}$

19. $\begin{vmatrix} 5 & 8 \\ 2 & -1 \end{vmatrix} = \begin{vmatrix} 5 & -2 \\ 2 & -5 \end{vmatrix}$

20. $\begin{vmatrix} 13 & 5 \\ 6 & 1 \end{vmatrix} = \begin{vmatrix} -2 & 5 \\ 3 & 1 \end{vmatrix}$

21. $\begin{vmatrix} 2 & 5 & 8 \\ 1 & 0 & 2 \\ 4 & 3 & 5 \end{vmatrix} = \begin{vmatrix} 2 & 5 & 8 \\ 1 & 0 & 2 \\ 7 & 3 & 11 \end{vmatrix}$

22. $2\begin{vmatrix} 4 & 2 & -1 \\ m & 2n & 3p \\ 5 & 1 & 0 \end{vmatrix} = \begin{vmatrix} 4 & 2 & -1 \\ 2m & 4n & 6p \\ 5 & 1 & 0 \end{vmatrix}$

23. $\begin{vmatrix} 3 & 5 & 0 \\ 2 & 1 & 3 \\ -5 & 1 & 6 \end{vmatrix} = \begin{vmatrix} 3 & 5 & 0 \\ 2+3k & 1+5k & 3+0k \\ -5 & 1 & 6 \end{vmatrix}$

24. $\begin{vmatrix} -4 & 2 & 1 \\ 3 & 0 & 5 \\ -1 & 4 & -2 \end{vmatrix} = \begin{vmatrix} -4 & 2 & 1+(-4)k \\ 3 & 0 & 5+3k \\ -1 & 4 & -2+(-1)k \end{vmatrix}$

Use Property 6 to find the value of each of the following determinants. See Examples 7 and 8.

25. $\begin{vmatrix} 2 & 4 \\ 3 & 6 \end{vmatrix}$

26. $\begin{vmatrix} -5 & 10 \\ 6 & -12 \end{vmatrix}$

27. $\begin{vmatrix} 4 & 8 & 0 \\ -1 & -2 & 1 \\ 2 & 4 & 3 \end{vmatrix}$

28. $\begin{vmatrix} 6 & 8 & -12 \\ -1 & 0 & 2 \\ 4 & 0 & -8 \end{vmatrix}$

29. $\begin{vmatrix} 3 & 1 & 2 \\ 2 & 0 & 1 \\ 1 & 0 & -2 \end{vmatrix}$

30. $\begin{vmatrix} -2 & 2 & 3 \\ 0 & 2 & 1 \\ -1 & 4 & 0 \end{vmatrix}$

31. $\begin{vmatrix} -4 & 2 & 3 \\ 2 & 0 & 1 \\ 0 & 4 & 2 \end{vmatrix}$

32. $\begin{vmatrix} 6 & 3 & 2 \\ 1 & 0 & 2 \\ -1 & 4 & 1 \end{vmatrix}$

33. $\begin{vmatrix} 1 & 0 & 2 & 2 \\ 2 & 4 & 1 & -1 \\ 1 & -3 & 1 & 0 \\ 1 & 1 & 0 & 1 \end{vmatrix}$

34. $\begin{vmatrix} 2 & -1 & 1 & 0 \\ 1 & 1 & 0 & 1 \\ 0 & -1 & 1 & 1 \\ 1 & 2 & 1 & 2 \end{vmatrix}$

Use the determinant

$$\begin{vmatrix} a & b & c \\ d & e & f \\ g & h & j \end{vmatrix}$$

to prove the following properties of determinants by finding the values of the original determinant and the new determinant and comparing them.

35. Property 2

36. Property 3

37. Property 4

7.6 A Matrix Application*

The diagram in Figure 7.1 shows the roads connecting four cities. This information is shown in matrix A, where the entries represent the number of roads connecting two cities without passing through another city. For example, from the diagram we see that there are two roads connecting city 1 to city 4 without passing

---

*Hugh G. Campbell, *Matrices With Applications,* © 1968, pp. 50–51. Reprinted by permission of Prentice-Hall, Inc., Englewood Cliffs, N.J.

through either city 2 or 3. This information is entered in row 1, column 4 and again in row 4, column 1 of the matrix below.

$$
\begin{array}{c}
 & & City \\
 & & \begin{array}{cccc} \mathbf{1} & \mathbf{2} & \mathbf{3} & \mathbf{4} \end{array} \\
City \begin{array}{c} \mathbf{1} \\ \mathbf{2} \\ \mathbf{3} \\ \mathbf{4} \end{array} & & \begin{bmatrix} 0 & 1 & 2 & 2 \\ 1 & 0 & 1 & 0 \\ 2 & 1 & 0 & 1 \\ 2 & 0 & 1 & 0 \end{bmatrix} & = A
\end{array}
$$

There are no roads connecting each city to itself. Also, there is one road connecting cities 3 and 2.

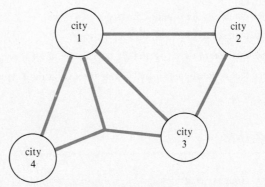

Figure 7.1

It can be shown that $A^2 (A^2 = A \cdot A)$ gives the number of ways to travel between any two cities by passing through exactly one other city, where opposite directions on the same road are considered to be different routes. Similarly, A^3 gives the number of ways to travel between any two cities by passing through exactly two cities. Also, $A + A^2$ represents the total number of ways to travel between two cities with at most one intermediate city.

Many other interpretations can be given to the diagram. For example, the lines could represent communication lines or mutual influence between people or nations.

7.6   EXERCISES

Using matrix A from the text, find A^2. Then answer the following questions.

1. How many ways are there to travel from city 1 to city 3 by passing through exactly one other city?

2. How many ways are there to travel from city 1 to city 3 by passing through at most one other city?

3. How many ways are there to travel from city 2 to city 4 by passing through exactly one other city?

4. How many ways are there to travel from city 2 to city 4 by passing through at most one other city?

Find A^3. Then answer the following questions.

5. How many ways are there to travel between cities 1 and 4 by passing through exactly two cities?

6. How many ways are there to travel between cities 1 and 4 by passing through at most two cities.

A small telephone system connects three cities. There are 4 lines between city 3 and city 2, 3 lines connecting city 3 and city 1, and 2 lines between city 1 and city 2.

7. Write a matrix B to represent this information.

8. Find B^2.

9. How many lines connect cities 1 and 2 through city 3?

10. How many lines connect cities 1 and 3 through city 2?

11. How many lines that connect cities 1 and 2 pass through at most one other city?

12. How many lines that connect cities 1 and 3 pass through at most one other city?

Chapter 7   Summary

Key Words	matrix, matrices	augmented matrix	column matrix
	order of a matrix	minor	scalar
	row matrix	cofactor	matrix row operations
	zero matrix	element	determinant
	identity matrix	square matrix	expansion

Addition of Matrices

The sum of two $m \times n$ matrices A and B is the $m \times n$ matrix $A + B$ in which each element is the sum of the corresponding elements of A and B.

Subtraction of Matrices

If A and B are two matrices of the same order, then
$$A - B = A + (-B).$$

Multiplication of Matrices

The product AB of an $m \times n$ matrix A and an $n \times k$ matrix B is found as follows. To get the ith row, jth column element of AB, multiply each element in the ith row of A by the corresponding element in the jth column of B. The sum of these products will give the element of row i, column j of AB.

Matrix Row Operations

The matrix row operations are:

1. interchanging any two rows of a matrix,

2. multiplying the elements of any row of a matrix by the same nonzero scalar k, and

3. adding a multiple of the elements of one row to the elements of another row.

Chapter 7   Review Exercises

Find the values of all variables in the following statements.

1. $\begin{bmatrix} 2 & z & 1 \\ m & 9 & -7 \end{bmatrix} = \begin{bmatrix} x & 5 & 1 \\ -8 & y & p \end{bmatrix}$

2. $\begin{bmatrix} 5 & -10 \\ a & 0 \end{bmatrix} = \begin{bmatrix} b-3 & c+2 \\ 4 & 0 \end{bmatrix}$

3. $\begin{bmatrix} 5 & x+2 \\ -6y & z \end{bmatrix} = \begin{bmatrix} a & 3x-1 \\ 5y & 9 \end{bmatrix}$

4. $\begin{bmatrix} 6+k & 2 & a+3 \\ -2+m & 3p & 2r \end{bmatrix} + \begin{bmatrix} 3-2k & 5 & 7 \\ 5 & 8p & 5r \end{bmatrix} = \begin{bmatrix} 5 & y & 6a \\ 2m & 11 & -35 \end{bmatrix}$

Perform each of the following operations whenever possible.

5. $\begin{bmatrix} 3 & -4 & 2 \\ 5 & -1 & 6 \end{bmatrix} + \begin{bmatrix} -3 & 2 & 5 \\ 1 & 0 & 4 \end{bmatrix}$

6. $\begin{bmatrix} 3 \\ 2 \\ 5 \end{bmatrix} - \begin{bmatrix} 8 \\ -4 \\ 6 \end{bmatrix} + \begin{bmatrix} 1 \\ 0 \\ 2 \end{bmatrix}$

7. $\begin{bmatrix} 2 & 5 & 8 \\ 1 & 9 & 2 \end{bmatrix} - \begin{bmatrix} 3 & 4 \\ 7 & 1 \end{bmatrix}$

8. $3\begin{bmatrix} 2 & 4 \\ -1 & 4 \end{bmatrix} - 2\begin{bmatrix} 5 & 8 \\ 2 & -2 \end{bmatrix}$

9. $-1\begin{bmatrix} 3 & -5 & 2 \\ 1 & 7 & -4 \end{bmatrix} + 5\begin{bmatrix} 0 & 2 \\ -1 & 3 \end{bmatrix}$

10. $10\begin{bmatrix} 2x+3y & 4x+y \\ x-5y & 6x+2y \end{bmatrix} + 2\begin{bmatrix} -3x-y & x+6y \\ 4x+2y & 5x-y \end{bmatrix}$

11. The speed limits in Italy, Britain, and the U.S. are 87 mph, 70 mph, and 55 mph, respectively. The corresponding fatalities per 100 million miles driven (in 1977) were 6.4, 4.0, and 3.3. Write this information as a 3 × 2 matrix and then as a 2 × 3 matrix.

12. In the years 1981 through 1985, a company had sales (in billions) of 13.0, 14.0, 14.2, 14.9, and 16.7. The company had income (in millions) of 1010, 1000, 979, 1079, and 1228. They declared dividends per common share (in dollars) of 1.32, 1.46, 1.54, 1.64, and 1.82. Write this information as a matrix in two ways.

Find each of the following matrix products, whenever possible.

13. $\begin{bmatrix} -3 & 4 \\ 2 & 8 \end{bmatrix}\begin{bmatrix} -1 & 0 \\ 2 & 5 \end{bmatrix}$

14. $\begin{bmatrix} 3 & 2 & -1 \\ 4 & 0 & 6 \end{bmatrix}\begin{bmatrix} -2 & 0 \\ 0 & 2 \\ 3 & 1 \end{bmatrix}$

15. $\begin{bmatrix} 1 & -2 & 4 & 2 \\ 0 & 1 & -1 & 8 \end{bmatrix}\begin{bmatrix} -1 \\ 2 \\ 0 \\ 1 \end{bmatrix}$

16. $\begin{bmatrix} 1 & 2 & 5 \\ -3 & 4 & 7 \\ 0 & 2 & -1 \end{bmatrix}\begin{bmatrix} 4 & 2 & 3 \\ 10 & -5 & 6 \end{bmatrix}$

17. $\begin{bmatrix} 4 & 2 & 3 \\ 10 & -5 & 6 \end{bmatrix}\begin{bmatrix} 1 & 2 & 5 \\ -3 & 4 & 7 \\ 0 & 2 & -1 \end{bmatrix}$

18. $\begin{bmatrix} 3 & -1 & 0 \end{bmatrix}\begin{bmatrix} 1 & 3 & 2 \\ 2 & -4 & 0 \\ 5 & 7 & 3 \end{bmatrix}$

Decide whether or not each of the following pairs of matrices are inverses.

19. $\begin{bmatrix} 2 & -3 \\ 1 & -2 \end{bmatrix}, \begin{bmatrix} 2 & -3 \\ 1 & -2 \end{bmatrix}$

20. $\begin{bmatrix} 1 & 0 \\ 2 & -3 \end{bmatrix}, \begin{bmatrix} 1 & 0 \\ 2/3 & -1/3 \end{bmatrix}$

21. $\begin{bmatrix} 2 & 0 & 6 \\ 0 & 1 & 0 \\ 1 & 0 & 1 \end{bmatrix}, \begin{bmatrix} -1 & 0 & 3/2 \\ 0 & 1 & 0 \\ 1/4 & 0 & -1 \end{bmatrix}$

22. $\begin{bmatrix} 1 & 0 & 2 \\ 0 & 2 & 4 \\ 0 & 0 & 1 \end{bmatrix}, \begin{bmatrix} 1 & 0 & -2 \\ 0 & 1/2 & -2 \\ 0 & 0 & 1 \end{bmatrix}$

Find the inverse, if it exists, for each of the following matrices.

23. $\begin{bmatrix} 2 & 1 \\ 5 & 3 \end{bmatrix}$

24. $\begin{bmatrix} -4 & 2 \\ 0 & 3 \end{bmatrix}$

25. $\begin{bmatrix} 2 & 0 \\ -1 & 5 \end{bmatrix}$

26. $\begin{bmatrix} 2 & 0 & 4 \\ 1 & -1 & 0 \\ 0 & 1 & -2 \end{bmatrix}$

27. $\begin{bmatrix} 2 & -1 & 0 \\ 1 & 0 & 1 \\ 1 & -2 & 0 \end{bmatrix}$

28. $\begin{bmatrix} 2 & 3 & 5 \\ -2 & -3 & -5 \\ 1 & 4 & 2 \end{bmatrix}$

Find each of the following determinants.

29. $\begin{vmatrix} -1 & 8 \\ 2 & 9 \end{vmatrix}$

30. $\begin{vmatrix} -2 & 4 \\ 0 & 3 \end{vmatrix}$

31. $\begin{vmatrix} -2 & 4 & 1 \\ 3 & 0 & 2 \\ -1 & 0 & 3 \end{vmatrix}$

32. $\begin{vmatrix} -1 & 2 & 3 \\ 4 & 0 & 3 \\ 5 & -1 & 2 \end{vmatrix}$

Solve each of the following determinant equations for x.

33. $\begin{vmatrix} -3 & 2 \\ 1 & x \end{vmatrix} = 5$

34. $\begin{vmatrix} 3x & 7 \\ -x & 4 \end{vmatrix} = 8$

35. $\begin{vmatrix} 2 & 5 & 0 \\ 1 & 3x & -1 \\ 0 & 2 & 0 \end{vmatrix} = 4$

36. $\begin{vmatrix} 6x & 2 & 0 \\ 1 & 5 & 3 \\ x & 2 & -1 \end{vmatrix} = 2x$

Explain why each of the following statements is true.

37. $\begin{vmatrix} 8 & 9 & 2 \\ 0 & 0 & 0 \\ 3 & 1 & 4 \end{vmatrix} = 0$

38. $\begin{vmatrix} 4 & 6 \\ 3 & 5 \end{vmatrix} = \begin{vmatrix} 4 & 3 \\ 6 & 5 \end{vmatrix}$

39. $\begin{vmatrix} 8 & 2 \\ 4 & 3 \end{vmatrix} = 2 \begin{vmatrix} 4 & 1 \\ 4 & 3 \end{vmatrix}$

40. $\begin{vmatrix} 4 & 6 & 2 \\ -3 & 8 & -5 \\ 4 & 6 & 2 \end{vmatrix} = 0$

41. $\begin{vmatrix} 5 & -1 & 2 \\ 3 & -2 & 0 \\ -4 & 1 & 2 \end{vmatrix} = \begin{vmatrix} 5 & -1 & 2 \\ 8 & -3 & 2 \\ -4 & 1 & 2 \end{vmatrix}$

42. $\begin{vmatrix} 8 & 2 & -5 \\ -3 & 1 & 4 \\ 2 & 0 & 5 \end{vmatrix} = - \begin{vmatrix} 8 & -5 & 2 \\ -3 & 4 & 1 \\ 2 & 5 & 0 \end{vmatrix}$

8

Systems of Equations and Inequalities

Many applications of mathematics require the solution of a large number of equations or inequalities having many variables. In this chapter we discuss methods of solving these systems of equations or inequalities.

8.1 Linear Systems with Two Variables

Any set of equations is a **system of equations.** The solution set of a system is the intersection of the solution sets of the individual equations. It is customary to write a system by listing its equations. For example, the system of equations $2x + y = 4$ and $x - y = 6$ is written as

$$2x + y = 4$$
$$x - y = 6.$$

A **first-degree equation in n unknowns** is any equation of the form

$$a_1x_1 + a_2x_2 + \cdots + a_nx_n = k,$$

where $a_1, a_2, \ldots, a_n$, and k are constants and $x_1, x_2, \ldots, x_n$ are variables. Such equations are also called linear equations. Generally, we discuss only systems of linear equations with two or three variables, although the methods used can be extended to systems with more variables.

Recall from Chapter 4 that the solution set of a linear equation in two variables is an infinite set of ordered pairs. Since the graph of such an equation is a straight line, there are three possibilities for the solution set of a system of two linear equations in two variables, as shown in Figure 8.1.

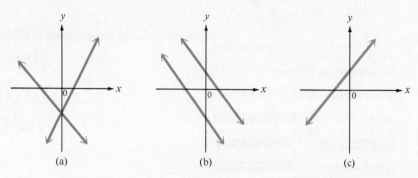

(a)   (b)   (c)

Figure 8.1

Graphs of a Linear System with Two Equations and Two Variables

1. The two graphs intersect in a single point. The coordinates of this point give the solution of the system. This is the most common case. See Figure 8.1(a).

2. The graphs are distinct parallel lines. When this is the case, the system is said to be **inconsistent.** That is, there is no solution common to both equations. The solution set of the linear system is empty. See Figure 8.1(b).

3. The graphs are the same line. In this case, the equations are said to be **dependent,** and any solution of one equation is also a solution of the other. See Figure 8.1(c).

Although the *number* of solutions of a linear system can often be seen from the graph of the equations of the system, it is usually difficult to determine an exact solution from the graph. One algebraic method of finding the solution of a system of two linear equations, called the **substitution method,** is illustrated in the following example.

EXAMPLE 1   Solve the system

$$2x + 5y = 21 \tag{1}$$
$$3x - 2y = -16. \tag{2}$$

Begin by solving one equation for one of the variables in terms of the other. Let us solve equation (1) for x.

$$2x + 5y = 21 \tag{1}$$
$$2x = 21 - 5y$$
$$x = \frac{21 - 5y}{2} \tag{3}$$

Now substitute this result for x into equation (2).

$$3x - 2y = -16 \qquad \textbf{(2)}$$

$$3\left(\frac{21 - 5y}{2}\right) - 2y = -16$$

To eliminate the fraction on the left, multiply both sides of the equation by 2 and then solve for y.

$$2 \cdot 3\left(\frac{21 - 5y}{2}\right) - 2 \cdot 2y = 2(-16)$$

$$3(21 - 5y) - 4y = -32$$

$$63 - 15y - 4y = -32$$

$$-19y = -95$$

$$y = 5$$

Substitute $y = 5$ back into equation (3) to find x.

$$x = \frac{21 - 5y}{2} \qquad \textbf{(3)}$$

$$x = \frac{21 - 5 \cdot 5}{2}$$

$$x = -2$$

The solution set for the system is $\{(-2, 5)\}$. Check by substituting -2 for x and 5 for y in each of the equations of the system. •

To solve a system of two equations by the **addition method,** multiply the equations on both sides by suitable numbers, so that when they are added, one variable is eliminated. The result is an equation in one variable which can be solved by the methods of Chapter 3. That solution is then substituted into one of the original equations, making it possible to solve for the other variable. In this process the given system is replaced by new systems that have the same solution set. Systems that have the same solution set are **equivalent systems.** The addition method is illustrated by the following examples.

EXAMPLE 2 Solve the system

$$3x - 4y = 1 \qquad \textbf{(4)}$$

$$2x + 3y = 12. \qquad \textbf{(5)}$$

Multiply both sides of equation (4) by -2 and both sides of equation (5) by 3 to get equations (6) and (7).

$$-6x + 8y = -2 \qquad \textbf{(6)}$$

$$6x + 9y = 36 \qquad \textbf{(7)}$$

Although this new system is not the same as the given system, it will have the same solution set.

Now add the two equations to eliminate x. Then solve the result for y.

$$-6x + 8y = -2$$
$$\underline{6x + 9y = 36}$$
$$17y = 34$$
$$y = 2$$

Substitute 2 for y in equation (4) or (5). We choose equation (4).

$$3x - 4(2) = 1$$
$$3x = 9$$
$$x = 3$$

The solution set of the given system is $\{(3, 2)\}$, which can be checked by substituting 3 for x and 2 for y in equation (5). ●

Since the method of solution shown in Example 2 results in the elimination of one variable from the system, it is also called the **elimination method** for solving a system.

EXAMPLE 3   Solve the system

$$3x - 2y = 4$$
$$-6x + 4y = 7$$

The variable x can be eliminated by multiplying both sides of the first equation by 2 and then adding.

$$6x - 4y = 8$$
$$\underline{-6x + 4y = 7}$$
$$0 = 15$$

Both variables were eliminated here, leaving the false statement $0 = 15$, a signal that these two equations have no solutions in common. The system is inconsistent, with an empty solution set.

The equations show that the graphs are parallel lines since they have the same slope but different y-intercepts. ●

EXAMPLE 4   Solve the system

$$-4x + y = 2$$
$$8x - 2y = -4.$$

Multiply both sides of the first equation by 2 and add.

$$-8x + 2y = 4$$
$$\underline{8x - 2y = -4}$$
$$0 = 0$$

This true statement, $0 = 0$, indicates that a solution of one equation is also a solution of the other, so the solution set is an infinite set of ordered pairs. The graphs of the equations are the same line, since the slopes and y-intercepts are equal. The two equations are dependent.   ●

Examples 3 and 4 suggest the following summary.

Inconsistent Systems and Systems of Dependent Equations

If $a_2 \neq 0$, $b_2 \neq 0$, and $c_2 \neq 0$, then the system of equations

$$a_1x + b_1y = c_1$$
$$a_2x + b_2y = c_2$$

is inconsistent if

$$\frac{a_1}{a_2} = \frac{b_1}{b_2} \neq \frac{c_1}{c_2},$$

and the graphs of the equations are parallel lines. The equations of the system are dependent if

$$\frac{a_1}{a_2} = \frac{b_1}{b_2} = \frac{c_1}{c_2},$$

and the graphs of the equations coincide.

Applications of mathematics often require the solution of a system of equations, as the next example shows.

EXAMPLE 5   The manager of a shoe store purchased 10 pairs of style 1501 shoes and 12 pairs of style 1470 shoes for $551. A later purchase of 4 pairs of style 1501 and 3 pairs of style 1470 cost $170. Find the cost per pair for each style.

A system of linear equations can be written from the information in the problem. To begin, let

$$x = \text{cost of a pair of style 1501 shoes}$$
$$y = \text{cost of a pair of style 1470 shoes.}$$

Since 10 pairs of 1501 and 12 pairs of 1470 shoes cost $551,

$$10x + 12y = 551. \tag{8}$$

Also,

$$4x + 3y = 170. \tag{9}$$

To solve this system by the addition method, multiply both sides of equation (9) by -4 and add the result to equation (8).

$$10x + 12y = 551$$
$$\underline{-16x - 12y = -680}$$
$$-6x \qquad\quad = -129$$
$$x \qquad\quad = 21.5$$

Substitute 21.5 for x in equation (8) or (9). Using (8),

$$10(21.5) + 12y = 551$$
$$215 + 12y = 551$$
$$12y = 336$$
$$y = 28.$$

Style 1501 shoes cost $21.50 a pair and style 1470 shoes cost $28 a pair.

Check by using the information of the original problem: 10 pairs of style 1501 and 12 pairs of style 1470 cost

$$10(21.50) + 12(28) = 215 + 336 = 551,$$

while 4 pairs of style 1501 and 3 pairs of style 1470 cost

$$4(21.50) + 3(28) = 86 + 84 = 170, \text{ as required.} \quad \bullet$$

8.1 EXERCISES

Use the substitution method to solve each system. Check your answers. See Example 1.

1. $y = 2x + 3$
$3x + 4y = 78$

2. $y = 4x - 6$
$2x + 5y = -8$

3. $3x - 2y = 12$
$5x + 2y = 4$

4. $8x + 3y = 2$
$5x - 6y = 17$

5. $2x - 3y + 1 = 0$
$4x + 3y - 25 = 0$

6. $3x - 5y = 0$
$2x - 3y - 3 = 0$

7. $\dfrac{x}{2} + \dfrac{y}{3} = \dfrac{7}{6}$

$\dfrac{2x}{3} - \dfrac{3y}{2} = \dfrac{-7}{3}$

8. $\dfrac{3x}{4} - \dfrac{y}{2} = 4$

$\dfrac{x}{3} + \dfrac{5y}{4} = \dfrac{-7}{6}$

9. $4.12x - 3.73y = 21.6$
$6.54x + 1.71y = 26.76$

10. $2.01x + 5.27y = 20.32$
$7.93x - 4.68y = -47.19$

Use the addition method to solve each system. Check your answers. (Hint for Exercises 27–30: let $1/x = t$, $1/y = u$.) See Examples 2–4.

11. $x + y = 9$
$2x - y = 0$

12. $4x + y = 9$
$3x - y = 5$

13. $5x + 3y = 7$
$7x - 3y = -19$

14. $2x + 7y = -8$
$-2x + 3y = -12$

15. $3x + 2y = 5$
$6x + 4y = 8$

16. $9x - 5y = 1$
$-18x + 10y = 1$

17. $2x - 3y = -7$
$5x + 4y = 17$

18. $4x + 3y = -1$
$2x + 5y = 3$

19. $5x + 7y = 6$
$10x - 3y = 46$

20. $12x - 5y = 9$
$3x - 8y = -18$

21. $4x - y = 9$
$-8x + 2y = -18$

22. $3x + 5y + 2 = 0$
$9x + 15y + 6 = 0$

23. $\dfrac{x}{2} + \dfrac{y}{3} = 8$

$\dfrac{2x}{3} + \dfrac{3y}{2} = 17$

24. $\dfrac{x}{5} + 3y = 31$

$2x - \dfrac{y}{5} = 8$

25. $\dfrac{3x}{2} - \dfrac{y}{3} = 5$

$\dfrac{5x}{2} + \dfrac{2y}{3} = 12$

26. $\dfrac{4x}{5} + \dfrac{y}{4} = -2$

$\dfrac{x}{5} + \dfrac{y}{8} = 0$

27. $\dfrac{2}{x} + \dfrac{1}{y} = \dfrac{3}{2}$

$\dfrac{3}{x} - \dfrac{1}{y} = 1$

28. $\dfrac{1}{x} + \dfrac{3}{y} = \dfrac{16}{5}$

$\dfrac{5}{x} + \dfrac{4}{y} = 5$

29. $\dfrac{2}{x} + \dfrac{1}{y} = 11$

$\dfrac{3}{x} - \dfrac{5}{y} = 10$

30. $\dfrac{2}{x} + \dfrac{3}{y} = 18$

$\dfrac{4}{x} - \dfrac{5}{y} = -8$

31. $.05x - .02y = -.18$
$.04x + .06y = -.22$

32. $.08x + .03y = -.24$
$.04x - .01y = -.32$

33. $.6x + .3y = .087$
$.5x - .4y = .378$

34. $.7x - .5y = -.884$
$.1x + .3y = .13$

Write each problem as a system of equations and then solve. See Example 5.

35. Find values of a and b so that the line with equation $ax + by = 5$ passes through the points $(-2, 1)$ and $(-1, -2)$.

36. Find m and b so that the line $y = mx + b$ passes through $(4, 6)$ and $(-5, -3)$.

37. At the Mustang Ranch, 6 goats and 5 sheep sell for $305, while 2 goats and 9 sheep cost $285. Find the cost of a goat and a sheep.

38. Linda Ramirez is a building contractor. If she hires 7 day laborers and 2 concrete finishers, her payroll for the day is $692, while 1 day laborer and 5 concrete finishers cost $476. Find the daily wage charge of each type of worker.

39. During summer vacation Hector and Ann earned a total of $2176. Hector worked 8 days less than Ann and earned $4 per day less. Find the number of days he worked and the daily wage made, if the total number of days worked by both was 72.

40. A bank teller has ten-dollar bills and twenty-dollar bills. He has 25 more twenties than tens. The value of the bills is $2900. How many of each kind does he have?

41. Mike Karelius plans to invest $30,000 he won in a lottery. With part of the money he buys a mutual fund, paying 9% a year. The rest he invests in utility bonds paying 10% per year. The first year his investments bring a return of $2820. How much is invested at each rate?

42. How much lowfat milk that is 3% butterfat should be mixed with milk that is 18% butterfat to get 25 gal of a lowfat milk that is 4.8% butterfat?

The break-even point for a company is the point where its costs equal its revenue. If both costs and revenue are expressed as linear equations, the break-even point is the solution of a linear system. In each of the following exercises, C represents the cost to produce x items, and R represents the revenue from the sale of x items. Use the substitution method to find the break-even point in each case, that is, the point where C = R.

43. $C = 1.5x + 252$
$R = 5.5x$

44. $C = 2.5x + 90$
$R = 3x$

45. $C = 20x + 10,000$
$R = 30x - 11,000$

46. $C = 4x + 125$
$R = 9x - 200$

47. $C = 18x + 240$
$R = 20x - 500$

48. $C = 4x + 48$
$R = 6x - 80$

*Usually, as the price of an item goes up, demand for the item goes down and the supply of the item goes up. Changes in gasoline prices illustrate this situation. The price where supply and demand are equal is called the **equilibrium price,** and the resulting supply or demand is called the **equilibrium supply** or **equilibrium demand.** In each of the following exercises p is the price of an item, while x represents the supply in one equation and the demand in the other. Find the equilibrium price and the equilibrium supply/demand.*

49. $p = 80 - \dfrac{3}{5}x$

$p = \dfrac{2}{5}x$

50. $p = 630 - \dfrac{3}{4}x$

$p = \dfrac{3}{4}x$

51. $3p = 84 - 2x$
$3p - x = 0$

52. $4p + x = 80$
$3p - 2x = 5$

53. $5p + 3x = 75$
$3x - 10p = 0$

54. $3p + 2x = 315$
$5x - 6p = 0$

8.2 Linear Systems with Three Variables

A solution of a linear equation $ax + by + cz = k$ with three variables is an **ordered triple** (x, y, z). For example, $(1, 2, -4)$ is a solution of $2x + 5y - 3z = 24$. The solution set of such an equation is an infinite set of ordered triples. It is shown in geometry that the graph of a linear equation in three variables is a plane in three-dimensional space. Considering the possible intersections of the planes representing three equations in three unknowns shows that the solution set of such a system may be either a single ordered triple (x, y, z), an infinite set of ordered triples (dependent equations), or empty (an inconsistent system). See Figure 8.2.

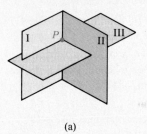 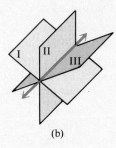 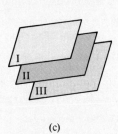

(a) (b) (c)

Figure 8.2

A system with three or more variables can be solved with repeated use of the addition method, as shown in the next example.

EXAMPLE 1 Solve the system

$$2x + y - z = 2 \tag{1}$$
$$x + 3y + 2z = 1 \tag{2}$$
$$x + y + z = 2. \tag{3}$$

Several steps are needed to obtain an equation in one variable. First, eliminate the same variable from each of two pairs of equations. Then eliminate another variable from the two resulting equations. Suppose we choose to eliminate x first. Multiply both sides of equation (2) by -2 and add to equation (1).

$$2x + y - z = 2$$
$$\underline{-2x - 6y - 4z = -2}$$
$$ - 5y - 5z = 0 \tag{4}$$

The variable x must be eliminated again from a different pair of equations, say (2) and (3). Multiply both sides of (2) by -1, then add the result to equation (3).

$$\begin{array}{r} -x - 3y - 2z = -1 \\ \underline{x + y + z = 2} \\ -2y - z = 1 \end{array} \qquad \text{(5)}$$

Now solve the system formed by the two equations (4) and (5). To eliminate z, multiply both sides of equation (4) by -1 and both sides of equation (5) by 5 and add.

$$\begin{array}{r} 5y + 5z = 0 \\ \underline{-10y - 5z = 5} \\ -5y \qquad = 5 \\ y = -1 \end{array}$$

Substitute y into equation (5), to find z. (Equation (4) could have been used.)

$$\begin{array}{r} -2(-1) - z = 1 \\ 2 - z = 1 \\ -z = -1 \\ z = 1 \end{array}$$

Now use equation (3) and the values for y and z to find x. (Either equation (1) or (2) could have been used.)

$$\begin{array}{r} x + y + z = 2 \\ x - 1 + 1 = 2 \\ x = 2 \end{array}$$

Thus the solution set of the system is $\{(2, -1, 1)\}$.   ●

EXAMPLE 2   Solve the system

$$\begin{array}{rl} x + 2y + z = 4 & \text{(6)} \\ 3x - y - 4z = -9. & \text{(7)} \end{array}$$

Geometrically, we want to locate the intersection of the two planes given by equations (6) and (7). The intersection of two different nonparallel planes is a line. Thus there will be an infinite number of ordered triples in the solution set, representing the points on the line of intersection. To describe these ordered triples, proceed as follows.

To eliminate x, multiply both sides of equation (6) by -3 and add to equation (7). (Either y or z could have been eliminated instead.)

$$\begin{array}{r} -3x - 6y - 3z = -12 \\ \underline{3x - y - 4z = -9} \\ -7y - 7z = -21 \qquad \text{(8)} \end{array}$$

Now solve equation (8) for z.

$$-7y - 7z = -21$$
$$-7z = 7y - 21$$
$$z = -y + 3$$

Solve equation (6) for x and substitute $-y + 3$ for z in the result.

$$x + 2y + z = 4$$
$$x = -2y - z + 4$$
$$x = -2y - (-y + 3) + 4$$
$$x = -y + 1$$

The system has an infinite number of solutions. For any value of y, the value of z is given by $-y + 3$ and x equals $-y + 1$. For example, if $y = 1$, then $x = -1 + 1 = 0$ and $z = -1 + 3 = 2$, giving the solution $(0, 1, 2)$. Verify that another solution is $(-1, 2, 1)$.

Had equation (8) been solved for y instead of z, the solution would have had a different form but would have led to the same set of solutions. The form of the solution set found above, $\{(-y + 1, y, -y + 3)\}$, is said to have **y arbitrary.** ●

Applications with three unknowns often require the solution of a system of three linear equations. As an example, the equation of the parabola $y = ax^2 + bx + c$ that passes through three given points can be found by solving a system of three equations with three unknowns.

EXAMPLE 3   Find the equation of the parabola $y = ax^2 + bx + c$ that passes through $(2, 4)$, $(-1, 1)$, and $(-2, 5)$.

Since the three points lie on the graph of the equation $y = ax^2 + bx + c$, they must satisfy the equation. Substituting each ordered pair into the equation gives three equations with three variables.

$$4 = a(2)^2 + b(2) + c \qquad \text{or} \qquad 4 = 4a + 2b + c \qquad \textbf{(9)}$$
$$1 = a(-1)^2 + b(-1) + c \qquad \text{or} \qquad 1 = a - b + c \qquad \textbf{(10)}$$
$$5 = a(-2)^2 + b(-2) + c \qquad \text{or} \qquad 5 = 4a - 2b + c \qquad \textbf{(11)}$$

This system can be solved by the addition method. First eliminate c using equations (9) and (10). Then eliminate c using equations (10) and (11).

$$
\begin{array}{ll}
4 = 4a + 2b + c & 1 = a - b + c \\
\underline{-1 = -a + b - c} & \underline{-5 = -4a + 2b - c} \\
3 = 3a + 3b & -4 = -3a + b
\end{array}
$$

Now solve the system of two equations in two variables by eliminating a, as shown at the top of the next page.

$$3 = 3a + 3b$$

$$-4 = -3a + b$$

$$\overline{-1 = \qquad 4b}$$

$$-\frac{1}{4} = \qquad b$$

Find a by substituting $-1/4$ for b in $3 = 3a + 3b$, which is equivalent to $1 = a + b$.

$$1 = a + b$$

$$1 = a - \frac{1}{4}$$

$$\frac{5}{4} = a$$

Finally, find c by substituting $a = 5/4$ and $b = -1/4$ in equation (10).

$$1 = a - b + c$$

$$1 = \frac{5}{4} - \left(-\frac{1}{4}\right) + c$$

$$1 = \frac{6}{4} + c = \frac{3}{2} + c$$

$$-\frac{1}{2} = c$$

The equation of the parabola is

$$4y = 5x^2 - x - 2 \qquad \text{or} \qquad y = \frac{5}{4}x^2 - \frac{1}{4}x - \frac{1}{2}. \quad \bullet$$

8.2   EXERCISES

Use the addition method to solve each of the following systems of three equations in three unknowns. Check your answers. See Example 1.

1. $x + y + z = 2$
  $2x + y - z = 5$
  $x - y + z = -2$

2. $2x + y + z = 9$
  $-x - y + z = 1$
  $3x - y + z = 9$

3. $x + 3y + 4z = 14$
  $2x - 3y + 2z = 10$
  $3x - y + z = 9$

4. $4x - y + 3z = -2$
  $3x + 5y - z = 15$
  $-2x + y + 4z = 14$

5. $x + 2y + 3z = 8$
  $3x - y + 2z = 5$
  $-2x - 4y - 6z = 5$

6. $3x - 2y - 8z = 1$
  $9x - 6y - 24z = -2$
  $x - y + z = 1$

7. $x + 4y - z = 6$
  $2x - y + z = 3$
  $3x + 2y + 3z = 16$

8. $4x - 3y + z = 9$
  $3x + 2y - 2z = 4$
  $x - y + 3z = 5$

9. $5x + y - 3z = -6$
  $2x + 3y + z = 5$
  $-3x - 2y + 4z = 3$

10. $2x - 5y + 4z = -35$
$5x + 3y - z = 1$
$x + y + z = 1$

11. $x - 3y - 2z = -3$
$3x + 2y - z = 12$
$-x - y + 4z = 3$

12. $x + y + z = 3$
$3x - 3y - 4z = -1$
$x + y + 3z = 11$

13. $2x + 6y - z = 6$
$4x - 3y + 5z = -5$
$6x + 9y - 2z = 11$

14. $8x - 3y + 6z = -2$
$4x + 9y + 4z = 18$
$12x - 3y + 8z = -2$

15. $\dfrac{1}{x} + \dfrac{1}{y} - \dfrac{1}{z} = \dfrac{1}{4}$
$\dfrac{2}{x} - \dfrac{1}{y} + \dfrac{3}{z} = \dfrac{9}{4}$
$-\dfrac{1}{x} - \dfrac{2}{y} + \dfrac{4}{z} = 1$

16. $\dfrac{3}{x} + \dfrac{2}{y} - \dfrac{1}{z} = \dfrac{11}{6}$
$\dfrac{1}{x} - \dfrac{1}{y} + \dfrac{3}{z} = -\dfrac{11}{12}$
$\dfrac{2}{x} + \dfrac{1}{y} + \dfrac{1}{z} = \dfrac{7}{12}$

17. $\dfrac{2}{x} - \dfrac{2}{y} + \dfrac{1}{z} = -1$
$\dfrac{4}{x} + \dfrac{1}{y} - \dfrac{2}{z} = -9$
$\dfrac{1}{x} + \dfrac{1}{y} - \dfrac{3}{z} = -9$

18. $\dfrac{5}{x} - \dfrac{1}{y} - \dfrac{2}{z} = -6$
$-\dfrac{1}{x} + \dfrac{3}{y} - \dfrac{3}{z} = -12$
$\dfrac{2}{x} - \dfrac{1}{y} - \dfrac{1}{z} = 6$

19. $1.2x + 3y - 4.8z = -5.28$
$-3x + 5.8y + 2.4z = -4.16$
$.7x - .5y + .1z = .75$

20. $-.6x + 2.5y - 3.2z = -1.25$
$1.1x + .4y + .8z = 1.22$
$-.3x - .7y - .4z = -.91$

21. $5.3x - 4.7y + 5.9z = 1.14$
$-2.5x + 3.2y - 1.4z = 7.22$
$2.25x - 2.88y + 1.26z = 4.88$

22. $7.77x - 8.61y + 4.2z = 15.96$
$11.6x - 4.9y + .8z = 12.4$
$3.7x - 4.1y + 2z = 7.6$

Solve each of the following systems in terms of the arbitrary variable x. See Example 2.

23. $x - 2y + 3z = 6$
$2x - y + 2z = 5$

24. $3x + 4y - z = 13$
$x + y + 2z = 15$

25. $5x - 4y + z = 9$
$x + y = 15$

26. $x - y + z = -6$
$4x + y + z = 7$

27. $3x - 5y - 4z = -7$
$y - z = -13$

28. $3x - 2y + z = 15$
$x + 4y - z = 11$

Work the following word problems. For Exercises 29 and 30, see Example 3.

29. Find a, b, and c so that the graph of the equation $y = ax^2 + bx + c$ passes through the points $(2, 3)$, $(-1, 0)$, and $(-2, 2)$.

30. Find a, b, and c so that $(2, 14)$, $(0, 0)$, and $(-1, -1)$ lie on the graph of $y = ax^2 + bx + c$.

31. Find the equation of the circle $x^2 + y^2 + ax + by + c = 0$ that passes through the points $(2, 1)$, $(-1, 0)$, $(3, 3)$.

32. Find the equation of the circle $x^2 + y^2 + ax + by + c = 0$ that passes through the points $(4, 2)$, $(-5, -2)$, and $(0, 3)$.

33. A cashier has a total of 30 bills, made up of ones, fives, and twenties. The number of twenties is 9 more than the number of ones. The total value of the money is $351. How many of each type of bill are there?

34. A wine merchant wishes to get 300 liters of wine that sells for $3 per liter. She wishes to mix wines selling for $4.50, $1.50, and $2.25, respectively. She must use twice as much of the $2.25 wine as the $1.50 wine. How many liters of each should she use?

35. A glue company needs to make some glue that it can sell for $120 per barrel. It wants to use 150 barrels of glue worth $100 per barrel, along with some glue worth $150 per barrel, and glue worth $190 per barrel. It must use the same number of barrels of $150 and $190 glue. How much of the $150 and $190 glue will be needed? How many barrels of $120 glue will be produced?

36. The Rolling Rocks sell three kinds of tickets, "up close," "middle," and "farther back." "Up close" tickets cost $2 more than "middle" tickets, while "middle" tickets cost $1 more than "farther back" tickets. Twice the cost of an "up close" ticket is $1 more than 3 times the cost of a "farther back" seat. Find the price of each kind of ticket.

37. The perimeter of a triangle is 33 cm. The longest side is 3 cm longer than the medium side. The medium side is twice the shortest side. Find the length of each side of the triangle.

38. The sum of the angles of a triangle is 180°. The largest angle is 10° more than twice the smallest. The third angle is 50° less than the sum of the other two angles. Find the measures of each of the three angles.

39. Jay Davis wins $100,000 in a lottery. He invests part of the money in real estate with an annual return of 10% and another part in a money market account at 9% interest. He invests the third part, which amounts to $20,000 less than the sum of the other two parts, in certificates of deposit at 7.5%. The total annual interest on the money is $8900. How much was invested in each way?

40. Betsy Mackay invests $10,000, received from her grandmother, in three ways. With one part, she buys mutual funds which offer a return of 8% per year. The second part, which amounts to twice the first, is used to buy government bonds at 9% per year. She puts the rest in the bank at 5% annual interest. The first year her investments bring a return of $830. How much did she invest in each way?

8.3   Solution of Linear Systems by Matrices

We have seen how to solve linear systems of equations by the elimination method. This section describes how to solve these systems by matrix methods. Matrix methods are particularly suitable for computer solutions of larger systems of equations having many unknowns. To begin, consider a system of three equations and three unknowns such as

$$a_1 x + b_1 y + c_1 z = d_1$$
$$a_2 x + b_2 y + c_2 z = d_2$$
$$a_3 x + b_3 y + c_3 z = d_3.$$

Write the coefficients as a 3×3 matrix,

$$\begin{bmatrix} a_1 & b_1 & c_1 \\ a_2 & b_2 & c_2 \\ a_3 & b_3 & c_3 \end{bmatrix}.$$

The resulting matrix is called the **coefficient matrix** of the system. Join the column of constants to the coefficient matrix to get the *augmented matrix* of the system,

$$\begin{bmatrix} a_1 & b_1 & c_1 & d_1 \\ a_2 & b_2 & c_2 & d_2 \\ a_3 & b_3 & c_3 & d_3 \end{bmatrix}.$$

(The bar is used only to separate the constants from the coefficients.)

The rows of this augmented matrix can be treated the same as the equations of a system of equations, since the augmented matrix is actually a short form of the system. Any transformation of the matrix which will result in an equivalent system is permitted. Operations which produce such transformations, called matrix row operations, were first mentioned in Chapter 7. We repeat them here for reference.

For any real number k and any augmented matrix of a system of linear equations, the following operations will produce the matrix of an equivalent system:

1. **interchanging any two rows of a matrix,**
2. **multiplying the elements of any row of a matrix by the same nonzero scalar k, and**
3. **adding a multiple of the elements of one row to the elements of another row.**

If the word "row" is replaced by "equation," it can be seen that the three transformations also apply to a system of equations, so that a system of equations can be solved by transforming its corresponding matrix into the matrix of an equivalent, simpler system.

The **Gauss-Jordan method** is a systematic method of using the matrix row operations to solve a system of equations. The equations should all be in the same form, with variable terms in the same order on the left, and the constant term on the right. The next example illustrates the method.

EXAMPLE 1   Use the Gauss-Jordan method to solve the linear system

$$3x - 4y = 1$$
$$5x + 2y = 19.$$

The equations are already in the proper form, so begin with the augmented matrix

$$\begin{bmatrix} 3 & -4 & 1 \\ 5 & 2 & 19 \end{bmatrix}.$$

The goal is to transform this augmented matrix into one in which the value of the variables will be easy to see. That is, since each column in the matrix represents the coefficients of one variable, we want to transform the matrix so that the augmented matrix is of the form

$$\begin{bmatrix} 1 & 0 & | & k \\ 0 & 1 & | & j \end{bmatrix}$$

for real numbers k and j. Once the augmented matrix is in this form, the matrix can be rewritten as a linear system to get

$$x = k$$
$$y = j.$$

The necessary transformations are performed as follows. (Note the similarity to the method of finding matrix inverses discussed in Chapter 7.) It is best to work in columns, beginning in each column with the element that is to become 1. In the augmented matrix,

$$\begin{bmatrix} 3 & -4 & | & 1 \\ 5 & 2 & | & 19 \end{bmatrix},$$

there is a 3 in the first row, first column position. To get a 1 in this position, use operation 2 and multiply each entry in the first row by 1/3.

$$\begin{bmatrix} 1 & -4/3 & | & 1/3 \\ 5 & 2 & | & 19 \end{bmatrix}$$

To get 0 in the second row, first column, multiply each element of the first row by -5 and add the result to the corresponding element in the second row, using operation 3.

$$\begin{bmatrix} 1 & -4/3 & | & 1/3 \\ 0 & 26/3 & | & 52/3 \end{bmatrix}$$

To get 1 in the second row, second column, multiply each element of the second row by 3/26, using operation 2.

$$\begin{bmatrix} 1 & -4/3 & | & 1/3 \\ 0 & 1 & | & 2 \end{bmatrix}$$

Finally, get 0 in the first row, second column by multiplying each element of the second row by 4/3 and adding the result to the corresponding element in the first row.

$$\begin{bmatrix} 1 & 0 & | & 3 \\ 0 & 1 & | & 2 \end{bmatrix}$$

This last matrix leads to the system

$$x = 3$$
$$y = 2,$$

which gives the solution set $\{(3, 2)\}$. This solution could have been read directly from the third column of the final matrix.   ●

EXAMPLE 2   Use the Gauss-Jordan method to solve the system

$$x - y + 5z = -6$$
$$3x + 3y - z = 10$$
$$x + 3y + 2z = 5.$$

Since the system is in proper form, begin by writing the augmented matrix of the linear system.

$$\begin{bmatrix} 1 & -1 & 5 & -6 \\ 3 & 3 & -1 & 10 \\ 1 & 3 & 2 & 5 \end{bmatrix}$$

The final matrix is to be of the form

$$\begin{bmatrix} 1 & 0 & 0 & m \\ 0 & 1 & 0 & n \\ 0 & 0 & 1 & p \end{bmatrix},$$

where m, n, and p are real numbers. This final form of the matrix gives the solution: $x = m$, $y = n$, and $z = p$.

We already have 1 in the first row, first column. To get a 0 in the second row of the first column, multiply each element in the first row by -3 and add the result to the corresponding element in the second row, using operation 3.

$$\begin{bmatrix} 1 & -1 & 5 & -6 \\ 0 & 6 & -16 & 28 \\ 1 & 3 & 2 & 5 \end{bmatrix}$$

Now, to change the last element in the first column to 0, use operation 3 and multiply each element of the first row by -1, then add the results to the corresponding elements of the third row.

$$\begin{bmatrix} 1 & -1 & 5 & -6 \\ 0 & 6 & -16 & 28 \\ 0 & 4 & -3 & 11 \end{bmatrix}$$

The same procedure is used to transform the second and third columns. For both of these columns perform the additional step of getting 1 in the appropriate position of each column.

$$\begin{bmatrix} 1 & -1 & 5 & -6 \\ 0 & 1 & -8/3 & 14/3 \\ 0 & 4 & -3 & 11 \end{bmatrix}$$ Second row multiplied by 1/6

$$\begin{bmatrix} 1 & 0 & 7/3 & -4/3 \\ 0 & 1 & -8/3 & 14/3 \\ 0 & 4 & -3 & 11 \end{bmatrix}$$ Second row added to first row

$$\begin{bmatrix} 1 & 0 & 7/3 & -4/3 \\ 0 & 1 & -8/3 & 14/3 \\ 0 & 0 & 23/3 & -23/3 \end{bmatrix}$$ -4 times second row added to third row

$$\begin{bmatrix} 1 & 0 & 7/3 & -4/3 \\ 0 & 1 & -8/3 & 14/3 \\ 0 & 0 & 1 & -1 \end{bmatrix}$$ Third row multiplied by 3/23

$$\begin{bmatrix} 1 & 0 & 0 & 1 \\ 0 & 1 & -8/3 & 14/3 \\ 0 & 0 & 1 & -1 \end{bmatrix}$$ $-7/3$ times third row added to first row

$$\begin{bmatrix} 1 & 0 & 0 & 1 \\ 0 & 1 & 0 & 2 \\ 0 & 0 & 1 & -1 \end{bmatrix}$$ 8/3 times third row added to second row

The linear system associated with this final matrix is

$$x = 1$$
$$y = 2$$
$$z = -1,$$

and the solution set is $\{(1, 2, -1)\}$.   ●

EXAMPLE 3   Use the Gauss-Jordan method to solve the system

$$x + y = 2$$
$$2x + 2y = 5.$$

Write the augmented matrix,

$$\begin{bmatrix} 1 & 1 & 2 \\ 2 & 2 & 5 \end{bmatrix}.$$

Multiply the elements in the first row by -2 and add the result to the corresponding elements in the second row.

$$\begin{bmatrix} 1 & 1 & 2 \\ 0 & 0 & 1 \end{bmatrix}$$

The next step would be to get a 1 in the second row, second column. Because of the zeros, it is impossible to go further. Since the second row corresponds to the equation

$$0x + 0y = 1$$

which has no solution, the system is inconsistent.   ●

EXAMPLE 4   Use the Gauss-Jordan method to solve the system

$$2x - 5y + 3z = 1$$
$$x + 4y - 2z = 8.$$

Recall from Section 8.1 that a system with two equations and three variables has an infinite number of solutions. The Gauss-Jordan method can be used to indicate the solution with one arbitrary variable. Start with the augmented matrix

$$\begin{bmatrix} 2 & -5 & 3 & | & 1 \\ 1 & 4 & -2 & | & 8 \end{bmatrix}.$$

Exchange rows to get a 1 in the first row, first column position.

$$\begin{bmatrix} 1 & 4 & -2 & | & 8 \\ 2 & -5 & 3 & | & 1 \end{bmatrix}$$

Now multiply each element in the first row by -2 and add to the corresponding element in the second row.

$$\begin{bmatrix} 1 & 4 & -2 & | & 8 \\ 0 & -13 & 7 & | & -15 \end{bmatrix}$$

Multiply each element in the second row by $-1/13$.

$$\begin{bmatrix} 1 & 4 & -2 & | & 8 \\ 0 & 1 & -7/13 & | & 15/13 \end{bmatrix}$$

Multiply each element in the second row by -4 and add to the corresponding element in the first row.

$$\begin{bmatrix} 1 & 0 & 2/13 & | & 44/13 \\ 0 & 1 & -7/13 & | & 15/13 \end{bmatrix}$$

This is as far as we can go with the Gauss-Jordan method. The equations which correspond to the final matrix are

$$x + \frac{2}{13}z = \frac{44}{13} \quad \text{and} \quad y - \frac{7}{13}z = \frac{15}{13}.$$

Solve these equations for x and y, respectively.

$$x + \frac{2}{13}z = \frac{44}{13} \qquad\qquad y - \frac{7}{13}z = \frac{15}{13}$$

$$x = \frac{44}{13} - \frac{2}{13}z \qquad\qquad y = \frac{15}{13} + \frac{7}{13}z$$

$$x = \frac{44 - 2z}{13} \qquad\qquad y = \frac{15 + 7z}{13}$$

The solution can now be written with z arbitrary, as

$$\left\{ \left(\frac{44 - 2z}{13}, \frac{15 + 7z}{13}, z \right) \right\}. \quad \bullet$$

Although we have shown examples with only two variables or three variables, the Gauss-Jordan method can be extended to solve linear systems with more variables. As the number of variables increases, the process quickly becomes very tedious and the opportunity for error increases. The method, however, is suitable for use by computers and a fairly large system can be solved quickly in that way.

8.3   EXERCISES   1 - 39

Write the augmented matrix for each of the following systems. Do not solve the system.

1. $2x + 3y = 11$
$\quad x + 2y = 8$

2. $3x + 5y = -13$
$\quad 2x + 3y = -9$

3. $x + 5y = 6$
$\quad x + 2y = 8$

4. $2x + 7y = 1$
$\quad 5x \qquad = -15$

5. $2x + y + z = 3$
$\quad 3x - 4y + 2z = -7$
$\quad x + y + z = 2$

6. $4x - 2y + 3z = 4$
$\quad 3x + 5y + z = 7$
$\quad 5x - y + 4z = 7$

7. $x + y \qquad = 2$
$\quad 2y + z = -4$
$\qquad z = 2$

8. $x \qquad = 6$
$\quad y + 2z = 2$
$\quad x \qquad - 3z = 6$

9. $x \qquad = 5$
$\quad y \qquad = -2$
$\qquad z = 3$

10. $x \qquad = 8$
$\quad y + z = 6$
$\qquad z = 2$

Write the system of equations associated with each of the following augmented matrices.
Do not try to solve.

11. $\begin{bmatrix} 1 & 0 & | & 2 \\ 0 & 1 & | & 3 \end{bmatrix}$

12. $\begin{bmatrix} 1 & 0 & | & 5 \\ 0 & 1 & | & -3 \end{bmatrix}$

13. $\begin{bmatrix} 2 & 1 & | & 1 \\ 3 & -2 & | & -9 \end{bmatrix}$

14. $\begin{bmatrix} 1 & -5 & | & -18 \\ 6 & 2 & | & 20 \end{bmatrix}$

15. $\begin{bmatrix} 1 & 0 & 0 & | & 2 \\ 0 & 1 & 0 & | & 3 \\ 0 & 0 & 1 & | & -2 \end{bmatrix}$

16. $\begin{bmatrix} 1 & 0 & 1 & | & 4 \\ 0 & 1 & 0 & | & 2 \\ 0 & 0 & 1 & | & 3 \end{bmatrix}$

17. $\begin{bmatrix} 3 & 2 & 1 & | & 1 \\ 0 & 2 & 4 & | & 22 \\ -1 & -2 & 3 & | & 15 \end{bmatrix}$

18. $\begin{bmatrix} 2 & 1 & 3 & | & 12 \\ 4 & -3 & 0 & | & 10 \\ 5 & 0 & -4 & | & -11 \end{bmatrix}$

Use the Gauss-Jordan method to solve each of the following systems of equations. See Examples 1–3.

19. $x + y = 5$
$\quad x - y = -1$

20. $x + 2y = 5$
$\quad 2x + y = -2$

21. $x + y = -3$
$\quad 2x - 5y = -6$

22. $3x - 2y = 4$
$\quad 3x + y = -2$

23. $2x - 3y = 10$
$\quad 2x + 2y = 5$

24. $4x + y = 5$
$\quad 2x + y = 3$

25. $2x - 5y = 10$
$\quad 4x - 5y = 15$

26. $4x - y = 3$
$\quad -2x + 3y = 1$

27. $2x - 3y = 2$
$\quad 4x - 6y = 1$

28. $x + 2y = 1$
$\quad 2x + 4y = 3$

29. $6x - 3y = 1$
$\quad -12x + 6y = -2$

30. $x - y = 1$
$\quad -x + y = -1$

31. $x + y \qquad = -1$
$\quad y + z = 4$
$\quad x \qquad + z = 1$

32. $x \qquad - z = -3$
$\quad y + z = 9$
$\quad x \qquad + z = 7$

33. $x + y - z = 6$
$\quad 2x - y + z = -9$
$\quad x - 2y + 3z = 1$

34. $x + 3y - 6z = 7$
$\quad 2x - y + 2z = 0$
$\quad x + y + 2z = -1$

35.
$$-x + y \quad\quad = -1$$
$$y - z = 6$$
$$x \quad\quad + z = -1$$

36.
$$x + y \quad\quad = 1$$
$$2x \quad\quad - z = 0$$
$$y + 2z = -2$$

37.
$$2x - y + 3z = 0$$
$$x + 2y - z = 5$$
$$2y + z = 1$$

38.
$$4x + 2y - 3z = 6$$
$$x - 4y + z = -4$$
$$-x \quad\quad + 2z = 2$$

39.
$$x - 2y + z = 5$$
$$2x + y - z = 2$$
$$-2x + 4y - 2z = 2$$

40.
$$3x + 5y - z = 4$$
$$4x - y + 2z = 1$$
$$-6x - 10y + 2z = 0$$

Use the Gauss-Jordan method to solve each of the following systems.

41.
$$3x + 2y \quad\quad - w = 0$$
$$2x \quad\quad + z + 2w = 5$$
$$x + 2y - z \quad\quad = -2$$
$$2x - y + z + w = 2$$

42.
$$x + 3y - 2z - w = 9$$
$$4x + y + z + 2w = 2$$
$$-3x - y + z - w = -5$$
$$x - y - 3z - 2w = 2$$

43.
$$3.5x + 2.9y = 12.91$$
$$1.7x - 3.8y = -6.23$$

44.
$$-4.3x + 1.1y = 10.1$$
$$4.8x - 2.7y = -17.31$$

45.
$$9.1x + 2.3y - z = -3.06$$
$$4.7x + y - 3.8z = -21.15$$
$$5.1y - 4.7z = -31.77$$

46.
$$7.2x + 4.8y + 3.1z = -14.96$$
$$3x - y + 11.5z = -15.6$$
$$1.1x + 2.4y \quad\quad = -1.24$$

Solve each of the following word problems by first setting up a system of equations. Use the Gauss-Jordan method.

47. Juanita deposits some money in a bank account paying 8% per year. She uses some additional money, amounting to 1/3 the amount placed in the bank, to buy bonds paying 5% per year. With the balance of her funds she buys a 9% certificate of deposit. The first year her investments bring a return of $830. If the total of the investments is $10,000, how much is invested at each rate?

48. To get the necessary funds for a planned expansion, a small company took out three loans totaling $25,000. The company was able to borrow some of the money at 8%. They borrowed $2000 more than 1/2 the amount of the 8% loan at 10%, and the rest at 9%. The total annual interest was $2220. How much did they borrow at each rate?

49. One day a service station sold 400 gal of unleaded gasoline, 150 gal of regular gasoline, and 130 gal of premium gasoline for a total of $842.90. The next day 380 gal of unleaded, 170 gal of regular, and 150 gal of premium were sold for $866.90. The difference in price per gallon of unleaded and regular is three times the difference in price per gallon of premium and regular. How much does this station charge for each type of gasoline?

50. A chemist has three solutions containing acid: some 5% solution, some 15% solution, and some 25% solution. She needs to mix some of each to get 50 liters of 20% solution. Since she has a lot of the 5% solution, she wants to use twice as much of that as the 15% solution. How much of each solution should she use?

Solve each of the following systems by the Gauss-Jordan method. Let z be the arbitrary variable. See Example 4.

51.
$$x - 3y + 2z = 10$$
$$2x - y - z = 8$$

52.
$$3x + y - z = 12$$
$$x + 2y + z = 10$$

53.
$$x + 2y - z = 0$$
$$3x - y + z = 6$$
$$-2x - 4y + 2z = 0$$

54. $3x + 5y - z = 0$
$4x - y + 2z = 1$
$-6x - 10y + 2z = 0$

55. $x - 2y + z = 5$
$-2x + 4y - 2z = 2$
$2x + y - z = 2$

56. $3x + 6y - 3z = 12$
$-x - 2y + z = 16$
$x + y - 2z = 20$

57. At rush hours, substantial traffic congestion is encountered at the traffic intersections shown in the figure.

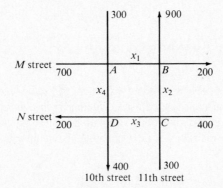

The city wishes to improve the signals at these corners so as to speed the flow of traffic. The traffic engineers first gather data. As the figure shows, 700 cars per hour come down M Street to intersection A: 300 cars per hour come to intersection A on 10th Street. A total of x_1 of these cars leave A on M Street, while x_4 cars leave A on 10th Street. The number of cars entering A must equal the number leaving, so that

$$x_1 + x_4 = 700 + 300$$

or
$$x_1 + x_4 = 1000. \qquad (1)$$

For intersection B, x_1 cars enter B on M Street, and x_2 cars enter B on 11th Street. The figure shows that 900 cars leave B on 11th while 200 leave on M. We have

$$x_1 + x_2 = 900 + 200$$

$$x_1 + x_2 = 1100. \qquad (2)$$

At intersection C, 400 cars enter on N Street, 300 on 11th Street, while x_2 leave on 11th Street and x_3 leave on N Street. This gives

$$x_2 + x_3 = 400 + 300$$

$$x_2 + x_3 = 700. \qquad (3)$$

Finally, intersection D has x_3 cars entering on N and x_4 entering on 10th. There are 400 leaving D on 10th and 200 leaving on N.

(a) Set up an equation for intersection D.

(b) Use equations (1), (2), (3), and your answer to part (a) to set up a system of equations. Solve the system by the Gauss-Jordan method.

(c) Since you got a row of zeros, the system of equations is dependent and does not have a unique solution. Solve equation (1) for x_1 and substitute the result into equation (2).

(d) Solve equation (1) for x_4. What is the largest possible value of x_1 so that x_4 is not negative?

(e) Your answer to part (c) should have been $x_2 - x_4 = 100$. Solve the equation for

x_4 and then find the smallest possible value of x_2 so that x_4 is not negative.

(f) Your answer to part (a) should have been $x_3 + x_4 = 600$. Solve the equation for x_4 and then find the largest possible values of x_3 and x_4 so that neither is negative.

(g) From your answers to parts (d) – (f), what is the maximum value of x_4 so that all the equations are satisfied and all variables are nonnegative? Of x_3? Of x_2? Of x_1?

8.4 Solution of Linear Systems by Inverses

Matrix equations are solved in much the same way as real number equations. Properties of matrices are used, together with a multiplication property of equality for matrices allowing both sides of a matrix equation to be multiplied by any appropriate matrix. The next example shows how this method works.

EXAMPLE 1   Given

$$A = \begin{bmatrix} 2 & 2 \\ -1 & -2 \end{bmatrix} \quad \text{and} \quad B = \begin{bmatrix} 2 \\ 3 \end{bmatrix},$$

find a matrix X so that $AX = B$.

Since A is 2×2 and B is 2×1, matrix X will be a 2×1 matrix, like B. To solve the matrix equation $AX = B$, use the fact that, if A^{-1} exists, $A^{-1}A = I$ and $IX = X$ as follows.

$$
\begin{array}{ll}
AX = B & \text{Given} \\
A^{-1}(AX) = A^{-1}B & \text{Multiplication Property of Equality} \\
(A^{-1}A)X = A^{-1}B & \text{Associative Property of Multiplication} \\
IX = A^{-1}B & \text{Multiplication Inverse Property} \\
X = A^{-1}B & \text{Multiplication Identity Property}
\end{array}
$$

In using the multiplication property of equality for matrices, be careful to multiply on the left on both sides of the equation. Unlike multiplication of real numbers, multiplication of matrices is not commutative. To solve for X, first find A^{-1} using the method introduced in Chapter 7.

Verify that

$$A^{-1} = \begin{bmatrix} 1 & 1 \\ -1/2 & -1 \end{bmatrix}.$$

As shown above,

$$X = A^{-1}B = \begin{bmatrix} 1 & 1 \\ -1/2 & -1 \end{bmatrix} \begin{bmatrix} 2 \\ 3 \end{bmatrix} = \begin{bmatrix} 5 \\ -4 \end{bmatrix}.$$

Check: $AX = \begin{bmatrix} 2 & 2 \\ -1 & -2 \end{bmatrix} \begin{bmatrix} 5 \\ -4 \end{bmatrix} = \begin{bmatrix} 2 \\ 3 \end{bmatrix} = B.$ ●

This method can be used to solve linear systems by first writing the system as a matrix equation $AX = B$, where X is a matrix of the variables of the system. The next example shows how to do this.

EXAMPLE 2   Use the inverse of the coefficient matrix to solve the system

$$2x - 3y = 4$$
$$x + 5y = 2.$$

To represent the system as a matrix equation, use one matrix for the coefficients, one for the variables, and one for the constants, as follows.

$$A = \begin{bmatrix} 2 & -3 \\ 1 & 5 \end{bmatrix}, \quad X = \begin{bmatrix} x \\ y \end{bmatrix}, \quad \text{and} \quad B = \begin{bmatrix} 4 \\ 2 \end{bmatrix}$$

The system can then be written in matrix form as the equation $AX = B$, since

$$AX = \begin{bmatrix} 2 & -3 \\ 1 & 5 \end{bmatrix} \begin{bmatrix} x \\ y \end{bmatrix} = \begin{bmatrix} 2x - 3y \\ x + 5y \end{bmatrix} = \begin{bmatrix} 4 \\ 2 \end{bmatrix} = B.$$

To solve the system, first find A^{-1} using the method discussed in Chapter 7.

$$A^{-1} = \begin{bmatrix} 5/13 & 3/13 \\ -1/13 & 2/13 \end{bmatrix}$$

Next, find the product $A^{-1}B$.

$$A^{-1}B = \begin{bmatrix} 5/13 & 3/13 \\ -1/13 & 2/13 \end{bmatrix} \begin{bmatrix} 4 \\ 2 \end{bmatrix} = \begin{bmatrix} 2 \\ 0 \end{bmatrix}$$

Since $X = A^{-1}B$,

$$X = \begin{bmatrix} x \\ y \end{bmatrix} = \begin{bmatrix} 2 \\ 0 \end{bmatrix},$$

and the solution set of the system is $\{(2, 0)\}$.   ●

EXAMPLE 3   Use the inverse of the coefficient matrix to solve the system

$$2x - 5y = 20$$
$$3x + 4y = 7.$$

The needed matrices are

$$A = \begin{bmatrix} 2 & -5 \\ 3 & 4 \end{bmatrix}, \quad X = \begin{bmatrix} x \\ y \end{bmatrix}, \quad \text{and} \quad B = \begin{bmatrix} 20 \\ 7 \end{bmatrix}.$$

Begin by finding A^{-1}.

$$A^{-1} = \begin{bmatrix} 4/23 & 5/23 \\ -3/23 & 2/23 \end{bmatrix}$$

Now find $A^{-1}B$.

$$A^{-1}B = \begin{bmatrix} 4/23 & 5/23 \\ -3/23 & 2/23 \end{bmatrix} \begin{bmatrix} 20 \\ 7 \end{bmatrix} = \begin{bmatrix} 5 \\ -2 \end{bmatrix}$$

Then, since $X = A^{-1}B$,

$$X = \begin{bmatrix} x \\ y \end{bmatrix} = \begin{bmatrix} 5 \\ -2 \end{bmatrix},$$

from which $x = 5$, $y = -2$, with solution set $\{(5, -2)\}$.   ●

8.4   EXERCISES

Write the matrix of coefficients, A, the matrix of variables, X, and the matrix of constants, B, for each of the following systems. Do not solve.

1. $x + y = 8$
$2x - y = 4$

2. $2x + y = 9$
$x + 3y = 17$

3. $4x + 5y = 7$
$2x + 3y = 5$

4. $2x + 3y = -2$
$5x + 4y = -12$

5. $x + y + z = 9$
$2x + y + 3z = 17$
$5x - 2y + 2z = 16$

6. $3x + y - 5z = 8$
$2x + 6y - 3z = 6$
$x + 5y - 9z = -12$

Solve each of the following systems by using the inverse of the coefficient matrix. See Examples 2 and 3.

7. $-x + y = 1$
$2x - y = 1$

8. $x + y = 5$
$x - y = -1$

9. $2x - y = -8$
$3x + y = -2$

10. $x + 3y = -12$
$2x - y = 11$

11. $2x + 3y = -10$
$3x + 4y = -12$

12. $2x - 3y = 10$
$2x + 2y = 5$

13. $2x - 5y = 10$
$4x - 5y = 15$

14. $2x - 3y = 2$
$4x - 6y = 1$

15. $5x + 2y = 20$
$3x + y = 12$

16. $x + 2y = 5$
$2x + y = -2$

17. $3x - 2y = 9$
$3x + y = -2$

18. $9x - y = 12$
$-18x + 2y = 9$

19. $6x - 3y = 1$
$-12x + 6y = -2$

20. $x - 3y = 1$
$3x + 4y = 29$

Use the inverse of the coefficient matrix to solve each of the following systems of equations. The inverses of the coefficient matrices for the first four problems were found in the text and exercises in Section 7.3.

21. $x + z = 3$
$2x - 2y - z = -2$
$3x = 3$

22. $x + 3y + 3z = 11$
$-x = -2$
$-4x - 4y - 3z = -10$

23. $3x + 6y + 3z = 12$
$6x + 4y - 2z = -4$
$y - z = -3$

24. $2x + 4z = 14$
$3x + y + 5z = 19$
$-x + y - 2z = -7$

25. $x + 3y + z = 2$
$x - 2y + 3z = -3$
$2x - 3y - z = 34$

26. $x + y - z = 6$
$2x - y + z = -9$
$x - 2y + 3z = 1$

27. $x + 3y - 6z = 7$
$2x - y + 2z = 0$
$x + y + 2z = -1$

28. $x - 2y + z = 5$
$-2x + 4y - 2z = 2$
$2x + y - z = 2$

Solve each of the following systems of four equations in four unknowns. The inverses of the coefficient matrices were found in Section 7.3.

29.
$$x - 2y + 3z \qquad = -4$$
$$y - z + w = 32$$
$$-2x + 2y - 2z + 4w = 16$$
$$2y - 3z + w = -8$$

30.
$$x + y \qquad + 2w = 3$$
$$2x - y + z - w = 3$$
$$3x + 3y + 2z - 2w = 5$$
$$x + 2y + z \qquad = 3$$

Solve the following systems using the inverse of the coefficient matrix.

31.
$$.9x + .3y = .51$$
$$1.2x - .7y = -.2$$

32.
$$.32x - .41y = .0115$$
$$.12x + .15y = .0195$$

33.
$$.1x + .1y \qquad = .08$$
$$- .1y + .3z = .04$$
$$.2x \qquad - .1z = .03$$

34.
$$1.2x \qquad - .3z = .01491$$
$$4x + 2.3y \qquad = .05675$$
$$4.8y - 3.1z = .12727$$

In Exercises 35 and 36, write a system of equations and solve it by using the inverse of the coefficient matrix.

35. Midtown Manufacturing Company makes two products, plastic plates and plastic cups. Both require time on two machines: plates require 1 hr on machine A and 2 hr on machine B; cups require 3 hr on machine A and 1 hr on machine B. Both machines operate 15 hr a day. How many of each product can be produced in a day under these conditions?

36. A company produces two models of bicycles, model 201 and model 301. Model 201 requires 2 hr of assembly time and model 301 requires 3 hr of assembly time. The parts for model 201 cost $25 per bike and the parts for model 301 cost $30 per bike. If the company has a total of 34 hr of assembly time and $365 available per day for these two models, how many of each can be made in a day?

8.5 Nonlinear Systems of Equations

A **nonlinear system of equations** is one in which at least one of the equations is not a first-degree equation. Since nonlinear systems vary, depending upon the type of equations in the system, different methods are required for different systems. Although the substitution method discussed in Section 8.1 can often be used, it may not be the most efficient approach to the solution. This section illustrates some of the methods best suited to solving certain types of nonlinear systems.

Some nonlinear systems can be solved by the addition method.

EXAMPLE 1   Solve the system

$$x^2 + y^2 = 4 \qquad (1)$$
$$2x^2 - y^2 = 8. \qquad (2)$$

Add the two equations to eliminate y^2.

$$x^2 + y^2 = 4$$
$$\underline{2x^2 - y^2 = 8}$$
$$3x^2 \quad\quad = 12$$
$$x^2 = 4$$

$$x = 2 \quad \text{or} \quad x = -2.$$

Substituting into equation (1) gives the corresponding values of y.

$$2^2 + y^2 = 4 \quad \text{or} \quad (-2)^2 + y^2 = 4$$
$$y^2 = 0 \quad\quad\quad\quad\quad y^2 = 0$$
$$y = 0 \quad\quad\quad\quad\quad y = 0$$

The solution set of the system is $\{(2, 0), (-2, 0)\}$. The graph shown in Figure 8.3 verifies this result. ●

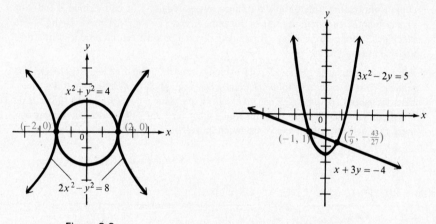

Figure 8.3   Figure 8.4

The substitution method is particularly useful in solving a nonlinear system that includes a linear equation. This method is illustrated by the next example.

EXAMPLE 2   Use the substitution method to solve the system

$$3x^2 - 2y = 5 \tag{3}$$
$$x + 3y = -4. \tag{4}$$

Equation (4) can be solved for x and the resulting expression substituted into equation (3) as shown below.

$$x + 3y = -4$$
$$x = -3y - 4 \tag{5}$$

Substituting this value of x into equation (3) gives

$$3x^2 - 2y = 5 \tag{3}$$
$$3(-3y - 4)^2 - 2y = 5$$
$$3(9y^2 + 24y + 16) - 2y = 5$$
$$27y^2 + 70y + 43 = 0.$$

Use the quadratic formula to solve for y.

$$y = \frac{-70 \pm \sqrt{4900 - 4644}}{54}$$

$$y = -1 \quad \text{or} \quad y = -43/27$$

Substitute these values for y into equation (4) to find the corresponding values for x.

$$x + 3(-1) = -4 \quad \text{or} \quad x + 3(-43/27) = -4$$

$$x = -1 \qquad\qquad\qquad x = \frac{7}{9}$$

The solution set is $\{(-1, -1), (7/9, -43/27)\}$. Figure 8.4 shows the graphs of both of the equations in the system.   ●

Sometimes a combination of the elimination method and the substitution method is effective in solving a system, as illustrated in Example 3.

EXAMPLE 3   Solve the system

$$x^2 + 3xy + y^2 = 22 \tag{6}$$
$$x^2 - xy + y^2 = 6. \tag{7}$$

Multiply both sides of equation (7) by -1. Then add the result to equation (6), as follows.

$$x^2 + 3xy + y^2 = 22$$
$$\underline{-x^2 + xy - y^2 = -6}$$
$$4xy \qquad\quad = 16 \tag{8}$$

Now solve equation (8) for either x or y and substitute the result into one of the given equations. We choose to solve for y.

$$y = \frac{4}{x}, \quad x \neq 0 \tag{9}$$

(The restriction $x \neq 0$ is included since if $x = 0$ there is no value of y which satisfies the system.) Substituting for y in equation (7) and simplifying gives

$$x^2 - x\left(\frac{4}{x}\right) + \left(\frac{4}{x}\right)^2 = 6.$$

Simplify the resulting equation as follows.

$$x^2 - x \left(\frac{4}{x}\right) + \left(\frac{4}{x}\right)^2 = 6$$

$$x^2 - 4 + \frac{16}{x^2} = 6$$

$$x^4 - 4x^2 + 16 = 6x^2$$

$$x^4 - 10x^2 + 16 = 0$$

$$(x^2 - 2)(x^2 - 8) = 0$$

$$x^2 = 2 \quad \text{or} \quad x^2 = 8$$

$$x = \sqrt{2} \quad \text{or} \quad x = -\sqrt{2} \quad \text{or} \quad x = 2\sqrt{2} \quad \text{or} \quad x = -2\sqrt{2}$$

Substitute these x values into equation (9) to find corresponding values of y.

If $x = \sqrt{2}$, $y = \dfrac{4}{\sqrt{2}} = 2\sqrt{2}$.

If $x = -\sqrt{2}$, $y = \dfrac{4}{-\sqrt{2}} = -2\sqrt{2}$.

If $x = 2\sqrt{2}$, $y = \dfrac{4}{2\sqrt{2}} = \sqrt{2}$.

If $x = -2\sqrt{2}$, $y = \dfrac{4}{-2\sqrt{2}} = -\sqrt{2}$.

The solution set of the system contains four ordered pairs:

$$\{(\sqrt{2}, 2\sqrt{2}), (-\sqrt{2}, -2\sqrt{2}), (2\sqrt{2}, \sqrt{2}), (-2\sqrt{2}, -\sqrt{2})\}. \quad \bullet$$

EXAMPLE 4   Solve the system

$$x^2 + y^2 = 16 \tag{10}$$

$$|x| + y = 4. \tag{11}$$

The substitution method is required here. From the definition of absolute value, equation (11) can be rewritten as $|x| = 4 - y$, giving

$$x = 4 - y \quad \text{or} \quad x = -(4 - y) = y - 4. \tag{12}$$

(Since $|x| \geq 0$ for all real x, $4 - y \geq 0$, or $4 \geq y$.) Substituting from either part of (12) into equation (10) gives the same result.

$$(4 - y)^2 + y^2 = 16 \quad \text{or} \quad (y - 4)^2 + y^2 = 16$$

Since $(4 - y)^2 = (y - 4)^2 = 16 - 8y + y^2$, either equation becomes

$$(16 - 8y + y^2) + y^2 = 16$$

$$2y^2 - 8y = 0$$

$$2y(y - 4) = 0$$

$$y = 0 \quad \text{or} \quad y = 4.$$

From equation (12),

$$\text{If } y = 0, \text{ then } x = 4 - 0 \quad \text{ or } \quad x = 0 - 4.$$
$$x = 4 \qquad\qquad\qquad x = -4$$
$$\text{If } y = 4, \text{ then } x = 4 - 4 \quad \text{ or } \quad x = 4 - 4.$$
$$x = 0$$

The solution set, $\{(4, 0), (-4, 0), (0, 4)\}$, includes the points of intersection shown in Figure 8.5. ●

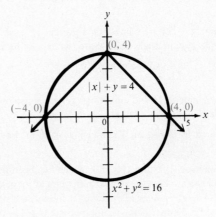

Figure 8.5

8.5 EXERCISES $1-35$ odd

Solve each of the following nonlinear systems of equations. See Examples 1–4.

1. $y = x^2$
$x + y = 2$

2. $y = -x^2 + 2$
$x - y = 0$

3. $y = (x - 1)^2$
$x - 3y = -1$

4. $y = (x + 3)^2$
$x + 2y = -2$

5. $y = x^2 + 4x$
$2x - y = -8$

6. $y = 6x + x^2$
$3x - 2y = 10$

7. $x^2 + y^2 = 13$
$2x + y = 1$

8. $2x^2 - y^2 = 7$
$y = 3x - 7$

9. $3x^2 + 2y^2 = 5$
$x - y = -2$

10. $x^2 + y^2 = 5$
$-3x + 4y = 2$

11. $x^2 + y^2 = 8$
$x^2 - y^2 = 0$

12. $x^2 + y^2 = 10$
$2x^2 - y^2 = 17$

13. $5x^2 - y^2 = 0$
$3x^2 + 4y^2 = 0$

14. $x^2 + y^2 = 4$
$2x^2 - 3y^2 = -12$

15. $3x^2 + y^2 = 3$
$4x^2 + 5y^2 = 4$

16. $x^2 + 2y^2 = 9$
$3x^2 - 4y^2 = 27$

17. $2x^2 + 3y^2 = 5$
$3x^2 - 4y^2 = -1$

18. $3x^2 + 5y^2 = 17$
$2x^2 - 3y^2 = 5$

19. $5x^2 - y^2 = 9$
$x^2 + y^2 = 45$

20. $x^2 + 3y^2 = 91$
$2x^2 - y^2 = 119$

21. $2x^2 + 2y^2 = 20$
$3x^2 + 3y^2 = 30$

22. $x^2 + y^2 = 4$
$5x^2 + 5y^2 = 28$

23. $9x^2 + 4y^2 = 1$
$x^2 + y^2 = 1$

24. $2x^2 - 3y^2 = 8$
$6x^2 + 5y^2 = 24$

25. $xy = 6$
$x + y = 5$

26. $xy = -4$
$2x + y = -7$

27. $xy = -15$
$4x + 3y = 3$

28. $xy = 8$
$3x + 2y = -16$

29. $2xy + 1 = 0$
$x + 16y = 2$

30. $-5xy + 2 = 0$
$x - 15y = 5$

31. $x^2 + 4y^2 = 25$
$xy = 6$

32. $5x^2 - 2y^2 = 6$
$xy = 2$

33. $x^2 + 2xy - y^2 = 14$
$\quad\ \ \ x^2 - y^2 = -16$

34. $3x^2 + xy + 3y^2 = 7$
$\quad\ \ \ x^2 + \quad\ \ \ y^2 = 2$

35. $x^2 - xy + y^2 = 5$
$\quad\ \ 2x^2 + xy - y^2 = 10$

36. $3x^2 + 2xy - y^2 = 9$
$\quad\ \ \ x^2 - xy + y^2 = 9$

37. $\quad\quad\ x = |y|$
$\quad\ x^2 + y^2 = 18$

38. $2x + |y| = 4$
$\quad\ x^2 + y^2 = 5$

39. $y = |x - 1|$
$\quad y = x^2 - 4$

40. $2x^2 - y^2 = 4$
$\quad\ \ |x| = |y|$

41. The sum of two numbers is 6. Their product is -16. Find the numbers.

42. The sum of the squares of two numbers is 100. One number is 2 more than the other. Find the numbers.

43. Does the straight line $3x - 2y = 9$ intersect the circle $x^2 + y^2 = 25$? (To find out, solve the system made up of these two equations.)

44. Do the parabola $y = x^2 + 4$ and the ellipse $2x^2 + y^2 - 4x - 4y = 0$ have any points in common?

45. The area of a right triangle is 216 sq cm. It has a hypotenuse of 30 cm. Find the lengths of the two legs.

46. A rectangle has an area of 240 sq m. The diagonal is 26 m. Find the length and width.

47. In skeet shooting, a person fires at a clay disc which has been thrown into the air. Suppose the disc follows a parabolic path with the height in meters at time t in seconds given by $y = -t^2 + 4t$, while the bullet follows the parabolic path $y = -2t^2 + 12t - 13$. Further, suppose the two paths are in the same plane with common axes. Find the number of seconds for the bullet to meet the disc.

48. In electronics, circuit gain is given by

$$G = \frac{Bt}{R + R_t},$$

where R is the value of a resistor, t is temperature, and B is a constant. The sensitivity of the circuit to temperature is given by

$$S = \frac{BR}{(R + R_t)^2}.$$

If $B = 3.7$ and t is $90°K$ (Kelvin), find the values of R and R_t that will make $G = .4$ and $S = .001$.

8.6 Systems of Inequalities and Linear Programming

As shown in Chapter 4, the solution set of an inequality in two variables is usually an infinite set whose graph is one or more regions of the coordinate plane. The solution set of a **system of inequalities,** such as

$$x + y < 3$$
$$x^2 < 2y,$$

is the intersection of the solution sets of its members. The solution is best visualized by its graph. To graph the solution set of the system, graph both inequalities on the same axes and identify the solution by shading heavily the region common to both graphs.

EXAMPLE 1   Graph the solution set of the system

$$x + y < 3$$
$$x^2 < 2y.$$

Figure 8.6 shows the graphs of both $x + y < 3$ and $x^2 < 2y$. The solution set of the system is the heavily shaded area. The points on the boundaries of $x + y < 3$ and $x^2 < 2y$ do not belong to the graph of the solution. For this reason, the boundaries are dashed lines. ●

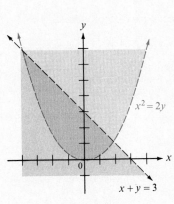

Figure 8.6

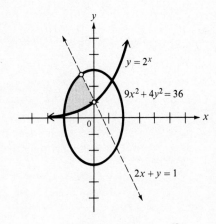

Figure 8.7

EXAMPLE 2   Graph the solution set of the system

$$y \geq 2^x$$
$$9x^2 + 4y^2 \leq 36$$
$$2x + y < 1.$$

The graph is obtained by graphing the three inequalities on the same axes and shading the region common to all three as shown in Figure 8.7. As the graph shows, two boundary lines are solid and one is dashed. The points where the dashed line intersects the solid lines are shown as open circles, since they do not belong to the solution set. ●

Linear Programming   An important application of mathematics to business and social science is called linear programming. **Linear programming** is used to find such things as minimum cost and maximum profit. The basic ideas of this technique are shown in the following example.

The Smith Company makes two products, tape decks and amplifiers. Each tape deck gives a profit of \$3, while each amplifier produces \$7. The company must manufacture at least one tape deck per day to satisfy one of its customers, but no more than five because of production problems. Also, the number of amplifiers produced cannot exceed six per day. As a further requirement, the number of tape decks cannot exceed the number of amplifiers. How many of each should the company manufacture in order to obtain the maximum profit?

To begin, translate the statements of the problem into symbols by assuming

$$x = \text{number of tape decks to be produced daily}$$
$$y = \text{number of amplifiers to be produced daily.}$$

According to the statement of the problem given above, the company must produce at least one tape deck (one or more), so

$$x \geq 1.$$

Since no more than 5 tape decks may be produced,

$$x \leq 5.$$

Since no more than 6 amplifiers may be made in one day,

$$y \leq 6.$$

The requirement that the number of tape decks may not exceed the number of amplifiers translates as

$$x \leq y.$$

The number of tape decks and of amplifiers cannot be negative, so

$$x \geq 0 \quad \text{and} \quad y \geq 0.$$

These restrictions, or **constraints,** that are placed on production form the system of inequalities

$$x \geq 1, \quad x \leq 5, \quad y \leq 6, \quad x \leq y, \quad x \geq 0, \quad y \geq 0.$$

To find the maximum possible profit that the company can make, subject to these constraints, sketch the graph of the solution of the system. See Figure 8.8. The only feasible values of x and y are those that satisfy all constraints; that is, the values which lie in the shaded region, called the **region of feasible solutions.**

Since each tape deck gives a profit of \$3, the daily profit from the production of x tape decks is $3x$ dollars. Also, the profit from the production of y amplifiers will be $7y$ dollars per day. The total daily profit is thus

$$\text{profit} = 3x + 7y.$$

The problem of the Smith Company may now be stated as follows: find values of x and y in the shaded region of Figure 8.8 that will produce the maximum possible value of $3x + 7y$.

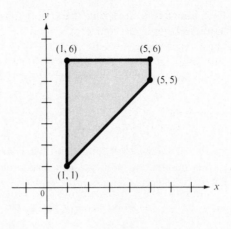

Figure 8.8

It can be shown that any optimum value (maximum or minimum) will always occur at a **vertex** (or corner) point of the region of feasible solutions. To locate the point (x, y) that gives the maximum profit, check the coordinates of the vertex points, shown in Figure 8.8 and listed below. Find the profit that corresponds to each coordinate pair and pick out the one that gives the maximum profit.

Point	Profit $= 3x + 7y$
$(1, 1)$	$3(1) + 7(1) = 10$
$(1, 6)$	$3(1) + 7(6) = 45$
$(5, 6)$	$3(5) + 7(6) = 57 \leftarrow$ Maximum
$(5, 5)$	$3(5) + 7(5) = 50$

The maximum profit of $57 is obtained when 5 tape decks and 6 amplifiers are produced each day.

EXAMPLE 3   Robin, who is ill, takes vitamin pills. Each day, she must have at least 16 units of vitamin A, at least 5 units of vitamin B_1, and at least 20 units of vitamin C. She can choose between red pills, costing 10¢ each, which contain 8 units of A, 1 of B_1, and 2 of C; and blue pills, costing 20¢ each, which contain 2 units of A, 1 of B_1, and 7 of C. How many of each pill should she buy in order to minimize her cost and yet fulfill her daily requirements?

Let x represent the number of red pills to buy, and let y represent the number of blue pills to buy. Then the cost in pennies per day is given by

$$\text{cost} = 10x + 20y.$$

Since Robin buys x of the 10¢ pills and y of the 20¢ pills, she gets vitamin A as follows: 8 units from each red pill and 2 units from each blue pill. Altogether, she gets $8x + 2y$ units of A per day. Since she must get at least 16 units,

$$8x + 2y \geq 16.$$

Each red pill and each blue pill supplies 1 unit of vitamin B_1. Robin needs at least 5 units per day, so

$$x + y \geq 5.$$

For vitamin C we get the inequality

$$2x + 7y \geq 20.$$

Also, $x \geq 0$ and $y \geq 0$, since Robin cannot buy negative numbers of the pills.

Again, we minimize total cost of the pills by finding the solution of the system of inequalities formed by the constraints. (See Figure 8.9.) The solution to this minimizing problem will also occur at a vertex point. Check the coordinates of the vertex points in the cost function to find the lowest cost.

Point	Cost $= 10x + 20y$	
(10, 0)	$10(10) + 20(0) = 100$	
(3, 2)	$10(3) + 20(2) = $ **70**	$\leftarrow$ Minimum
(1, 4)	$10(1) + 20(4) = 90$	
(0, 8)	$10(0) + 20(8) = 160$	

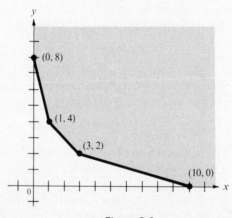

Figure 8.9

Robin's best bet is to buy 3 red pills and 2 blue ones, for a total cost of 70¢ per day. She receives minimum amounts of vitamins B_1 and C but an excess of vitamin A. Even with an excess of A, this is still the best buy.   ●

8.6   EXERCISES

Graph the solution set of each system of inequalities. See Examples 1 and 2.

1. $x + y \leq 4$
$\quad x - 2y \geq 6$

2. $2x + y > 2$
$\quad x - 3y < 6$

3. $4x + 3y < 12$
$\quad y + 4x > -4$

4. $3x + 5y \leq 15$
$\quad x - 3y \geq 9$

5. $x + y \leq 6$
$\quad 2x + 2y \geq 12$

6. $3x + 4y < 15$
$\quad 6x + 8y > 30$

7. $x + 2y \leq 4$
$\quad y \geq x^2 - 1$

8. $4x - 3y \leq 12$
$\quad y \leq x^2$

9. $y \leq -x^2$
$\quad y \geq x^2 - 6$

10. $x^2 + y^2 \leq 9$
$\quad\quad x \leq -y^2$

11. $x^2 - y^2 < 1$
$\quad -1 < y < 1$

12. $x^2 + y^2 \leq 36$
$\quad -4 \leq x \leq 4$

13. $2x^2 - y^2 > 4$
$\quad 2y^2 - x^2 > 4$

14. $y \geq x^2 + 4x + 4$
$\quad\quad y < -x^2$

15. $\dfrac{x^2}{16} + \dfrac{y^2}{9} \leq 1$
$\quad\dfrac{x^2}{4} - \dfrac{y^2}{16} \geq 1$

16. $\dfrac{x^2}{36} - \dfrac{y^2}{9} \geq 1$
$\quad\dfrac{x^2}{81} + y^2 \leq 1$

17. $x + y \leq 4$
$\quad x - y \leq 5$
$\quad 4x + y \leq -4$

18. $3x - 2y \geq 6$
$\quad x + y \leq -5$
$\quad\quad y \leq 4$

19. $2y + x \leq -5$
$\quad\quad y \geq 3 + x$
$\quad\quad x \geq 0$
$\quad\quad y \geq 0$

20. $2x + 3y \leq 12$
$\quad 2x + 3y > -6$
$\quad 3x + y < 4$
$\quad\quad\quad x \geq 0$
$\quad\quad\quad y \geq 0$

21. $\dfrac{x^2}{4} + \dfrac{y^2}{9} > 1$
$\quad x^2 - y^2 \geq 1$
$\quad -4 \leq x \leq 4$

22. $2x - 3y < 6$
$\quad 4x^2 + 9y^2 < 36$
$\quad\quad x \geq -1$

23. $y \geq 3^x$
$\quad y \geq 2$

24. $y \leq \left(\dfrac{1}{2}\right)^x$
$\quad y \geq 4$

25. $|x| \geq 2$
$\quad |y| \geq 4$
$\quad y < x^2$

26. $|x| + 2 \geq 4$
$\quad\quad |y| \leq 1$
$\quad\dfrac{x^2}{9} + \dfrac{y^2}{16} \leq 1$

27. $y \leq |x + 2|$
$\quad\dfrac{x^2}{16} - \dfrac{y^2}{9} \leq 1$

28. $y \leq \log x$
$\quad y \geq |x - 2|$

29. $y \geq |4 - x|$
$\quad y \geq |x|$

30. $|x + 2| < y$
$\quad |x| \leq 3$

In Exercises 31–34, use graphical methods to find values of x and y satisfying the given conditions. Find the value of the maximum or minimum.

31. Find $x \geq 0$ and $y \geq 0$ such that
$$2x + 3y \leq 6$$
$$4x + y \leq 6$$
and $5x + 2y$ is maximized.

32. Find $x \geq 0$ and $y \geq 0$ such that
$$x + y \leq 10$$
$$5x + 2y \geq 20$$
$$2y \geq x$$
and $x + 3y$ is minimized.

33. Find $x \geq 2$ and $y \geq 5$ such that
$$3x - y \geq 12$$
$$x + y \leq 15$$
and $2x + y$ is minimized.

34. Find $x \geq 10$ and $y \geq 20$ such that
$$2x + 3y \leq 100$$
$$5x + 4y \leq 200$$
and $x + 3y$ is maximized.

Solve each of the following linear programming problems. See Example 3.

35. Farmer Jones raises only pigs and geese. She wants to raise no more than 16 animals with no more than 12 geese. She spends $5 to raise a pig and $2 to raise a goose. She has $50 available for this purpose. Find the maximum profit she can make if she makes a profit of $8 per goose and $4 per pig.

36. A wholesaler of party goods wishes to display her products at a convention of social secretaries in such a way that she gets the maximum number of inquiries about her whistles and hats. Her booth at the convention has 12 sq m of floor space to be used for display purposes. A display unit for hats requires 2 sq m; and for whistles, 4 sq m. Experience tells the wholesaler that she should never have more than a total of 5 units of whistles and hats on display at one time. If she receives three inquiries for each unit of hats and two inquiries for each unit of whistles on display, how many of each should she display in order to get the maximum number of inquiries?

37. An office manager wants to buy some filing cabinets. She knows that cabinet #1 costs \$10 each, requires 6 sq ft of floor space, and holds 8 cu ft of files. Cabinet #2 costs \$20 each, requires 8 sq ft of floor space, and holds 12 cu ft. She can spend no more than \$140 due to budget limitations, while her office has room for no more than 72 sq ft of cabinets. She wants the maximum storage capacity within the limits imposed by funds and space. How many of each type of cabinet should she buy?

38. In a small town in South Carolina, zoning rules require that the window space (in square feet) in a house be at least one-sixth of the space used up by solid walls. The cost to heat the house is 2¢ for each square foot of solid walls and 8¢ for each square foot of windows. Find the maximum total area (windows plus walls) if \$16 is available to pay for heat.

39. The manufacturing process requires that oil refineries manufacture at least 2 gal of gasoline for each gallon of fuel oil. To meet the winter demand for fuel oil, at least 3 million gal a day must be produced. The demand for gasoline is no more than 6.4 million gal per day. If the price of gasoline is \$1.90 and the price of fuel oil is \$1.50 per gal, how much of each should be produced to maximize revenue?

40. Seall Manufacturing Company makes color television sets. It produces a bargain set that sells for \$100 profit and a deluxe set that sells for \$150 profit. On the assembly line the bargain set requires 3 hr, while the deluxe set takes 5 hr. The cabinet shop spends 1 hr on the cabinet for the bargain set and 3 hr on the cabinet for the deluxe set. Both sets require 2 hr of time for testing and packing. On a particular production run the Seall Company has available 3900 work hours on the assembly line, 2100 work hours in the cabinet shop, and 2200 work hours in the testing and packing department. How many sets of each type should it produce to make maximum profit? What is the maximum profit?

Appendix   Solution of Linear Systems by Determinants—Cramer's Rule

In Section 8.1, we discussed the elimination method for solving systems of equations. We can use this method to solve the general system of two equations in two variables,

$$a_1 x + b_1 y = c_1 \tag{1}$$
$$a_2 x + b_2 y = c_2. \tag{2}$$

To eliminate y and solve for x, multiply both sides of equation (1) by b_2 and equation (2) by $-b_1$ and then add.

$$a_1 b_2 x + b_1 b_2 y = c_1 b_2$$
$$\underline{-a_2 b_1 x - b_1 b_2 y = -c_2 b_1}$$
$$(a_1 b_2 - a_2 b_1)x \qquad\qquad = c_1 b_2 - c_2 b_1$$
$$x = \frac{c_1 b_2 - c_2 b_1}{a_1 b_2 - a_2 b_1}$$

To solve for y, multiply both sides of equation (1) by $-a_2$ and equation (2) by a_1 and add.

$$-a_1a_2x - a_2b_1y = -a_2c_1$$
$$\underline{a_1a_2x + a_1b_2y = a_1c_2}$$
$$(a_1b_2 - a_2b_1)y = a_1c_2 - a_2c_1$$
$$y = \frac{a_1c_2 - a_2c_1}{a_1b_2 - a_2b_1}$$

Both numerators and the common denominator of these values for x and y can be written as determinants, since

$$c_1b_2 - c_2b_1 = \begin{vmatrix} c_1 & b_1 \\ c_2 & b_2 \end{vmatrix}, \qquad a_1c_2 - a_2c_1 = \begin{vmatrix} a_1 & c_1 \\ a_2 & c_2 \end{vmatrix},$$

$$\text{and } a_1b_2 - a_2b_1 = \begin{vmatrix} a_1 & b_1 \\ a_2 & b_2 \end{vmatrix}.$$

The solutions for x and y can be written with determinants as

$$x = \frac{\begin{vmatrix} c_1 & b_1 \\ c_2 & b_2 \end{vmatrix}}{\begin{vmatrix} a_1 & b_1 \\ a_2 & b_2 \end{vmatrix}} \qquad \text{and} \qquad y = \frac{\begin{vmatrix} a_1 & c_1 \\ a_2 & c_2 \end{vmatrix}}{\begin{vmatrix} a_1 & b_1 \\ a_2 & b_2 \end{vmatrix}}.$$

This result, known as **Cramer's rule,** is summarized below.

Cramer's Rule for 2 × 2 Systems

Given the system

$$a_1x + b_1y = c_1$$
$$a_2x + b_2y = c_2,$$

with $a_1b_2 - a_2b_1 \neq 0$, then

$$x = \frac{\begin{vmatrix} c_1 & b_1 \\ c_2 & b_2 \end{vmatrix}}{\begin{vmatrix} a_1 & b_1 \\ a_2 & b_2 \end{vmatrix}} \qquad \text{and} \qquad y = \frac{\begin{vmatrix} a_1 & c_1 \\ a_2 & c_2 \end{vmatrix}}{\begin{vmatrix} a_1 & b_1 \\ a_2 & b_2 \end{vmatrix}}.$$

For convenience, the three determinants in the solution are written as

$$\begin{vmatrix} a_1 & b_1 \\ a_2 & b_2 \end{vmatrix} = D, \qquad \begin{vmatrix} c_1 & b_1 \\ c_2 & b_2 \end{vmatrix} = D_x, \qquad \text{and} \qquad \begin{vmatrix} a_1 & c_1 \\ a_2 & c_2 \end{vmatrix} = D_y.$$

Then

$$x = \frac{D_x}{D} \text{ and } y = \frac{D_y}{D}, \text{ where } D \neq 0.$$

To use Cramer's rule, the system must first be written in the form shown above. The elements of D are then the coefficients of the variables in the given system. The elements of D_x are obtained by replacing the coefficients of x in the coefficient matrix by the respective constants; D_y is defined similarly. If $D = 0$ the system does not have a unique solution. It is either dependent or inconsistent.

EXAMPLE 1 Use Cramer's rule to solve the system

$$5x + 7y = -1$$
$$6x + 8y = 1.$$

By Cramer's rule, $x = D_x/D$ and $y = D_y/D$. Now evaluate D, D_x, and D_y.

$$D = \begin{vmatrix} 5 & 7 \\ 6 & 8 \end{vmatrix} = 5(8) - 6(7) = -2$$

$$D_x = \begin{vmatrix} -1 & 7 \\ 1 & 8 \end{vmatrix} = (-1)(8) - (1)(7) = -15$$

$$D_y = \begin{vmatrix} 5 & -1 \\ 6 & 1 \end{vmatrix} = 5(1) - (6)(-1) = 11$$

Thus $x = -15/(-2) = 15/2$ and $y = 11/(-2) = -11/2$. The solution set is $\{(15/2, -11/2)\}$, as can be verified by substituting within the given system. ●

Cramer's rule can be generalized to systems of three equations in three variables (or n equations in n variables).

Cramer's Rule for 3 × 3 Systems

Given the system

$$a_1x + b_1y + c_1z = d_1$$
$$a_2x + b_2y + c_2z = d_2$$
$$a_3x + b_3y + c_3z = d_3,$$

with

$$D_x = \begin{vmatrix} d_1 & b_1 & c_1 \\ d_2 & b_2 & c_2 \\ d_3 & b_3 & c_3 \end{vmatrix}, \qquad D_y = \begin{vmatrix} a_1 & d_1 & c_1 \\ a_2 & d_2 & c_2 \\ a_3 & d_3 & c_3 \end{vmatrix},$$

$$D_z = \begin{vmatrix} a_1 & b_1 & d_1 \\ a_2 & b_2 & d_2 \\ a_3 & b_3 & d_3 \end{vmatrix}, \qquad D = \begin{vmatrix} a_1 & b_1 & c_1 \\ a_2 & b_2 & c_2 \\ a_3 & b_3 & c_3 \end{vmatrix} \neq 0,$$

then

$$x = \frac{D_x}{D}, \qquad y = \frac{D_y}{D}, \qquad \text{and} \qquad z = \frac{D_z}{D}.$$

EXAMPLE 2   Use Cramer's rule to solve the system

$$x + y - z + 2 = 0$$
$$2x - y + z + 5 = 0$$
$$x - 2y + 3z - 4 = 0.$$

To use Cramer's rule, the system must be rewritten in the form

$$x + y - z = -2$$
$$2x - y + z = -5$$
$$x - 2y + 3z = 4.$$

Verify that the required determinants are as follows:

$$D = \begin{vmatrix} 1 & 1 & -1 \\ 2 & -1 & 1 \\ 1 & -2 & 3 \end{vmatrix} = -3, \qquad D_x = \begin{vmatrix} -2 & 1 & -1 \\ -5 & -1 & 1 \\ 4 & -2 & 3 \end{vmatrix} = 7,$$

$$D_y = \begin{vmatrix} 1 & -2 & -1 \\ 2 & -5 & 1 \\ 1 & 4 & 3 \end{vmatrix} = -22, \qquad D_z = \begin{vmatrix} 1 & 1 & -2 \\ 2 & -1 & -5 \\ 1 & -2 & 4 \end{vmatrix} = -21.$$

Thus

$$x = \frac{D_x}{D} = \frac{7}{-3} = -\frac{7}{3}, \quad y = \frac{D_y}{D} = \frac{-22}{-3} = \frac{22}{3}, \quad \text{and} \quad z = \frac{D_z}{D} = \frac{-21}{-3} = 7,$$

so the solution set is $\{(-7/3, 22/3, 7)\}$.   ●

EXAMPLE 3   Use Cramer's rule to solve the system

$$2x - 3y + 4z = 10$$
$$6x - 9y + 12z = 24$$
$$x + 2y - 3z = 5.$$

Find D, D_x, D_y, and D_z.

$$D = \begin{vmatrix} 2 & -3 & 4 \\ 6 & -9 & 12 \\ 1 & 2 & -3 \end{vmatrix} = 2 \begin{vmatrix} -9 & 12 \\ 2 & -3 \end{vmatrix} - 6 \begin{vmatrix} -3 & 4 \\ 2 & -3 \end{vmatrix} + 1 \begin{vmatrix} -3 & 4 \\ -9 & 12 \end{vmatrix}$$

$$= 2(3) - 6(1) + 1(0)$$

$$= 0$$

Since Cramer's rule requires $D \neq 0$, it is a good idea to calculate D first. When $D = 0$, the system is either dependent or inconsistent. Use the elimination method or the Gauss-Jordan method to tell which. Verify that this system is inconsistent.   ●

APPENDIX EXERCISES 1-27

Use Cramer's rule to solve each of the following systems of equations. If D = 0, use another method to determine the solution. See Example 1.

1. $x + y = 4$
$2x - y = 2$

2. $3x + 2y = -4$
$2x - y = -5$

3. $4x + 3y = -7$
$2x + 3y = -11$

4. $4x - y = 0$
$2x + 3y = 14$

5. $5x + 4y = 10$
$3x - 7y = 6$

6. $3x + 2y = -4$
$5x - y = 2$

7. $2x - 3y = -5$
$x + 5y = 17$

8. $x + 9y = -15$
$3x + 2y = 5$

9. $3x + 2y = 4$
$6x + 4y = 8$

10. $1.5x + 3y = 5$
$2x + 4y = 3$

11. $12x + 8y = 3$
$15x + 10y = 9$

12. $15x - 10y = 5$
$9x + 6y = 3$

Use Cramer's rule to solve each of the following systems of equations. If D = 0, use another method to complete the solution. See Examples 2 and 3.

13. $4x - y + 3z = -3$
$3x + y + z = 0$
$2x - y + 4z = 0$

14. $5x + 2y + z = 15$
$2x - y + z = 9$
$4x + 3y + 2z = 13$

15. $2x - y + 4z = -2$
$3x + 2y - z = -3$
$x + 4y + 2z = 17$

16. $x + y + z = 4$
$2x - y + 3z = 4$
$4x + 2y - z = -15$

17. $4x - 3y + z = -1$
$5x + 7y + 2z = -2$
$3x - 5y - z = 1$

18. $2x - 3y + z = 8$
$-x - 5y + z = -4$
$3x - 5y + 2z = 12$

19. $x + 2y + 3z = 4$
$4x + 3y + 2z = 1$
$-x - 2y - 3z = 0$

20. $2x - y + 3z = 1$
$-2x + y - 3z = 2$
$5x - y + z = 2$

21. $-2x - 2y + 3z = 4$
$5x + 7y - z = 2$
$2x + 2y - 3z = -4$

22. $-3x + 2y - 2z = 4$
$4x + y + z = 5$
$3x - 2y + 2z = 1$

23. $2x + 3y = 13$
$2y - z = 5$
$x + 2z = 4$

24. $3x - z = -10$
$y + 4z = 8$
$x + 2z = -1$

25. $5x - y = -4$
$3x + 2z = 4$
$4y + 3z = 22$

26. $3x + 5y = -7$
$2x + 7z = 2$
$4y + 3z = -8$

27. $x + 2y = 10$
$3x + 4z = 7$
$-y - z = 1$

28. $5x - 2y = 3$
$4y + z = 8$
$x + 2z = 4$

Use Cramer's rule to solve each of the following systems.

29. $x + z = 0$
$y + 2z + w = 0$
$2x - w = 0$
$x + 2y + 3z = -2$

30. $x - y + z + w = 6$
$2y - w = -7$
$x - z = -1$
$y + w = 1$

31. $.4x - .6y = .4$
$.3x + .2y = -.22$

32. $-.5x + .4y = .43$
$.1x + .2y = .11$

33. $.4x - .2y + .3z = -.08$
$-.5x + .3y - .7z = .19$
$.8x - .7y + .1z = -.21$

34. $-.6x + .2y - .5z = -.16$
$.8x - .3y - .2z = -.48$
$.4x + .6y - .8z = -.8$

For the following two exercises, use the system of equations

$$a_1x + b_1y = c_1$$
$$a_2x + b_2y = c_2.$$

35. Assume $D_x = 0$ and $D_y = 0$. Show that if $c_1c_2 \neq 0$, then $D = 0$, and the equations are consistent (not inconsistent).

36. Assume $D = 0$, $D_x = 0$, and $a_1a_2 \neq 0$. Show that $D_y = 0$.

Chapter 8 Summary

Key Words

system of equations
inconsistent system
dependent equations
addition method
elimination method
y arbitrary
Gauss-Jordan method
system of inequalities
constraints

first-degree equation in n unknowns
substitution method
equivalent systems
ordered triple
coefficient matrix
nonlinear system of equations
linear programming
region of feasible solutions
Cramer's rule

Matrix Row Operations

For any real number k and any augmented matrix of a system of linear equations, the following operations will result in the matrix of an equivalent system:

1. **interchanging any two rows of a matrix,**

2. **multiplying the elements of any row of a matrix by the same nonzero scalar k, and**

3. **adding a multiple of the elements of one row to the elements of another row.**

Cramer's Rule for 2 × 2 Systems

Given the system

$$a_1x + b_1y = c_1$$
$$a_2x + b_2y = c_2,$$

with $a_1b_2 - a_2b_1 \neq 0$, then

$$x = \frac{\begin{vmatrix} c_1 & b_1 \\ c_2 & b_2 \end{vmatrix}}{\begin{vmatrix} a_1 & b_1 \\ a_2 & b_2 \end{vmatrix}} \quad \text{and} \quad y = \frac{\begin{vmatrix} a_1 & c_1 \\ a_2 & c_2 \end{vmatrix}}{\begin{vmatrix} a_1 & b_1 \\ a_2 & b_2 \end{vmatrix}}.$$

Cramer's Rule for 3 × 3 Systems

Given the system

$$a_1x + b_1y + c_1z = d_1$$
$$a_2x + b_2y + c_2z = d_2$$
$$a_3x + b_3y + c_3z = d_3,$$

with

$$D_x = \begin{vmatrix} d_1 & b_1 & c_1 \\ d_2 & b_2 & c_2 \\ d_3 & b_3 & c_3 \end{vmatrix}, \quad D_y = \begin{vmatrix} a_1 & d_1 & c_1 \\ a_2 & d_2 & c_2 \\ a_3 & d_3 & c_3 \end{vmatrix},$$

$$D_z = \begin{vmatrix} a_1 & b_1 & d_1 \\ a_2 & b_2 & d_2 \\ a_3 & b_3 & d_3 \end{vmatrix}, \quad D = \begin{vmatrix} a_1 & b_1 & c_1 \\ a_2 & b_2 & c_2 \\ a_3 & b_3 & c_3 \end{vmatrix} \neq 0,$$

then

$$x = \frac{D_x}{D}, \quad y = \frac{D_y}{D}, \quad \text{and} \quad z = \frac{D_z}{D}.$$

Chapter 8   Review Exercises   *NO EVEN ONES ON TEST*

Use the substitution method to solve each of the following linear systems. Identify any systems with dependent equations or any inconsistent systems.

1. $4x - 3y = -1$
$3x + 5y = 50$

2. $10x + 3y = 8$
$5x - 4y = 26$

3. $7x - 10y = 11$ *won't be on test*
$\dfrac{3x}{2} - 5y = 8$

must write in $3(3)$

4. $\dfrac{x}{2} + \dfrac{2y}{3} = -8$
$\dfrac{3x}{4} + \dfrac{y}{3} = 0$

5. $\dfrac{x}{2} - \dfrac{y}{5} = \dfrac{11}{10}$
$2x - \dfrac{4y}{5} = \dfrac{22}{5}$

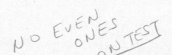

6. $4x + 5y - 5 = 0$
$3x + 7y + 6 = 0$

Use the addition method to solve each of the following linear systems. Identify any systems with dependent equations or any inconsistent systems.

7. $3x - 5y = -18$
$2x + 7y = 19$

8. $6x + 5y = 53$
$4x - 3y = 29$

9. $\dfrac{2}{3}x - \dfrac{3}{4}y = 13$
$\dfrac{1}{2}x + \dfrac{2}{3}y = -5$

10. $3x + 7y = 10$
$18x + 42y = 50$

11. $\dfrac{1}{x} + \dfrac{1}{y} = \dfrac{7}{10}$
$\dfrac{3}{x} - \dfrac{5}{y} = \dfrac{1}{2}$

12. $.9x - .2y = .8$
$.3x + .7y = 4.1$

Solve each of the following problems by writing a system of equations and then solving it.

13. A student bought some candy bars, paying 25¢ each for some and 50¢ each for others. The student bought a total of 22 bars, paying a total of $8.50. How many of each kind of bar did he buy?

14. Ink worth $25 a bottle is to be mixed with ink worth $18 per bottle to get 12 bottles of ink worth $20 each. How much of each type of ink should be used?

15. Donna Sharp wins $50,000 in a lottery. She invests part of the money at 6%, twice as much at 7%, with $10,000 more than the amount invested at 6% invested at 9%. Total annual interest is $3800. How much is invested at each rate?

16. The sum of three numbers is 23. The second number is 3 more than the first. The sum of the first and twice the third is 4. Find the three numbers.

17. A gold merchant has some 12 carat gold (12/24 pure gold), and some 22 carat gold (22/24 pure). How many grams of each should be mixed to get 25 g of 15 carat gold?

18. A chemist has some 40% acid solution and some 60% solution. How many liters of each should be used to get 40 liters of a 45% solution?

Use the addition method to solve each of the following linear systems.

19. $2x - 3y + z = -5$
$x + 4y + 2z = 13$
$5x + 5y + 3z = 14$

20. $x - 3y = 12$
$2y + 5z = 1$
$4x + z = 25$

21. $x + y - z = 5$
$2x + y + 3z = 2$
$4x - y + 2z = -1$

22. $5x - 3y + 2z = -5$
$2x + 2y - z = 4$
$4x - y + z = -1$

Solve each of the following problems by writing a system of equations and then solving it.

23. Three kinds of tea worth $4.60, $5.75, and $6.50 per lb are to be mixed to get 20 lb of tea worth $5.25 per lb. The amount of $4.60 tea used is to be equal to the total amount of the other two kinds together. How many pounds of each tea should be used?

24. A 5% solution of a drug is to be mixed with some 15% solution and some 10% solution to get 20 ml of 8% solution. The amount of 5% solution used must be 2 ml more than the sum of the other two solutions. How many milliliters of each solution should be used?

25. The cashier at an amusement park has a total of $2480, made up of fives, tens, and twenties. The total number of bills is 290, and the value of the tens is $60 more than the value of the twenties. How many of each type of bill does the cashier have?

26. Ms. Levy invests $50,000 three ways—at 8%, 8 1/2%, and 11%. In total, she receives $4436.25 per year in interest. The interest from the 11% investment is $80 more than the interest on the 8% investment. Find the amount she has invested at each rate.

Find solutions for the following systems in terms of the arbitrary variable z.

27. $3x - 4y + z = 2$
$2x + y - 4z = 1$

28. $2x + 3y - z = 5$
$-x + 2y + 4z = 8$

Use the Gauss-Jordan method to solve each of the following systems.

29. $2x + 3y = 10$
$-3x + y = 18$

30. $5x + 2y = -10$
$3x - 5y = -6$

31. $3x + y = -7$
$x - y = -5$

32. $x \quad - z = -3$
$y + z = 6$
$2x \quad - 3z = -9$

33. $2x - y + 4z = -1$
$-3x + 5y - z = 5$
$2x + 3y + 2z = 3$

34. $5x - 8y + z = 1$
$3x - 2y + 4z = 3$
$10x - 16y + 2z = 3$

Use the method of matrix inverses to solve each of the following.

35. $x + y = 4$
$2x + 3y = 10$

36. $5x - 3y = -2$
$2x + 7y = -9$

37. $2x + y = 5$
$3x - 2y = 4$

38. $x - 2y = 7$
$3x + y = 7$

39. $x + y + z = 1$
$2x - y \quad = -2$
$3y + z = 2$

40. $x \quad = -3$
$y + z = 6$
$2x \quad - 3z = -9$

41. $3x - 2y + 4z = 1$
$4x + y - 5z = 2$
$-6x + 4y - 8z = -2$

42. $x + 2y \quad = -1$
$3y - z = -5$
$x + 2y - z = -3$

Solve each of the following nonlinear systems of equations.

43. $y = x^2 - 1$
$x + y = 1$

44. $x^2 + y^2 = 2$
$3x + y = 4$

45. $x^2 + 2y^2 = 22$
$2x^2 - y^2 = -1$

46. $x^2 - 4y^2 = 19$
$x^2 + y^2 = 29$

47. $xy = 4$
$x - 6y = 2$

48. $x^2 + 2xy + y^2 = 4$
$x = 3y - 2$

49. Do the circle $x^2 + y^2 = 144$ and the line $x + 2y = 8$ have any points in common? If so, what are they?

50. Find a value of b so that the straight line $3x - y = b$ touches the circle $x^2 + y^2 = 25$ in only one point.

51. The difference of the squares of two real numbers is 64. Their product is 60. Find the numbers.

Graph the solution of each of the following systems of inequalities.

52. $x + y \le 6$
$2x - y \ge 3$

53. $x - 3y \ge 6$
$y^2 \le 16 - x^2$

54. $9x^2 + 16y^2 \ge 144$
$x^2 - y^2 \ge 16$

55. A bakery makes both cakes and cookies. Each batch of cakes requires 2 hr in the oven and 3 hr in the decorating room. Each batch of cookies needs 1 1/2 hr in the oven and 2/3 hr in the decorating room. The oven is available no more than 16 hr a day, while the decorating room can be used no more than 12 hr per day. Set up a system of inequalities expressing the given information, and then graph the system.

56. A candy company has 100 kg of chocolate-covered nuts and 125 kg of chocolate-covered raisins to be sold as two different mixtures. One mix will contain 1/2 nuts and 1/2 raisins, while the other mix will contain 1/3 nuts and 2/3 raisins. Set up a system of inequalities expressing the given information, and then graph the system.

Solve the following linear programming problems.

57. Find $x \geq 0$ and $y \geq 0$ such that

$$3x + 2y \leq 12$$
$$5x + y \geq 5$$

and $2x + 4y$ is maximized.

58. Find $x \geq 0$ and $y \geq 0$ such that

$$x + y \leq 50$$
$$2x + y \geq 20$$
$$x + 2y \geq 30$$

and $4x + 2y$ is minimized.

Solve each of the following systems by Cramer's rule. Identify any systems with dependent equations or any inconsistent systems.

59. $3x + y = -1$
$\quad\; 5x + 4y = 10$

60. $3x + 7y = 2$
$\quad\;\; 5x - y = -22$

61. $3x + 2y + z = 2$
$\quad\; 4x - y + 3z = -16$
$\quad\;\; x + 3y - z = 12$

62. $\quad 5x - 2y - z = 8$
$\quad -5x + 2y + z = -8$
$\quad\;\;\; x - 4y - 2z = 0$

9

Zeros of Polynomials

Recall from Chapter 5 that a *polynomial function of degree n,* or simply a *polynomial,* is an expression of the form

$$P(x) = a_n x^n + a_{n-1} x^{n-1} + \ldots + a_1 x + a_0,$$

where n is a nonnegative integer and $a_n \neq 0$. In Chapter 5 the coefficients a_0, a_1, . . . , a_n were all real numbers. In this chapter, we extend the definition to include coefficients that are complex numbers.* We use $P(x)$ instead of $f(x)$ in this chapter to emphasize that the function is a polynomial function. The number a_n is the **leading coefficient** of $P(x)$. If all the coefficients of a polynomial expression are 0, the polynomial is 0 and is called the **zero polynomial.** The zero polynomial has no degree. However, a polynomial of the form $P(x) = a_0$ for a nonzero complex number a_0 has degree 0.

The values of x that satisfy $P(x) = 0$ are called the **zeros** of $P(x)$. Methods for finding zeros of first-and second-degree polynomials were given in Chapter 3. In this chapter several theorems are presented that help to find, or at least approximate, the zeros of polynomials of higher degree.

---

*Remember, *complex numbers* are real numbers, imaginary numbers like $5i$ or $-i\sqrt{3}$, or numbers such as $4 - 2i$ or $6 + i\sqrt{7}$.

9.1   Polynomial Division

In many polynomial division problems the divisor is a first-degree binomial of the form $x - k$, where the coefficient of x is 1. A shortcut for this type of division problem is developed by omitting all variables and writing only coefficients. Zero is used to represent the coefficient of any missing powers of the variables. Since the coefficient of x in the division is always 1 in problems of this type, the 1 can be left out also. Using these omissions, the division problem on the left below is written in the shortened form shown on the right.

$$
\begin{array}{r}
3x^2 + 10x + 40 \\
x - 4\overline{)3x^3 - 2x^2 + 0x - 150} \\
\underline{3x^3 - 12x^2} \\
10x^2 \\
\underline{10x^2 - 40x} \\
40x \\
40x - 160 \\
\underline{} \\
10
\end{array}
\qquad
\begin{array}{r}
3 \quad 10 \quad 40 \\
-4\overline{)3 - 2 \quad\; 0 - 150} \\
\mathbf{3} - 12 \\
10 \\
\mathbf{10} - 40 \\
40 \\
\mathbf{40} - 160 \\
\underline{} \\
10
\end{array}
$$

The boldface numbers are repetitions of the numbers directly above them and can be omitted also.

$$
\begin{array}{r}
3 \quad 10 \quad 40 \\
-4\overline{)3 - 2 \quad\; 0 - 150} \\
- 12 \\
10 \\
- 40 \\
40 \\
- 160 \\
10
\end{array}
$$

The entire problem can now be condensed, with the top row of numbers omitted, since it duplicates the bottom row.

$$
\begin{array}{r}
-4\overline{)3 - 2 \quad\; 0 - 150} \\
\underline{- 12 - 40 - 160} \\
3 \quad 10 \quad 40 \quad 10
\end{array}
$$

We obtained the bottom row by subtracting -12, -40, and -160 from the corresponding terms above. For reasons that will become clear later in this section, change the -4 at the left to 4, which also changes the sign of the numbers in the second row, and then *add*.

$$
\begin{array}{r}
4\overline{)3 - 2 \quad\; 0 - 150} \\
\underline{12 \quad 40 \quad 160} \\
3 \quad 10 \quad 40 \quad 10
\end{array}
$$

Read the quotient and remainder from the bottom row. The quotient will have degree one less than the polynomial being divided.

$$\frac{3x^3 - 2x^2 - 150}{x - 4} = 3x^2 + 10x + 40 + \frac{10}{x - 4}$$

This shortcut process is called **synthetic division.**

In summary, to use synthetic division to divide a polynomial by a binomial of the form $x - k$, begin by writing the coefficients of the polynomial in decreasing powers of the variable, using 0 as the coefficient of any missing powers. The number k is written to the left in the same row. In the example above, $x - k$ is $x - 4$, so k is 4. Next bring down the leading coefficient of the polynomial, 3 in the example above, as the first number in the last row. Multiply the 3 by 4 to get the first number in the second row, 12. Add 12 to -2; this gives 10, the second number in the third row. Multiply 10 by 4 to get 40, the next number in the second row. Add 40 to 0 to get the third number in the third row, and so on. This process of multiplying each result in the third row by k and adding the product to the number in the next column is repeated until there is a number in the last row for each coefficient in the first row. To avoid incorrect results, it is very important to use a 0 for any missing terms when setting up the division.

EXAMPLE 1   Use synthetic division to divide $5m^3 - 6m^2 - 28m - 2$ by $m + 2$.
Begin by writing

$$-2\overline{)5 \quad -6 \quad -28 \quad -2}.$$

The 2 is changed to -2 since k is found by writing $m + 2$ as $m - (-2)$. Next, bring down the 5.

$$
\begin{array}{r}
-2\overline{)5 \quad -6 \quad -28 \quad -2} \\
\hline
5
\end{array}
$$

Now, multiply -2 by 5 to get -10, and add it to the -6 in the first row. The result is -16.

$$
\begin{array}{r}
-2\overline{)5 \quad\; -6 \quad -28 \quad\;\; 2} \\
-10 \\
\hline
5 \quad -16
\end{array}
$$

Next, $(-2)(-16) = 32$. Add this to the -28 in the first row.

$$
\begin{array}{r}
-2\overline{)5 \quad\; -6 \quad -28 \quad -2} \\
-10 \quad\;\; 32 \\
\hline
5 \quad -16 \quad\;\;\; 4
\end{array}
$$

Finally, $(-2)(4) = -8$, giving

$$
\begin{array}{r}
-2\overline{)5 \quad\; -6 \quad -28 \quad\; -2} \\
-10 \quad\;\; 32 \quad\; -8 \\
\hline
5 \quad -16 \quad\;\;\; 4 \quad -10.
\end{array}
$$

The coefficients of the quotient polynomial and the remainder are read directly from the bottom row. Since the degree of the quotient will always be one less than the degree of the polynomial to be divided,

$$\frac{5m^3 - 6m^2 - 28m - 2}{m + 2} = 5m^2 - 16m + 4 + \frac{-10}{m + 2}. \quad \bullet$$

The result of the division in Example 1 can be written as

$$5m^3 - 6m^2 - 28m - 2 = (m + 2)(5m^2 - 16m + 4) + (-10)$$

by multiplying both sides by the denominator $m + 2$. The following theorem is a generalization of the division process illustrated above.

Division Algorithm

> For any polynomial $P(x)$ and any complex number k, there exists a unique polynomial $Q(x)$ and number r such that
>
> $$P(x) = (x - k)Q(x) + r.$$

For example, in the synthetic division above,

$$5m^3 - 6m^2 - 28m - 2 = (m + 2)(5m^2 - 16m + 4) + (-10)$$
$$P(x) \qquad = (x - k) \qquad Q(x) \qquad + \quad r.$$

By the division algorithm, $P(x) = (x - k)Q(x) + r$. This equality is true for all complex values of x, so it is true for $x = k$. Thus

$$P(k) = (k - k)Q(k) + r$$
$$P(k) = r.$$

This proves the following theorem.

Remainder Theorem

> If the polynomial $P(x)$ is divided by $x - k$, then the remainder is $P(k)$.

As an example of this theorem, we have seen that the polynomial $P(m) = 5m^3 - 6m^2 - 28m - 2$ can be written as

$$P(m) = (m + 2)(5m^2 - 16m + 4) + (-10).$$

To find $P(-2)$, substitute -2 for m.

$$P(-2) = (-2 + 2)[5(-2)^2 - 16(-2) + 4] + (-10)$$
$$= 0\,[5(-2)^2 - 16(-2) + 4] + (-10)$$
$$= -10$$

By the remainder theorem, instead of replacing m by -2 to find $P(-2)$, divide $P(m)$ by $m + 2$ using synthetic division. Then $P(-2)$ is the remainder, -10.

EXAMPLE 2   Let $P(x) = -x^4 + 3x^2 - 4x - 5$. Find $P(-2)$.
Use the remainder theorem and synthetic division.

$$
\begin{array}{r|rrrrr}
-2) & -1 & 0 & 3 & -4 & -5 \\
    &   & 2 & -4 & 2 & 4 \\
\hline
    & -1 & 2 & -1 & -2 & -1
\end{array}
$$

The remainder when $P(x)$ is divided by $x - (-2) = x + 2$ is -1, so $P(-2) = -1$. •

The remainder theorem gives a quick way to decide if a number k is a zero of a polynomial $P(x)$. Use synthetic division to find $P(k)$; if the remainder is zero, then $P(k) = 0$ and k is a zero of $P(x)$. When $P(k) = 0$, the number k is also called a **root** or **solution** of the equation $P(x) = 0$.

EXAMPLE 3   Decide whether or not the given number is a zero of the given polynomial.
(a) $2; P(x) = x^3 - 4x^2 + 9x - 10$
Use synthetic division.

$$
\begin{array}{r|rrrr}
2) & 1 & -4 & 9 & -10 \\
   &   & 2 & -4 & 10 \\
\hline
   & 1 & -2 & 5 & 0
\end{array}
$$

Since the remainder is 0, $P(2) = 0$, and 2 is a zero of the polynomial $P(x) = x^3 - 4x^2 + 9x - 10$.

(b) $-4; P(x) = x^4 + x^2 - 3x + 1$
Remember to use a coefficient of 0 for the missing x^3 term in the synthetic division.

$$
\begin{array}{r|rrrrr}
-4) & 1 & 0 & 1 & -3 & 1 \\
    &   & -4 & 16 & -68 & 284 \\
\hline
    & 1 & -4 & 17 & -71 & 285
\end{array}
$$

The remainder is not 0, so -4 is not a zero of $P(x) = x^4 + x^2 - 3x + 1$. In fact, $P(-4) = 285$. •

9.1 EXERCISES

Use synthetic division to perform each of the following divisions. See Example 1.

1. $\dfrac{x^3 + 2x^2 - 17x - 10}{x + 5}$

2. $\dfrac{a^4 + 4a^3 + 2a^2 + 9a + 4}{a + 4}$

3. $\dfrac{m^4 - 3m^3 - 4m^2 + 12m}{m - 2}$

4. $\dfrac{p^4 - 3p^3 - 5p^2 + 2p - 16}{p + 2}$

5. $\dfrac{3x^3 - 11x^2 - 20x + 3}{x - 5}$

6. $\dfrac{4p^3 + 8p^2 - 16p - 9}{p + 3}$

7. $\dfrac{4m^3 - 3m - 2}{m + 1}$

8. $\dfrac{3q^3 - 4q + 2}{q - 1}$

9. $\dfrac{x^5 + 3x^4 + 2x^3 + 2x^2 + 3x + 1}{x + 2}$

10. $\dfrac{m^6 - 3m^4 + 2m^3 - 6m^2 - 5m + 3}{m + 2}$

11. $\dfrac{\frac{1}{3}x^3 - \frac{2}{9}x^2 + \frac{1}{27}x + 1}{x - \frac{1}{3}}$

12. $\dfrac{x^3 + x^2 + \frac{1}{2}x + \frac{1}{8}}{x + \frac{1}{2}}$

13. $\dfrac{9z^3 - 8z^2 + 7z - 2}{z - 2}$

14. $\dfrac{-11p^4 + 2p^3 - 8p^2 - 4}{p + 1}$

15. $\dfrac{y^3 - 1}{y - 1}$

16. $\dfrac{r^5 - 1}{r - 1}$

17. $\dfrac{x^4 - 1}{x - 1}$

18. $\dfrac{x^7 + 1}{x + 1}$

Express each polynomial in the form $P(x) = (x - k)Q(x) + r$ for the given value of k.

19. $P(x) = x^3 + x^2 + x - 8; \quad k = 1$

20. $P(x) = 2x^3 + 3x^2 + 4x - 10; \quad k = -1$

21. $P(x) = -x^3 + 2x^2 + 4; \quad k = -2$

22. $P(x) = -4x^3 + 2x^2 - 3x - 10; \quad k = 2$

23. $P(x) = x^4 - 3x^3 + 2x^2 - x + 5; \quad k = 3$

24. $P(x) = 2x^4 + x^3 - 15x^2 + 3x; \quad k = -3$

25. $P(x) = 3x^4 + 4x^3 - 10x^2 + 15; \quad k = -1$

26. $P(x) = -5x^4 + x^3 + 2x^2 + 3x + 1; \quad k = 1$

For each of the following polynomials, use the remainder theorem to find P(k). See Example 2.

27. $k = 5; \quad P(x) = -x^2 + 2x + 7$

28. $k = -3; \quad P(x) = 3x^2 + 8x + 5$

29. $k = 3; \quad P(x) = x^2 - 4x + 5$

30. $k = -2; \quad P(x) = x^2 + 5x + 6$

31. $k = -1; \quad P(x) = x^3 - 4x^2 + 2x + 1$

32. $k = 2; \quad P(x) = 2x^3 - 3x^2 - 5x + 4$

33. $k = 2 + i; \quad P(x) = x^2 - 5x + 1$

34. $k = 3 - 2i; \quad P(x) = x^2 - x + 3$

35. $k = 1 - i; \quad P(x) = x^3 + x^2 - x + 1$

36. $k = 2 - 3i; \quad P(x) = x^3 + 2x^2 + x - 5$

Use synthetic division to decide whether or not the given number is a zero of the given polynomial. See Example 3.

37. $2; \quad P(x) = x^2 + 2x - 8$

38. $-1; \quad P(m) = m^2 + 4m - 5$

39. $2; \quad P(g) = g^3 - 3g^2 + 4g - 4$

40. $-3; \quad P(m) = m^3 + 2m^2 - m + 6$

41. $4; \quad P(r) = 2r^3 - 6r^2 - 9r + 4$

42. $-6; \quad P(y) = 2y^3 + 9y^2 - 16y + 12$

43. $1 - i; \quad P(y) = y^2 - 2y + 2$

44. $3 - 2i; \quad P(r) = r^2 - 6r + 13$

45. $2 + i; \quad P(k) = k^2 + 3k + 4$

46. $1 - 2i; \quad P(z) = z^2 - 3z + 5$

47. $i; \quad P(x) = x^3 + 2ix^2 + 2x + i$

48. $-i; \quad P(p) = p^3 - ip^2 + 3p + 5i$

49. $1 + i; \quad P(p) = p^3 + 3p^2 - p + 1$

50. $2 - i; \quad P(r) = 2r^3 - r^2 + 3r - 5$

9.2 Factoring Polynomials

By the remainder theorem, if $P(k) = 0$, then the remainder when $P(x)$ is divided by $x - k$ is zero. This means that $x - k$ is a factor of $P(x)$. Conversely, if $x - k$ is a factor of $P(x)$, then $P(k)$ must equal 0. This is summarized in the following theorem.

Factor Theorem

> The polynomial $x - k$ is a factor of the polynomial $P(x)$ if and only if $P(k) = 0$.

EXAMPLE 1

Is $x - 1$ a factor of $P(x) = 2x^4 + 3x^2 - 5x + 7$?

By the factor theorem, $x - 1$ will be a factor of $P(x)$ only if $P(1) = 0$. Use synthetic division and the remainder theorem to decide.

$$
\begin{array}{r|rrrr}
1) & 2 & 0 & 3 & -5 & 7 \\
  &   & 2 & 2 & 5 & 0 \\
\hline
  & 2 & 2 & 5 & 0 & 7
\end{array}
$$

Since the remainder is 7, $P(1) = 7$, not 0, so $x - 1$ is not a factor of $P(x)$. ●

EXAMPLE 2

Is $x - i$ a factor of $P(x) = 3x^3 + (-4 - 3i)x^2 + (5 + 4i)x - 5i$?

The only way $x - i$ can be a factor of $P(x)$ is for $P(i)$ to be 0. To see if this is the case, use synthetic division.

$$
\begin{array}{r|rrrr}
i) & 3 & -4 - 3i & 5 + 4i & -5i \\
  &   & 3i & -4i & 5i \\
\hline
  & 3 & -4 & 5 & 0
\end{array}
$$

Since the remainder is 0, $P(i) = 0$, and $x - i$ is a factor of $P(x)$. The quotient $3x^2 - 4x + 5$ gives the other factor, so $P(x)$ can be factored as

$$P(x) = (x - i)(3x^2 - 4x + 5). \quad ●$$

The next theorem says that every polynomial of degree 1 or more has a zero, so every such polynomial can be factored. This theorem was first proved by the mathematician K. F. Gauss in his doctoral thesis in 1799 when he was 22 years old. Although many proofs of this result have been given, all of them involve mathematics beyond the algebra in this book, so no proof is included here.

Fundamental Theorem of Algebra

> Every polynomial of degree 1 or more has at least one complex zero.

From the fundamental theorem, if $P(x)$ is of degree 1 or more then there is some number k such that $P(k) = 0$. By the factor theorem, then

$$P(x) = (x - k) \cdot Q(x)$$

for some polynomial $Q(x)$. The fundamental theorem and the factor theorem can be used to factor $Q(x)$ in the same way. Assuming that $P(x)$ has degree n and repeating this process n times gives

$$P(x) = a(x - k_1)(x - k_2) \ldots (x - k_n),$$

where a is the leading coefficient of $P(x)$. Each of these factors leads to a zero of $P(x)$, so $P(x)$ has the n zeros $k_1, k_2, k_3, \ldots, k_n$.

This result is used to prove the next theorem. The proof is left for the exercises. See Exercise 57.

Theorem

> A polynomial of degree n has at most n distinct zeros.

The theorem says that there exist *at most* n distinct zeros. For example, the polynomial function $P(x) = x^3 + 3x^2 + 3x + 1 = (x + 1)^3$ is of degree 3 but has only one zero, -1. Actually, the zero -1 occurs three times, since there are three factors of $x + 1$, and is called a **zero of multiplicity 3.**

EXAMPLE 3   Find a polynomial $P(x)$ of degree 3 that satisfies the following conditions.

(a) Zeros of -1, 2, and 4; $P(1) = 3$

These three zeros give $x - (-1) = x + 1$, $x - 2$, and $x - 4$ as factors of $P(x)$. Since $P(x)$ is to be of degree 3, these are the only possible factors by the theorem just above. Therefore, $P(x)$ has the form

$$P(x) = a(x + 1)(x - 2)(x - 4)$$

for some real number a. To find a, use the fact that $P(1) = 3$.

$$P(1) = a(1 + 1)(1 - 2)(1 - 4) = 3$$
$$a(2)(-1)(-3) = 3$$
$$6a = 3$$
$$a = \frac{1}{2}$$

Thus, $$P(x) = \frac{1}{2}(x + 1)(x - 2)(x - 4),$$

or $$P(x) = \frac{1}{2}x^3 - \frac{5}{2}x^2 + x + 4.$$

(b) -2 is a zero of multiplicity 3; $P(-1) = 4$

The polynomial $P(x)$ has the form

$$P(x) = a(x + 2)(x + 2)(x + 2)$$
$$= a(x + 2)^3.$$

Since $P(-1) = 4$,

$$P(-1) = a(-1 + 2)^3 = 4,$$
$$a(1)^3 = 4$$

or
$$a = 4,$$

and $P(x) = 4(x + 2)^3 = 4x^3 + 24x^2 + 48x + 32.$ •

The remainder theorem can be used to show that both $2 + i$ and $2 - i$ are zeros of $P(x) = x^3 - x^2 - 7x + 15$. It is not just coincidence that both $2 + i$ and its conjugate $2 - i$ are zeros of this polynomial. If $a + bi$ is a zero of a polynomial function with *real* coefficients, then so is $a - bi$. This is given as the next theorem. The proof is left for the exercises. See Exercise 58.

**Conjugate
Zeros Theorem**

> If $P(x)$ is a polynomial having only real coefficients and if $a + bi$ is a zero of $P(x)$, then the conjugate $a - bi$ is also a zero of $P(x)$.

The requirement that the polynomial have only real coefficients is very important. For example, $P(x) = x - (1 + i)$ has $1 + i$ as a zero, but the conjugate $1 - i$ is not a zero.

EXAMPLE 4

Find a polynomial of lowest degree having real coefficients and zeros 3 and $2 + i$.

The complex number $2 - i$ also must be a zero, so the polynomial has at least three zeros, 3, $2 + i$, and $2 - i$. For the polynomial to be of lowest degree these must be the only zeros. By the factor theorem there must be three factors, $x - 3$, $x - (2 + i)$, and $x - (2 - i)$. A polynomial of lowest degree is

$$P(x) = (x - 3)[x - (2 + i)][x - (2 - i)]$$
$$= (x - 3)(x - 2 - i)(x - 2 + i)$$
$$= x^3 - 7x^2 + 17x - 15.$$

Other polynomials, such as $2(x^3 - 7x^2 + 17x - 15)$ or $\sqrt{5}(x^3 - 7x^2 + 17x - 15)$, for example, also satisfy the given conditions on zeros. The information on zeros given in the problem is not enough to give a specific value for the leading coefficient. •

The theorem on conjugate zeros is important in helping predict the number of real zeros of polynomials with real coefficients. A polynomial with real coefficients of odd degree n, where $n \geq 1$, must have at least one real zero (since zeros of the form $a + bi$, where $b \neq 0$, occur in conjugate pairs.) On the other hand, a polynomial with real coefficients of even degree n need have no real zeros but may have up to n real zeros.

EXAMPLE 5   Find all zeros of $P(x) = x^4 - 7x^3 + 18x^2 - 22x + 12$, given that $1 - i$ is a zero.

Since $1 - i$ is a zero, by the conjugate zeros theorem $1 + i$ is also a zero. To find the remaining zeros, first divide the original polynomial by $1 - i$ and then divide the quotient by $1 + i$, as follows.

$$
\begin{array}{r|rrrrr}
1 - i) & 1 & -7 & 18 & -22 & 12 \\
       &   & 1 - i & -7 + 5i & 16 - 6i & -12 \\
\hline
       & 1 & -6 - i & 11 + 5i & -6 - 6i & 0 \\
\end{array}
$$

$$
\begin{array}{r|rrrr}
1 + i) & 1 & -6 - i & 11 + 5i & -6 - 6i \\
       &   & 1 + i & -5 - 5i & 6 + 6i \\
\hline
       & 1 & -5 & 6 & 0 \\
\end{array}
$$

Now find the zeros of the quadratic polynomial $x^2 - 5x + 6$. Factoring the polynomial shows that the zeros are 2 and 3, so the four zeros of $P(x)$ are $1 - i$, $1 + i$, 2, and 3.   •

EXAMPLE 6   Factor $P(x)$ into linear factors (factors of the form $ax - b$), given that k is a zero of $P(x)$.

(a) $P(x) = 6x^3 + 19x^2 + 2x - 3; k = -3$

Since $k = -3$ is a zero of $P(x)$, $x - (-3) = x + 3$ is a factor. Use synthetic division to divide $P(x)$ by $x + 3$.

$$
\begin{array}{r|rrrr}
-3) & 6 & 19 & 2 & -3 \\
    &   & -18 & -3 & 3 \\
\hline
    & 6 & 1 & -1 & 0 \\
\end{array}
$$

The quotient is $6x^2 + x - 1$, so

$$P(x) = (x + 3)(6x^2 + x - 1).$$

Factor $6x^2 + x - 1$ as $(2x + 1)(3x - 1)$ to get

$$P(x) = (x + 3)(2x + 1)(3x - 1),$$

where all factors are linear.

(b) $P(x) = 3x^3 + (-1 + 3i)x^2 + (-12 + 5i)x + 4 - 2i; k = 2 - i$

One factor is $x - (2 - i)$ or $x - 2 + i$. Divide $P(x)$ by $x - (2 - i)$.

$$
\begin{array}{r|rrrr}
2 - i) & 3 & -1 + 3i & -12 + 5i & 4 - 2i \\
       &   & 6 - 3i & 10 - 5i & -4 + 2i \\
\hline
       & 3 & 5 & -2 & 0 \\
\end{array}
$$

By the division algorithm,

$$P(x) = (x - 2 + i)(3x^2 + 5x - 2).$$

Use the quadratic formula to factor $3x^2 + 5x - 2$; $a = 3$, $b = 5$, and $c = -2$.

$$x = \frac{-b \pm \sqrt{b^2 - 4ac}}{2a}$$

$$= \frac{-5 \pm \sqrt{5^2 - 4(3)(-2)}}{2(3)} = \frac{-5 \pm \sqrt{49}}{6}$$

From this result,

$$x = \frac{-5 + 7}{6} = \frac{1}{3} \quad \text{or} \quad x = \frac{-5 - 7}{6} = -2.$$

These two solutions lead to the factors $(x - 1/3)$ and $(x + 2)$, so a linear factored form of $P(x)$ is

$$P(x) = a(x - 2 + i)\left(x - \frac{1}{3}\right)(x - 2).$$

Choose $a = 3$ since $P(x)$ has a leading coefficient of 3.

$$P(x) = 3(x - 2 + i)\left(x - \frac{1}{3}\right)(x - 2)$$

Multiply the 3 and the factor $x - 1/3$ to get

$$P(x) = (x - 2 + i)(3x - 1)(x - 2). \quad \bullet$$

9.2 EXERCISES

Use the factor theorem to decide whether or not the second polynomial is a factor of the first. See Example 1.

1. $4x^2 + 2x + 42$; $x - 3$
2. $-3x^2 - 4x + 2$; $x + 2$
3. $x^3 + 2x^2 - 3$; $x - 1$
4. $2x^3 + x + 2$; $x + 1$
5. $3x^3 - 12x^2 - 11x - 20$; $x - 5$
6. $4x^3 + 6x^2 - 5x - 2$; $x + 2$
7. $2x^4 + 5x^3 - 2x^2 + 5x + 3$; $x + 3$
8. $5x^4 + 16x^3 - 15x^2 + 8x + 16$; $x + 4$

For each of the following, find a polynomial of degree 3 with real coefficients which satisfies the given conditions. See Example 3.

9. Zeros of -3, -1, and 4; $P(2) = 5$
10. Zeros of 1, -1, and 0; $P(2) = -3$
11. Zeros of -2, 1, and 0; $P(-1) = -1$
12. Zeros of 2, 5, and -3; $P(1) = -4$
13. Zeros of 3, i, and $-i$; $P(2) = 50$
14. Zeros of -2, i, and $-i$; $P(-3) = 30$

For each of the following, find a polynomial of lowest degree with real coefficients having the given zeros. See Example 4.

15. $5 + i$ and $5 - i$
16. $3 - 2i$ and $3 + 2i$
17. 2, $1 - i$ and $1 + i$
18. -3, $2 - i$, and $2 + i$
19. $1 + \sqrt{2}$, $1 - \sqrt{2}$, and 1
20. $1 - \sqrt{3}$, $1 + \sqrt{3}$, and -2
21. $2 + i$, $2 - i$, 3, and -1
22. $3 + 2i$, $3 - 2i$, -1, and 2

23. 2 and $3 + i$

24. -1 and $4 - 2i$

25. $1 - \sqrt{2}, 1 + \sqrt{2}$, and $1 - i$

26. $2 + \sqrt{3}, 2 - \sqrt{3}$, and $2 + 3i$

27. $2 - i, 3 + 2i$

28. $5 + i, 4 - i$

29. $4, 1 - 2i, 3 + 4i$

30. $-1, 1 + \sqrt{2}, 1 - \sqrt{2}, 1 + 4i$

31. $1 + 2i$, 2 (multiplicity 2)

32. $2 + i, -3$ (multiplicity 2)

For each of the following polynomials, one zero is given. Find all others. See Example 5.

33. $P(x) = x^3 - x^2 - 4x - 6$; 3

34. $P(x) = x^3 - 5x^2 + 17x - 13$; 1

35. $P(x) = x^3 + x^2 - 4x - 24$; $-2 + 2i$

36. $P(x) = x^3 + x^2 - 20x - 50$; $-3 + i$

37. $P(x) = 2x^3 - 2x^2 - x - 6$; 2

38. $P(x) = 2x^3 - 5x^2 + 6x - 2$; $1 + i$

39. $P(x) = x^4 + 5x^2 + 4$; $-i$

40. $P(x) = x^4 + 10x^3 + 27x^2 + 10x + 26$; i

41. $P(x) = x^4 - 3x^3 + 6x^2 + 2x - 60$; $1 + 3i$

42. $P(x) = x^4 - 6x^3 - x^2 + 86x + 170$; $5 + 3i$

43. $P(x) = x^4 - 6x^3 + 15x^2 - 18x + 10$; $2 + i$

44. $P(x) = x^4 + 8x^3 + 26x^2 + 72x + 153$; $-3i$

Factor $P(x)$ into linear factors given that k is a zero of $P(x)$.

45. $P(x) = 2x^3 - 3x^2 - 17x + 30$; $k = 2$

46. $P(x) = 2x^3 - 3x^2 - 5x + 6$; $k = 1$

47. $P(x) = 6x^3 + 25x^2 + 3x - 4$; $k = -4$

48. $P(x) = 8x^3 + 50x^2 + 47x - 15$; $k = -5$

49. $P(x) = x^3 + (7 - 3i)x^2 + (12 - 21i)x - 36i$; $k = 3i$

50. $P(x) = 2x^3 + (3 + 2i)x^2 + (1 + 3i)x + i$; $k = -i$

51. $P(x) = 2x^3 + (3 - 2i)x^2 + (-8 - 5i)x + 3 + 3i$; $k = 1 + i$

52. $P(x) = 6x^3 + (19 - 6i)x^2 + (16 - 7i)x + 4 - 2i$; $k = -2 + i$

53. Show that -2 is a zero of multiplicity 2 of $P(x) = x^4 + 2x^3 - 7x^2 - 20x - 12$ and find all other complex zeros. Then write $P(x)$ in factored form.

54. Show that -1 is a zero of multiplicity 3 of $P(x) = x^5 + 9x^4 + 33x^3 + 55x^2 + 42x + 12$, and find all other complex zeros. Then write $P(x)$ in factored form.

55. The displacement at time t of a particle moving along a straight line is given by

$$s(t) = t^3 - 2t^2 - 5t + 6,$$

where t is in seconds and s is measured in centimeters. The displacement is 0 after 1 sec has elapsed. At what other times is the displacement 0?

56. For headphone radios, the cost function (in thousands of dollars) is given by

$$C(x) = 2x^3 - 9x^2 + 17x - 4,$$

and the revenue function (in thousands of dollars) is given by $R(x) = 5x$, where x is the number of items (in hundred thousands) produced. Cost equals revenue if 200,000 items are produced ($x = 2$). Find all other break-even points.

Prove each of the following statements.

57. A polynomial of degree n has at most n distinct zeros. (*Hint:* use an indirect proof, where you assume the statement is true and show that it leads to a contradiction.)

58. If $a + bi$ is a zero of a polynomial $P(x)$ with all real coefficients, then $a - bi$ is also a zero of $P(x)$.

9.3 Rational Zeros of Polynomial Functions

We know by the fundamental theorem of algebra that every polynomial of degree 1 or more has a zero. However, the fundamental theorem merely says that such a zero exists. It gives no help at all in identifying zeros. Other theorems can be used to find any rational zeros of polynomials with rational coefficients or to find decimal approximations of any irrational zeros.

The next theorem gives a useful method for finding a set of possible zeros of a polynomial with integer coefficients.

Rational Zeros Theorem

> Let $P(x) = a_n x^n + a_{n-1} x^{n-1} + \ldots + a_1 x + a_0$, $a_n \neq 0$, be a polynomial with integer coefficients. If p/q is a rational number written in lowest terms and if p/q is a zero of $P(x)$, then p is a factor of the constant term a_0 and q is a factor of the leading coefficient a_n.

To prove this theorem, $P(p/q) = 0$ since p/q is a zero of $P(x)$, so

$$a_n(p/q)^n + a_{n-1}(p/q)^{n-1} + \ldots + a_1(p/q) + a_0 = 0.$$

This can also be written as

$$a_n(p^n/q^n) + a_{n-1}(p^{n-1}/q^{n-1}) + \ldots + a_1(p/q) + a_0 = 0$$

Multiply both sides of this last result by q^n and add $-a_0 q^n$ to both sides.

$$a_n p^n + a_{n-1} p^{n-1} q + \ldots + a_1 p q^{n-1} = -a_0 q^n$$

Factoring out p gives

$$p(a_n p^{n-1} + a_{n-1} p^{n-2} q + \ldots + a_1 q^{n-1}) = -a_0 q^n.$$

This result shows that $-a_0 q^n$ equals the product of the two factors, p and $(a_n p^{n-1} + \ldots + a_1 q^{n-1})$. For this reason, p must be a factor of $-a_0 q^n$. Since we have assumed that p/q is written in lowest terms, p and q have no common factor other than 1, so p is not a factor of q^n. Thus p must be a factor of a_0. In a similar way we can show that q is a factor of a_n.

EXAMPLE 1   Find all rational zeros of $P(x) = 2x^4 - 11x^3 + 14x^2 - 11x + 12$.

If p/q is to be a rational zero of $p(x)$, by the rational zeros theorem p must be a factor of $a_0 = 12$ and q must be a factor of $a_4 = 2$. The possible values of p are ± 1, ± 2, ± 3, ± 4, ± 6, or ± 12, while q must be ± 1 or ± 2. The possible rational zeros are found by forming all possible quotients of the form p/q; any rational zero of $P(x)$ will come from the list

$$\pm 1, \ \pm 1/2, \ \pm 2, \ \pm 3, \ \pm 3/2, \ \pm 4, \ \pm 6, \ \text{or} \ \pm 12.$$

Though we are not sure that any of these numbers are zeros, if $P(x)$ has any rational zeros, they will be in the list above. These proposed zeros can be checked by synthetic division. Doing so, we find that 4 is a zero.

$$4 \overline{)\, 2 \quad -11 \quad 14 \quad -11 \quad 12}$$
$$\underline{\qquad 8 \quad -12 \quad 8 \quad -12}$$
$$2 \quad -3 \quad 2 \quad -3 \quad 0$$

As a fringe benefit of this calculation, we can now look for zeros of the simpler polynomial $Q(x) = 2x^3 - 3x^2 + 2x - 3$. Any rational zero of $Q(x)$ will have a numerator of ± 3 or ± 1, with a denominator of ± 1 or ± 2. Thus, any rational zeros of $Q(x)$ will come from the list

$$\pm 3, \; \pm 3/2, \; \pm 1, \; \pm 1/2.$$

Again use synthetic division and trial and error to find that 3/2 is a zero.

$$3/2 \overline{)\, 2 \quad -3 \quad 2 \quad -3}$$
$$\underline{\qquad\qquad 3 \quad 0 \quad 3}$$
$$2 \quad 0 \quad 2 \quad 0$$

The quotient is $2x^2 + 2$, which, by the quadratic formula, has i and $-i$ as zeros. They are, however, imaginary zeros. The rational zeros of the polynomial function

$$P(x) = 2x^4 - 11x^3 + 14x^2 - 11x + 12$$

are 4 and 3/2. ●

EXAMPLE 2 Find all rational zeros of $P(x) = 6x^4 + 7x^3 - 12x^2 - 3x + 2$, and factor the polynomial.

For a rational number p/q to be a zero of $P(x)$, p must be a factor of $a_0 = 2$ and q must be a factor of $a_4 = 6$. Thus, p can be ± 1 or ± 2 and q can be ± 1, ± 2, ± 3, or ± 6. The rational zeros, p/q, must be from the following list.

$$\pm 1, \; \pm 2, \; \pm 1/2, \; \pm 1/3, \; \pm 1/6, \; \pm 2/3$$

Check 1 first because it is easy.

$$1 \overline{)\, 6 \quad 7 \quad -12 \quad -3 \quad 2}$$
$$\underline{\qquad 6 \quad 13 \quad 1 \quad -2}$$
$$6 \quad 13 \quad 1 \quad -2 \quad 0$$

The 0 remainder shows that 1 is a zero. Now use the quotient polynomial $6x^3 + 13x^2 + x - 2$ and synthetic division to find that -2 is also a zero.

$$-2 \overline{)\, 6 \quad 13 \quad 1 \quad -2}$$
$$\underline{\qquad\; -12 \quad -2 \quad 2}$$
$$6 \quad 1 \quad -1 \quad 0$$

The new quotient polynomial is $6x^2 + x - 1$. Use the quadratic formula or factor to solve the equation $6x^2 + x - 1 = 0$. The remaining two zeros are 1/3 and $-1/2$.

Factor the polynomial $P(x)$ in the following way. Since the four zeros of $P(x) = 6x^4 + 7x^3 - 12x^2 - 3x + 2$ are 1, -2, 1/3, and $-1/2$, the factors are $x - 1, x + 2, x - 1/3,$ and $x + 1/2,$ and $P(x) = a(x - 1)(x + 2)(x - 1/3)$ $(x + 1/2)$. We know $a = 6,$ so

$$
\begin{aligned}
P(x) &= 6(x - 1)(x + 2)(x - 1/3)(x + 1/2) \\
&= (x - 1)(x + 2)(3)(x - 1/3)(2)(x + 1/2) \\
&= (x - 1)(x + 2)(3x - 1)(2x + 1). \quad \bullet
\end{aligned}
$$

To find any rational zeros of a polynomial with fractional coefficients, first multiply the polynomial by a number that will clear it of all fractional coefficients. Then use the rational zeros theorem.

EXAMPLE 3   Find all rational zeros of

$$
P(x) = x^4 - \frac{1}{6}x^3 + \frac{2}{3}x^2 - \frac{1}{6}x - \frac{1}{3}.
$$

We want to find the values of x that make $P(x) = 0,$ or

$$
x^4 - \frac{1}{6}x^3 + \frac{2}{3}x^2 - \frac{1}{6}x - \frac{1}{3} = 0.
$$

Multiply both sides by 6 to eliminate all fractions. This gives

$$
6x^4 - x^3 + 4x^2 - x - 2 = 0.
$$

This polynomial will have the same zeros as $P(x)$. The possible rational zeros are of the form p/q where p is $\pm 1,$ or $\pm 2,$ and q is $\pm 1, \pm 2, \pm 3,$ or ± 6. Then p/q may be

$$
\pm 1, \pm 2, \pm 1/2, \pm 1/3, \pm 1/6, \text{ or } \pm 2/3.
$$

Use synthetic division to find that $-1/2$ and $2/3$ are zeros.

$$
\begin{array}{r|rrrrr}
-\frac{1}{2}) & 6 & -1 & 4 & -1 & -2 \\
 & & -3 & 2 & -3 & 2 \\
\hline
 & 6 & -4 & 6 & -4 & 0
\end{array}
$$

$$
\begin{array}{r|rrrr}
\frac{2}{3}) & 6 & -4 & 6 & -4 \\
 & & 4 & 0 & 4 \\
\hline
 & 6 & 0 & 6 & 0
\end{array}
$$

The final quotient is $Q(x) = 6x^2 + 6 = 6(x^2 + 1)$. This polynomial can be solved by the quadratic formula; its zeros are i and $-i$. Since these zeros are imaginary numbers, there are just two rational zeros: $-1/2$ and $2/3$. $\bullet$

9.3 EXERCISES

Give all possible rational zeros for the following polynomials.

1. $P(x) = 6x^3 + 17x^2 - 31x - 1$

2. $P(x) = 15x^3 + 61x^2 + 2x - 1$

3. $P(x) = 12x^3 + 20x^2 - x - 2$

4. $P(x) = 12x^3 + 40x^2 + 41x + 3$

5. $P(x) = 2x^3 + 7x^2 + 12x - 8$

6. $P(x) = 2x^3 + 20x^2 + 68x - 40$

7. $P(x) = x^4 + 4x^3 + 3x^2 - 10x + 50$

8. $P(x) = x^4 - 2x^3 + x^2 + 18$

Find all rational zeros of the following polynomials. See Examples 1 and 2.

9. $P(x) = x^3 - 2x^2 - 13x - 10$

10. $P(x) = x^3 + 5x^2 + 2x - 8$

11. $P(x) = x^3 + 6x^2 - x - 30$

12. $P(x) = x^3 - x^2 - 10x - 8$

13. $P(x) = x^3 + 9x^2 - 14x - 24$

14. $P(x) = x^3 + 3x^2 - 4x - 12$

15. $P(x) = x^4 + 9x^3 + 21x^2 - x - 30$

16. $P(x) = x^4 + 4x^3 - 7x^2 - 34x - 24$

Find the rational zeros of the following polynomials; then write each polynomial in factored form. See Example 2.

17. $P(x) = 6x^3 + 17x^2 - 31x - 12$

18. $P(x) = 15x^3 + 61x^2 + 2x - 8$

19. $P(x) = 12x^3 + 20x^2 - x - 6$

20. $P(x) = 12x^3 + 40x^2 + 41x + 12$

21. $P(x) = 2x^3 + 7x^2 + 12x - 8$

22. $P(x) = 2x^3 + 20x^2 + 68x - 40$

23. $P(x) = x^4 + 4x^3 + 3x^2 - 10x + 50$

24. $P(x) = x^4 - 2x^3 + x^2 + 18$

25. $P(x) = x^4 + 2x^3 - 13x^2 - 38x - 24$

26. $P(x) = 6x^4 + x^3 - 7x^2 - x + 1$

27. $P(x) = 3x^4 + 4x^3 - x^2 + 4x - 4$

28. $P(x) = x^4 + 8x^3 + 16x^2 - 8x - 17$

29. $P(x) = x^5 + 3x^4 - 5x^3 - 11x^2 + 12$

30. $P(x) = 4x^5 + 4x^4 - 37x^3 - 37x^2 + 9x + 9$

Find all rational zeros of the following polynomials. See Example 3.

31. $P(x) = x^3 - \frac{4}{3}x^2 - \frac{13}{3}x - 2$

32. $P(x) = x^3 + x^2 - \frac{16}{9}x + \frac{4}{9}$

33. $P(x) = x^4 + \frac{1}{4}x^3 + \frac{11}{4}x^2 + x - 5$

34. $P(x) = \frac{10}{7}x^4 - x^3 - 7x^2 + 5x - \frac{5}{7}$

35. $P(x) = \frac{1}{3}x^5 + x^4 - \frac{5}{3}x^3 - \frac{11}{3}x^2 + 4$

36. $P(x) = x^5 + x^4 - \frac{37}{4}x^2 + \frac{9}{4}x + \frac{9}{4}$

37. Show that $P(x) = x^2 - 2$ has no rational zeros, so $\sqrt{2}$ must be irrational.

38. Show that $P(x) = x^2 - 5$ has no rational zeros, so $\sqrt{5}$ must be irrational.

39. Show that $P(x) = x^4 + 5x^2 + 4$ has no rational zeros.

40. Show that $P(x) = x^5 - 3x^3 + 5$ has no rational zeros.

41. Show that any integer zeros of a polynomial function must be factors of the constant term a_0.

9.4 Approximate Zeros of Polynomials

Every polynomial of degree 1 or more has a zero. However, we do not know whether or not a polynomial has real zeros. Even if it does have real zeros, we often have no way to find them. In this section we discuss methods of approximating any real zeros a polynomial function may have. These methods work well using a computer or a calculator.

Much of our work in locating real zeros uses the following result, which is related to the fact that graphs of polynomial functions are unbroken curves, with no gaps or sudden jumps. The proof requires advanced methods, so it is not given here.

**Intermediate
Value Theorem
for Polynomials**

If $P(x)$ is a polynomial function with only real coefficients, and if for real numbers a and b, $P(a)$ and $P(b)$ are opposite in sign, then there exists at least one real zero between a and b.

As Figure 9.1 suggests, if $P(a)$ and $P(b)$ are opposite in sign, then there must be some number c, between a and b, such that $P(c) = 0$. This helps to identify the intervals where the zeros of the polynomial are located.

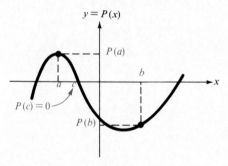

Figure 9.1

EXAMPLE 1   Does $P(x) = x^3 - 2x^2 - x + 1$ have any real zeros between 2 and 3?
Use synthetic division to find $P(2)$ and $P(3)$.

$$\begin{array}{r} 2\overline{)1 \quad -2 \quad -1 \quad 1} \\ \underline{ \quad 2 \quad 0 \quad -2} \\ 1 \quad 0 \quad -1 \quad -1 \end{array} \qquad \begin{array}{r} 3\overline{)1 \quad -2 \quad -1 \quad 1} \\ \underline{ \quad 3 \quad 3 \quad 6} \\ 1 \quad 1 \quad 2 \quad 7 \end{array}$$

Since $P(2)$ is negative but $P(3)$ is positive, there must be a real zero between 2 and 3.   •

The intermediate value theorem for polynomials is helpful in limiting the search for real zeros to a smaller and smaller interval. In Example 1 we used the theorem to find that there is a real zero between 2 and 3. The theorem could then be used repeatedly to express the zero more accurately.

The next theorem gives another method for narrowing the search for real zeros. It depends on the fact that in any polynomial function, as $|x|$ gets larger and larger, so does $|y|$. This is illustrated by the typical polynomial functions shown in Figure 9.2. The signal that this has occurred is found in a row of the synthetic division, as stated in the next theorem.

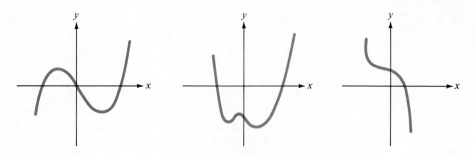

Figure 9.2

Boundedness Theorem

Let $P(x)$ be a polynomial with real coefficients and with a positive leading coefficient. If $P(x)$ is divided synthetically by $x - c$ and

(a) if $c > 0$ and all numbers in the bottom row of the synthetic division are nonnegative, then $P(x)$ has no zero greater than c;

(b) if $c < 0$ and the numbers in the bottom row of the synthetic division alternate in sign (with 0 considered positive or negative, as needed), then $P(x)$ has no zero less than c.

EXAMPLE 2

Approximate the real zeros of $P(x) = x^4 - 6x^3 + 8x^2 + 2x - 1$.

First, check for rational zeros. The only possible rational zeros are ± 1.

$$
\begin{array}{r|rrrr}
1) & 1 & -6 & 8 & 2 & -1 \\
  &   & 1 & -5 & 3 & 5 \\
\hline
  & 1 & -5 & 3 & 5 & 4 \\
\end{array}
$$

$$
\begin{array}{r|rrrr}
-1) & 1 & -6 & 8 & 2 & -1 \\
   &   & -1 & 7 & -15 & 13 \\
\hline
   & 1 & -7 & 15 & -13 & 12 \\
\end{array}
$$

Neither 1 nor -1 is a zero, so the polynomial has no rational zeros. We must now search in some consistent way for the location of irrational real zeros using the two theorems in this section.

The leading coefficient of $P(x)$ is positive and the numbers in the last row of the second synthetic division above alternate in sign. Since $-1 < 0$, by the boundedness theorem -1 is less than any zero of $P(x)$. [We already know that -1 is not equal to a zero of $P(x)$.] Also, $P(-1) = 12 > 0$. By substitution, or synthetic division, $P(0) = -1 < 0$. Thus there is at least one real zero between -1 and 0.

Let us try $c = -.5$. Divide $P(x)$ by $x + .5$.

$$
\begin{array}{r|rrrr}
-.5) & 1 & -6 & 8 & 2 & -1 \\
    &   & -.5 & 3.25 & -5.625 & 1.8125 \\
\hline
    & 1 & -6.5 & 11.25 & -3.625 & .8125 \\
\end{array}
$$

Since $P(-.5) > 0$ and $P(0) < 0$, there is a real zero between $-.5$ and 0.

Now try $c = -.4$.

$$
\begin{array}{r|rrrr}
-.4) & 1 & -6 & 8 & 2 & -1 \\
     &   & -.4 & 2.56 & -4.224 & .8896 \\
\hline
     & 1 & -6.4 & 10.56 & -2.224 & -.1104
\end{array}
$$

Since $P(-.5)$ is positive, but $P(-.4)$ is negative, there is a zero between $-.5$ and $-.4$. The value of $P(-.4)$ is closer to zero than $P(-.5)$, so it is probably safe to say that, to one decimal place of accuracy, $-.4$ is a real zero of $P(x)$. A more accurate result can be found, if desired, by continuing this process.

To find the remaining real zeros of $P(x)$, continue in the same way. Use synthetic division to find $P(1)$, $P(2)$, $P(3)$, and so on, until you observe a change in sign. It is helpful to use the shortened form of synthetic division shown below. Only the last row of the synthetic division is shown for each line. The first row of the chart is used for each division and the work in the second row of the division is done mentally.

x					$P(x)$
	1	-6	8	2	-1
-1	1	-7	15	-13	12
0	1	-6	8	2	-1
1	1	-5	3	5	4
2	1	-4	0	2	3
3	1	-3	-1	-1	-4
4	1	-2	0	2	7

←—— Zero between -1 and 0 (row -1 / 0)
←—— Zero between 0 and 1
←—— Zero between 2 and 3
←—— Zero between 3 and 4

Since the polynomial is degree 4, there are no more than 4 zeros. Expand the table to find the real zeros to the nearest tenth. For example, for the zero between 0 and 1, we might work as follows. Start halfway between 0 and 1 with $x = .5$. Since $P(.5) > 0$ and $P(0) < 0$, try $x = .4$ next.

x				$P(x)$
	1	-6	8	2	-1
.5	1	-5.5	5.25	4.63	1.31
.4	1	-5.6	5.76	4.30	.72
.3	1	-5.7	6.29	3.89	.17
.2	1	-5.8	6.84	3.37	-.33

←—— Zero between .3 and .2

The value $P(.3) = .17$ is closer to 0 than $P(.2) = -.33$, so to the nearest tenth, the zero is .3. Use synthetic division to verify that the remaining two zeros are approximately 2.4 and 3.7. ●

Descartes' rule of signs, stated below, gives a useful, practical test for finding the number of positive or negative real zeros of a given polynomial. The terms of the polynomial are assumed to be in descending order.

Descartes' Rule of Signs

Let $P(x)$ be a polynomial with real coefficients.

(a) The number of positive real zeros of $P(x)$ is either equal to the number of variations in sign occurring in the coefficients of $P(x)$, or else is less than the number of variations by a positive even integer.

(b) The number of negative real zeros of $P(x)$ either equals the number of variations in sign of $P(-x)$, or else is less than the number of variations by a positive even integer.

In the theorem, the number of variations in sign of the coefficients of $P(x)$ or $P(-x)$ refers to changes from positive to negative in successive terms of the polynomial. Missing terms (those with 0 coefficients) are counted as no change in sign and can be ignored.

For the purposes of this theorem, zeros of multiplicity k count as k zeros. For example,

$$P(x) = (x - 1)^4 = \underset{1}{+\ x^4} \underset{2}{-4x^3} \underset{3}{+\ 6x^2} \underset{4}{-\ 4x} + 1$$

has 4 changes of sign. By Descartes' rule of signs, $P(x)$ has 4 positive real zeros. In this case, each of the 4 positive real zeros is 1.

The polynomial in Example 2,

$$P(x) = x^4 - 6x^3 + 8x^2 + 2x - 1,$$

has 3 variations in sign:

$$\underset{1}{+\ x^4} \underset{2}{-\ 6x^3} + 8x^2 \underset{3}{+\ 2x - 1}$$

Thus, by Descartes' rule of signs, $P(x)$ has either 3 or $3 - 2 = 1$ positive real zeros. We found in Example 2 that $P(x)$ has 3 positive real zeros. Since $P(x)$ is of degree 4 and has 3 positive real zeros, it must have 1 negative real zero, which corresponds to the result above. This could be verified with part (b) of Descartes' rule of signs. Since

$$P(-x) = (-x)^4 - 6(-x)^3 + 8(-x)^2 + 2(-x) - 1$$
$$= x^4 + 6x^3 + 8x^2 - 2x - 1$$

has only one variation in sign, $P(x)$ has only one negative real zero, which again corresponds to the result above.

EXAMPLE 3   Find the number of positive and negative real zeros of

$$Q(x) = x^5 + 5x^4 + 3x^2 + 2x + 1.$$

The polynomial $Q(x)$ has no variations in sign and so has no positive real zeros. Here

$$Q(-x) = -x^5 + 5x^4 + 3x^2 - 2x + 1,$$

with three variations in sign, so $Q(x)$ has either 3 or 1 negative real zeros. The other zeros are complex numbers.   ●

If a polynomial of degree 3 or more cannot be factored, the only way it can be graphed is to find many points to plot. The theorems studied in this chapter are helpful in deciding which points to plot.

EXAMPLE 4   Graph the function $P(x) = 8x^3 - 12x^2 + 2x + 1$.

By Descartes' rule of signs, $P(x)$ has 2 or 0 positive real zeros and 1 negative real zero. To find some ordered pairs belonging to the graph, use synthetic division to evaluate, say $P(3)$, in hopes of finding a number greater than all zeros of $P(x)$.

$$
\begin{array}{r|rrrr}
3 & 8 & -12 & 2 & 1 \\
  &   & 24 & 36 & 114 \\
\hline
  & 8 & 12 & 38 & 115
\end{array}
$$

Since $P(3) = 115$, the point $(3, 115)$ belongs to the graph. Also, since the bottom row is all positive, there are no zeros greater than 3. Now find $P(-1)$.

$$
\begin{array}{r|rrrr}
-1 & 8 & -12 & 2 & 1 \\
   &   & -8 & 20 & -22 \\
\hline
   & 8 & -20 & 22 & -21
\end{array}
$$

From this result, the point $(-1, -21)$ belongs to the graph. Since the signs in the last row alternate, -1 is less than any zero of $P(x)$. Find several points between -1 and 3 in order to sketch the graph using the shortened form of synthetic division shown in the chart below.

x				$P(x)$	Ordered pair	
	8	-12	2	1		
3	8	12	38	115	$(3, 115)$	
2	8	4	10	21	$(2, 21)$	← Zero
1	8	-4	-2	-1	$(1, -1)$	← Zero
0	8	-12	2	1	$(0, 1)$	← Zero
-1	8	-20	22	-21	$(-1, -21)$	

We have located all three real zeros. Since the numbers in the row for $x = 2$ are all positive and $2 > 0$, 2 is greater than or equal to any zero of $P(x)$ and the top row of the chart is not needed.

By the intermediate value theorem, there is a zero between 0 and 1 and between -1 and 0, as well as between 1 and 2. To get the graph, plot the points from the chart and then draw a continuous curve through them, as in Figure 9.3 below. •

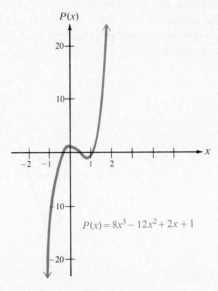

$P(x)$

$P(x) = 8x^3 - 12x^2 + 2x + 1$

Figure 9.3

EXAMPLE 5

Graph $P(x) = 3x^4 - 14x^3 + 54x - 3$.

Use Descartes' rule of signs to see that there are 3 positive real zeros or 1 positive real zero and 1 negative real zero. To find points to plot, use synthetic division to make a table like the one shown below. Start with $x = 0$ and work up through the positive integers until a row with all positive numbers is found. Then work down through the negative integers until a row with alternating signs is found.

x					$P(x)$	Ordered Pair
	3	-14	0	54	-3	
5	3	1	5	79	392	$(5, 392)$
4	3	-2	-8	22	85	$(4, 85)$
3	3	-5	-15	9	24	$(3, 24)$
2	3	-8	-16	22	41	$(2, 41)$
1	3	-11	-11	43	40	$(1, 40)$
0	3	-14	0	54	-3	$(0, -3)$
-1	3	-17	17	37	-40	$(-1, -40)$
-2	3	-20	40	-26	49	$(-2, 49)$

Since the row in the chart for $x = 5$ contains all positive numbers, the polynomial has no zero greater than 5. Also, since the row for $x = -2$ has numbers which alternate in sign, there is no zero less than -2. By the changes in sign of $P(x)$, the polynomial has zeros between 0 and 1 and between -2 and -1. Plotting the points found above and drawing a continuous curve through them gives the graph in Figure 9.4. ●

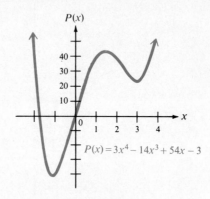

Figure 9.4

9.4 EXERCISES

Use the intermediate value theorem for polynomials to show that the following polynomials have a real zero between the numbers given. See Example 1.

1. $P(x) = 3x^2 - 2x - 6$; 1 and 2

2. $P(x) = x^3 + x^2 - 5x - 5$; 2 and 3

3. $P(x) = 2x^3 - 8x^2 + x + 16$; 2 and 2.5

4. $P(x) = 3x^3 + 7x^2 - 4$; 1/2 and 1

5. $P(x) = 2x^4 - 4x^2 + 3x - 6$; 2 and 1.5

6. $P(x) = x^4 - 4x^3 - x + 1$; 1 and .3

7. $P(x) = -3x^4 - x^3 + 2x^2 + 4$; 1 and 1.2

8. $P(x) = x^5 + 2x^4 + x^3 + 3$; -1.3 and -2

Show that the real zeros of each of the following polynomial functions satisfy the given conditions.

9. $P(x) = x^4 - x^3 + 3x^2 - 8x + 8$; no real zero greater than 2

10. $P(x) = 2x^5 - x^4 + 2x^3 - 2x^2 + 4x - 4$; no real zero greater than 1

11. $P(x) = x^4 + x^3 - x^2 + 3$; no real zero less than -2

12. $P(x) = x^5 + 2x^3 - 2x^2 + 5x + 5$; no real zero less than -1

13. $P(x) = 3x^4 + 2x^3 - 4x^2 + x - 1$; no real zero greater than 1

14. $P(x) = 3x^4 + 2x^3 - 4x^2 + x - 1$; no real zero less than -2

15. $P(x) = x^5 - 3x^3 + x + 2$; no real zero greater than 2

16. $P(x) = x^5 - 3x^3 + x + 2$; no real zero less than -3

For each of the following polynomials, (a) Find the number of positive and negative real zeros. See Example 3. (b) Approximate each zero as a decimal to the nearest tenth. See Example 2.

17. $P(x) = x^3 + 3x^2 - 2x - 6$

18. $P(x) = x^3 + x^2 - 5x - 5$

19. $P(x) = x^3 - 4x^2 - 5x + 14$

20. $P(x) = x^3 + 9x^2 + 34x + 13$

21. $P(x) = x^3 + 6x - 13$

22. $P(x) = 4x^3 - 3x^2 + 4x - 5$

23. $P(x) = 4x^4 - 8x^3 + 17x^2 - 2x - 14$

24. $P(x) = 3x^4 - 4x^3 - x^2 + 8x - 2$

25. $P(x) = -x^4 + 2x^3 + 3x^2 + 6$

26. $P(x) = -2x^4 - x^2 + x - 5$

The following polynomials have zeros in the given intervals. Approximate these zeros to the nearest hundredth.

27. $P(x) = x^4 + x^3 - 6x^2 - 20x - 16$,   [3.2, 3.3] and [−1.4, −1.1]

28. $P(x) = x^4 - 2x^3 - 2x^2 - 18x + 5$,   [.2, .4] and [3.7, 3.8]

29. $P(x) = x^4 - 4x^3 - 20x^2 + 32x + 12$;   [−4, −3], [−1, 0], [1, 2], and [6, 7]

30. $P(x) = x^4 - 4x^3 - 44x^2 + 160x - 80$;   [−7, −6], [0, 1], [2, 3], and [7, 8]

Graph each of the following polynomials. See Examples 4 and 5.

31. $P(x) = x^3 - 7x - 6$

32. $P(x) = x^3 + x^2 - 4x - 4$

33. $P(x) = x^4 - 5x^2 + 6$

34. $P(x) = x^3 - 3x^2 - x + 3$

35. $P(x) = 6x^3 + 11x^2 - x - 6$

36. $P(x) = x^4 - 2x^2 - 8$

37. $P(x) = -x^3 + 6x^2 - x - 14$

38. $P(x) = 6x^4 - x^3 - 23x^2 - 4x + 12$

Chapter 9 Summary

Key Words	leading coefficient zeros of a polynomial root of an equation	zero polynomial synthetic division multiplicity

Division Algorithm   For any polynomial $P(x)$ and any complex number k, there exists a unique polynomial $Q(x)$ and number r such that

$$P(x) = (x - k)Q(x) + r.$$

Remainder Theorem   If the polynomial $P(x)$ is divided by $x - k$, then the remainder is $P(k)$.

Factor Theorem   The polynomial $x - k$ is a factor of the polynomial $P(x)$ if and only if $P(k) = 0$.

Fundamental Theorem of Algebra   Every polynomial of degree 1 or more has at least one complex zero.

Conjugate Zeros Theorem   If $P(x)$ is a polynomial having only real coefficients and if $a + bi$ is a zero of $P(x)$, then the conjugate $a - bi$ is also a zero of $P(x)$.

Rational Zeros Theorem

Let $P(x) = a_n x^n + a_{n-1} x^{n-1} + \cdots + a_1 x + a_0$, $a_n \neq 0$, be a polynomial with integer coefficients. If p/q is a rational number written in lowest terms and if p/q is a zero of $P(x)$, then p is a factor of the constant term a_0 and q is a factor of the leading coefficient a_n.

Intermediate Value Theorem for Polynomials

If $P(x)$ is a polynomial function with only real coefficients, and if for real numbers a and b, $P(a)$ and $P(b)$ are opposite in sign, then there exists at least one real zero between a and b.

Boundedness Theorem

Let $P(x)$ be a polynomial with real coefficients and with a positive leading coefficient. If $P(x)$ is divided synthetically by $x - c$, and

(a) if $c > 0$ and all numbers in the bottom row of the synthetic division are nonnegative, then $P(x)$ has no zero greater than c;

(b) if $c < 0$ and the numbers in the bottom row of the synthetic division alternate in sign (with 0 considered positive or negative, as needed), then $P(x)$ has no zero less than c.

Descartes' Rule of Signs

Let $P(x)$ be a polynomial with real coefficients.

(a) The number of positive real zeros of $P(x)$ is either equal to the number of variations in sign occurring in the coefficients of $P(x)$, or else is less than the number of variations by a positive even integer.

(b) The number of negative real zeros of $P(x)$ either equals the number of variations in sign of $P(-x)$, or else is less than the number of variations by a positive even integer.

Chapter 9   Review Exercises

Use synthetic division to perform each division.

1. $(2x^3 - x^2 - x - 6) \div (x + 1)$

2. $(3t^3 + 10t^2 - 9t - 4) \div (t + 4)$

3. $(4m^3 + m^2 - 12) \div (m - 2)$

4. $(5y^3 + 24y^2 + 25y - 6) \div (y + 3)$

Express each polynomial in the form $P(x) = (x - k)Q(x) + r$ for the given value of k.

5. $P(z) = 2z^3 + 15z^2 + 30z + 20$; $k = -2$

6. $P(r) = r^3 + 6r^2 - r - 35$; $k = -5$

7. $P(m) = 2m^3 - 13m^2 + 17m + 15$; $k = 4$

8. $P(x) = 3x^3 + 2x^2 - 19x + 16$; $k = 2$

Use synthetic division to find $P(2)$ for each of the following.

9. $P(x) = x^3 + 3x^2 - 5x + 1$

10. $P(x) = 2x^3 - 4x^2 + 3x - 10$

11. $P(x) = 5x^4 - 12x^2 + 2x - 8$

12. $P(x) = x^5 - 3x^2 + 2x - 4$

Find a polynomial of lowest degree having the following zeros.

13. $-1, 4, 7$

14. $8, 2, 3$

15. $-\sqrt{7}, \sqrt{7}, 2, -1$

16. $1 + \sqrt{5}, 1 - \sqrt{5}, -4, 1$

17. Is -1 a zero of $P(x) = 2x^4 + x^3 - 4x^2 + 3x + 1$?

18. Is -2 a zero of $P(x) = 2x^4 + x^3 - 4x^2 + 3x + 1$?

19. Is $x + 1$ a factor of $P(x) = x^3 + 2x^2 + 3x - 1$?

20. Is $x + 1$ a factor of $P(x) = 2x^3 - x^2 + x + 4$?

21. Find a polynomial of degree 3 with -2, 1, and 4 as zeros, and $P(2) = 16$.

22. Find a polynomial of degree 4 with real coefficients having 1, -1, and $3i$ as zeros, and $P(2) = 39$.

23. Find a lowest-degree polynomial with real coefficients having zeros 2, -2, and $-i$.

24. Find a lowest-degree polynomial with real coefficients having zeros 2, -3, and $5i$.

25. Find a polynomial of lowest degree with real coefficients having -3 and $1 - i$ as zeros.

26. Find all zeros of $P(x) = x^4 - 3x^3 - 8x^2 + 22x - 24$, given that $1 - i$ is a zero, and factor $P(x)$.

27. Find all zeros of $P(x) = x^4 - 6x^3 + 14x^2 - 24x + 40$, given that $3 + i$ is a zero, and factor $P(x)$.

28. Find all zeros of $P(x) = x^4 + x^3 - x^2 + x - 2$, given that 1 is a zero, and factor $P(x)$.

Find all rational zeros of the following.

29. $P(x) = 2x^3 - 9x^2 - 6x + 5$

30. $P(x) = 3x^3 - 10x^2 - 27x + 10$

31. $P(x) = x^3 - \dfrac{17}{6}x^2 - \dfrac{13}{3}x - \dfrac{4}{3}$

32. $P(x) = 8x^4 - 14x^3 - 29x^2 - 4x + 3$

33. $P(x) = -x^4 + 2x^3 - 3x^2 + 11x - 7$

34. $P(x) = x^5 - 3x^4 - 5x^3 + 15x^2 + 4x - 12$

35. Use a polynomial to show that $\sqrt{11}$ is irrational.

36. Show that $P(x) = x^3 - 9x^2 + 2x - 5$ has no rational zeros.

Show that the following polynomials have real zeros satisfying the given conditions.

37. $P(x) = 3x^3 - 8x^2 + x + 2$, zero in $[-1, 0]$ and $[2, 3]$

38. $P(x) = 4x^3 - 37x^2 + 50x + 60$, zero in $[2, 3]$ and $[7, 8]$

39. $P(x) = x^3 + 2x^2 - 22x - 8$, zero in $[-1, 0]$ and $[-6, -5]$

40. $P(x) = 2x^4 - x^3 - 21x^2 + 51x - 36$ has no real zero greater than 4

41. $P(x) = 6x^4 + 13x^3 - 11x^2 - 3x + 5$ has no zero greater than 1 or less than -3

Approximate the real zeros of each of the following as decimals to the nearest tenth.

42. $P(x) = 2x^3 - 11x^2 - 2x + 2$

43. $P(x) = x^4 - 4x^3 - 5x^2 + 14x - 15$

44. **(a)** Find the number of positive and negative zeros of $P(x) = x^3 + 3x^2 - 4x - 2$.
   (b) Show that $P(x)$ has a zero between -4 and -3. Approximate this zero to the nearest tenth.

Graph each of the following polynomials

45. $P(x) = 2x^3 - 11x^2 - 2x + 2$ (See Exercise 42.)

46. $P(x) = x^4 - 4x^3 - 5x^2 + 14x - 15$ (See Exercise 43.)

47. $P(x) = x^3 + 3x^2 - 4x - 2$

48. $P(x) = 2x^4 - 3x^3 + 4x^2 + 5x - 1$

10

Sequences

A *sequence* is a list of numbers. Familiar sequences include the sequence of positive even integers,

$$2, 4, 6, 8, 10, \ldots ,$$

and the sequence of powers of 2,

$$2, 4, 8, 16, 32, 64, 128, \ldots .$$

In this chapter we study sequences, giving special attention to two types of sequences, arithmetic and geometric. We also look at mathematical induction, a method which is used, among other things, to prove theorems about sequences.

10.1 Sequences

A **sequence** is a function having as domain a set of positive integers. Instead of using $f(x)$ notation to indicate a sequence, it is customary to use a_n, where n is used to represent a positive integer. The elements in the range of a sequence, called the **terms** of the sequence, are $a_1, a_2, a_3, \ldots .$ The **general term,** or **nth term,** of the sequence is a_n.

EXAMPLE 1 Write the first five terms for each of the following sequences.

(a) $a_n = \dfrac{n + 1}{n + 2}$

Replacing n, in turn, with 1, 2, 3, 4, and 5 gives

$$\frac{2}{3}, \frac{3}{4}, \frac{4}{5}, \frac{5}{6}, \frac{6}{7}.$$

(b) $a_n = (-1)^n \cdot n$

Replace n, in turn, with 1, 2, 3, 4, and 5, to get

$$a_1 = (-1)^1 \cdot 1 = -1$$
$$a_2 = (-1)^2 \cdot 2 = 2$$
$$a_3 = (-1)^3 \cdot 3 = -3$$
$$a_4 = (-1)^4 \cdot 4 = 4$$
$$a_5 = (-1)^5 \cdot 5 = -5.$$

(c) $b_n = \dfrac{(-1)^n}{2^n}$

Here $b_1 = -1/2$, $b_2 = 1/4$, $b_3 = -1/8$, $b_4 = 1/16$, and $b_5 = -1/32$. ●

A sequence is a **finite sequence** if the domain is the set $\{1, 2, 3, 4, \ldots, n\}$, where n is a positive integer. An **infinite sequence** has the set of all positive integers as its domain.

EXAMPLE 2   The sequence of positive even integers,

$$2, 4, 6, 8, 10, 12, 14, \ldots,$$

is infinite, but the sequence of days in June,

$$1, 2, 3, 4, \ldots, 29, 30,$$

is finite. ●

EXAMPLE 3   Find the first four terms for the sequence defined as follows: $a_1 = 4$, and for $n > 1$, $a_n = 2 \cdot a_{n-1} + 1$.
We know $a_1 = 4$. Since $a_n = 2 \cdot a_{n-1} + 1$,

$$a_2 = 2 \cdot a_1 + 1 = 2 \cdot 4 + 1 = 9$$
$$a_3 = 2 \cdot a_2 + 1 = 2 \cdot 9 + 1 = 19$$
$$a_4 = 2 \cdot a_3 + 1 = 2 \cdot 19 + 1 = 39. ●$$

The definition of this sequence is an example of a **recursive definition,** one in which each term is defined as an expression involving the previous term.
The sum of the terms of a sequence is called a **series.** A compact shorthand notation can be used to write a series. For example,

$$1 + 5 + 9 + 13 + 17 + 21$$

is the sum of the first six terms in the sequence with general term $a_n = 4n - 3$.

The series $1 + 5 + 9 + 13 + 17 + 21$ with general term $a_n = 4n - 3$ can be written as

$$\sum_{i=1}^{6} (4i - 3).$$

The Greek letter sigma, Σ, is used to mean "sum." To evaluate this sum, replace i in $4i - 3$ first by 1, then 2, then 3, until finally i is replaced with 6.

$$\sum_{i=1}^{6} (4i - 3) = (4 \cdot 1 - 3) + (4 \cdot 2 - 3) + (4 \cdot 3 - 3)$$
$$+ (4 \cdot 4 - 3) + (4 \cdot 5 - 3) + (4 \cdot 6 - 3)$$
$$= 1 + 5 + 9 + 13 + 17 + 21 = 66$$

Here i is called the **index of summation.** Do not confuse this use of i with the use of i to represent an imaginary number.

EXAMPLE 4

Evaluate each of the following sums.

(a) $\displaystyle\sum_{i=1}^{4} i^2(i + 1) = 1^2(1 + 1) + 2^2(2 + 1) + 3^2(3 + 1) + 4^2(4 + 1)$

$$= 1 \cdot 2 + 4 \cdot 3 + 9 \cdot 4 + 16 \cdot 5$$
$$= 2 + 12 + 36 + 80$$
$$= 130$$

(b) $\displaystyle\sum_{i=3}^{6} \frac{i + 1}{i - 2} = \frac{3 + 1}{3 - 2} + \frac{4 + 1}{4 - 2} + \frac{5 + 1}{5 - 2} + \frac{6 + 1}{6 - 2}$

$$= \frac{4}{1} + \frac{5}{2} + \frac{6}{3} + \frac{7}{4} = \frac{41}{4} \quad \bullet$$

In the next few sections, we develop formulas that can be used to evaluate some of these sums without finding each term individually.

In statistics, the **mean** of a set of numbers $x_1, x_2, x_3, \ldots, x_n$ is defined as

$$\bar{x} = \frac{\displaystyle\sum_{i=1}^{n} x_i}{n}.$$

That is, to find the mean, add all the members of the set of numbers and divide the sum by the number of members.

EXAMPLE 5

Find the mean for the set of numbers 10, 12, 18, 24, 20, 22, 14.

Here $n = 7$, since there are seven numbers in the list.

$$\sum_{i=1}^{7} x_i = 10 + 12 + 18 + 24 + 20 + 22 + 14 = 120$$

Thus, the mean is

$$\bar{x} = \frac{\sum_{i=1}^{7} x_i}{7} = \frac{120}{7} = 17.1,$$

to the nearest tenth.   ●

10.1   EXERCISES

Write the first five terms of each of the following sequences. See Example 1.

1. $2_n = 6n + 4$ **2.** $a_n = 3n - 2$ **3.** $a_n = 2^n$ **4.** $a_n = 3^n$

5. $a_n = (-1)^{n+1}$ **6.** $a_n = (-1)^n(n + 2)$ **7.** $a_n = \dfrac{2n}{n + 3}$ **8.** $a_n = \dfrac{-4}{n + 5}$

9. $a_n = \dfrac{8n - 4}{2n + 1}$ **10.** $a_n = \dfrac{-3n + 6}{n + 1}$ **11.** $a_n = (-2)^n(n)$

12. $a_n = (-1/2)^n(n^{-1})$ **13.** $a_n = \dfrac{n^2 + 1}{n^2 + 2}$ **14.** $a_n = \dfrac{(-1)^{n-1}(n + 1)}{n + 2}$

15. $a_n = x^n$ **16.** $a_n = n \cdot x^{-n}$

Find the first ten terms for the sequences defined as follows. See Example 3.

17. $a_1 = 4, a_n = a_{n-1} + 5$ **18.** $a_1 = -3, a_n = a_{n-1} + 2$ **19.** $a_1 = -9, a_n = a_{n-1} - 4$

20. $a_1 = -8, a_n = a_{n-1} - 7$ **21.** $a_1 = 2, a_n = 2 \cdot a_{n-1}$ **22.** $a_1 = -3, a_n = 2 \cdot a_{n-1}$

23. $a_1 = 1, a_2 = 1, a_n = a_{n-1} + a_{n-2}$ (The Fibonacci sequence)

24. $a_1 = 3, a_2 = 2, a_n = a_{n-1} - a_{n-2}$

Evaluate each of the following sums. See Example 4.

25. $\displaystyle\sum_{i=1}^{5} (2i + 1)$ **26.** $\displaystyle\sum_{i=1}^{6} (3i - 2)$ **27.** $\displaystyle\sum_{i=1}^{9} i$ **28.** $\displaystyle\sum_{i=1}^{4} \frac{1}{i}$

29. $\displaystyle\sum_{i=1}^{5} (i + 1)^{-1}$ **30.** $\displaystyle\sum_{i=1}^{4} i^i$ **31.** $\displaystyle\sum_{i=1}^{3} i^{-i}$ **32.** $\displaystyle\sum_{i=1}^{4} (i + 1)^2$

33. $\displaystyle\sum_{i=1}^{6} (-1)^i \cdot i$ **34.** $\displaystyle\sum_{i=1}^{7} (-1)^{i+1} \cdot i^2$ **35.** $\displaystyle\sum_{i=3}^{7} i/2$ **36.** $\displaystyle\sum_{i=5}^{9} (i + 1)/3$

37. $\displaystyle\sum_{i=5}^{8} (i^2 + 1)/2$ **38.** $\displaystyle\sum_{i=3}^{5} (i + 3)/i$ **39.** $\displaystyle\sum_{i=0}^{4} 2^i$ **40.** $\displaystyle\sum_{i=0}^{4} 3^{-i}$

Write out each of the following. Do not try to evaluate.

41. $\displaystyle\sum_{i=1}^{4} x^i$ **42.** $\displaystyle\sum_{i=3}^{7} x^{-i}$ **43.** $\displaystyle\sum_{i=1}^{5} i \cdot x^i$ **44.** $\displaystyle\sum_{i=1}^{5} i^2/x^i$

Find the mean for each of the following sets of numbers. See Example 5.

45. 10, 9, 6, 12, 8, 7, 7, 7, 8, 6   **46.** 124, 132, 129, 135, 130

47. 28.1, 29.2, 31.5, 26.0, 28.9, 30.4   **48.** .002, .005, .001, .003, .004, .002

10.2   Arithmetic Sequences

A sequence in which each term after the first is obtained by adding a fixed number to the previous term is called an **arithmetic sequence** (or **arithmetic progression**). The fixed number that is added is called the **common difference.** The sequence

$$5, 9, 13, 17, 21, \ldots$$

is an arithmetic sequence since each term after the first is obtained by adding 4 to the previous term. That is,

$$9 = 5 + 4$$
$$13 = 9 + 4$$
$$17 = 13 + 4$$
$$21 = 17 + 4,$$

and so on. The common difference is 4.

EXAMPLE 1   Find the common difference, d, for the arithmetic sequence

$$-9, -7, -5, -3, -1, \ldots.$$

Since this sequence is arithmetic, d can be found by choosing any two adjacent terms and subtracting the first from the second. If we choose -7 and -5,

$$d = -5 - (-7) = 2.$$

Choosing -9 and -7 would give $d = -7 - (-9) = 2$, the same result.   ●

EXAMPLE 2   Write the first five terms for each of the following arithmetic sequences.

(a) The first term is 7, and each succeeding term is found by adding -3 to the preceding term.

Here $a_1 = 7$
$$a_2 = 7 + (-3) = 4$$
$$a_3 = 4 + (-3) = 1$$
$$a_4 = 1 + (-3) = -2$$
$$a_5 = -2 + (-3) = -5.$$

(b) $a_1 = -12, d = 5$

We have $a_1 = -12$
$$a_2 = -12 + d = -12 + 5 = -7$$
$$a_3 = -7 + d = -7 + 5 = -2$$
$$a_4 = -2 + d = -2 + 5 = 3$$
$$a_5 = 3 + d = 3 + 5 = 8.$$ ●

If a_1 is the first term of an arithmetic sequence and d is the common difference, then the terms of the sequence are given by

$$a_1 = a_1$$
$$a_2 = a_1 + d$$
$$a_3 = a_2 + d = a_1 + d + d = a_1 + 2d$$
$$a_4 = a_3 + d = a_1 + 2d + d = a_1 + 3d$$
$$a_5 = a_1 + 4d$$
$$a_6 = a_1 + 5d,$$

and, by this pattern, $a_n = a_1 + (n - 1)d$.

nth Term of an Arithmetic Sequence

> In an arithmetic sequence with first term a_1 and common difference d, the nth term, a_n, is given by
> $$a_n = a_1 + (n - 1)d.$$

EXAMPLE 3   Find a_{13} and a_n for the arithmetic sequence

$$-3, 1, 5, 9, \ldots$$

Here $a_1 = -3$ and $d = 1 - (-3) = 4$. To find a_{13}, substitute 13 for n in the formula above.

$$a_{13} = a_1 + (13 - 1)d$$
$$a_{13} = -3 + (12)4$$
$$a_{13} = -3 + 48$$
$$a_{13} = 45$$

To find a_n, substitute values for a_1 and d in the formula for a_n.

$$a_n = -3 + (n - 1) \cdot 4$$
$$a_n = -3 + 4n - 4$$
$$a_n = 4n - 7 \quad \bullet$$

EXAMPLE 4   Find a_{18} and a_n for the arithmetic sequence having $a_2 = 9$ and $a_3 = 15$.
Find d: $d = a_3 - a_2 = 15 - 9 = 6$.

Since
$$a_2 = a_1 + d,$$
$$9 = a_1 + 6 \qquad \text{and} \qquad a_1 = 3.$$

Then
$$a_{18} = 3 + (18 - 1) \cdot 6$$
$$a_{18} = 105$$

and
$$a_n = 3 + (n - 1) \cdot 6$$
$$a_n = 3 + 6n - 6$$
$$a_n = 6n - 3. \quad \bullet$$

EXAMPLE 5

Suppose that an arithmetic sequence has $a_8 = -16$ and $a_{16} = -40$. Find a_1.

Since $a_8 = a_1 + (8 - 1)d$, replacing a_8 with -16 gives $-16 = a_1 + 7d$ or $a_1 = -16 - 7d$. Similarly, $-40 = a_1 + 15d$ or $a_1 = -40 - 15d$. From these two equations, using the substitution method from Chapter 8, $-16 - 7d = -40 - 15d$, so $d = -3$. To find a_1, substitute -3 for d in $-16 = a_1 + 7d$:

$$-16 = a_1 + 7d$$
$$-16 = a_1 + 7(-3)$$
$$a_1 = 5. \quad \bullet$$

EXAMPLE 6

A child building a tower with blocks uses 15 for the first row. Each row has 2 blocks less than the previous row. If there are 8 rows in the tower, how many blocks are used for the top row?

The number of blocks in each row forms an arithmetic sequence with $a_1 = 15$ and $d = -2$. We want to find a_n for $n = 8$. Using the formula

$$a_n = a_1 + (n - 1)d$$

gives

$$a_8 = 15 + (8 - 1)(-2) = 1.$$

There is just one block in the top row. $\bullet$

Suppose a man borrows $3000 and agrees to pay $100 per month plus interest of 1% per month on the unpaid balance until the loan is paid off. The first month he pays $100 to reduce the loan plus interest of $(.01)3000 = 30$ dollars. The second month he pays another $100 toward the loan and interest of $(.01)2900 = 29$ dollars. Since the loan is reduced by $100 each month, his interest payments decrease by $(.01)100 = 1$ dollar each month, forming the arithmetic sequence

$$30, 29, 28, \ldots , 3, 2, 1.$$

The total amount of interest paid is given by the sum of the terms of this sequence. We shall develop a formula to find this sum without adding all thirty numbers directly, and then we will return to the problem.

To find this formula, suppose a sequence has terms $a_1, a_2, a_3, a_4 \ldots$. Then S_n is defined as the sum of the first n terms of the sequence. That is,

$$S_n = a_1 + a_2 + a_3 + \ldots + a_n.$$

To find a formula for S_n, write the sum of the first n terms as follows.

$$S_n = a_1 + [a_1 + d] + [a_1 + 2d] + \ldots + [a_1 + (n - 1)d]$$

Next, write this same sum in reversed order.

$$S_n = [a_1 + (n - 1)d] + [a_1 + (n - 2)d] + \ldots + [a_1 + d] + a_1$$

Now add the corresponding sides of these two equations.

$$S_n + S_n = (a_1 + [a_1 + (n - 1)d]) + ([a_1 + d] + [a_1 + (n - 2)d])$$
$$+ \ldots + ([a_1 + (n - 1)d] + a_1)$$

From this,

$$2S_n = [2a_1 + (n - 1)d] + [2a_1 + (n - 1)d] + \ldots + [2a_1 + (n - 1)d].$$

Since there are n of the $[2a_1 + (n - 1)d]$ terms on the right,

$$2S_n = n[2a_1 + (n - 1)d]$$

$$S_n = \frac{n}{2} [2a_1 + (n - 1)d].$$

Since $a_n = a_1 + (n - 1)d$, we also have $S_n = \frac{n}{2} [a_1 + a_1 + (n - 1)d]$, or

$$S_n = \frac{n}{2} (a_1 + a_n).$$

This work with sums of arithmetic sequences is summarized below.

Sum of the First n Terms of an Arithmetic Sequence

> If an arithmetic sequence has first term a_1 and common difference d, then the sum of the first n terms, S_n, is given by
>
> $$S_n = \frac{n}{2} (a_1 + a_n)$$
>
> or $\qquad$ $$S_n = \frac{n}{2} [2a_1 + (n - 1)d].$$

The first formula is used when the first and last terms are known; otherwise the second formula is used.

Either one of these formulas can be used to find the amount of interest the man above will pay on the $3000 loan. In the sequence of interest payments $a_1 = 30$, $d = -1$, $n = 30$, and $a_n = 1$. Choosing the first formula,

$$S_n = \frac{n}{2} (a_1 + a_n),$$

gives $\qquad$ $$S_{30} = \frac{30}{2} (30 + 1) = 15(31) = 465,$$

so the man will pay a total of $465 interest over 30 months.

EXAMPLE 7 $\quad$ Find S_{12} for the arithmetic sequence

$$-9, -5, -1, 3, 7, \ldots.$$

Using $a_1 = -9$ and $d = 4$ in the second formula,

$$S_n = \frac{n}{2} [2a_1 + (n - 1)d],$$

gives

$$S_{12} = \frac{12}{2} [2(-9) + 11(4)] = 6(-18 + 44) = 156. \quad \bullet$$

EXAMPLE 8   Find the sum of the first 60 positive integers.
Here $n = 60$, $a_1 = 1$, and $a_{60} = 60$. We want S_{60}:

$$S_{60} = \frac{60}{2}(1 + 60) = 30 \cdot 61 = 1830. \quad \bullet$$

EXAMPLE 9   The sum of the first 17 terms of an arithmetic sequence is 187. If $a_{17} = -13$, find a_1 and d.
Use the formula for S_n, with $n = 17$, to find a_1.

$$S_{17} = \frac{17}{2}(a_1 + a_{17})$$

$$187 = \frac{17}{2}(a_1 - 13)$$

$$374 = 17(a_1 - 13)$$

$$22 = a_1 - 13$$

$$a_1 = 35$$

Since $a_{17} = a_1 + (17 - 1)d$,

$$-13 = 35 + 16d$$

$$-48 = 16d$$

$$d = -3. \quad \bullet$$

Any sum of the form

$$\sum_{i=1}^{n} (mi + p),$$

where m and p are real numbers, represents the sum of the terms of an arithmetic sequence having first term

$$a_1 = m(1) + p = m + p$$

and common difference $d = m$. These sums can be evaluated by the formulas in this section, as shown by the next example.

EXAMPLE 10   Find each of the following sums.

(a) $\displaystyle\sum_{i=1}^{10} (4i + 8)$

This sum represents the sum of the first ten terms of the arithmetic sequence having

$$a_1 = 4 \cdot 1 + 8 = 12,$$

$$n = 10,$$

and

$$a_n = a_{10} = 4 \cdot 10 + 8 = 48.$$

Thus

$$\sum_{i=1}^{10} (4i + 8) = S_{10} = \frac{10}{2}(12 + 48) = 5(60) = 300.$$

(b) $\displaystyle\sum_{i=1}^{15} (9 - i) = S_{15} = \frac{15}{2} [8 + (-6)] = \frac{15}{2}(2) = 15$ ●

10.2   EXERCISES

For each of the following arithmetic sequences, write the indicated number of terms. See Example 2.

1. $a_1 = 4, d = 2, n = 5$

2. $a_1 = 6, d = 8, n = 4$

3. $a_2 = 9, d = -2, n = 4$

4. $a_3 = 7, d = -4, n = 4$

5. $a_3 = -2, d = -4, n = 4$

6. $a_2 = -12, d = -6, n = 5$

7. $a_3 = 6, a_4 = 12, n = 6$

8. $a_5 = 8, a_6 = 5, n = 6$

For each of the following sequences that are arithmetic, find d and a_n. See Examples 1 and 3.

9. $12, 17, 22, 27, 32, 37, \ldots$

10. $8, 17, 26, 35, 44, 53, \ldots$

11. $18, 15, 12, 9, 6, \ldots$

12. $30, 24, 18, 12, \ldots$

13. $-19, -12, -5, 2, 9, \ldots$

14. $-30, -20, -12, -6, -2, \ldots$

15. $x, x + m, x + 2m, x + 3m, x + 4m, \ldots$

16. $k + p, k + 2p, k + 3p, k + 4p, \ldots$

17. $2z + m, 2z, 2z - m, 2z - 2m, 2z - 3m, \ldots$

18. $3r - 4z, 3r - 3z, 3r - 2z, 3r - z, 3r, 3r + z, \ldots$

Find a_8 and a_n for each of the following arithmetic sequences. See Example 4.

19. $a_1 = 5, d = 2$

20. $a_1 = -3, d = -4$

21. $a_3 = 2, d = 1$

22. $a_4 = 5, d = -2$

23. $a_1 = 8, a_2 = 6$

24. $a_1 = 6, a_2 = 3$

25. $a_{10} = 6, a_{12} = 15$

26. $a_{15} = 8, a_{17} = 2$

27. $a_1 = x, a_2 = x + 3$

28. $a_2 = y + 1, d = -3$

29. $a_6 = 2m, a_7 = 3m$

30. $a_5 = 4p + 1, a_7 = 6p + 7$

Find the sum of the first ten terms for each of the following arithmetic sequences. See Example 7.

31. $a_1 = 8, d = 3$

32. $a_1 = -9, d = 4$

33. $a_3 = 5, a_4 = 8$

34. $a_2 = 9, a_4 = 13$

35. $5, 9, 13, \ldots$

36. $8, 6, 4, \ldots$

37. $3\frac{1}{2}, 5, 6\frac{1}{2}, \ldots$

38. $2\frac{1}{2}, 3\frac{3}{4}, 5, \ldots$

39. $a_1 = 10, a_{10} = 5\frac{1}{2}$

40. $a_1 = -8, a_{10} = -5/4$

41. $a_1 = 9.428, d = -1.723$

42. $a_1 = -3.119, d = 2.422$

43. $a_4 = 2.556, a_5 = 3.004$

44. $a_7 = 11.192, a_9 = 4.812$

The terms of the following sums represent arithmetic sequences. Use the formulas of this section to evaluate each sum. See Example 10.

45. $\displaystyle\sum_{i=1}^{3} (i + 4)$

46. $\displaystyle\sum_{i=1}^{5} (i - 8)$

47. $\displaystyle\sum_{i=1}^{10} (2i + 3)$

48. $\displaystyle\sum_{i=1}^{15} (5i - 9)$

49. $\displaystyle\sum_{i=1}^{12} (7i + 2)$

50. $\displaystyle\sum_{i=1}^{10} (8i - 3)$

51. $\displaystyle\sum_{i=1}^{12} (-5 - 8i)$

52. $\displaystyle\sum_{i=1}^{19} (-3 - 4i)$

53. $\displaystyle\sum_{i=1}^{1000} i$

54. $\displaystyle\sum_{i=1}^{2000} i$

55. $\displaystyle\sum_{i=6}^{15} (4i - 2)$

56. $\displaystyle\sum_{i=9}^{20} (8i + 3)$

57. $\displaystyle\sum_{i=7}^{12} (6 - 2i)$

58. $\displaystyle\sum_{i=4}^{17} (-3 - 5i)$

Find a_1 for each of the following arithmetic sequences. See Examples 5 and 9.

59. $a_9 = 47, a_{15} = 77$

60. $a_{10} = 50, a_{20} = 110$

61. $a_{15} = 168, a_{16} = 180$

62. $a_{10} = -54, a_{17} = -89$

63. $S_{20} = 1090, a_{20} = 102$

64. $S_{31} = 5580, a_{31} = 360$

65. $S_{12} = -108, a_{12} = -19$

66. $S_{25} = 650, a_{25} = 62$

67. Find the sum of all the integers from 51 to 71.

68. Find the sum of all the integers from -8 to 30.

69. If a clock strikes the proper number of bongs each hour on the hour, how many bongs will it bong in a month of 30 days?

70. A stack of telephone poles has 30 in the bottom row, 29 in the next, and so on, with one pole in the top row. How many poles are in the stack?

71. A sky diver falls 10 m during the first second, 20 m during the second, 30 m during the third, and so on. How many meters will the diver fall during the tenth second? During the first ten seconds?

72. Deepwell Drilling Company charges a flat $100 set-up charge, plus $5 for the first foot, $6 for the second, $7 for the third, and so on. Find the total charge for a 70-foot well.

73. An object falling under the force of gravity falls about 16 ft the first second, 48 ft during the second, 80 ft during the third second, and so on. How far would the object fall during the eighth second? What is the total distance the object would fall in eight seconds?

74. The population of a city was 49,000 five years ago. Each year the zoning commission permits an increase of 580 in the population. What will the population be five years from now?

75. A super slide of uniform slope is to be built on a level piece of land. There are to be twenty equally spaced supports, with the longest support 15 m long and the shortest 2 m long. Find the total length of all the supports.

76. How much material would be needed for the rungs of a ladder of 31 rungs, if the rungs taper uniformly from 18 in to 28 in?

Explain why each of the following sequences is arithmetic.

77. log 2, log 4, log 8, log 16, log 32, . . .

78. log 12, log 36, log 108, log 324, . . .

10.3   Geometric Sequences

A **geometric sequence** (or **geometric progression**) is a sequence in which each term after the first is obtained by multiplying the preceding term by a constant nonzero real number, called the **common ratio.** An example of a geometric sequence is 2, 8, 32, 128, . . . in which the first term is 2 and the common ratio is 4.

If the common ratio of a geometric sequence is r, then by the definition of a geometric sequence,

$$\frac{a_{n+1}}{a_n} = r$$

for every positive integer n. By this definition, the common ratio can be found by choosing any term except the first and dividing it by the preceding term.

As we said, the geometric sequence 2, 8, 32, 128, . . . has $r = 4$. Notice that

$$8 = 2 \cdot 4$$
$$32 = 8 \cdot 4 = (2 \cdot 4) \cdot 4 = 2 \cdot 4^2$$
$$128 = 32 \cdot 4 = (2 \cdot 4^2) \cdot 4 = 2 \cdot 4^3.$$

To generalize this result, assume that a geometric sequence has first term a_1 and common ratio r. The second term can be written as $a_2 = a_1 r$, the third as $a_3 = a_2 r = (a_1 r)r = a_1 r^2$, and so on. Following this pattern, the nth term is $a_n = a_1 r^{n-1}$.

nth Term of a Geometric Sequence

> In the geometric sequence with first term a_1 and common ratio r, the nth term is
>
> $$a_n = a_1 r^{n-1}.$$

EXAMPLE 1

Find a_5 and a_n for each of the following geometric sequences.

(a) 4, 12, 36, 108, . . .

The first term, a_1, is 4. To find r, choose any term except the first and divide it by the preceding term. For example,

$$r = 36/12 = 3.$$

To find the fifth term, a_5, start with $a_n = a_1 r^{n-1}$ and replace n with 5, r with 3, and a_1 with 4.

$$a_5 = 4 \cdot (3)^{5-1} = 4 \cdot 3^4 = 324$$

Also, $a_n = 4 \cdot 3^{n-1}.$

(b) 64, 32, 16, 8, . . .

Here $r = 8/16 = 1/2$, and $a_1 = 64$, so

$$a_5 = 64\left(\frac{1}{2}\right)^{5-1} = 64\left(\frac{1}{16}\right) = 4.$$

Also, $$a_n = 64\left(\frac{1}{2}\right)^{n-1}. \quad \bullet$$

EXAMPLE 2   Find a_1 and r for each of the following geometric sequences.

(a) The third term is 20 and the sixth term is -160.

Use the formula for the nth term of a geometric sequence.

$$\text{For } n = 3, \qquad a_3 = a_1 r^2 = 20;$$
$$\text{for } n = 6, \qquad a_6 = a_1 r^5 = -160.$$

Since $a_1 r^2 = 20$, then $a_1 = 20/r^2$. Substitute this in the second equation.

$$a_1 r^5 = -160$$
$$\left(\frac{20}{r^2}\right) r^5 = -160$$
$$20 r^3 = -160$$
$$r^3 = -8$$
$$r = -2.$$

Since $a_1 r^2 = 20$ and $r = -2$, then $a_1 = 5$.

(b) $a_5 = 15$ and $a_7 = 375$

First substitute $n = 5$ and then $n = 7$ into $a_n = a_1 r^{n-1}$.

$$a_5 = a_1 r^4 = 15 \qquad \text{and} \qquad a_7 = a_1 r^6 = 375$$

Solve the first equation for a_1 to get $a_1 = 15/r^4$. Then substitute for a_1 in the second equation.

$$a_1 r^6 = 375$$
$$\frac{15}{r^4} \cdot r^6 = 375$$
$$15 r^2 = 375$$
$$r^2 = 25$$
$$r = \pm 5$$

Either 5 or -5 can be used for r. To find a_1, use

$$a_1 = \frac{15}{r^4}.$$

Replace r with ± 5.

$$a_1 = \frac{15}{(\pm 5)^4} = \frac{15}{625} = \frac{3}{125}$$

There are two sequences that satisfy the given conditions: one with $a_1 = 3/125$ and $r = 5$; the other with $a_1 = 3/125$ and $r = -5$. ●

EXAMPLE 3   An insect population is growing in such a way that each generation is 1.5 times as large as the previous generation. Suppose there were 100 insects in the first generation. How many would there be in the fourth generation?

The population can be written as a geometric sequence with a_1 as the first-generation population, a_2 the second-generation population, and so on. Then the fourth-generation population will be a_4. Using the formula for a_n, where $n = 4$, $r = 1.5$, and $a_1 = 100$, gives

$$a_4 = a_1 r^3 = 100(1.5)^3 = 100(3.375) = 337.5.$$

In the fourth generation, the population will number about 338 insects.   ●

In applications of geometric sequences, it is often necessary to know the sum of the first n terms of a sequence. For example, we might want to know the total number of insects in the four generations of the population just discussed.

To find a formula for the sum of the first n terms of a geometric sequence, S_n, first write the series as

$$S_n = a_1 + a_2 + a_3 + \cdots + a_n.$$

This can also be written as

$$S_n = a_1 + a_1 r + a_1 r^2 + \cdots + a_1 r^{n-1}. \tag{1}$$

If $r = 1$, $S_n = na_1$, which is a correct formula for this case. If $r \neq 1$, multiply both sides of (1) by r, obtaining

$$rS_n = a_1 r + a_1 r^2 + a_1 r^3 + \cdots + a_1 r^n. \tag{2}$$

If (2) is subtracted from (1),

$$S_n - rS_n = a_1 - a_1 r^n$$

or

$$S_n(1 - r) = a_1(1 - r^n),$$

which finally gives

$$S_n = \frac{a_1(1 - r^n)}{1 - r}, \quad (r \neq 1).$$

This discussion is summarized below.

Sum of the First n Terms of a Geometric Sequence

If a geometric sequence has first term a_1 and common ratio r, then the sum of the first n terms, S_n, is given by

$$S_n = \frac{a_1(1 - r^n)}{1 - r}, \quad r \neq 1.$$

This formula can be used to find the total insect population in Example 3 over the four-generation period. With $n = 4$, $a_1 = 100$, and $r = 1.5$,

$$S_4 = \frac{100(1 - 1.5^4)}{1 - 1.5} = \frac{100(1 - 5.0625)}{-.5} = 812.5,$$

so the total population for the four generations will amount to about 813 insects.

EXAMPLE 4   **(a)** Find S_5 for the geometric sequence

$$3, 6, 12, 24, 48.$$

Here $a_1 = 3$ and $r = 2$. Using the formula above,

$$S_5 = \frac{3(1 - 2^5)}{1 - 2} = \frac{3(1 - 32)}{-1} = \frac{3(-31)}{-1} = 93.$$

(b) Find S_4 for the sequence

$$10, 2, \frac{2}{5}, \ldots .$$

Here $a_1 = 10$, $r = 1/5$, and $n = 4$. Using the formula for S_n,

$$S_4 = \frac{10\left[1 - \left(\frac{1}{5}\right)^4\right]}{1 - \frac{1}{5}}$$

$$= \frac{10\left(1 - \frac{1}{625}\right)}{\frac{4}{5}}$$

$$= 10\left(\frac{624}{625}\right) \cdot \frac{5}{4}$$

$$S_4 = \frac{312}{25}. \quad \bullet$$

A sum of the form

$$\sum_{i=1}^{n} m \cdot p^i$$

represents the sum of the terms of a geometric sequence having first term

$$a_1 = m \cdot p^1 = mp$$

and common ratio $r = p$. These sums can be found by the formula for S_n given above.

EXAMPLE 5   Find each of the following sums.

(a) $\displaystyle\sum_{i=1}^{6} 2 \cdot 3^i$

In this sum, $a_1 = 2 \cdot 3^1 = 6$ and $r = 3$. Thus,

$$\sum_{i=1}^{6} 2 \cdot 3^i = S_6 = \frac{6(1 - 3^6)}{1 - 3} = \frac{6(1 - 729)}{-2} = \frac{6(-728)}{-2} = 2184.$$

(b) $\displaystyle\sum_{i=1}^{5} \left(\frac{3}{4}\right)^i = S_5 = \frac{\dfrac{3}{4}\left[1 - \left(\dfrac{3}{4}\right)^5\right]}{1 - \dfrac{3}{4}}$

$$= \frac{\dfrac{3}{4}\left(1 - \dfrac{243}{1024}\right)}{\dfrac{1}{4}}$$

$$= 3\left(\frac{781}{1024}\right) = \frac{2343}{1024} \quad \bullet$$

10.3 EXERCISES

For each of the following, write the terms of the geometric sequence that satisfies the given conditions.

1. $a_1 = 2$, $r = 3$, $n = 4$ **2.** $a_1 = 4$, $r = 2$, $n = 5$

3. $a_1 = 1/2$, $r = 4$, $n = 4$ **4.** $a_1 = 2/3$, $r = 6$, $n = 3$

5. $a_1 = -2$, $r = -3$, $n = 4$ **6.** $a_1 = -4$, $r = 2$, $n = 5$

7. $a_1 = 3125$, $r = 1/5$, $n = 7$ **8.** $a_1 = 729$, $r = 2/3$, $n = 5$

9. $a_1 = -1$, $r = -1$, $n = 6$ **10.** $a_1 = -1$, $r = 1$, $n = 5$

11. $a_3 = 6$, $a_4 = 12$, $n = 5$ **12.** $a_2 = 9$, $a_3 = 3$, $n = 4$

Find a_5 and a_n for each of the following geometric sequences. See Example 1.

13. $a_1 = 4$, $r = 3$ **14.** $a_1 = 8$, $r = 4$

15. $a_1 = -2$, $r = 3$ **16.** $a_1 = -5$, $r = 4$

17. $a_1 = -3$, $r = -5$ **18.** $a_1 = -4$, $r = -2$

19. $a_2 = 3$, $r = 2$ **20.** $a_3 = 6$, $r = 3$

21. $a_4 = 64$, $r = -4$ **22.** $a_4 = 81$, $r = -3$

For each of the following sequences that are geometric, find r and a_n. See Examples 1 and 2.

23. 6, 12, 24, 48, . . .   **24.** 4, 16, 64, 256, . . .

25. 3/4, 3/2, 3, 6, 12, . . .   **26.** 5/6, 5/3, 10/3, 20/3, 40/3, . . .

27. -4, 2, -1, 1/2, . . .   **28.** 49, -7, 1, $-1/7$, . . .

29. 18, 20, 24, 32, 48, . . .   **30.** -7, -5, -3, -1, 1, 3, . . .

31. $a_3 = 9$, $r = 2$ **32.** $a_5 = 6$, $r = 1/2$

33. $a_3 = -2$, $r = 3$ **34.** $a_2 = 5$, $r = 1/2$

Find the sum of the first five terms for each of the following geometric sequences. See Example 4.

35. 3, 6, 12, 24, . . .   **36.** 5, 20, 80, 320, . . .

37. 12, -6, 3, $-3/2$, . . .

38. 18, -3, 1/2, $-1/12$, . . .

39. 9, -6, 4, . . .

40. 50, -10, 2, . . .

41. $a_1 = 4$, $r = 2$

42. $a_1 = 3$, $r = 3$

43. $a_2 = 1/3$, $r = 3$

44. $a_2 = -1$, $r = 2$

45. $a_1 = 8.423$, $r = 2.859$

46. $a_1 = -3.772$, $r = -1.553$

47. $a_1 = -5$, $r = .2033$

48. $a_1 = 4$, $r = -.8765$

Find each of the following sums. See Example 5.

49. $\displaystyle\sum_{i=1}^{4} 2^i$

50. $\displaystyle\sum_{i=1}^{6} 3^i$

51. $\displaystyle\sum_{i=1}^{4} (-3)^i$

52. $\displaystyle\sum_{i=1}^{4} (-3^i)$

53. $\displaystyle\sum_{i=1}^{8} 64(1/2)^i$

54. $\displaystyle\sum_{i=1}^{6} 81(2/3)^i$

55. $\displaystyle\sum_{i=1}^{4} (2/5)^i$

56. $\displaystyle\sum_{i=1}^{4} (-3/4)^i$

57. $\displaystyle\sum_{i=1}^{4} 6(3/2)^i$

58. $\displaystyle\sum_{i=1}^{5} 9(5/3)^i$

59. $\displaystyle\sum_{i=3}^{6} 2^i$

60. $\displaystyle\sum_{i=4}^{7} 3^i$

61. $\displaystyle\sum_{i=1}^{4} 5(3/5)^{i-1}$

62. $\displaystyle\sum_{i=1}^{5} 256(-3/4)^{i-1}$

63. Suppose you could save \$1 on January 1, \$2 on January 2, \$4 on January 3, and so on. What amount would you save on January 31?

64. What would be the total amount of your savings during January? (*Hint:* $2^{31} =$ 2,147,483,648.)

65. Richland Oil has a well that produced \$4,000,000 of income its first year. Each year thereafter, the well produced half as much as it did the previous year. What total amount of income would be produced by the well in 6 years?

66. Fruit and vegetable dealer Olive Greene paid 10¢ per lb for 10,000 lb of onions. Each week the price she charges increases by .1¢ per lb, while her onions lose 5% of their weight. If she sells all the onions after 6 weeks, does she make or lose money? How much?

67. The final step in processing a black and white photographic print is to immerse the print in a chemical called "fixer." The print is then washed in running water. Under certain conditions, 98% of the fixer in a print will be removed with 15 min of washing. How much of the original fixer would then be left after one hour?

68. A scientist has a vat containing 100 liters of a pure chemical. Twenty liters are drained and replaced with water. After complete mixing, 20 liters of the mixture are drained and replaced with water. What will be the strength of the mixture after nine such drainings?

69. The half-life of a radioactive substance is the time it takes for half the substance to decay. Suppose that the half-life of a substance is 3 yr and that 10^{15} molecules of the substance are present initially. How many molecules will be present after 15 yr?

70. Each year a machine loses 20% of the value it had at the beginning of the year. Find the value of a machine at the end of 6 yr if it cost \$100,000 new.

A sequence of equal payments made at equal periods of time is called an **annuity**. *Find the total value of each of the following annuities.*

71. Payments of $1000 at the end of each year for 9 years at 8% interest compounded annually. (*Hint:* the first payment earns interest for 8 years and amounts to $(1.08)^8$; the second amounts to $(1.08)^7$, and so on. The amounts form a geometric sequence with $r = 1.08$.)

72. Payments of $800 at the end of each year for 12 years at 10% interest compounded annually.

73. Payments of $2430 at the end of each year for 10 years at 11% interest compounded annually.

74. Payments of $1500 at the end of each year for 6 years at 12% interest compounded annually.

Find a_1 and r for each of the following geometric sequences. See Example 2.

75. $a_2 = 6$, $a_6 = 486$

76. $a_3 = -12$, $a_6 = 96$

77. $a_2 = 64$, $a_8 = 1$

78. $a_2 = 100$, $a_5 = 1/10$

Explain why the following sequences are geometric.

79. $\log 6, \log 6^2, \log 6^4, \log 6^8, \ldots$

80. $\log 2, \log 4, \log 16, \log 256, \ldots$

10.4   Sums of Infinite Geometric Sequences

In the last section the sum of the first *n* terms of a geometric sequence was given as

$$S_n = \sum_{i=1}^{n} a_i = \frac{a_1(1 - r^n)}{1 - r},$$

where a_1 is the first term and r $(r \neq 1)$ is the common ratio. Now we extend this discussion to include infinite geometric sequences such as the infinite sequence

$$2, 1, \frac{1}{2}, \frac{1}{4}, \frac{1}{8}, \frac{1}{16}, \ldots$$

with first term 2 and common ratio 1/2. Using the formula above gives the following sequence.

$$S_1 = 2, \; S_2 = 3, \; S_3 = \frac{7}{2}, \; S_4 = \frac{15}{4}, \; S_5 = \frac{31}{8}, \; S_6 = \frac{63}{16}$$

These sums seem to be getting closer and closer to the number 4. In fact, by selecting a value of *n* large enough, we can make S_n as close as desired to 4. This is expressed as

$$\lim_{n \to \infty} S_n = 4.$$

(Read: ''the limit of S_n as n increases without bound is 4.'') For no value of n is $S_n = 4$. However, if n is large enough, then S_n is as close to 4 as desired.*

Since

$$\lim_{n \to \infty} S_n = 4,$$

the number 4 is called the *sum* of the infinite geometric sequence

$$2, 1, \frac{1}{2}, \frac{1}{4}, \ldots$$

or

$$2 + 1 + \frac{1}{2} + \frac{1}{4} + \frac{1}{8} + \ldots = 4.$$

EXAMPLE 1   Find $1 + \dfrac{1}{3} + \dfrac{1}{9} + \dfrac{1}{27} + \ldots$.

Use the formula for the first n terms of a geometric sequence to get

$$S_1 = 1, S_2 = \frac{4}{3}, S_3 = \frac{13}{9}, S_4 = \frac{40}{27},$$

and in general

$$S_n = \frac{1\left[1 - \left(\frac{1}{3}\right)^n\right]}{1 - \frac{1}{3}}.$$

The chart below shows the value of $(1/3)^n$ for larger and larger values of n.

n	1	10	100	200
$\left(\dfrac{1}{3}\right)^n$	$\dfrac{1}{3}$	1.69×10^{-5}	1.94×10^{-48}	3.76×10^{-96}

As n gets larger and larger, $(1/3)^n$ gets closer and closer to 0. That is,

$$\lim_{n \to \infty} \left(\frac{1}{3}\right)^n = 0,$$

making it reasonable that

$$\lim_{n \to \infty} S_n = \lim_{n \to \infty} \frac{1\left[1 - \left(\frac{1}{3}\right)^n\right]}{1 - \frac{1}{3}} = \frac{1(1 - 0)}{1 - \frac{1}{3}} = \frac{1}{\frac{2}{3}} = \frac{3}{2}.$$

Hence,

$$1 + \frac{1}{3} + \frac{1}{9} + \frac{1}{27} + \ldots = \frac{3}{2}. \quad \bullet$$

---

*These phrases ''large enough'' and ''as close as desired'' are not nearly precise enough for mathematicians; much of a standard calculus course is devoted to making them more precise.

If a geometric sequence has a first term a_1 and a common ratio r, then

$$S_n = \frac{a_1(1 - r^n)}{1 - r}$$

for every positive integer n. If $-1 < r < 1$, then $\lim_{n \to \infty} r^n = 0$, and

$$\lim_{n \to \infty} S_n = \frac{a_1(1 - 0)}{1 - r} = \frac{a_1}{1 - r}.$$

This quotient, $a_1/(1 - r)$, is called the **sum of an infinite geometric sequence.** The limit $\lim_{n \to \infty} S_n$ is often expressed as S_∞ or $\sum_{i=1}^{\infty} a_i$. These results lead to the following definition.

Sum of an Infinite Geometric Sequence

> The sum of an infinite geometric sequence with first term a_1 and common ratio r, where $-1 < r < 1$, is given by
>
> $$S_\infty = \lim_{n \to \infty} S_n = \sum_{i=1}^{\infty} a_i = \frac{a_1}{1 - r}.$$

EXAMPLE 2   **(a)** Find the sum

$$-\frac{3}{4} + \frac{3}{8} - \frac{3}{16} + \frac{3}{32} - \frac{3}{64} + \ldots .$$

The first term is $a_1 = -3/4$. To find r, divide any two adjacent terms. For example,

$$r = \frac{-\dfrac{3}{16}}{\dfrac{3}{8}} = -\frac{1}{2}.$$

Since $-1 < r < 1$, the formula in this section applies, and

$$S_\infty = \frac{a_1}{1 - r} = \frac{-\dfrac{3}{4}}{1 - \left(-\dfrac{1}{2}\right)} = -\frac{1}{2}.$$

(b) $\displaystyle\sum_{i=1}^{\infty} \left(\frac{3}{5}\right)^i = \frac{\dfrac{3}{5}}{1 - \dfrac{3}{5}} = \frac{3}{2}$ ●

The formula in this section can be used to convert repeating decimals (which represent rational numbers) to fractions of the form p/q, where p and q are integers, with $q \neq 0$.

EXAMPLE 3   Write each repeating decimal in the form p/q, where p and q are integers.

(a) .090909 . . .

This decimal can be written as

$$.090909 . . . = .09 + .0009 + .000009 + . . . ,$$

which is the sum of the terms of an infinite geometric sequence having $a_1 = .09$ and $r = .01$. The sum of this sequence is given by

$$S_\infty = \frac{a_1}{1 - r} = \frac{.09}{1 - .01} = \frac{.09}{.99} = \frac{1}{11}.$$

Thus, .090909 . . . $= 1/11$.

(b) 2.5121212 . . .

Write the number as

$$2.5121212 . . . = 2.5 + .012 + .00012 + .0000012 + ...$$
$$= 2.5 + (.012 + .00012 + .0000012 + . . .).$$

Beginning with the second term, this is the sum of the terms of an infinite geometric sequence with $a_1 = .012$ and $r = .01$, so 2.5121212 . . . can be written

$$2.5 + \frac{.012}{1 - .01} = 2.5 + \frac{.012}{.990} = 2.5 + \frac{2}{165} = \frac{829}{330}. \quad \bullet$$

10.4 EXERCISES

Find r for each of the following infinite geometric sequences. Identify any whose sums would exist. See Example 2.

1. 9, 18, 36, 72, 144, . . .

2. 3, 9, 27, 81, . . .

3. 10, 100, 1000, 10,000, . . .

4. $-8, -4, -2, -1, -1/2, . . .$

5. 12, 6, 3, 3/2, . . .

6. $-8, -16, -32, -64, . . .$

7. 1, 1.1, 1.21, 1.331, . . .

8. $-1, 1.2, -1.44, 1.728, . . .$

9. $1, -.9, .81, -.729, . . .$

10. 1, .7, .49, .343, . . .

Find each of the following sums by using the formula of this section where it applies. See Example 2.

11. $16 + 4 + 1 + . . .$

12. $81 + 27 + 9 + 3 + 1 + . . .$

13. $100 + 10 + 1 + . . .$

14. $128 + 64 + 32 + . . .$

15. $90 + 30 + 10 + . . .$

16. $25 + 5 + 1 + . . .$

17. $256 - 128 + 64 - 32 + 16 - . . .$

18. $120 - 60 + 30 - 15 + . . .$

19. $108 - 36 + 12 - 4 + . . .$

20. $10,000 - 1000 + 100 - 10 + 1 - . . .$

21. $\dfrac{3}{4} + \dfrac{3}{8} + \dfrac{3}{16} + \ldots$

22. $\dfrac{4}{5} + \dfrac{2}{5} + \dfrac{1}{5} + \ldots$

23. $3 - \dfrac{3}{2} + \dfrac{3}{4} - \ldots$

24. $9 - 3 + 1 - \ldots$

25. $\dfrac{1}{3} - \dfrac{2}{9} + \dfrac{4}{27} - \dfrac{8}{81} + \ldots$

26. $1 + \dfrac{1}{1.01} + \dfrac{1}{(1.01)^2} + \ldots$

27. $\dfrac{1}{36} + \dfrac{1}{30} + \dfrac{1}{25} + \ldots$

28. $1 + \dfrac{1}{2^2} + \dfrac{1}{2^4} + \ldots$

29. $\displaystyle\sum_{i=1}^{\infty} (1/4)^i$

30. $\displaystyle\sum_{i=1}^{\infty} (9/10)^i$

31. $\displaystyle\sum_{i=1}^{\infty} (1.2)^i$

32. $\displaystyle\sum_{i=1}^{\infty} (1.001)^i$

33. $\displaystyle\sum_{i=1}^{\infty} (-1/4)^i$

34. $\displaystyle\sum_{i=1}^{\infty} (.3)^i$

35. $\displaystyle\sum_{i=1}^{\infty} 1/5^i$

36. $\displaystyle\sum_{i=1}^{\infty} (-1/2)^i$

37. $\displaystyle\sum_{i=1}^{\infty} 10^{-i}$

38. $\displaystyle\sum_{i=1}^{\infty} 4^{-i}$

39. $\displaystyle\sum_{i=1}^{\infty} (1/2)^{-i}$

40. $\displaystyle\sum_{i=1}^{\infty} (3/4)^{-i}$

Express each repeating decimal in the form p/q, where p and q are integers. See Example 3.

41. $0.55555\ldots$

42. $0.33333\ldots$

43. $0.121212\ldots$

44. $0.858585\ldots$

45. $0.313131\ldots$

46. $0.909090\ldots$

47. $0.508508508\ldots$

48. $0.613613613\ldots$

49. $0.3455555\ldots$

50. $0.5688888\ldots$

51. Mitzi drops a ball from a height of 10 m and notices that on each bounce the ball returns to about 3/4 of its previous height. About how far will the ball travel before it comes to rest? (*Hint:* consider the sum of two sequences.)

52. A sugar factory receives an order for 1000 units of sugar. The production manager thus orders production of 1000 units of sugar. He forgets, however, that the production of sugar requires some sugar (to prime the machines, for example), and so he ends up with only 900 units of sugar. He then orders an additional 100 units, and receives only 90 units. A further order for 10 units produces 9 units. Finally seeing he is wrong, the manager decides to try mathematics. He views the production process as an infinite geometric progression with $a_1 = 1000$ and $r = .1$. Using this, find the number of units of sugar that he should have ordered originally.

53. After a person pedaling a bicycle removes his or her feet from the pedals, the wheel rotates 400 times the first minute. As it continues to slow down, each minute it rotates only 3/4 as many times as in the previous minute. How many times will the wheel rotate before coming to a complete stop?

54. A pendulum bob swings through an arc 40 cm long on its first swing. Each swing thereafter, it swings only 80% as far as on the previous swing. How far will it swing altogether before coming to a complete stop?

55. A sequence of equilateral triangles is constructed. The first triangle has sides 2 m in length. To get the second triangle, midpoints of the sides of the original triangle are connected. If this process could be continued indefinitely, what would be the total perimeter of all the triangles?

56. What would be the total area of all the triangles in Exercise 55, disregarding the overlapping?

10.5 Mathematical Induction

Many results in mathematics are claimed to be true for any positive integer. Any of these results could be checked for $n = 1$, $n = 2$, $n = 3$, and so on, but since the set of positive integers is infinite it would be impossible to check every possible case. For example, let S_n represent the statement that the sum of the first n positive integers is $n(n + 1)/2$,

$$S_n: 1 + 2 + 3 + \ldots + n = \frac{n(n + 1)}{2}.$$

The truth of this statement can be checked quickly for the first few values of n.

If $n = 1$, S_1 is $\qquad 1 = \dfrac{1(1 + 1)}{2}$, a true statement.

If $n = 2$, S_2 is $\qquad 1 + 2 = \dfrac{2(2 + 1)}{2}$, a true statement.

If $n = 3$, S_3 is $\qquad 1 + 2 + 3 = \dfrac{3(3 + 1)}{2}$, a true statement.

If $n = 4$, S_4 is $\quad 1 + 2 + 3 + 4 = \dfrac{4(4 + 1)}{2}$, a true statement.

We could continue in this way as long as we wanted, yet we could never prove that S_n is true for *every* positive integer value of n. To prove that such statements are true for every positive integer value of n, use the following principle.

Principle of Mathematical Induction

> Let S_n be a statement concerning the positive integer n. Suppose that
>
> **1.** S_1 is true;
>
> **2.** when S_n is true for the positive integer k, then S_n is also true for the integer $k + 1$ (that is, the truth of S_k implies the truth of S_{k+1});
>
> then S_n is true for every positive integer value of n.

A proof by mathematical induction can be explained as follows. By (1) above, the statement is true when $n = 1$. By (2) above, the fact that the statement is true for $n = 1$ implies that it is true for $n = 1 + 1 = 2$. Using (2) again, the statement is thus true for $2 + 1 = 3$, for $3 + 1 = 4$, for $4 + 1 = 5$, and so on. Continuing in this way shows that the statement must be true for *every* positive integer, no matter how large.

The situation is similar to that of a number of dominoes lined up as shown in Figure 10.1. If the first domino is pushed over, it pushes the next, which pushes the next, and so on until all are down.

Figure 10.1

Another example of the principle of mathematical induction might be an infinite ladder. Suppose the rungs are spaced so that, whenever you are on a rung, you know you can move to the next rung. Then *if* you can get to the first rung, you can go as high up the ladder as you wish.

Two separate steps are required for a proof by mathematical induction:

Step 1 Prove that the statement is true for $n = 1$.

Step 2 Assume that S_n is true for the positive integer k. Show that this implies that S_n is also true for the positive integer $k + 1$.

In the next example we use mathematical induction to prove the statement S_n discussed above.

EXAMPLE 1   Let S_n represent the statement

$$1 + 2 + 3 + \ldots + n = \frac{n(n + 1)}{2}.$$

Prove that S_n is true for every positive integer n.

The proof by mathematical induction is as follows.

Step 1 Show that the statement is true when $n = 1$. If $n = 1$, S_1 becomes

$$1 = \frac{1(1 + 1)}{2},$$

which is true.

Step 2 Assume that S_n is true for the statement $n = k$. That is, assume that

$$1 + 2 + 3 + \ldots + k = \frac{k(k + 1)}{2}$$

is true. Show that this implies the truth of S_{k+1}, where S_{k+1} is the statement

$$1 + 2 + 3 + \ldots + k + (k + 1) = \frac{(k + 1)[(k + 1) + 1]}{2}.$$

We have assumed that

$$1 + 2 + 3 + \ldots + k = \frac{k(k + 1)}{2}.$$

Add $k + 1$ to both sides of the known equation.

$$1 + 2 + 3 + \ldots + k + (k + 1) = \frac{k(k + 1)}{2} + (k + 1)$$

Factoring on the right gives

$$= (k + 1)\left(\frac{k}{2} + 1\right)$$

$$= (k + 1)\left(\frac{k + 2}{2}\right)$$

$$1 + 2 + 3 + \ldots + k + (k + 1) = \frac{(k + 1)[(k + 1) + 1]}{2}.$$

This final result is the statement we wished to establish for $n = k + 1$; therefore the truth of S_k implies the truth of S_{k+1}. The two steps required for a proof by mathematical induction have now been completed, so our statement S_n is true for every positive integer value of n. ●

EXAMPLE 2

Prove: $4 + 7 + 10 + \ldots + (3n + 1) = \dfrac{n(3n + 5)}{2}$.

Step 1 Show that the statement is true when $n = 1$. If $n = 1$, S_n becomes

$$4 = \frac{1(3 \cdot 1 + 5)}{2},$$

a true statement.

Step 2 Assume that S_k is true. That is, assume that

$$4 + 7 + 10 + \ldots + (3k + 1) = \frac{k(3k + 5)}{2}$$

is true. Show that this implies the truth of S_{k+1}, where S_{k+1} is

$$4 + 7 + 10 + \ldots + (3k + 1) + [3(k + 1) + 1]$$
$$= \frac{(k + 1)[3(k + 1) + 5]}{2}.$$

We have assumed that

$$4 + 7 + 10 + \ldots + (3k + 1) = \frac{k(3k + 5)}{2}.$$

Adding $[3(k + 1) + 1]$ to both sides gives

$$4 + 7 + 10 + \ldots + (3k + 1) + [3(k + 1) + 1]$$
$$= \frac{k(3k + 5)}{2} + [3(k + 1) + 1].$$

Clear the parentheses in the new term on the right side of the equals sign and simplify.

$$= \frac{k(3k + 5)}{2} + 3k + 3 + 1$$

$$= \frac{k(3k + 5)}{2} + 3k + 4$$

Now combine the two terms on the right.

$$= \frac{k(3k + 5)}{2} + \frac{2(3k + 4)}{2}$$

$$= \frac{k(3k + 5) + 2(3k + 4)}{2}$$

$$= \frac{3k^2 + 5k + 6k + 8}{2}$$

$$= \frac{3k^2 + 11k + 8}{2}$$

$$= \frac{(k + 1)(3k + 8)}{2}$$

Finally, $4 + 7 + 10 + \ldots + (3k + 1) + [3(k + 1) + 1]$

$$= \frac{(k + 1)[3(k + 1) + 5]}{2}.$$

This final result is the statement we want for S_{k+1}. Therefore the truth of S_k implies the truth of S_{k+1}. The two steps required for a proof by mathematical induction are completed, so the general statement S_n is true for every positive integer value of n. ●

EXAMPLE 3   Prove that if x is a real number between 0 and 1, then for every positive integer n,

$$0 < x^n < 1.$$

Here S_1 is: if $0 < x < 1$, then $0 < x^1 < 1$, which is true. Assume now that S_k is true:

$$\text{if } 0 < x < 1, \text{ then } 0 < x^k < 1.$$

We must now show that this implies the truth of S_{k+1}. Multiply all members of $0 < x^k < 1$ by x to get

$$x \cdot 0 < x \cdot x^k < x \cdot 1.$$

(Here we use the fact that $0 < x$.) Simplify to get

$$0 < x^{k+1} < x.$$

Since $x < 1$,

$$x^{k+1} < x < 1$$

and $$0 < x^{k+1} < 1.$$

By this work, the truth of S_k implies the truth of S_{k+1}, so the given statement is true for every positive integer n.   ●

10.5 EXERCISES

Use the method of mathematical induction to prove the following statements. Assume that n is a positive integer. See Examples 1–3.

1. $2 + 4 + 6 \ldots + 2n = n(n + 1)$

2. $1 + 3 + 5 + \ldots + (2n - 1) = n^2$

3. $3 + 6 + 9 \ldots + 3n = \dfrac{3n(n + 1)}{2}$

4. $5 + 10 + 15 + \ldots + 5n = \dfrac{5n(n + 1)}{2}$

5. $2 + 4 + 8 + \ldots + 2^n = 2^{n+1} - 2$

6. $3 + 3^2 + 3^3 + \ldots + 3^n = \dfrac{3(3^n - 1)}{2}$

7. $1^2 + 2^2 + 3^2 + \ldots + n^2 = \dfrac{n(n + 1)(2n + 1)}{6}$

8. $1^3 + 2^3 + 3^3 + \ldots + n^3 = \dfrac{n^2(n + 1)^2}{4}$

9. $5 \cdot 6 + 5 \cdot 6^2 + 5 \cdot 6^3 + \ldots + 5 \cdot 6^n = 6(6^n - 1)$

10. $7 \cdot 8 + 7 \cdot 8^2 + 7 \cdot 8^3 + \ldots + 7 \cdot 8^n = 8(8^n - 1)$

11. $\dfrac{1}{1 \cdot 2} + \dfrac{1}{2 \cdot 3} + \dfrac{1}{3 \cdot 4} + \ldots + \dfrac{1}{n(n + 1)} = \dfrac{n}{n + 1}$

12. $\dfrac{1}{1 \cdot 4} + \dfrac{1}{4 \cdot 7} + \dfrac{1}{7 \cdot 10} + \ldots + \dfrac{1}{(3n - 2)(3n + 1)} = \dfrac{n}{3n + 1}$

13. $\dfrac{1}{2} + \dfrac{1}{2^2} + \dfrac{1}{2^3} + \ldots + \dfrac{1}{2^n} = 1 - \dfrac{1}{2^n}$

14. $\dfrac{4}{5} + \dfrac{4}{5^2} + \dfrac{4}{5^3} + \ldots + \dfrac{4}{5^n} = 1 - \dfrac{1}{5^n}$

15. $x^{2n} + x^{2n-1}y + \ldots + xy^{2n-1} + y^{2n} = \dfrac{x^{2n+1} - y^{2n+1}}{x - y}$

16. $x^{2n-1} + x^{2n-2}y + \ldots + xy^{2n-2} + y^{2n-1} = \dfrac{x^{2n} - y^{2n}}{x - y}$

17. $(a^m)^n = a^{mn}$ (Assume that a and m are constant.)

18. $(ab)^n = a^n b^n$ (Assume that a and b are constant.)

19. If $a > 1$, then $a^n > 1$.

20. If $a > 1$, then $a^n > a^{n-1}$.

21. If $0 < a < 1$, then $a^n < a^{n-1}$.

22. $2^n > n^2$

23. $4^n > n^4$, for $n \geq 7$

24. What is wrong with the following proof by mathematical induction?

Prove: Any natural number equals the next natural number; that is, $n = n + 1$.

Proof. To begin, we assume the statement true for some natural number $n = k$:

$$k = k + 1.$$

We must now show that the statement is true for $n = k + 1$. If we add 1 to both sides, we have

$$k + 1 = k + 1 + 1$$
$$k + 1 = k + 2.$$

Hence, if the statement is true for $n = k$, it is also true for $n = k + 1$. Thus, the theorem is proved.

25. Suppose that n straight lines (with $n \geq 2$) are drawn in a plane, where no two lines are parallel and no three lines pass through the same point. Show that the number of points of intersection of the lines is $(n^2 - n)/2$.

26. The series of sketches below starts with an equilateral triangle having sides of length 1. In the following steps, equilateral triangles are constructed on each side of the preceding figure. The lengths of the sides of these new triangles are 1/3 the length of the sides of the preceding triangles. Develop a formula for the number of sides of the nth figure. Use mathematical induction to prove your answer.

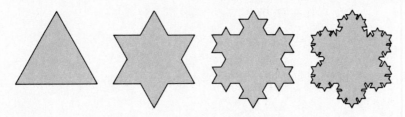

27. Find the perimeter of the nth figure in Exercise 26.

28. Show that the area of the nth figure in Exercise 26 is

$$\sqrt{3} \left[\frac{2}{5} - \frac{3}{20} \left(\frac{4}{9} \right)^{n-1} \right].$$

Chapter 10   Summary

Key Words	sequence	terms
	general term	nth term
	finite sequence	infinite sequence
	series	index of summation
	mean	common difference
	arithmetic sequence	arithmetic progression
	geometric sequence	geometric progression
	common ratio	sum of an infinite geometric sequence

nth Term of an Arithmetic Sequence	In an arithmetic sequence with first term a_1 and common difference d, the nth term, a_n, is given by $$a_n = a_1 + (n - 1)d.$$

Sum of the First n Terms of an Arithmetic Sequence

If an arithmetic sequence has first term a_1 and common difference d, then the sum of the first n terms, S_n, is given by

$$S_n = \frac{n}{2}(a_1 + a_n)$$

or

$$S_n = \frac{n}{2}[2a_1 + (n - 1)d].$$

nth Term of a Geometric Sequence

In the geometric sequence with first term a_1 and common ratio r, the nth term is

$$a_n = a_1 r^{n-1}.$$

Sum of the First n Terms of a Geometric Sequence

If a geometric sequence has first term a_1 and common ratio r, then the sum of the first n terms, S_n, is given by

$$S_n = \frac{a_1(1 - r^n)}{1 - r}, \quad r \neq 1.$$

Sum of an Infinite Geometric Sequence

The sum of an infinite geometric sequence with first term a_1 and common ratio r, where $-1 < r < 1$, is given by

$$S_\infty = \lim_{n \to \infty} S_n = \sum_{i=1}^{\infty} a_i = \frac{a_1}{1 - r}.$$

Principle of Mathematical Induction

Let S_n be a statement concerning the positive integer n. Suppose that

1. S_1 is true;

2. when S_n is true for the positive integer k, then S_n is also true for the integer $k + 1$ (that is, the truth of S_k implies the truth of S_{k+1});

then S_n is true for every positive integer value of n.

Chapter 10   Review Exercises

Write the first five terms for each of the following sequences.

1. $a_n = \dfrac{n}{n + 1}$ **2.** $a_n = (-2)^n$ **3.** $a_n = 2(n + 3)$

4. $a_n = n(n + 1)$ **5.** $a_1 = 5, a_n = a_{n-1} - 3$ **6.** $a_1 = -2, a_n = 3a_{n-1}$

7. $a_1 = 5, a_2 = 3, a_n = a_{n-1} - a_{n-2}$ for $n \geq 2$

8. $b_1 = -2, b_2 = 2, b_3 = -4, b_n = -2 \cdot b_{n-2}$ if n is even, and $b_n = 2 \cdot b_{n-2}$ if n is odd.

9. Arithmetic, $a_1 = 6$, $d = -4$

10. Arithmetic, $a_3 = 9$, $a_4 = 7$

11. Arithmetic, $a_1 = 3 - \sqrt{5}$, $a_2 = 4$

12. Arithmetic, $a_3 = \pi$, $a_4 = 0$

13. Geometric, $a_1 = 4$, $r = 2$

14. Geometric, $a_4 = 8$, $r = 1/2$

15. Geometric, $a_1 = -3$, $a_2 = 4$

16. Geometric, $a_3 = 8$, $a_5 = 72$

17. A certain arithmetic sequence has $a_6 = -4$ and $a_{17} = 51$. Find a_1 and a_{20}.

18. For a given geometric sequence, $a_1 = 4$ and $a_5 = 324$. Find a_6.

Find a_8 for each of the following arithmetic sequences.

19. $a_1 = 6$, $d = 2$

20. $a_1 = -4$, $d = 3$

21. $a_1 = 6x - 9$, $a_2 = 5x + 1$

22. $a_3 = 11m$, $a_5 = 7m - 4$

Find S_{12} for each of the following arithmetic sequences.

23. $a_1 = 2$, $d = 3$ 24. $a_2 = 6$, $d = 10$ 25. $a_1 = -4k$, $d = 2k$

Find a_5 for each of the following geometric sequences.

26. $a_1 = 3$, $r = 2$ 27. $a_2 = 3125$, $r = 1/5$ 28. $a_1 = 6$, $a_3 = 24$

29. $a_1 = 5x$, $a_2 = x^2$ 30. $a_2 = \sqrt{6}$, $a_4 = 6\sqrt{6}$

Find S_4 for each of the following geometric sequences.

31. $a_1 = 1$, $r = 2$ 32. $a_1 = 3$, $r = 3$ 33. $a_1 = 2k$, $a_2 = -4k$

Evaluate each of the following sums that exist.

34. $18 + 9 + 9/2 + 9/4 + \ldots$

35. $20 + 15 + 45/4 + 135/16 + \ldots$

36. $-5/6 + 5/9 - 10/27 + \ldots$

37. $1/16 + 1/8 + 1/4 + 1/2 + \ldots$

38. $.9 + .09 + .009 + .0009 + \ldots$

Convert each of the following repeating decimals to rational numbers.

39. $.6666\ldots$ 40. $2.1333\ldots$ 41. $.512512512\ldots$ 42. $.2499999\ldots$

Evaluate each of the following sums that exist.

43. $\displaystyle\sum_{i=1}^{4} \frac{2}{i}$ 44. $\displaystyle\sum_{i=1}^{7} (-1)^{i+1} \cdot 6$ 45. $\displaystyle\sum_{i=4}^{8} 3i(2i - 5)$ 46. $\displaystyle\sum_{i=1}^{6} i(i + 2)$

47. $\displaystyle\sum_{i=1}^{4} \frac{i + 1}{i}$ 48. $\displaystyle\sum_{i=1}^{12} (8i + 2)$ 49. $\displaystyle\sum_{i=1}^{10,000} i$ 50. $\displaystyle\sum_{i=1}^{6} 4 \cdot 3^i$

51. $\displaystyle\sum_{i=1}^{4} 8 \cdot 2^i$ 52. $\displaystyle\sum_{i=1}^{\infty} \left(\frac{5}{8}\right)^i$ 53. $\displaystyle\sum_{i=1}^{\infty} -10\left(\frac{5}{2}\right)^i$ 54. $\displaystyle\sum_{i=1}^{\infty} 6\left(\frac{2}{3}\right)^i$

55. A stack of canned goods in a market display requires 15 cans on the bottom, 13 in the next layer, 11 in the next layer, and so on. How many cans are needed for the display?

56. Gale Stockdale borrows $3000 at simple interest of 12% per year. He will repay the loan in monthly payments of $130, $129, $128, and so on. If he makes 30 payments, what is the total amount required to pay off the loan plus the interest?

57. On a certain production line, during their first week, new employees turn out 5/4 as many items each day as on the previous day. If a new employee produces 48 items the first day, how many will she produce on the fifth day of work?

58. The half-life of a radioactive substance is 20 years. If 600 g are present at the start, how much will be left after 100 years?

Use mathematical induction to prove that each of the following is true for every positive integer n.

59. $2 + 6 + 10 + 14 + \ldots + (4n - 2) = 2n^2$

60. $2^2 + 4^2 + 6^2 + \ldots + (2n)^2 = \dfrac{2n(n + 1)(2n + 1)}{3}$

61. $2 + 2^2 + 2^3 + \ldots + 2^n = 2(2^n - 1)$

62. $1 \cdot 4 + 2 \cdot 9 + 3 \cdot 16 + \ldots + n(n + 1)^2 = \dfrac{n(n + 1)(n + 2)(3n + 5)}{12}$

63. $1^3 + 3^3 + 5^3 + \ldots + (2n - 1)^3 = n^2(2n^2 - 1)$

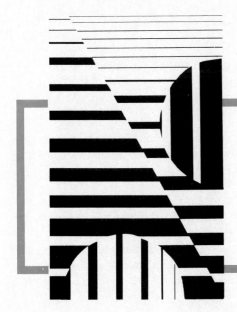

11

Counting and Probability

How many different poker hands can be dealt from a 52-card deck? How many license plates can be made using 2 letters followed by 4 numbers? These are some of the questions answered by *counting theory,* which gives methods of counting the number of ways a certain event can occur. *Probability theory* uses these methods of counting to give the probability of events.

11.1 Permutations

If there are 3 roads from Albany to Baker and 2 roads from Baker to Creswich, in how many ways can one travel from Albany to Creswich by way of Baker? For each of the 3 roads from Albany to Baker, there are 2 different roads from Baker to Creswich, so that there are $3 \cdot 2 = 6$ different ways to make the trip, as shown in the **tree diagram** in Figure 11.1.

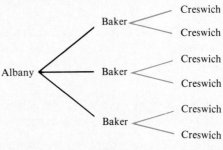

Figure 11.1

This example illustrates the following property of counting.

Fundamental Principle of Counting

> If one event can occur in *m* ways and a second event can occur in *n* ways, then both events can occur in *mn* ways, provided the outcome of the first event does not influence the outcome of the second.

The fundamental principle of counting can be extended to any number of events, provided the outcome of no one event influences the outcome of another. Such events are called **independent events.**

EXAMPLE 1

A restaurant offers a choice of 3 salads, 5 main dishes, and 2 desserts. Use the fundamental principle of counting to find the number of different 3-course meals that can be selected.

Three independent events are involved: selecting a salad, selecting a main dish, and selecting a dessert. The first event can occur in 3 ways, the second event can occur in 5 ways, and the third event can occur in 2 ways; thus there are

$$3 \cdot 5 \cdot 2 = 30 \text{ possible meals.} \quad \bullet$$

EXAMPLE 2

Janet Branson has 5 different books that she wishes to arrange on her desk. How many different arrangements are possible?

Five events are involved: selecting a book for the first spot, selecting a book for the second spot, and so on. Here the outcome of the first event *does* influence the outcome of the other events (since one book has already been chosen). For the first spot Branson has 5 choices, for the second spot 4 choices, for the third spot 3 choices, and so on. Now use the fundamental principle of counting to find that there are

$$5 \cdot 4 \cdot 3 \cdot 2 \cdot 1 \text{ or } 120 \text{ different arrangements.} \quad \bullet$$

In using the fundamental principle of counting we often encounter such products as $5 \cdot 4 \cdot 3 \cdot 2 \cdot 1$ from Example 2. For convenience in writing these products, use the symbol *n*! (read ''*n* factorial''), which is defined as follows for any counting number *n*.

$$n! = n(n - 1)(n - 2)(n - 3) \ldots (2)(1)$$

Thus $5 \cdot 4 \cdot 3 \cdot 2 \cdot 1$ is written as 5!. Also, $3! = 3 \cdot 2 \cdot 1 = 6$. By the definition of *n*!, $n[(n - 1)!] = n!$ for all natural numbers $n \geq 2$. It is convenient to have this relation hold also for $n = 1$, so, by definition,

$$0! = 1.$$

EXAMPLE 3

Suppose Branson wishes to place only 3 of the 5 books on her desk. How many arrangements of 3 books are possible?

She again has 5 ways to fill the first spot, 4 ways to fill the second spot, and 3 ways to fill the third. Since she wants to use only 3 books, there are only 3 spots to be filled (3 events) instead of 5, with

$$5 \cdot 4 \cdot 3 = 60 \text{ arrangements.} \quad \bullet$$

The number 60 in the example above is called the number of *permutations* of 5 things taken 3 at a time, written $P(5, 3) = 60$. In Example 2 we found the number of ways of arranging 5 elements from a set of 5 elements, written $P(5, 5) = 120$.

Permutations of n Elements n at a Time

If $P(n, n)$ denotes the number of permutations of n elements taken n at a time, then

$$P(n, n) = n!$$

A **permutation** of n elements taken r at a time is the number of ways of arranging r elements taken from a set of n elements $(r \le n)$. Generalizing from the examples above,

$$P(n, r) = n(n - 1)(n - 2) \cdots (n - r + 1)$$
$$= \frac{n(n - 1)(n - 2) \cdots (n - r + 1)(n - r)(n - r - 1) \cdots (2)(1)}{(n - r)(n - r - 1) \cdots (2)(1)}$$
$$= \frac{n!}{(n - r)!}.$$

Summarizing, we have the following result.

Permutations of n Elements r at a Time

If $P(n, r)$ denotes the number of permutations of n elements taken r at a time, then

$$P(n, r) = \frac{n!}{(n - r)!}.$$

EXAMPLE 4   Suppose 8 people enter an event in a swim meet. In how many ways could the gold, silver, and bronze prizes be awarded?

Using the fundamental principle of counting, there are 3 choices to be made giving $8 \cdot 7 \cdot 6 = 336$. However, we can also use the formula for $P(n, r)$ to get the same result.

$$P(8, 3) = \frac{8!}{5!} = \frac{8 \cdot 7 \cdot 6 \cdot 5 \cdot 4 \cdot 3 \cdot 2 \cdot 1}{5 \cdot 4 \cdot 3 \cdot 2 \cdot 1} = 8 \cdot 7 \cdot 6 = 336. \quad \bullet$$

EXAMPLE 5   In how many ways can 6 students be seated in a row of 6 desks?

$$P(6, 6) = 6! = 6 \cdot 5 \cdot 4 \cdot 3 \cdot 2 \cdot 1 = 720 \quad \bullet$$

11.1 EXERCISES

Evaluate each of the following. See Examples 4 and 5.

1. $P(7, 7)$ 2. $P(5, 3)$ 3. $P(6, 5)$ 4. $P(4, 2)$

5. $P(10, 2)$ 6. $P(8, 2)$ 7. $P(8, 3)$ 8. $P(11, 4)$

9. $P(7, 1)$ 10. $P(18, 0)$ 11. $P(9, 0)$ 12. $P(14, 1)$

13. In how many ways can 8 people be seated in a row of 8 seats?

14. In how many ways can 7 out of 10 people be assigned to 7 seats?

15. In how many ways can 5 players be assigned to the 5 positions on a basketball team, assuming that any player can play any position? In how many ways can 10 players be assigned to the 5 positions?

16. A couple has narrowed down their choice of names for a new baby to 3 first names and 5 middle names. How many different first and middle name arrangements are possible?

17. How many different homes are available if a builder offers a choice of 5 basic plans, 3 roof styles, and 2 types of siding?

18. An automaker produces 7 models, each available in 6 colors, with 4 upholstery fabrics and 5 interior colors. How many varieties of the auto are available?

19. A concert is to consist of five works: two modern, two romantic, and one classical. In how many ways can the program be arranged?

20. If the program in Exercise 19 must be shortened to 3 works chosen from the 5, how many arrangements are possible?

21. In Exercise 19, how many different programs are possible if the 2 modern works are to be played first, then the 2 romantic, and then the classical?

22. How many 4-letter radio station call letters can be made if the first letter must be K or W and no letter may be repeated? How many if repeats are allowed?

23. How many of the 4-letter call letters in Exercise 22 with no repeats end in K?

24. How many 7-digit telephone numbers are possible if the first digit cannot be zero and
    (a) only odd digits can be used;
    (b) the number must be a multiple of 10 (that is, must end in zero);
    (c) the number must be a multiple of 100;
    (d) the first three digits are 481;
    (e) no repetitions are allowed?

25. (a) How many different license numbers consisting of 3 letters followed by 3 digits are possible?
    (b) How many are possible if either the 3 digits or the 3 letters can come first?

26. A business school gives courses in typing, shorthand, transcription, business English, technical writing, and accounting. How many ways can a student arrange his program if he takes 3 courses?

27. How many different license plate numbers can be formed using 3 letters followed by 3 digits if no repeats are allowed? How many if there are no repeats and either letters or numbers come first?

28. Show that $P(n, n - 1) = P(n, n)$.

11.2   Combinations

In the last section we discussed a method for finding the number of ways we can arrange r elements taken from a set of n elements. Sometimes, however, we are not interested in the arrangement (or order) of the elements.

For example, suppose three people (Ms. Opelka, Mr. Adams, and Ms. Jacobs) apply for 2 identical jobs. Ignoring all other factors, in how many ways can the personnel officer select 2 people from the 3 applicants? Here the arrangement or order of the people is unimportant. Selecting Ms. Opelka and Mr. Adams is the same as selecting Mr. Adams and Ms. Opelka. Therefore, there are only 3 ways to select 2 of the 3 applicants:

> Ms. Opelka and Mr. Adams;
>
> Ms. Opelka and Ms. Jacobs;
>
> Mr. Adams and Ms. Jacobs.

These three choices are called the combinations of 3 elements taken 2 at a time. A **combination** of n elements taken r at a time is the number of ways in which r elements can be chosen from n elements.

Each combination of r elements forms $r!$ permutations. Therefore, we can find the number of combinations of n elements taken r at a time by dividing the number of permutations, $P(n, r)$, by $r!$ to get

$$\frac{P(n, r)}{r!}$$

combinations. This expression can be rewritten as follows.

$$\frac{P(n, r)}{r!} = \frac{\dfrac{n!}{(n - r)!}}{r!} = \frac{n!}{(n - r)!r!}$$

The symbol $\binom{n}{r}$ is used to represent the number of combinations of n things taken r at a time. With this symbol the results above are stated as follows.

Combinations of n Elements r at a Time

> If $\binom{n}{r}$ represents the number of combinations of n things taken r at a time, then
>
> $$\binom{n}{r} = \frac{n!}{(n - r)!r!}.$$

In the discussion above, we found that $\binom{3}{2} = 3$. The same result can be found using the formula.

$$\binom{3}{2} = \frac{3!}{(3 - 2)!2!} = \frac{3 \cdot 2 \cdot 1}{1 \cdot 2 \cdot 1} = 3$$

EXAMPLE 1 How many different committees of 3 people can be chosen from a group of 8 people?

We are not interested in the order in which the members of the committee are chosen, so use combinations to get

$$\binom{8}{3} = \frac{8!}{5!3!} = \frac{8 \cdot 7 \cdot 6 \cdot 5 \cdot 4 \cdot 3 \cdot 2 \cdot 1}{5 \cdot 4 \cdot 3 \cdot 2 \cdot 1 \cdot 3 \cdot 2 \cdot 1} = 56. \quad \bullet$$

EXAMPLE 2 From a group of 30 employees, 3 are to be selected to work on a special project.

(a) In how many different ways can the employees be selected?

Here we wish to know the number of 3-element combinations from a set of 30 elements. (We want combinations, not permutations, because we do not care about order within the group of 3.) Using the formula gives

$$\binom{30}{3} = \frac{30!}{27!3!} = 4060.$$

There are 4060 ways to select the project group.

(b) In how many different ways can the group of 3 be selected if it has already been decided that a certain employee must work on the project?

Since one employee has already been selected to work on the project, the problem is reduced to selecting 2 more employees from the 29 employees that are left:

$$\binom{29}{2} = \frac{29!}{27!2!} = 406.$$

In this case, the project group can be selected in 406 different ways. $\quad \bullet$

EXAMPLE 3 Find $\binom{n}{n}$.

Use the formula with $r = n$.

$$\binom{n}{n} = \frac{n!}{(n-n)!n!} = \frac{n!}{0!n!} = \frac{n!}{1 \cdot n!} = 1$$

This is reasonable since there is only one way of selecting a group of n objects from a set of n objects when order is disregarded. $\quad \bullet$

EXAMPLE 4 Find $\binom{n}{1}$.

Again use the formula.

$$\binom{n}{1} = \frac{n!}{(n-1)!1!} = \frac{n \cdot (n-1)!}{(n-1)!1!} = n$$

This result agrees with what we already know: there are n distinct one-element subsets of an n-element set. $\quad \bullet$

The results of Examples 3 and 4 can be stated as properties and applied directly to special cases. Thus, from Example 3,

$$\binom{5}{5} = 1, \qquad \binom{8}{8} = 1,$$

and so on. Also, from Example 4,

$$\binom{6}{1} = 6 \quad \text{and} \quad \binom{9}{1} = 9.$$

EXAMPLE 5

Show that $\binom{5}{3} = \binom{5}{2}$.

From the formula,

$$\binom{5}{3} = \frac{5!}{3!2!} \quad \text{and} \quad \binom{5}{2} = \frac{5!}{2!3!}$$

Since multiplication is commutative,

$$\binom{5}{3} = \binom{5}{2}. \quad \bullet$$

Example 5 shows that there are the same number of ways to select three things from a group of five as to select two things from a group of five. This makes sense, because each time you choose a set of two, you are also choosing a set of three, the three not selected.

Example 5 is a special case of the following result.

If $\binom{n}{r}$ represents the number of combinations of n things taken r at a time, then

$$\binom{n}{r} = \binom{n}{n-r}.$$

The proof of this result is left as an exercise.

11.2 EXERCISES

Evaluate each of the following. See Examples 1 and 2.

1. $\binom{6}{5}$ **2.** $\binom{4}{2}$ **3.** $\binom{8}{5}$ **4.** $\binom{10}{2}$

5. $\binom{15}{4}$ **6.** $\binom{9}{3}$ **7.** $\binom{10}{7}$ **8.** $\binom{10}{3}$

9. $\binom{14}{1}$ **10.** $\binom{20}{2}$ **11.** $\binom{18}{0}$ **12.** $\binom{13}{0}$

13. A club has 30 members. If a committee of 4 is selected at random, how many committees are possible?

14. How many different samples of 3 apples can be drawn from a crate of 25 apples?

15. Three students are to be selected from a group of 12 students to participate in a special class. In how many ways can this be done? In how many ways can the group that will not participate be selected?

16. Hal's Hamburger Heaven sells hamburgers with cheese, relish, lettuce, tomato, mustard, or ketchup. How many different hamburgers can be made using any 3 of the extras?

17. How many different 2-card hands can be dealt from a deck of 52 cards?

18. How many different 13-card bridge hands can be dealt from a deck of 52 cards?

19. Five cards are marked with the numbers 1, 2, 3, 4, and 5, shuffled, and 2 cards are then drawn. How many different 2-card combinations are possible?

20. If a bag contains 15 marbles, how many samples of 2 marbles can be drawn from it? How many samples of 4 marbles?

21. In Exercise 20, if the bag contains 3 yellow, 4 white, and 8 blue marbles, how many samples of 2 can be drawn in which both marbles are blue?

22. In Exercise 14, if it is known that there are 5 rotten apples in the crate:
    (a) How many samples of 3 could be drawn in which all 3 are rotten?
    (b) How many samples of 3 could be drawn in which there are 1 rotten apple and 2 good apples?

23. A city council is composed of 5 liberals and 4 conservatives. Three members are to be selected randomly as delegates to a convention.
    (a) How many delegations are possible?
    (b) How many delegations could have all liberals?
    (c) How many delegations could have 2 liberals and 1 conservative?
    (d) If 1 member of the council serves as mayor, how many delegations are possible which include the mayor?

24. Seven workers decide to send a delegation of 2 to their supervisor to discuss their grievances.
    (a) How many different delegations are possible?
    (b) If it is decided that a certain employee must be in the delegation, how many different delegations are possible?
    (c) If there are 2 women and 5 men in the group, how many delegations would include a woman?

A poker hand is made up of 5 cards drawn at random from a deck of 52 cards. Any 5 cards in one suit are called a flush. The 5 highest cards, that is, the A, K, Q, J, and 10, of any one suit are called a royal flush. Use combinations to set up each of the following. Do not evaluate.

25. Find the total number of all possible poker hands.

26. How many royal flushes in hearts are possible?

27. How many royal flushes in any of the four suits are possible?

28. How many flushes in hearts are possible?

29. How many flushes in any of the four suits are possible?

30. How many combinations of 3 aces and 2 eights are possible?

Solve the following problems by using either combinations or permutations.

31. How many ways can the letters of the word TOUGH be arranged?

32. If Matthew has 8 courses to choose from, how many ways can he arrange his schedule if he must pick 4 of them?

33. How many samples of 3 pineapples can be drawn from a crate of 12?

34. Velma specializes in making different vegetable soups with carrots, celery, beans, peas, mushrooms, and potatoes. How many different soups can she make using any 4 ingredients?

35. Show that $\binom{n}{r} = \binom{n}{n-r}$.

36. Find $\binom{n}{n-1}$.

11.3 Probability

The study of probability has become increasingly popular because of the wide range of practical applications. In this section, we introduce the basic ideas of probability.

Consider an experiment that has one or more possible **outcomes,** each of which is equally likely to occur. For example, the experiment of tossing a coin has two equally likely possible outcomes: landing heads up (*H*) or landing tails up (*T*). Also, the experiment of rolling a die has 6 equally likely outcomes: landing so the face that is up shows 1, 2, 3, 4, 5, or 6 points.

The set *S* of all possible outcomes of a given experiment is called the **sample space** for the experiment. (In this text all sample spaces are finite.) A sample space for the experiment of tossing a coin consists of the outcomes *H* and *T*. This sample space can be written in set notation as

$$S = \{H,\ T\}.$$

Similarly, a sample space for the experiment of rolling a single die is

$$S = \{1,\ 2,\ 3,\ 4,\ 5,\ 6\}.$$

Any subset of the sample space is called an **event.** In the experiment with the die, for example, "the number showing is a three" is an event, say E_1, such that $E_1 = \{3\}$. "The number showing is greater than three" is also an event, say E_2, such that $E_2 = \{4,\ 5,\ 6\}$. To represent the number of outcomes which belong to event *E*, we use $n(E)$. Then $n(E_1) = 1$ and $n(E_2) = 3$.

The **probability** of an event *E*, written $P(E)$, is the ratio of *the number of outcomes in sample space S that belong to event E* to *the total number of outcomes in sample space S.*

Probability of Event E

$$P(E) = \frac{n(E)}{n(S)}$$

To use this definition to find the probability of the event E_1 given above, start with the sample space for the experiment, $S = \{1, 2, 3, 4, 5, 6\}$, and the desired event, $E_1 = \{3\}$. Since $n(E_1) = 1$ and since there are 6 outcomes in the sample space,

$$P(E_1) = \frac{n(E_1)}{n(S)} = \frac{1}{6}.$$

EXAMPLE 1   A single die is rolled. Write the following events in set notation and give the probability for each event.

(a) E_3: the number showing is even
   Use the definition above. Since $E_3 = \{2, 4, 6\}$, $n(E_3) = 3$. As shown above, $n(S) = 6$, so

$$P(E_3) = \frac{3}{6} = \frac{1}{2}.$$

(b) E_4: the number showing is greater than 4
   Again $n(S) = 6$. Event $E_4 = \{5, 6\}$, with $n(E_4) = 2$. By the definition,

$$P(E_4) = \frac{2}{6} = \frac{1}{3}.$$

(c) $E_5 = \{1, 2, 3, 4, 5, 6\}$ and $P(E_5) = \frac{6}{6} = 1$

(d) $E_6 = \emptyset$ and $P(E_6) = \frac{0}{6} = 0$.   ●

In part (c), $E_5 = S$. Therefore the event E_5 is certain to occur every time the experiment is performed. An event which is certain to occur, such as E_5, always has a probability of 1. On the other hand, $E_6 = \emptyset$ and $P(E_6)$ is 0. The probability of an impossible event, such as E_6, is always 0, since none of the outcomes in the sample space satisfy the event. For any event E, $P(E)$ is between 0 and 1 inclusive.

The set of all outcomes in the sample space which do *not* belong to event E is called the **complement** of E, written E'. For example, in the experiment of drawing a single card from a standard deck of 52 cards, let E be the event "the card is an ace." Then E' is the event "the card is not an ace." From the definition of E', for any event E,

$$E \cup E' = S \quad \text{and} \quad E \cap E' = \emptyset.$$

Probability concepts can be illustrated using Venn diagrams. The rectangle in Figure 11.2 represents the sample space in an experiment. The area inside the circle represents event E, while the area inside the rectangle, but outside the circle, represents event E'.

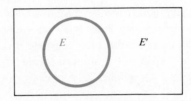

Figure 11.2

EXAMPLE 2 In the experiment of drawing a card from a well-shuffled deck, find the probability of events E, the card is an ace, and E'.

Since there are four aces in the deck of 52 cards, $n(E) = 4$ and $n(S) = 52$. Therefore, $P(E) = \dfrac{n(E)}{n(S)} = \dfrac{4}{52} = \dfrac{1}{13}$. Of the 52 cards, 48 are not aces, so

$$P(E') = \frac{n(E')}{n(S)} = \frac{48}{52} = \frac{12}{13}. \quad \bullet$$

In Example 2, $P(E) + P(E') = (1/13) + (12/13) = 1$. This is always true for any event E and its complement E'. That is,

$$P(E) + P(E') = 1.$$

This can be restated as

$$P(E) = 1 - P(E') \qquad \text{or} \qquad P(E') = 1 - P(E).$$

These two equations suggest an alternate way to compute the probability of an event. For example, if it is known that $P(E) = 1/10$, then

$$P(E') = 1 - \frac{1}{10} = \frac{9}{10}.$$

Sometimes probability statements are expressed in terms of odds, a comparison of $P(E)$ with $P(E')$. The **odds** in favor of an event E are expressed as the ratio of $P(E)$ to $P(E')$ or as the fraction $P(E)/P(E')$. For example, if the probability of rain can be established as $1/3$, the odds that it will rain are

$$P(\text{rain}) \text{ to } P(\text{no rain}) = \frac{1}{3} \text{ to } \frac{2}{3} = \frac{1/3}{2/3} = \frac{1}{2} \qquad \text{or} \qquad 1 \text{ to } 2.$$

On the other hand, the odds that it will not rain are 2 to 1 (or 2/3 to 1/3). If we know that the odds in favor of an event are, say, 3 to 5, then the probability of the event is 3/8, while the probability of the complement of the event is 5/8. If the odds favoring event E are m to n, then

$$P(E) = \frac{m}{m + n} \quad \text{and} \quad P(E') = \frac{n}{m + n}.$$

We now consider the probability of a **compound event** which involves an *alternative*, such as E or F, where E and F are events. For example, in the experiment of rolling a die, suppose H is the event "the result is a 3," and K is the event "the result is an even number." What is the probability of "the result is a 3 or an even number"? We have

$$H = \{3\} \qquad\qquad P(H) = \frac{1}{6}$$

$$K = \{2, 4, 6\} \qquad\qquad P(K) = \frac{3}{6} = \frac{1}{2}$$

$$H \text{ or } K = \{2, 3, 4, 6\} \quad P(H \text{ or } K) = \frac{4}{6} = \frac{2}{3}.$$

Notice that $P(H) + P(K) = P(H \text{ or } K)$.

Before we assume that this relationship is true in general, let us consider another event for this experiment, "the result is a 2," event G. Now

$$G = \{2\} \qquad\qquad P(G) = \frac{1}{6}$$

$$K = \{2, 4, 6\} \qquad\qquad P(K) = \frac{3}{6} = \frac{1}{2}$$

$$K \text{ or } G = \{2, 4, 6\} \qquad P(K \text{ or } G) = \frac{3}{6} = \frac{1}{2}.$$

In this case $P(K) + P(G) \neq P(K \text{ or } G)$.

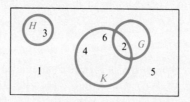

Figure 11.3

As Figure 11.3 shows, the difference in the two examples above comes from the fact that events H and K cannot occur simultaneously. Such events are called **mutually exclusive events.** We see that $H \cap K = \emptyset$, which is true for any two mutually exclusive events. Events K and G, however, can occur simultaneously.

Both are satisfied if the result of the roll is a 2, the element in their intersection $(K \cap G = \{2\})$. This example suggests the following property.

Probability of
Alternate
Events

> For any events E and F,
>
> $$P(E \text{ or } F) = P(E \cup F) = P(E) + P(F) - P(E \cap F).$$

EXAMPLE 3   One card is drawn from a well-shuffled deck of 52 cards. What is the probability of the following outcomes?

(a) The card is an ace or a spade.

The events "drawing an ace" and "drawing a spade" are not mutually exclusive since it is possible to draw the ace of spades, an outcome satisfying both events. The probability is

$$P(\text{ace or spade}) = P(\text{ace}) + P(\text{spade}) - P(\text{ace and spade})$$

$$= \frac{4}{52} + \frac{13}{52} - \frac{1}{52} = \frac{16}{52} = \frac{4}{13}.$$

(b) The card is a three or a king.

"Drawing a 3" and "drawing a king" are mutually exclusive events because it is impossible to draw one card that is both a 3 and a king.

$$P(3 \text{ or } K) = P(3) + P(K) - P(3 \text{ and } K)$$

$$= \frac{4}{52} + \frac{4}{52} - 0 = \frac{8}{52} = \frac{2}{13} \quad \bullet$$

EXAMPLE 4   For the experiment consisting of one roll of a pair of dice, find the probability that the sum of the points showing is at most 4.

We can rewrite "at most 4" as "2 or 3 or 4." (A sum of 1 is meaningless here.) Then

$$P(\text{at most 4}) = P(2 \text{ or } 3 \text{ or } 4)$$
$$= P(2) + P(3) + P(4), \tag{1}$$

since the events represented by "2," "3," and "4" are mutually exclusive.

The sample space for this experiment includes the 36 possible pairs of numbers from 1 to 6: (1, 1), (1, 2), (1, 3), (1, 4), (1, 5), (1, 6), (2, 1), (2, 2), and so on. The pair (1, 1) is the only one with a sum of 2, so $P(2) = 1/36$. Also $P(3) = 2/36$ since both (1, 2) and (2, 1) give a sum of 3. The pairs, (1, 3), (2, 2), and (3, 1) have a sum of 4, so $P(4) = 3/36$. Substituting into equation (1) above gives

$$P(\text{at most 4}) = \frac{1}{36} + \frac{2}{36} + \frac{3}{36} = \frac{6}{36} = \frac{1}{6}. \quad \bullet$$

The properties of probability discussed so far are summarized below.

Properties of Probability

For any events E and F:

1. $0 \leq P(E) \leq 1$

2. $P(\text{a certain event}) = 1$

3. $P(\text{an impossible event}) = 0$

4. $P(E') = 1 - P(E)$

5. $P(E \text{ or } F) = P(E \cup F) = P(E) + P(F) - P(E \cap F).$

11.3 EXERCISES

Write a sample space with equally likely outcomes for each of the following experiments.

1. A two-headed coin is tossed once.

2. Two ordinary coins are tossed.

3. Three ordinary coins are tossed.

4. Slips of paper marked with the numbers 1, 2, 3, 4, and 5 are placed in a box. After mixing well, two slips are drawn.

5. An unprepared student takes a three-question true/false quiz in which he guesses the answer to all three questions.

6. A die is rolled and then a coin is tossed.

Write the following events in set notation and give the probability of each event. See Example 1.

7. In the experiment described in Exercise 2:
   (a) both coins show the same face;
   (b) at least one coin turns up heads.

8. In Exercise 1:
   (a) the result of the toss is heads;
   (b) the result of the toss is tails.

9. In Exercise 4:
   (a) both slips are marked with even numbers;
   (b) both slips are marked with odd numbers;
   (c) both slips are marked with the same number
   (d) one slip is marked with an odd number, the other with an even number.

10. In Exercise 5:
   (a) the student gets all three answers correct;
   (b) he gets all three answers wrong;
   (c) he get exactly two answers correct;
   (d) he gets at least one answer correct.

11. A marble is drawn at random from a box containing 3 yellow, 4 white, and 8 blue marbles. Find the following probabilities.
  (a) A yellow marble is drawn,
  (b) A blue marble is drawn.
  (c) A black marble is drawn.
  (d) What are the odds in favor of drawing a yellow marble?
  (e) What are the odds against drawing a blue marble?

12. A baseball player with a batting average of .300 comes to bat. What are the odds in favor of his getting a hit?

13. In Exercise 4, what are the odds that the sum of the numbers on the two slips of paper is 5?

14. If the odds that it will rain are 4 to 5, what is the probability of rain?

15. If the odds that a candidate will win an election are 3 to 2, what is the probability that the candidate will lose?

16. A card is drawn from a well-shuffled deck of 52 cards. Find the probability that the card is
  (a) a 9;   (b) black;   (c) a black 9;   (d) a heart;
  (e) the 9 of hearts;   (f) a face card (K, Q, J of any suit).

17. Mrs. Elliott invites 10 relatives to a party: her mother, two uncles, three brothers, and four cousins. If the chances of any one guest arriving first are equally likely, find the following probabilities.
  (a) The first guest is an uncle or a brother.
  (b) The first guest is a brother or cousin.
  (c) The first guest is a brother or her mother.

18. One card is drawn from a standard deck of 52 cards. What is the probability that the card is
  (a) a 9 or a 10?   (b) red or a 3?   (c) a heart or black?
  (d) less than a 4? (Consider aces as 1's.)

19. Two dice are rolled. Find the probability of the following events.
  (a) The sum of the points is at least 10.
  (b) The sum of the points is either 7 or at least 10.
  (c) The sum of the points is 2 or the dice both show the same number.

20. A student estimates that his probability of getting an A in a certain course is .4; a B, .3; a C, .2; and a D, .1.
  (a) Assuming that only the grades A, B, C, D, and F are possible, what is the probability that he will fail the course?
  (b) What is the probability that he will receive a grade of C or better?
  (c) What is the probability that he will receive at least a B in the course?
  (d) What is the probability that he will get at most a C in the course?

21. If a marble is drawn from a bag containing 3 yellow, 4 white, and 8 blue marbles, what is the probability of the following?
  (a) The marble is either yellow or white.
  (b) It is either yellow or blue.
  (c) It is either blue or white.

22. The table shows the probability that a customer of a department store will make a purchase in the indicated price range.

Cost	Probability
Below $5	.25
$5–$19.99	.37
$20–$39.99	.11
$40–$69.99	.09
$70–$99.99	.07
$100–$149.99	.08
$150 or more	.03

Find the probability that a customer makes a purchase that is
(a) less than $20; **(b)** $40 or more; **(c)** more than $99.99; **(d)** less than $100.

11.4 The Binomial Theorem

In this section combinations are used as an aid in evaluating expressions such as $(x + y)^5$. To see how this is done, first list some expansions of $(x + y)^n$, for various positive integer values of n. For example,

$$(x + y)^1 = x + y$$
$$(x + y)^2 = x^2 + 2xy + y^2$$
$$(x + y)^3 = x^3 + 3x^2y + 3xy^2 + y^3$$
$$(x + y)^4 = x^4 + 4x^3y + 6x^2y^2 + 4xy^3 + y^4$$
$$(x + y)^5 = x^5 + 5x^4y + 10x^3y^2 + 10x^2y^3 + 5xy^4 + y^5$$

and so on. Studying these results reveals a pattern that can be used to write a general expression for $(x + y)^n$.

First, notice that each expression begins with x raised to the same power as the binomial itself. That is, the expansion of $(x + y)^1$ has a first term of x^1, $(x + y)^2$ has a first term of x^2, $(x + y)^3$ has a first term of x^3, and so on. Also, the last term in each expansion is y to the same power as the binomial. Thus the expansion of $(x + y)^n$ should begin with the term x^n and end with the term y^n.

Also, the exponents on x decrease by one in each term after the first, while the exponents on y, beginning with y in the second term, increase by one in each succeeding term. That is, the *variables* in the terms of the expansion of $(x + y)^n$ have the following pattern.

$$x^n, \ x^{n-1}y, \ x^{n-2}y^2, \ x^{n-3}y^3, \ \ldots \ , \ xy^{n-1}, \ y^n$$

This pattern suggests that the sum of the exponents on x and y in each term is n. For example, in the third term in the list above, the variable is $x^{n-2}y^2$, and the sum of the exponents is $n - 2 + 2 = n$.

Now examine the pattern for the *coefficients* in the terms of the expansions shown on the preceding page. In the product

$$(x + y)^5 = (x + y)(x + y)(x + y)(x + y)(x + y) \qquad (*)$$

the term x occurs 5 times, once in each factor. To get the first term of the expansion, form the product of these five x's to get x^5. The product x^5 can occur in just this one way, by taking an x from each factor in equation (*), so that the coefficient of x^5 is 1.

The term with x^4y occurs in more than one way. For example, we can take the x's from the first four factors and the y from the last factor, or we may take the x's from the last four factors and the y from the first factor, and so on. Since there are five factors in equation (*) from which four x's must be selected for the term x^4y, there are a total of $\binom{5}{4}$ ways in which this can be done. Thus, the term x^4y occurs in $\binom{5}{4}$, or 5, ways and has coefficient 5. In the same way, combinations are used to find coefficients for each of the terms in the expansion $(x + y)^5$.

$$(x + y)^5 = x^5 + \binom{5}{4} x^4y + \binom{5}{3} x^3y^2 + \binom{5}{2} x^2y^3 + \binom{5}{1} xy^4 + y^5$$

The coefficient 1 of the first term could be written as $\binom{5}{5}$ and of the last term as $\binom{5}{0}$ to complete the pattern if desired.

Generalizing, the coefficient for a term in the expansion of $(x + y)^n$ in which the variable is $x^{n-r}y^r$ is $\binom{n}{n-r}$. The **general binomial expansion** is given by the following theorem.

Binomial Theorem

> For any positive integer n,
>
> $$(x + y)^n = x^n + \binom{n}{n-1}x^{n-1}y + \binom{n}{n-2}x^{n-2}y^2 + \binom{n}{n-3}x^{n-3}y^3$$
>
> $$+ \ldots + \binom{n}{n-r}x^{n-r}y^r + \ldots + \binom{n}{1}xy^{n-1} + y^n.$$

The binomial theorem can be proved using mathematical induction (discussed in Section 10.5). Note that the expansion of $(x + y)^n$ has $n + 1$ terms.

EXAMPLE 1

Write out the binomial expansion of $(x + y)^9$.

Using the binomial theorem,

$$(x + y)^9 = x^9 + \binom{9}{8} x^8y + \binom{9}{7} x^7y^2 + \binom{9}{6} x^6y^3 + \binom{9}{5} x^5y^4$$

$$+ \binom{9}{4} x^4y^5 + \binom{9}{3} x^3y^6 + \binom{9}{2} x^2y^7 + \binom{9}{1} xy^8 + y^9.$$

Now evaluate each of the combinations.

$$(x + y)^9 = x^9 + \frac{9!}{8!1!} x^8 y + \frac{9!}{7!2!} x^7 y^2 + \frac{9!}{6!3!} x^6 y^3 + \frac{9!}{5!4!} x^5 y^4$$

$$+ \frac{9!}{4!5!} x^4 y^5 + \frac{9!}{3!6!} x^3 y^6 + \frac{9!}{2!7!} x^2 y^7 + \frac{9!}{1!8!} xy^8 + y^9$$

$$= x^9 + 9x^8 y + 36x^7 y^2 + 84x^6 y^3 + 126x^5 y^4 + 126x^4 y^5$$

$$+ 84x^3 y^6 + 36x^2 y^7 + 9xy^8 + y^9 \quad \bullet$$

EXAMPLE 2   Expand $\left(a - \dfrac{b}{2} \right)^5$.

Again, use the binomial theorem.

$$\left(a - \frac{b}{2} \right)^5 = a^5 + \binom{5}{4} a^4 \left(-\frac{b}{2} \right) + \binom{5}{3} a^3 \left(-\frac{b}{2} \right)^2 + \binom{5}{2} a^2 \left(-\frac{b}{2} \right)^3$$

$$+ \binom{5}{1} a \left(-\frac{b}{2} \right)^4 + \left(-\frac{b}{2} \right)^5$$

$$= a^5 + 5a^4 \left(-\frac{b}{2} \right) + 10a^3 \left(-\frac{b}{2} \right)^2 + 10a^2 \left(-\frac{b}{2} \right)^3$$

$$+ 5a \left(-\frac{b}{2} \right)^4 + \left(-\frac{b}{2} \right)^5$$

$$= a^5 - \frac{5}{2} a^4 b + \frac{5}{2} a^3 b^2 - \frac{5}{4} a^2 b^3 + \frac{5}{16} ab^4 - \frac{1}{32} b^5 \quad \bullet$$

An easy way to find the coefficients of a binomial expansion for small powers of n is to use **Pascal's triangle,** shown below. This triangular array is named for the famous seventeenth-century mathematician Blaise Pascal (1623–1662), one of the first to use it extensively.

Pascal's Triangle

```
                          1       1
                      1       2       1
                  1       3       3       1
              1       4       6       4       1
          1       5      10      10       5       1
      1       6      15      20      15       6       1
  1       7      21      35      35      21       7       1
1       8      28      56      70      56      28       8       1
  .   .   .       .       .       .       .   .   .
      .           .       .       .           .
```

The pattern can be extended indefinitely. The nth row of the triangle gives the coefficients of $(x + y)^n$. For example, the second row gives the coefficients of $(x + y)^2$, the third row gives the coefficients of $(x + y)^3$, and so on. Each row begins and ends with 1. Each of the other numbers can be found by adding the two numbers above it. For example, in the center of the fourth row, the entry 6 is the sum of the two 3s above it.

rth Term   We can also write any single term of a binomial expansion. For example, suppose we want to find just the 7th term of $(x + y)^9$. The 7th term would have y raised to the 6th power (since y has the power 1 in the second term, the power 2 in the third term, and so on). The exponents on x and y in each term must have a sum of 9, so the exponent on x in the 7th term is $9 - 6 = 3$. Thus, the 7th term should be

$$\frac{9!}{3!6!} x^3 y^6.$$

In Example 1 above, we did, indeed, find just that for the 7th term. This suggests the next theorem.

rth Term of the Binomial Expansion

> The rth term of the binomial expansion of $(x + y)^n$, where $n \geq r - 1$, is
>
> $$\binom{n}{n - (r - 1)} x^{n-(r-1)} y^{r-1}.$$

EXAMPLE 3   Find the seventh term of $(a + 2b)^{10}$.

In the seventh term $2b$ has an exponent of 6, while a has an exponent of $10 - 6$, or 4. Thus the seventh term is

$$\binom{10}{4} a^4 (2b)^6 = 210 a^4 (64 b^6) = 13{,}440 a^4 b^6. \quad \bullet$$

The binomial expansion can be used to find powers of certain numbers, as suggested by the next example.

EXAMPLE 4   Use the binomial theorem to find $(1.01)^8$ correct to three decimal places.

Write 1.01 as $1 + .01$. Then use the binomial theorem.

$$(1.01)^8 = (1 + .01)^8$$
$$= 1^8 + \binom{8}{7}(1)^7(.01) + \binom{8}{6}(1)^6(.01)^2 + \ldots + (.01)^8$$
$$= 1 + 8(1)(.01) + 28(1)(.0001) + \ldots + (.0000000000000001)$$
$$= 1 + .08 + .0028 + \ldots + .0000000000000001$$
$$(1.01)^8 \approx 1.083$$

Since we require only three decimal places of accuracy, there is no need to evaluate all the terms of the expansion. Additional terms, however, would give the answer to more decimal places of accuracy. $\bullet$

11.4 EXERCISES

Write out the binomial expansion for each of the following. See Examples 1 and 2.

1. $(x + y)^6$

2. $(m + n)^4$

3. $(p - q)^5$

4. $(a - b)^7$

5. $(r^2 + s)^5$

6. $(m + n^2)^4$

7. $(p + 2q)^4$

8. $(3r - s)^6$

9. $(7p + 2q)^4$

10. $(4a - 5b)^5$

11. $(3x - 2y)^6$

12. $(7k - 9j)^4$

13. $\left(\dfrac{m}{2} - 1\right)^6$

14. $\left(3 + \dfrac{y}{3}\right)^5$

15. $\left(2p + \dfrac{q}{3}\right)^4$

16. $\left(\dfrac{r}{6} - \dfrac{m}{2}\right)^3$

For each of the following write the indicated term of the binomial expansion. See Example 3.

17. 5th term of $(m - 2p)^{12}$

18. 4th term of $(3x + y)^6$

19. 6th term of $(x + y)^9$

20. 12th term of $(a - b)^{15}$

21. 9th term of $(2m + n)^{10}$

22. 7th term of $(3r - 5s)^{12}$

23. 17th term of $(p^2 + q)^{20}$

24. 10th term of $(2x^2 + y)^{14}$

25. 8th term of $(x^3 + 2y)^{14}$

26. 13th term of $(a + 2b^3)^{12}$

Use the binomial expansion to evaluate each of the following to three decimal places. See Example 4.

27. $(1.10)^{10}$

28. $(0.99)^{15}$ (*Hint:* $0.99 = 1 - 0.01$.)

29. $(1.99)^8$

30. $(2.99)^3$

31. $(3.02)^6$

32. $(1.01)^9$

33. Prove the binomial theorem by mathematical induction.

In calculus it is shown that

$$(1 + x)^n = 1 + nx + \frac{n(n - 1)}{2!} x^2 + \frac{n(n - 1)(n - 2)}{3!} x^3 + \cdots$$

for any real number n (not just positive integers) and any real number x, where $|x| < 1$. This result, a generalized binomial theorem, may be used to find approximate values of powers or roots. For example,

$$\sqrt[4]{630} = (625 + 5)^{1/4}$$

$$= \left[625\left(1 + \frac{5}{625}\right)\right]^{1/4}$$

$$= 625^{1/4}\left(1 + \frac{5}{625}\right)^{1/4}$$

34. Use the expression given above for $(1 + x)^n$ to approximate $(1 + 5/625)^{1/4}$ to the nearest thousandth. Then approximate $\sqrt[4]{630}$.

35. Approximate $\sqrt[3]{9.42}$, using this method.

36. Approximate $(1.02)^{-3}$.

37. Approximate $1/(1.04)^5$.

Chapter 11   Summary

Key Words

tree diagram
permutation
outcomes
event
complement
compound event
general binomial expansion

independent events
combination
sample space
probability
odds
mutually exclusive events
Pascal's triangle

Fundamental Principle of Counting

If one event can occur in m ways and a second event can occur in n ways, then both events can occur in mn ways, provided the outcome of the first event does not influence the outcome of the second.

Permutations of n Elements n at a Time

If $P(n, n)$ denotes the number of permutations of n elements taken n at a time, then

$$P(n, n) = n!$$

Permutations of n Elements r at a time

If $P(n, r)$ denotes the number of permutations of n elements taken r at a time, then

$$P(n, r) = \frac{n!}{(n - r)!}.$$

Combinations of n Elements r at a Time

If $\binom{n}{r}$ represents the number of combinations of n things taken r at a time, then

$$\binom{n}{r} = \frac{n!}{(n - r)!r!}.$$

Probability of Event E

$$P(E) = \frac{n(E)}{n(S)}$$

Properties of Probability

For any events E and F:

1. $0 \le P(E) \le 1$

2. $P(\text{a certain event}) = 1$

3. $P(\text{an impossible event}) = 0$

4. $P(E') = 1 - P(E)$

5. $P(E \text{ or } F) = P(E \cup F) = P(E) + P(F) - P(E \cap F).$

Binomial Theorem

For any positive integer n,

$$(x + y)^n = x^n + \binom{n}{n - 1} x^{n-1}y + \binom{n}{n - 2} x^{n-2}y^2 + \binom{n}{n - 3} x^{n-3}y^3$$

$$+ \ldots + \binom{n}{n - r} x^{n-r}y^r + \ldots + \binom{n}{1} xy^{n-1} + y^n.$$

Chapter 11 Review Exercises

Find each of the following.

1. $P(5, 5)$

2. $P(9, 2)$

3. $P(6, 0)$

4. $\begin{pmatrix} 8 \\ 3 \end{pmatrix}$

5. $\begin{pmatrix} 10 \\ 5 \end{pmatrix}$

6. $\begin{pmatrix} 6 \\ 0 \end{pmatrix}$

7. $\begin{pmatrix} 7 \\ 1 \end{pmatrix}$

8. Two people are planning their wedding. They can select from 2 different chapels, 4 soloists, 3 organists, and 2 ministers. How many different wedding services will be possible?

9. John Jacobs, who is furnishing his apartment, wants to buy a new couch. He can select from 5 different styles, each available in 3 different fabrics, with 6 color choices. How many different couches are available?

10. Four students are to be assigned to 4 different summer jobs. Each student is qualified for all 4 jobs. In how many ways can the jobs be assigned?

11. A student body council consists of a president, vice-president, secretary-treasurer, and 3 representatives at large. Three members are to be selected to attend a conference.
(a) How many different such delegations are possible?
(b) How many if the president must attend?

12. Nine football teams are competing for 1st, 2nd, and 3rd place titles in a statewide tournament. In how many ways can the winners be determined?

13. How many different license plates can be formed with a letter, followed by 3 digits, and then 3 letters? How many such license plates have no repeats?

Solve each of the following equations for n.

14. $P(n, 3) = 8 \cdot P(n - 1, 2)$

15. $30 \cdot P(n, 2) = P(n + 2, 4)$

16. $\begin{pmatrix} n + 2 \\ 4 \end{pmatrix} = 15 \cdot \begin{pmatrix} n \\ 4 \end{pmatrix}$

17. $\begin{pmatrix} n \\ n - 2 \end{pmatrix} = 66$

A card is drawn from a standard deck of 52 cards. Find the probability that each of the following is drawn.

18. A black king

19. A face card or an ace

20. An ace or a diamond

21. A card that is not a diamond

22. A card that is not a diamond or not black

A sample shipment of 5 swimming pool filters is chosen. The probability of exactly 0, 1, 2, 3, 4, or 5 filters being defective is given in the following table.

Number defective	0	1	2	3	4	5
Probability	.31	.25	.18	.12	.08	.06

Find the probability that the following numbers of filters are defective.

23. No more than 3

24. At least 2

Use the binomial theorem to expand each of the following.

25. $(x + 2y)^4$

26. $(3z - 5w)^3$

27. $\left(\dfrac{k}{2} - g\right)^5$

Find the indicated term or terms for each of the following expansions.

28. Fifth term of $(3x - 2y)^6$

29. Eighth term of $(2m + n^2)^{12}$

30. First four terms of $(3 + x)^{16}$

31. Last three terms of $(2m - 3n)^{15}$

Use the binomial expansion to evaluate the following numbers to three decimal places.

32. $(1.02)^{12}$

33. $(.98)^7$

Show that each of the following is true for positive integers n and r with $r \le n$.

34. $P(n, n - 1) = P(n, n)$

35. $P(n, 1) = n$

36. $P(n, 0) = 1$

37. $\dbinom{n}{n} = 1$

38. $\dbinom{n}{0} = 1$

39. $\dbinom{n}{n - 1} = n$

40. $\dbinom{n}{n - r} = \dbinom{n}{r}$

TABLE 1   THE METRIC SYSTEM

METRIC TO ENGLISH

To convert from	To	Multiply by
meters	yards	1.094
meters	feet	3.281
meters	inches	39.37
kilometers	miles	.6214
grams	pounds	.00220
kilograms	pounds	2.20
liters	quarts	1.06
liters	gallons	.264

ENGLISH TO METRIC

To convert from	To	Multiply by
yards	meters	.9144
feet	meters	.3048
inches	meters	.0254
miles	kilometers	1.6093
pounds	grams	454
pounds	kilograms	.454
quarts	liters	.946
gallons	liters	3.785

TABLE 2   SELECTED POWERS OF NUMBERS

n	n^2	n^3	n^4	n^5	n^6
2	4	8	16	32	64
3	9	27	81	243	729
4	16	64	256	1024	4096
5	25	125	625	3125	15,625
6	36	216	1296	7776	46,656
7	49	343	2401	16,807	117,649
8	64	512	4096	32,768	262,144
9	81	729	6561	59,049	531,441
10	100	1000	10,000	100,000	1,000,000
11	121	1331	14,641	161,051	1,771,561
12	144	1728	20,736	248,832	2,985,984

Table 3   Squares and Square Roots   **459**

TABLE 3   SQUARES AND SQUARE ROOTS

n	n^2	$\sqrt{n}$	$\sqrt{10n}$	n	n^2	$\sqrt{n}$	$\sqrt{10n}$
1	1	1.000	3.162	51	2601	7.141	22.583
2	4	1.414	4.472	52	2704	7.211	22.804
3	9	1.732	5.477	53	2809	7.280	23.022
4	16	2.000	6.325	54	2916	7.348	23.238
5	25	2.236	7.071	55	3025	7.416	23.452
6	36	2.449	7.746	56	3136	7.483	23.664
7	49	2.646	8.367	57	3249	7.550	23.875
8	64	2.828	8.944	58	3364	7.616	24.083
9	81	3.000	9.487	59	3481	7.681	24.290
10	100	3.162	10.000	60	3600	7.746	24.495
11	121	3.317	10.488	61	3721	7.810	24.698
12	144	3.464	10.954	62	3844	7.874	24.900
13	169	3.606	11.402	63	3969	7.937	25.100
14	196	3.742	11.832	64	4096	8.000	25.298
15	225	3.873	12.247	65	4225	8.062	25.495
16	256	4.000	12.649	66	4356	8.124	25.690
17	289	4.123	13.038	67	4489	8.185	25.884
18	324	4.243	13.416	68	4624	8.246	26.077
19	361	4.359	13.784	69	4761	8.307	26.268
20	400	4.472	14.142	70	4900	8.367	26.458
21	441	4.583	14.491	71	5041	8.426	26.646
22	484	4.690	14.832	72	5184	8.485	26.833
23	529	4.796	15.166	73	5329	8.544	27.019
24	576	4.899	15.492	74	5476	8.602	27.203
25	625	5.000	15.811	75	5625	8.660	27.386
26	676	5.099	16.125	76	5776	8.718	27.568
27	729	5.196	16.432	77	5929	8.775	27.749
28	784	5.292	16.733	78	6084	8.832	27.928
29	841	5.385	17.029	79	6241	8.888	28.107
30	900	5.477	17.321	80	6400	8.944	28.284
31	961	5.568	17.607	81	6561	9.000	28.460
32	1024	5.657	17.889	82	6724	9.055	28.636
33	1089	5.745	18.166	83	6889	9.110	28.810
34	1156	5.831	18.439	84	7056	9.165	28.983
35	1225	5.916	18.708	85	7225	9.220	29.155
36	1296	6.000	18.974	86	7396	9.274	29.326
37	1369	6.083	19.235	87	7569	9.327	29.496
38	1444	6.164	19.494	88	7744	9.381	29.665
39	1521	6.245	19.748	89	7921	9.434	29.833
40	1600	6.325	20.000	90	8100	9.487	30.000
41	1681	6.403	20.248	91	8281	9.539	30.166
42	1764	6.481	20.494	92	8464	9.592	30.332
43	1849	6.557	20.736	93	8649	9.644	30.496
44	1936	6.633	20.976	94	8836	9.695	30.659
45	2025	6.708	21.213	95	9025	9.747	30.822
46	2116	6.782	21.448	96	9216	9.798	30.984
47	2209	6.856	21.679	97	9409	9.849	31.145
48	2304	6.928	21.909	98	9604	9.899	31.305
49	2401	7.000	22.136	99	9801	9.950	31.464
50	2500	7.071	22.361	100	10000	10.000	31.623

TABLE 4   COMMON LOGARITHMS

n	0	1	2	3	4	5	6	7	8	9
1.0	.0000	.0043	.0086	.0128	.0170	.0212	.0253	.0294	.0334	.0374
1.1	.0414	.0453	.0492	.0531	.0569	.0607	.0645	.0682	.0719	.0755
1.2	.0792	.0828	.0864	.0899	.0934	.0969	.1004	.1038	.1072	.1106
1.3	.1139	.1173	.1206	.1239	.1271	.1303	.1335	.1367	.1399	.1430
1.4	.1461	.1492	.1523	.1553	.1584	.1614	.1644	.1673	.1703	.1732
1.5	.1761	.1790	.1818	.1847	.1875	.1903	.1931	.1959	.1987	.2014
1.6	.2041	.2068	.2095	.2122	.2148	.2175	.2201	.2227	.2253	.2279
1.7	.2304	.2330	.2355	.2380	.2405	.2430	.2455	.2480	.2504	.2529
1.8	.2553	.2577	.2601	.2625	.2648	.2672	.2695	.2718	.2742	.2765
1.9	.2788	.2810	.2833	.2856	.2878	.2900	.2923	.2945	.2967	.2989
2.0	.3010	.3032	.3054	.3075	.3096	.3118	.3139	.3160	.3181	.3201
2.1	.3222	.3243	.3263	.3284	.3304	.3324	.3345	.3365	.3385	.3404
2.2	.3424	.3444	.3464	.3483	.3502	.3522	.3541	.3560	.3579	.3598
2.3	.3617	.3636	.3655	.3674	.3692	.3711	.3729	.3747	.3766	.3784
2.4	.3802	.3820	.3838	.3856	.3874	.3892	.3909	.3927	.3945	.3962
2.5	.3979	.3997	.4014	.4031	.4048	.4065	.4082	.4099	.4116	.4133
2.6	.4150	.4166	.4183	.4200	.4216	.4232	.4249	.4265	.4281	.4298
2.7	.4314	.4330	.4346	.4362	.4378	.4393	.4409	.4425	.4440	.4456
2.8	.4472	.4487	.4502	.4518	.4533	.4548	.4564	.4579	.4594	.4609
2.9	.4624	.4639	.4654	.4669	.4683	.4698	.4713	.4728	.4742	.4757
3.0	.4771	.4786	.4800	.4814	.4829	.4843	.4857	.4871	.4886	.4900
3.1	.4914	.4928	.4942	.4955	.4969	.4983	.4997	.5011	.5024	.5038
3.2	.5051	.5065	.5079	.5092	.5105	.5119	.5132	.5145	.5159	.5172
3.3	.5185	.5198	.5211	.5224	.5237	.5250	.5263	.5276	.5289	.5302
3.4	.5315	.5328	.5340	.5353	.5366	.5378	.5391	.5403	.5416	.5428
3.5	.5441	.5453	.5465	.5478	.5490	.5502	.5514	.5527	.5539	.5551
3.6	.5563	.5575	.5587	.5599	.5611	.5623	.5635	.5647	.5658	.5670
3.7	.5682	.5694	.5705	.5717	.5729	.5740	.5752	.5763	.5775	.5786
3.8	.5798	.5809	.5821	.5832	.5843	.5855	.5866	.5877	.5888	.5899
3.9	.5911	.5922	.5933	.5944	.5955	.5966	.5977	.5988	.5999	.6010
4.0	.6021	.6031	.6042	.6053	.6064	.6075	.6085	.6096	.6107	.6117
4.1	.6128	.6138	.6149	.6160	.6170	.6180	.6191	.6201	.6212	.6222
4.2	.6232	.6243	.6253	.6263	.6274	.6284	.6294	.6304	.6314	.6325
4.3	.6335	.6345	.6355	.6365	.6375	.6385	.6395	.6405	.6415	.6425
4.4	.6435	.6444	.6454	.6464	.6474	.6484	.6493	.6503	.6513	.6522
4.5	.6532	.6542	.6551	.6561	.6571	.6580	.6590	.6599	.6609	.6618
4.6	.6628	.6637	.6646	.6656	.6665	.6675	.6684	.6693	.6702	.6712
4.7	.6721	.6730	.6739	.6749	.6758	.6767	.6776	.6785	.6794	.6803
4.8	.6812	.6821	.6830	.6839	.6848	.6857	.6866	.6875	.6884	.6893
4.9	.6902	.6911	.6920	.6928	.6937	.6946	.6955	.6964	.6972	.6981
5.0	.6990	.6998	.7007	.7016	.7024	.7033	.7042	.7050	.7059	.7067
5.1	.7076	.7084	.7093	.7101	.7110	.7118	.7126	.7135	.7143	.7152
5.2	.7160	.7168	.7177	.7185	.7193	.7202	.7210	.7218	.7226	.7235
5.3	.7243	.7251	.7259	.7267	.7275	.7284	.7292	.7300	.7308	.7316
5.4	.7324	.7332	.7340	.7348	.7356	.7364	.7372	.7380	.7388	.7396
n	0	1	2	3	4	5	6	7	8	9

Table 4   Common Logarithms   **461**

TABLE 4   COMMON LOGARITHMS (CONTINUED)

n	0	1	2	3	4	5	6	7	8	9
5.5	.7404	.7412	.7419	.7427	.7435	.7443	.7451	.7459	.7466	.7474
5.6	.7482	.7490	.7497	.7505	.7513	.7520	.7528	.7536	.7543	.7551
5.7	.7559	.7566	.7574	.7582	.7589	.7597	.7604	.7612	.7619	.7627
5.8	.7634	.7642	.7649	.7657	.7664	.7672	.7679	.7686	.7694	.7701
5.9	.7709	.7716	.7723	.7731	.7738	.7745	.7752	.7760	.7767	.7774
6.0	.7782	.7789	.7796	.7803	.7810	.7818	.7825	.7832	.7839	.7846
6.1	.7853	.7860	.7868	.7875	.7882	.7889	.7896	.7903	.7910	.7917
6.2	.7924	.7931	.7938	.7945	.7952	.7959	.7966	.7973	.7980	.7987
6.3	.7993	.8000	.8007	.8014	.8021	.8028	.8035	.8041	.8048	.8055
6.4	.8062	.8069	.8075	.8082	.8089	.8096	.8102	.8109	.8116	.8122
6.5	.8129	.8136	.8142	.8149	.8156	.8162	.8169	.8176	.8182	.8189
6.6	.8195	.8202	.8209	.8215	.8222	.8228	.8235	.8241	.8248	.8254
6.7	.8261	.8267	.8274	.8280	.8287	.8293	.8299	.8306	.8312	.8319
6.8	.8325	.8331	.8338	.8344	.8351	.8357	.8363	.8370	.8376	.8382
6.9	.8388	.8395	.8401	.8407	.8414	.8420	.8426	.8432	.8439	.8445
7.0	.8451	.8457	.8463	.8470	.8476	.8482	.8488	.8494	.8500	.8506
7.1	.8513	.8519	.8525	.8531	.8537	.8543	.8549	.8555	.8561	.8567
7.2	.8573	.8579	.8585	.8591	.8597	.8603	.8609	.8615	.8621	.8627
7.3	.8633	.8639	.8645	.8651	.8657	.8663	.8669	.8675	.8681	.8686
7.4	.8692	.8698	.8704	.8710	.8716	.8722	.8727	.8733	.8739	.8745
7.5	.8751	.8756	.8762	.8768	.8774	.8779	.8785	.8791	.8797	.8802
7.6	.8808	.8814	.8820	.8825	.8831	.8837	.8842	.8848	.8854	.8859
7.7	.8865	.8871	.8876	.8882	.8887	.8893	.8899	.8904	.8910	.8915
7.8	.8921	.8927	.8932	.8938	.8943	.8949	.8954	.8960	.8965	.8971
7.9	.8976	.8982	.8987	.8993	.8998	.9004	.9009	.9015	.9020	.9025
8.0	.9031	.9036	.9042	.9047	.9053	.9058	.9063	.9069	.9074	.9079
8.1	.9085	.9090	.9096	.9101	.9106	.9112	.9117	.9122	.9128	.9133
8.2	.9138	.9143	.9149	.9154	.9159	.9165	.9170	.9175	.9180	.9186
8.3	.9191	.9196	.9201	.9206	.9212	.9217	.9222	.9227	.9232	.9238
8.4	.9243	.9248	.9253	.9258	.9263	.9269	.9274	.9279	.9284	.9289
8.5	.9294	.9299	.9304	.9309	.9315	.9320	.9325	.9330	.9335	.9340
8.6	.9345	.9350	.9355	.9360	.9365	.9370	.9375	.9380	.9385	.9390
8.7	.9395	.9400	.9405	.9410	.9415	.9420	.9425	.9430	.9435	.9440
8.8	.9445	.9450	.9455	.9460	.9465	.9469	.9474	.9479	.9484	.9489
8.9	.9494	.9499	.9504	.9509	.9513	.9518	.9523	.9528	.9533	.9538
9.0	.9542	.9547	.9552	.9557	.9562	.9566	.9571	.9576	.9581	.9586
9.1	.9590	.9595	.9600	.9605	.9609	.9614	.9619	.9624	.9628	.9633
9.2	.9638	.9643	.9647	.9652	.9657	.9661	.9666	.9671	.9675	.9680
9.3	.9685	.9689	.9694	.9699	.9703	.9708	.9713	.9717	.9722	.9727
9.4	.9731	.9736	.9741	.9745	.9750	.9754	.9759	.9763	.9768	.9773
9.5	.9777	.9782	.9786	.9791	.9795	.9800	.9805	.9809	.9814	.9818
9.6	.9823	.9827	.9832	.9836	.9841	.9845	.9850	.9854	.9859	.9863
9.7	.9868	.9872	.9877	.9881	.9886	.9890	.9894	.9899	.9903	.9908
9.8	.9912	.9917	.9921	.9926	.9930	.9934	.9939	.9943	.9948	.9952
9.9	.9956	.9961	.9965	.9969	.9974	.9978	.9983	.9987	.9991	.9996
n	0	1	2	3	4	5	6	7	8	9

TABLE 5A   NATURAL LOGARITHMS AND POWERS OF e

x	e^x	e^{-x}	$\ln x$	x	e^x	e^{-x}	$\ln x$
0.00	1.00000	1.00000		1.60	4.95302	0.20189	0.4700
0.01	1.01005	0.99004	-4.6052	1.70	5.47394	0.18268	0.5306
0.02	1.02020	0.98019	-3.9120	1.80	6.04964	0.16529	0.5878
0.03	1.03045	0.97044	-3.5066	1.90	6.68589	0.14956	0.6419
0.04	1.04081	0.96078	-3.2189	2.00	7.38905	0.13533	0.6931
0.05	1.05127	0.95122	-2.9957				
0.06	1.06183	0.94176	-2.8134	2.10	8.16616	0.12245	0.7419
0.07	1.07250	0.93239	-2.6593	2.20	9.02500	0.11080	0.7885
0.08	1.08328	0.92311	-2.5257	2.30	9.97417	0.10025	0.8329
0.09	1.09417	0.91393	-2.4079	2.40	11.02316	0.09071	0.8755
0.10	1.10517	0.90483	-2.3026	2.50	12.18248	0.08208	0.9163
				2.60	13.46372	0.07427	0.9555
0.11	1.11628	0.89583	-2.2073	2.70	14.87971	0.06720	0.9933
0.12	1.12750	0.88692	-2.1203	2.80	16.44463	0.06081	1.0296
0.13	1.13883	0.87810	-2.0402	2.90	18.17412	0.05502	1.0647
0.14	1.15027	0.86936	-1.9661	3.00	20.08551	0.04978	1.0986
0.15	1.16183	0.86071	-1.8971				
0.16	1.17351	0.85214	-1.8326	3.50	33.11545	0.03020	1.2528
0.17	1.18530	0.84366	-1.7720	4.00	54.59815	0.01832	1.3863
0.18	1.19722	0.83527	-1.7148	4.50	90.01713	0.01111	1.5041
0.19	1.20925	0.82696	-1.6607				
				5.00	148.41316	0.00674	1.6094
0.20	1.22140	0.81873	-1.6094	5.50	224.69193	0.00409	1.7047
0.30	1.34985	0.74081	-1.2040				
0.40	1.49182	0.67032	-0.9163	6.00	403.42879	0.00248	1.7918
0.50	1.64872	0.60653	-0.6931	6.50	665.14163	0.00150	1.8718
0.60	1.82211	0.54881	-0.5108				
0.70	2.01375	0.49658	-0.3567	7.00	1096.63316	0.00091	1.9459
0.80	2.22554	0.44932	-0.2231	7.50	1808.04241	0.00055	2.0149
0.90	2.45960	0.40656	-0.1054	8.00	2980.95799	0.00034	2.0794
				8.50	4914.76884	0.00020	2.1401
1.00	2.71828	0.36787	0.0000				
1.10	3.00416	0.33287	0.0953	9.00	8130.08392	0.00012	2.1972
1.20	3.32011	0.30119	0.1823	9.50	13359.72683	0.00007	2.2513
1.30	3.66929	0.27253	0.2624				
1.40	4.05519	0.24659	0.3365	10.00	22026.46579	0.00005	2.3026
1.50	4.48168	0.22313	0.4055				

TABLE 5B   ADDITIONAL NATURAL LOGARITHMS

x	$\ln x$	x	$\ln x$	x	$\ln x$
11	2.3979	20	2.9957	70	4.2485
12	2.4849	25	3.2189	75	4.3175
13	2.5649	30	3.4012	80	4.3820
14	2.6391	35	3.5553	85	4.4427
		40	3.6889	90	4.4998
15	2.7081				
16	2.7726	45	3.8067	95	4.5539
17	2.8332	50	3.9120	100	4.6052
18	2.8904	55	4.0073		
19	2.9444	60	4.0943	1000	6.9078
		65	4.1744		

Answers to Selected Exercises

If you need further help with algebra, you might want to get a copy of the *Study Guide* and *Student Solutions Manual* designed to go along with this book. Your local college bookstore either has these books or can order them for you.

CHAPTER 1

Section 1.1 (page 5)

1. 12, 13, 14, 15, 16, 17, 18, 19, 20   **3.** 1, 1/2, 1/4, 1/8, 1/16, 1/32   **5.** 17, 22, 27, 32, 37, 42, 47
7. 8, 9, 10, 11, 12, 13, 14   **9.** Finite   **11.** Infinite   **13.** Infinite   **15.** Infinite   **17.** $\in$
19. $\not\subseteq$ **21.** $\in$ **23.** $\not\subseteq$ **25.** $\not\subseteq$ **27.** $\not\subseteq$ **29.** False   **31.** True   **33.** True
35. True   **37.** False   **39.** True   **41.** True   **43.** False   **45.** True   **47.** False
49. True   **51.** True   **53.** True   **55.** True   **57.** False   **59.** True   **61.** True
63. False   **65.** $\subseteq$ **67.** $\not\subseteq$ **69.** $\subseteq$ **71.** $\{0, 2, 4\}$
73. $\{0, 1, 2, 3, 4, 5, 6, 7, 8, 9, 11, 13\}$ **75.** M or $\{0, 2, 4, 6, 8\}$
77. $\{0, 1, 2, 3, 4, 5, 7, 9, 11, 13\}$ **79.** $\{0, 2, 4, 6, 8, 10, 12\}$ **81.** $\{10, 12\}$
83. $\varnothing$; $\varnothing$ and R are disjoint   **85.** N or $\{1, 3, 5, 7, 9, 11, 13\}$ **87.** R or $\{0, 1, 2, 3, 4\}$
89. $\{0, 1, 2, 3, 4, 6, 8\}$ **91.** R or $\{0, 1, 2, 3, 4\}$ **93.** $\varnothing$; Q' and $(N' \cap U)$ are disjoint
95. All students in this school who are not taking this course   **97.** All students in this school who are taking calculus and history   **99.** All students in this school who are taking this course or history
101. $\{1, 2, 3, 4, 5, 6\}$ **103.** $\{hhh, hht, hth, thh, htt, tht, tth, ttt\}$ **105.** X and Y are disjoint.
107. $X \subseteq Y$ **109.** $X = \varnothing$

Section 1.2 (page 13)

1. 1, 3   **3.** -6, $-12/4$ (or -3), 0, 1, 3   **5.** $-\sqrt{3}$, $\pi/2$, $\sqrt{12}$ **7.** Natural, whole, integer, rational, real   **9.** Integer, rational, real   **11.** Rational, real   **13.** Irrational, real
15. ($-\sqrt{36} = -6$) integer, rational, real   **17.** Meaningless   **19.** $\{4, 5, 6, 7, 8\}$
21. $\{0, 1, 2, 3\}$ **23.** $\{11, 13, 15, 17, \ldots\}$ **25.** -1 **27.** 21   **29.** -5 **31.** -2
33. 22   **35.** 0   **37.** 24   **39.** 2/3   **41.** $-25/36$ **43.** $-6/7$ **45.** Not a real number
47. -1 **49.** 2.22   **51.** -20 **53.** -54 **55.** $-1/2$ **57.** Not a real number
59. 2   **61.** 14/15   **63.** $-.095$ **65.** -2.463 **67.** $-.979$ **69.** $\subseteq$ **71.** $\cup$
73. $\cap$ **75.** $\cup$ **77.** $=$ or $\subseteq$ **79.** $\cup$ or $\cap$ **81.** (a) Multiply both sides of $a^2/b^2 = 2$ by b^2.   (b) $2b^2$ is a multiple of 2.   (c) a^2 is even because a^2 equals the even number $2b^2$; since the square of an odd number is odd, a must be even.   (d) Replace a with $2c$.   (e) $(2c)^2 = 4c^2$; divide both sides by 2.
(f) $2c^2$ is a multiple of 2.   (g) If $2c^2$ is even, b^2 is even, since $b^2 = 2c^2$.   (h) We have both a and b even, but we claimed that the fraction a/b was in lowest terms.

Section 1.3 (page 19)

1. True   **3.** True   **5.** False; the identity property says $8 \cdot 1 = 8$.   **7.** False; 0 is not a natural number.   **9.** True   **11.** False, since $4 + (7 \cdot 9) = 4 + 63 = 67$, while $(4 + 7) \cdot (4 + 9) = 11 \cdot 13 = 143$.   **13.** True   **15.** False; by the identity property, $2 + 0 = 2$.
17. Commutative   **19.** Commutative   **21.** Associative   **23.** Distributive   **25.** Inverse
27. Identity   **29.** Commutative and distributive   **31.** Distributive   **33.** $6x + 6y$
35. $-3z + 3y$ **37.** $-8r + k$ **39.** $ar + as - at$ **41.** $8(r - 2)$ **43.** $m(n + p)$
45. $36m$ **47.** $20y$ **49.** $-9r$ **51.** $8y - 4z + 12q$ **53.** $-28p + 7m - 35$
55. $2y/3 + 4z/9 - 5/3$ **57.** For example, $11 - 2 \neq 2 - 11$.   **59.** For example, $(12 - 3) - 1 \neq 12 - (3 - 1)$.   **61.** No; for example, $6 + (5 \cdot 4) \neq (6 + 5) \cdot (6 + 4)$.

Section 1.4 (page 26)

1. $-9, -4, -2, 3, 8$ **3.** $-|9|, -|-6|, |-8|$ **5.** $-5, -4, -2, -\sqrt{3}, \sqrt{6}, \sqrt{8}, 3$
7. $3/4, 7/5, \sqrt{2}, 22/15, 8/5$ **9.** $-|-8| - |-6|, -|-2| + (-3), -|3|, -|-2|, |-8 + 2|$ **11.** 8
13. 6 **15.** 4 **17.** 16 **19.** 8 **21.** -16 **23.** 6 **25.** -6 **27.** $8 - \sqrt{50}$
29. $5 - \sqrt{7}$ **31.** $\pi - 3$ **33.** $x - 4$ **35.** $8 - 2k$ **37.** $56 - 7m$
39. $8 + 4m$ or $4m + 8$ **41.** $y - x$ **43.** $3 + x^2$ **45.** $1 + p^2$ **47.** $-6 < 15$
49. $30 > 20$ **51.** $1 \leq 2$ **53.** $x < 3$ **55.** $k < -7$ **57.** $k > 25$ **59.** First part of multiplication property   **61.** Second part of multiplication property   **63.** First part of multiplication property   **65.** Triangle inequality, $|a + b| \leq |a| + |b|$ **67.** Property of absolute value, $|a| \cdot |b| = |ab|$
69. Property of absolute value, $|a/b| = |a|/|b|$ **71.** If $x = y$ or $x = -y$
73. If $y = 0$ or if $|x| \geq |y|$ and x and y have opposite signs   **75.** If $|x| \geq |y|$ and either both positive or both negative   **77.** -1 if $x < 0$ and 1 if $x > 0$

Chapter 1 Review Exercises (page 28)

1. False   **3.** False   **5.** True   **7.** False   **9.** False   **11.** The set has no elements
13. 1, 2, 3, 4   **15.** The set of all students at State University who are female chemistry majors
17. The set of all students at State University who are chemistry majors or out for track   **19.** The set of all students at State University who are not freshmen and are male   **21.** $\{-3, 0\}$
23. $\{-4, -2, -1, 2, 3, 5\}$ **25.** D, or $\{0, 1, 2, 3, 4, 5\}$ **27.** $\{-3, 0, 3, 5\}$ **29.** $\{2\}$
31. 0, 6   **33.** $-12, -6, -9/10, -\sqrt{4}, 0, 1/8, 6$ **35.** All are real numbers.   **37.** Whole numbers, integers, rational numbers, real numbers   **39.** Integers, rational numbers, real numbers
41. Rational numbers, real numbers   **43.** Meaningless   **45.** -15 **47.** 6   **49.** 32
51. $-37/18$ **53.** $-3/20$ **55.** 17   **57.** -32 **59.** -13 **61.** Commutative
63. Inverse   **65.** Distributive   **67.** Distributive   **69.** Identity   **71.** Distributive
73. $9p + 9q$ **75.** $-15 + 3r$ **77.** $a(b + c)$ **79.** $-7, -3, -2, 0, \pi, 8$
81. $-|3 - (-2)|, -|-2|, |6 - 4|, |8 + 1|$ **83.** -3 **85.** -1 **87.** $3 - \sqrt{8}$
89. $m - 3$ **91.** $4 - \pi$ **93.** Second part of multiplication property of inequality   **95.** Addition property of inequality   **97.** Property of absolute value, $|a/b| = |a|/|b|$ **99.** Property of absolute value, $|a \cdot b| = |a| \cdot |b|$ **101.** If $x \geq 0$ **103.** If $x \leq -y$

CHAPTER 2

Section 2.1 (page 37)

1. 1024   **3.** 1/243   **5.** $-1/64$ **7.** 8   **9.** $x^4/625$ **11.** $-1/8$ **13.** 1   **15.** -1
17. 1   **19.** 2^{10} or 1024   **21.** $x^{15}y^{12}$ **23.** p^8/q^2 **25.** z^9/y^9 **27.** $2^3a^9/b^9$ or $8a^9/b^9$
29. r^6/s^{15} **31.** $1/(ab^3)$ **33.** $1/(m^8n^4)$ **35.** $d^6/(2^2c^4)$ or $d^6/(4c^4)$ **37.** $x^6/(2^2y^4)$ or $x^6/(4y^4)$
39. $1/6^4$ or 1/1296   **41.** m **43.** $(3m)/n$ **45.** $(3^5m^{20}n^5)/(2^5x^{15})$ or $(243m^{20}n^5)/(32x^{15})$
47. $216m^9$ **49.** $x^2/(2y)$ **51.** $(4 + s)^6$ **53.** $x/2$ **55.** $1/(r^{10}s^8t^{12})$ **57.** $k^2/(p + q)^7$
59. 5   **61.** 6   **63.** 0   **65.** 73/8   **67.** 9/8   **69.** 1   **71.** 1/6   **73.** $-13/66$

75. 35/18　　**77.** 6.93×10^4　　**79.** 1.27×10^7　　**81.** 1×10^{-4}　　**83.** $7.92 \times$

85. 9.65×10^{-1}　　**87.** 3700　　**89.** 170,000,000　　**91.** .0000402　　**93.** .00000

95. 4.232　　**97.** 1.8×10^{-9}　　**99.** .002 or 2×10^{-3}　　**101.** 70,000 or 7×10^4

103. 4.86×10^7　　**105.** 1.29×10^{-4}　　**107.** -6.04×10^{11}　　**109.** -2.51×10

111. 3.2×10^4 hr, or about 3 yr and 8 mo　　**113.** About 8.33 min　　**115.** About 1.25

117. $2k$　　**119.** $30p^{3-r}$　　**121.** $15x^{p+2}$　　**123.** k^{-7} or $1/k^7$　　**125.** $1/(8b^y)$　　**127.** $-2m^{2-n}$

Section 2.2　(page 45)

1. Polynomial; degree 11; monomial　　**3.** Polynomial; degree 5; binomial　　**5.** Polynomial; degree 6;
binomial　　**7.** Polynomial; degree 4; trinomial　　**9.** Not a polynomial　　**11.** Not a polynomial
13. $x^2 - x + 3$　　**15.** $3x^3 - 3x^2 + 6x + 2$　　**17.** $9y^2 - 4y + 4$　　**19.** $p^3 - 7p^2 - p - 7$
21. $6a^2 + a - 1$　　**23.** $3b^2 - 8b - 1$　　**25.** $-14q^2 + 11q - 14$　　**27.** $4p^2 - 16$
29. $11y^3 - 18y^2 + 4y$　　**31.** $28r^2 + r - 2$　　**33.** $18p^2 - 27pq - 35q^2$　　**35.** $15x^2 - (7/3)x - 2/9$
37. $(6/25)y^2 + (11/40)yz + (1/16)z^2$　　**39.** $25r^2 - 4$　　**41.** $16x^2 - 9y^2$　　**43.** $36k^2 - 36k + 9$
45. $16m^2 + 16mn + 4n^2$　　**47.** $4p^2 - 12p + 9 + 4pq - 6q + q^2$　　**49.** $9q^2 + 30q + 25 - p^2$
51. $8z^3 - 12z^2 + 6z - 1$　　**53.** $125r^3 + 225r^2t^2 + 135rt^4 + 27t^6$　　**55.** $12x^5 + 8x^4 - 20x^3 + 4x^2$
57. $15m^4 - 10m^3 + 5m^2 - 5m$　　**59.** $-2z^3 + 7z^2 - 11z + 4$　　**61.** $27p^3 - 1$　　**63.** $8m^3 + 1$
65. $m^2 + mn - 2n^2 - 2km + 5kn - 3k^2$　　**67.** $a^2 - 2ab + b^2 + 4ac - 4bc + 4c^2$
69. $x^{-2} - 4x^{-1} + 4$　　**71.** $12m^{-2} - 5m^{-1}n^{-1} - 2n^{-2}$　　**73.** $2m^2 - 4m + 8$
75. $5x^3 + 10x^2 + 4x$　　**77.** $5y^2 - 3xy + 8x^2$　　**79.** $3y + 2$　　**81.** $5p - 1$
83. $3x + 2 + [-2/(4x - 5)]$　　**85.** $4z - 3 + [-5/(2z + 5)]$　　**87.** $x^2 - 3x + 2$
89. $5x^2 - 3x + 4 + 4/(3x + 4)$　　**91.** $3m^3 - 5m^2 + 7m - 1 + 3/(m - 3)$　　**93.** $x^2 + 2x + 3$
95. $2z^3 - z^2 - (1/2)z - 5/4 + [(-13/4)z + 17/4]/(2z^2 + z + 1)$
97. $2p^3 - 5p^2 - p + 1/3 + [(-7/3)p^2 - 7p + 7/3]/(3p^3 + p^2 - 1)$
99. $3x^2 - 4xy + y^2$　　**101.** $y^2 - 2yz + z^2$　　**103.** -8　　**105.** 1　　**107.** -24
109. $k^{2m} - 4$　　**111.** $b^{2r} + b^r - 6$　　**113.** $3p^{2x} - 5p^x - 2$　　**115.** $m^{2x} - 4m^x + 4$
117. $q^{2p} - 10q^pp^q + 25p^{2q}$　　**119.** $27k^{3a} - 54k^{2a} + 36k^a - 8$　　**121.** m　　**123.** $m + n$
125. $x^2 + y^2 + z^2 + 2xy + 2xz + 2yz$

Section 2.3　(page 51)

1. $12(m + 3)$　　**3.** $4(2k^2 + 1)$　　**5.** $4m(3n - 2)$　　**7.** $3r(4r^2 + 2r - 1)$
9. $2px(3x - 4x^2 - 6)$　　**11.** $9a^3b^2(5a - 7b^2 + 2ab)$　　**13.** $4k^2m^3(1 + 2k^2 - 3m)$
15. $2(a + b)(1 + 2m)$　　**17.** $(r + s)(a^3 + b^2)$　　**19.** $(y + 2)(3y + 2)$　　**21.** $(r + 3)(3r - 5)$
23. $(m - 1)(2m^2 - 7m + 7)$　　**25.** $25r^3(r - 2)$　or　$-25r^3(-r + 2)$
27. $12a^2(-a^3 + 2a - 3)$　or　$-12a^2(a^3 - 2a + 3)$
29. $36m^2(-m^2n^3 + 2n^5 - 4m^3)$　or　$-36m^2(m^2n^3 - 2n^5 + 4m^3)$
31. $(m + 7)(m + 2)$　　**33.** $(x + 5)(x - 1)$　　**35.** $(z + 4)(z + 5)$　　**37.** $(a + b)(a + 4b)$
39. $(s + 7t)(s - 5t)$　　**41.** $(y - 7z)(y + 3z)$　　**43.** $6(a - 10)(a + 2)$　　**45.** $3m(m + 1)(m + 3)$
47. $(4p - 1)(p + 1)$　　**49.** $(2z - 5)(z + 6)$　　**51.** $3(2r - 1)(2r + 5)$　　**53.** $(6r - 5s)(3r + 2s)$
55. $(4a - b)(5a - 2b)$　　**57.** $(3k - 2p)(2k + 3p)$　　**59.** $(2x + 1)(x - 3)$　　**61.** $(2a - 5)(a - 6)$
63. $(5y + 2)(3y - 1)$　　**65.** Prime　　**67.** $(5a + 3b)(a - 2b)$　　**69.** $(7m + 2n)(3m + n)$
71. $x^2(3 - x)^2$　　**73.** $z(2r - 3)(r + 1)$　　**75.** $2a^2(4a - b)(3a + 2b)$　　**77.** $4z^3(8z + 3a)(z - a)$
79. $(m + 4)(p + 3)$　　**81.** $(q + 2)(r + 2s)$　　**83.** $(2s + 3)(3t - 5)$　　**85.** $(x + y^2)(4 - m)$
87. $(t^3 + s^2)(r - p)$　　**89.** $(3p - 7)(2p + 5)$　　**91.** $(5z - 2x)(4z - 9x)$　　**93.** $(3 - m^2)(5 - r^2)$
95. $(1 - p)(3m^2 + 2)$　　**97.** $(m + n^p)(m - 2n^p)$　　**99.** $(3z^a + x^b)(2z^a - x^b)$
101. $(2x^n + 3y^n)(x^n - 13y^n)$　　**103.** $9.44(x^3 + 5.1x^2 - 3.2x + 4)$
105. $7.88(8.2z^4 - 3z^2 + z + 14.5)$　　**107.** $p^{-4}(1 - p^2)$
109. $4k^{-3}(3 + k - 2k^2) = 4k^{-3}(1 + k)(3 - 2k)$　　**111.** $25p^{-6}(4 - 2p^4 + 3p^8)$

$5p^{-6}(22 - 10p^4 + 25p^8)$

Section 2.4 (page 55)

1. $(p + 10)(p - 10)$ **3.** $(3a + 4)(3a - 4)$ **5.** $(12z + 11)(12z - 11)$
7. $(5s^2 + 3t)(5s^2 - 3t)$ **9.** $(xy + 5a)(xy - 5a)$ **11.** $(a + b + 4)(a + b - 4)$
13. $(3 + 2m - n)(3 - 2m + n)$ **15.** $(6q + 10a - b)(6q - 10a + b)$ **17.** $4xy$
19. $(p^2 + 25)(p + 5)(p - 5)$ **21.** $(3r^4 + 4s^5)(3r^4 - 4s^5)$ **23.** $25(2m^3 + p^2)(2m^3 - p^2)$
25. $(3m - 2)^2$ **27.** $(6a + 5)^2$ **29.** $(5k - 4p)^2$ **31.** $2(4a - 3b)^2$ **33.** $(2xy + 7)^2$
35. $(a - 3b - 3)^2$ **37.** $(q + 3 + p)(q + 3 - p)$
39. $(a + b + x + y)(a + b - x - y)$ **41.** $(2 - a)(4 + 2a + a^2)$
43. $(5x - 3)(25x^2 + 15x + 9)$ **45.** $(6p + 5q)(36p^2 - 30pq + 25q^2)$
47. $(2 + x)(2 - x)(4 + 2x + x^2)(4 - 2x + x^2)$ **49.** $(11pq + 5k^2)(121p^2q^2 - 55pqk^2 + 25k^4)$
51. $(3y^3 + 5z^2)(9y^6 - 15y^3z^2 + 25z^4)$ **53.** $125(p^2 - 2q)(p^4 + 2p^2q + 4q^2)$
55. $r(r^2 + 18r + 108)$ **57.** $(3 - m - 2n)(9 + 3m + 6n + m^2 + 4mn + 4n^2)$
59. $(5m^2 - 2z - 2x^2)(25m^4 + 10m^2z + 10m^2x^2 + 4z^2 + 8zx^2 + 4x^4)$
61. $(m^2 - 5)(m^2 + 2)$ **63.** $(3p^2 - 1)(2p^2 + 3)$ **65.** $9(7k - 3)(k + 1)$ **67.** $(2y - 3)^2$
69. $(3a - 7)^2$ **71.** $-3(10 - 9m)(1 + 2m)$ **73.** $(3a^{2k} + b^{4k})(3a^{2k} - b^{4k})$
75. $(4 + x^a + y^a)(4 - x^a - y^a)$ **77.** $(2y^a - 3)^2$ **79.** $(x^n - y^{2n})(x^{2n} + x^ny^{2n} + y^{4n})$
81. $(5r^{2k} + 6s^{4k})(25r^{4k} - 30r^{2k}s^{4k} + 36s^{8k})$ **83.** $[3(m + p)^k + 5][2(m + p)^k - 3]$
85. ± 14 **87.** ± 36 **89.** 9

Section 2.5 (page 62)

1. $\{x \mid x \neq -6\}$ **3.** $\{x \mid x \neq 3/5\}$ **5.** Set of all real numbers **7.** $5p/2$ **9.** $8/9$
11. $3/(t - 3)$ **13.** $(2x + 4)/x$ **15.** $(m - 2)/(m + 3)$ **17.** $(2m + 3)/(4m + 3)$
19. $(y + 2)/2$ **21.** $25p^2/9$ **23.** $2/9$ **25.** $3/10$ **27.** $5x/y$ **29.** $2(a + 4)/(a - 3)$
31. 1 **33.** $(k + 2)/(k + 3)$ **35.** $(m + 6)/(m + 3)$ **37.** $(m - 3)/(2m - 3)$
39. $(x^2 - 1)/x^2$ **41.** $(x + y)/(x - y)$ **43.** $(x^2 - xy + y^2)/(x^2 + xy + y^2)$ **45.** $14/r$
47. $19/(6k)$ **49.** $5/(12y)$ **51.** 1 **53.** $(6 + p)/(2p)$ **55.** $(8 - y)/(4y)$ **57.** $137/(30m)$
59. $(2y + 1)/[y(y + 1)]$ **61.** $3/[2a(a + b)]$ **63.** $-2/[(a + 1)(a - 1)]$
65. $(2m^2 + 2)/[(m - 1)(m + 1)]$ **67.** $4/(a - 2)$ **69.** $(3x + y)/(2x - y)$
71. $5/[(a - 2)(a - 3)(a + 2)]$ **73.** $(x - 11)/[(x + 4)(x - 4)(x - 3)]$
75. $a(a + 5)/[(a + 6)(a + 1)(a - 1)]$ **77.** $(6x^2 - 6x - 5)/[(x + 5)(x - 2)(x - 1)]$
79. $(p + 5)/[p(p + 1)]$ **81.** $-1/[x(x + h)]$ **83.** $(x + 1)/(x - 1)$ **85.** $-1/(x + 1)$
87. $(2 - b)(1 + b)/[b(1 - b)]$ **89.** $(m^3 - 4m - 1)/(m - 2)$ **91.** $(3z^2 - 75z + 8)(z + 1)/[5(z - 25)]$
93. $a + b$ **95.** -1 **97.** $1/(ab)$ **99.** $y(xy - 9)/(x^2y^2 - 9)$

Section 2.6 (page 68)

1. 2 **3.** 4 **5.** $1/9$ **7.** $27/8$ **9.** $1/25$ **11.** $16/9$ **13.** $256/81$ **15.** $4p^2$
17. $9x^4$ **19.** $5y^3z^4$ **21.** 2^2 or 4 **23.** $27^{1/3}$ or 3 **25.** 4^2 or 16 **27.** $m^{7/3}$
29. $(1 + n)^{5/4}$ **31.** $6yz^{2/3}$ **33.** $r^{3/14}s^{7/20}$ **35.** r **37.** $1/(12k^{5/2})$ **39.** $a^{2/3}b^2$
41. $h^{1/3}t^{1/5}/k^{2/5}$ **43.** 1 **45.** $x^{11/12}$ **47.** $16/y^{11/12}$ **49.** $1/(k + 5)^{1/2}$ **51.** $4z^2/(x^5y)$
53. r^{6+p} **55.** $m^{3/2}$ **57.** $x^{(2n^2-1)/n}$ **59.** $p^{(m+n+m^2)/(mn)}$ **61.** $y - 10y^2$
63. $-4k^{10/3} + 24k^{4/3}$ **65.** $x^2 - x$ **67.** $r - 2 + r^{-1}$ or $r - 2 + 1/r$
69. $x^{1/2} + x^{-1/2}$ or $x^{1/2} + 1/x^{1/2}$ **71.** $k^{3/4}(4k + 1)$ **73.** $z^{-1/2}(9 + 2z)$ **75.** $p^{-7/4}(p - 2)$
77. $(p + 4)^{-3/2}(p^2 + 9p + 21)$ **79.** 64 **81.** $\$64,000,000$ **83.** About $\$10,000,000$
85. 29 **87.** 177

Section 2.7 (page 74)

1. 5 **3.** 6 **5.** -5 **7.** $5\sqrt{2}$ **9.** $20\sqrt{5}$ **11.** $50\sqrt{2}$ **13.** $3\sqrt[3]{3}$ **15.** $5\sqrt[3]{2}$
17. $-2\sqrt[4]{2}$ **19.** $-3\sqrt{5}/5$ **21.** $-\sqrt[3]{12}/2$ **23.** $\sqrt[4]{24}/2$ **25.** $-24\sqrt{6}$ **27.** $32\sqrt[3]{2}$
29. $2x^2z^4\sqrt{2x}$ **31.** $2zx^2y\sqrt[3]{2z^2x^2y}$ **33.** $np^2\sqrt[4]{m^2n^3}$ **35.** $5rz\sqrt[6]{5^3r^2z^4}$ or $5rz\sqrt[6]{125r^2z^4}$

37. $\sqrt{6x}/(3x)$ **39.** $x^2y\sqrt{xy}/z$ **41.** $2\sqrt[3]{x}/x$ **43.** $-km\sqrt[3]{k^2}/r^2$ **45.** $h\sqrt[4]{9g^3hr^2}/(3r^2)$
47. $m\sqrt[3]{n^2}/n$ **49.** $2\sqrt[4]{x^3y^3}$ **51.** $\sqrt[6]{4}$ or $\sqrt[3]{2}$ **53.** $\sqrt[18]{x}$ **55.** $9\sqrt{3}$ **57.** $\sqrt{2}$
59. $-2\sqrt{7p}$ **61.** $2\sqrt[3]{2}$ **63.** $7\sqrt[3]{3}$ **65.** $2\sqrt{3}$ **67.** $13\sqrt[3]{4}/6$
69. $ab\sqrt{ab}(b - 2a^2 + b^3)$ **71.** -7 **73.** 10 **75.** $11 + 4\sqrt{6}$ **77.** $5\sqrt{6}$
79. $2\sqrt[3]{9} - 7\sqrt[3]{3} - 4$ **81.** $-3(1 + \sqrt{2})$ **83.** $(4\sqrt{3} - 3)/13$ **85.** $4(2 + \sqrt{y})/(4 - y)$
87. $p(\sqrt{p} - 2)/(p - 4)$ **89.** $a(\sqrt{a + b} + 1)/(a + b - 1)$ **91.** $-2x - 2\sqrt{x(x + 1)} - 1$
93. $3/(2\sqrt{3})$ **95.** $-1/[2(1 - \sqrt{2})]$ **97.** $x/(\sqrt{x} + x)$ **99.** $-1/(2x - 2\sqrt{x(x + 1)} + 1)$
101. 5.36×10^3 **103.** 1.40×10^{-1} **105.** 2.64×10^{-1} **107.** 9.93, or 9.93×10^0
109. $|m + n|$ **111.** $|z - 3x|$ **113.** 3.14626437

Chapter 2 Review Exercises (page 77)
1. $1/64$ **3.** $16/25$ **5.** $-10z^8$ **7.** $(-16p^7)/q$ **9.** $y^6/(36x^4z^4)$ **11.** $1/a^{23}$
13. $(36s)/r$ **15.** $1/[p^2(m + n)]$ **17.** $r^2/(s^4t^4)$ **19.** $-20/(m^5n^6)$ **21.** a^{5-p} **23.** $p^{2m}/9$
25. $-7m^2 - m - 3$ **27.** $8r^4 - 8r^2 + 11r$ **29.** $-r^5 + r^4 + 19r^2$ **31.** $16y^2 + 42y - 49$
33. $21z^2 - 5zy - 50y^2$ **35.** $9k^2 - 30km + 25m^2$ **37.** $125x^3 - 150x^2 + 60x - 8$
39. $15w^3 - 22w^2 + 11w - 2$ **41.** $a^2 - b^2 - ac + 3bc - 2c^2$ **43.** $p^{2q} - 2p^q - 3$
45. $6y^2 - 4y + 1$ **47.** $3m^3n/2 - m^2n^3 + 2m^6/n$ **49.** $3p + 8$ **51.** $4z - 1 + (-8)/(7z - 5)$
53. $y^2 - 4y + 8 + 4/(2y - 3)$ **55.** $5m - 7$ **57.** $k^3 - 2k^2 + 3$ **59.** $z(7z - 9z^2 + 1)$
61. $4p^3(3p^2 - 2p + 5)$ **63.** $(r + 7p)(r - 6p)$ **65.** $(3m + 1)(2m - 5)$
67. $(5a - b)(3a + 2b)$ **69.** $5m^3(3m - 5n)(2m + n)$ **71.** $(13y^2 + 1)(13y^2 - 1)$
73. $(2p - 5r^3)^2$ **75.** $8(y - 5z^2)(y^2 + 5yz^2 + 25z^4)$ **77.** $3(9r - 10)(2r + 1)$
79. $(r - 3s)(a + 5b)$ **81.** $(4m - 7 + 5a)(4m - 7 - 5a)$ **83.** $x/(2y)$ **85.** $1/[2k^2(k - 1)]$
87. $(x + 1)/(x + 4)$ **89.** $(3m + n)(3m - n)$ **91.** $37/(20y)$ **93.** $(x + 9)/[(x - 3)(x - 1)(x + 1)]$
95. $(q + p)/(pq - 1)$ **97.** $p^{1/3}$ **99.** $-14r^{17/12}$ **101.** $y^{1/2}$ **103.** k^{2-9p}
105. $10z^{7/3} - 4z^{1/3}$ **107.** $3p^2 + 3p^{3/2} - 5p - 5p^{1/2}$ **109.** $(2 + 5p)/p^{1/2}$
111. $(p + 2)/[2(p + 1)^{3/2}]$ **113.** $(2x^2 + 15)/(2x^2 + 5)^{4/3}$ **115.** $10\sqrt{2}$ **117.** $5\sqrt[4]{2}$
119. $\sqrt{21r}/(3r)$ **121.** $2^3y^4\sqrt{2m}/m^2$ or $8y^4\sqrt{2m}/m^2$ **123.** $\sqrt[12]{m}$ **125.** 66
127. $m(-9\sqrt{2m} + 5\sqrt{m})$ **129.** $6(3 + \sqrt{2})/7$ **131.** $k(\sqrt{k} + 3)/(k - 9)$

CHAPTER 3

Section 3.1 (page 85)
1. Identity **3.** Conditional **5.** Identity **7.** Identity **9.** Not equivalent
11. Equivalent **13.** Not equivalent **15.** Not equivalent **17.** $\{4\}$ **19.** $\{12\}$
21. $\{-2/7\}$ **23.** $\{-7/8\}$ **25.** $\{-1\}$ **27.** $\{3\}$ **29.** $\{4\}$ **31.** $\{12\}$ **33.** $\{3/4\}$
35. $\{-12/5\}$ **37.** $\emptyset$ **39.** $\{27/7\}$ **41.** $\{-59/6\}$ **43.** $\{-9/4\}$ **45.** $\emptyset$ **47.** $\{-4\}$
49. $\{-19/75\}$ **51.** $x = -3a + b$ **53.** $x = (3a + b)/(3 - a)$
55. $x = (3 - 3a)/(a^2 - a - 1)$ **57.** $x = 2a^2/(a^2 + 3)$ **59.** $x = (m + 4)/(5 + 2m)$
61. $68°F$ **63.** $15°C$ **65.** $37.8°C$ **67.** $104°F$ **69.** 13% **71.** $\$432$
73. $\$1500$ **75.** $\$205.41$ **77.** $\$66.50$ **81.** 6 **83.** -2 **85.** $\{.72\}$
87. $\{6.53\}$ **89.** $\{-13.26\}$

Section 3.2 (page 93)
1. $V = k/P$ **3.** $l = V/(wh)$ **5.** $g = (V - V_0)/t$ **7.** $g = 2s/t^2$ **9.** $B = 2A/h - b$
11. $r_1 = S/(2\pi h) - r_2$ **13.** $l = gt^2/(4\pi^2)$ **15.** $h = (S - 2\pi r^2)/(2\pi r)$ **17.** $f = AB(p + 1)/24$
19. $R = r_1r_2/(r_2 + r_1)$ **21.** 6 cm **23.** 7 quarters **25.** 2 liters **27.** $38/3$ kg
29. $400/3$ liters **31.** About 840 mi **33.** 15 min **35.** 35 kph **37.** $18/5$ hr **39.** 39 hr
41. $40/3$ hr **43.** $\$57.65$ **45.** $\$12,000$ at 13% and $\$8000$ at 16% **47.** $\$54,000$ at $8\ 1/2\%$ and
$\$6000$ at 16% **49.** $\$70,000$ for land that makes a profit and $\$50,000$ for land that produces a loss
51. $\$267.57$ **53.** No solution; given numbers are inconsistent **55.** $\$20,000$ at 12% and $\$40,000$ at 10%

Section 3.3   (page 101)

1. Imaginary and complex   **3.** Real and complex   **5.** Imaginary and complex   **7.** Complex
9. $0 + 10i$ **11.** $0 - 20i$ **13.** $0 - i\sqrt{39}$ **15.** $5 + 2i$ **17.** $-6 - 14i$ **19.** $9 - 5i\sqrt{2}$
21. -5 **23.** -4 **25.** 2 **27.** -2 **29.** $7 - i$ **31.** $2 + 0i$ **33.** $1 - 10i$
35. $-10 + 5i$ **37.** $8 - i$ **39.** $-14 + 2i$ **41.** $5 - 12i$ **43.** $-8 - 6i$ **45.** $13 + 0i$
47. $7 + 0i$ **49.** $0 + 25i$ **51.** $12 + 9i$ **53.** $0 + i$ **55.** $7/25 - (24/25)i$
57. $26/29 + (7/29)i$ **59.** $-2 + i$ **61.** $-2i$ **63.** $[(3 - 2\sqrt{5}) + (-2 - 3\sqrt{5})i]/13$
65. $5/2 + i$ **67.** $-16/65 - (37/65)i$ **69.** $27/10 + (11/10)i$ **71.** i **73.** i **75.** 1
77. $-i$ **79.** 1 **81.** i **83.** $a = 0$ or $b = 0$

Section 3.4   (page 110)

1. $\{\pm 4\}$ **3.** $\{\pm 3\sqrt{3}\}$ **5.** $\{3 \pm \sqrt{5}\}$ **7.** $\{(1 \pm 2\sqrt{3})/3\}$ **9.** $\{2, 3\}$ **11.** $\{-5/2, 10/3\}$
13. $\{-3/2, -1/4\}$ **15.** $\{3, 5\}$ **17.** $\{1 \pm \sqrt{5}\}$ **19.** $\{(-1 \pm i)/2\}$ **21.** $\{(1 \pm \sqrt{5})/2\}$
23. $\{(-1 \pm \sqrt{7})/2\}$ **25.** $\{3 \pm \sqrt{2}\}$ **27.** $\{(3 \pm i\sqrt{7})/2\}$ **29.** $\{(1 \pm i\sqrt{7})/4\}$
31. $\{1 \pm 2i\}$ **33.** $\{3, -1/4\}$ **35.** $\{3/2, 1\}$ **37.** $\{1.243, -.643\}$ **39.** $\{1.281, -.781\}$
41. $\{3.562, -.562\}$ **43.** $\{(\sqrt{2} \pm \sqrt{6})/2\}$ **45.** $\{\sqrt{2}, \sqrt{2}/2\}$ **47.** $\{(-i \pm i\sqrt{5})/2\}$
49. $\{(1 \pm \sqrt{2})/i\}$ or $\{-i \pm i\sqrt{2}\}$ **51.** $\{[(1 - i) \pm (\sqrt{7} + i\sqrt{7})]/4\}$ **53.** $\{1, (-1 \pm i\sqrt{3})/2\}$
55. $\{-3, (3 \pm 3i\sqrt{3})/2\}$ **57.** $\{4, -2 \pm 2i\sqrt{3}\}$ **59.** $\{-5/2, (5 \pm 5i\sqrt{3})/4\}$ **61.** 0; one
rational solution   **63.** 1; two rational solutions   **65.** 84; two irrational solutions   **67.** -23; two
complex solutions   **69.** 16, 18 or $-18, -16$ **71.** 12, -2 **73.** 50 m by 100 m   **75.** 9 ft
by 12 ft   **77.** 50 mph   **79.** 16.95 cm, 20.95 cm, 26.95 cm   **81.** ± 12 **83.** 121/4
85. 1, 9   **87.** $t = \pm \sqrt{2sg}/g$ **89.** $h = \pm \sqrt{d^4kL/L}$ or $\pm d^2\sqrt{kL}/L$
91. $t = \pm \sqrt{(s - s_0 - k)g}/g$ **93.** $R = (E^2 - 2Pr \pm E\sqrt{E^2 - 4Pr})/(2P)$
95. (a) $x = (y \pm \sqrt{8 - 11y^2})/4$ (b) $y = (x \pm \sqrt{6 - 11x^2})/3$

Section 3.5   (page 116)

1. $\{\pm\sqrt{3}, \pm i\sqrt{5}\}$ **3.** $\{\pm 1, \pm\sqrt{10}/2\}$ **5.** $\{4, 6\}$ **7.** $\{(-5 \pm \sqrt{21})/2\}$
9. $\{(-6 \pm 2\sqrt{3})/3, (-4 \pm \sqrt{2})/2\}$ **11.** $\{7/2, -1/3\}$ **13.** $\{-63, 28\}$ **15.** $\{\pm\sqrt{5}/5\}$
17. $\{\pm i\sqrt{3}, \pm i\sqrt{6}/3\}$ **19.** $\{1\}$ **21.** $\{1/2\}$ **23.** $\{-1\}$ **25.** $\{5\}$ **27.** $\{9\}$
29. $\{9\}$ **31.** $\{5/4\}$ **33.** $\{2, -2\}$ **35.** $\{2, -2/9\}$ **37.** $\{0, 9\}$ **39.** $\{-2\}$ **41.** $\{2\}$
43. $\{3/2\}$ **45.** $\{31\}$ **47.** $\{-3, 1\}$ **49.** $\{-27, 3\}$ **51.** $\{1/4, 1\}$ **53.** $\{0, 8\}$
55. $\{\pm 2.121, \pm 1.528\}$ **57.** $\{\pm.917, \pm.630i\}$ **59.** $h = (d/k)^2$ **61.** $L = P^2g/4$
63. $y = (a^{2/3} - x^{2/3})^{3/2}$

Section 3.6   (page 124)

1. $(-1, 4)$ **3.** $(-\infty, 0)$ **5.** $[1, 2)$ **7.** $(-\infty, -9)$

9. $-4 < x < 3$ **11.** $x \le -1$ **13.** $-2 \le x < 6$ **15.** $x \le -4$

17. $(-\infty, 4]$ **19.** $[-1, +\infty)$ **21.** $(-\infty, 6]$ **23.** $(-\infty, 4)$

25. $[-11/5, +\infty)$ **27.** $[1, 4]$ **29.** $(-6, -4)$ **31.** $(-16, 19]$

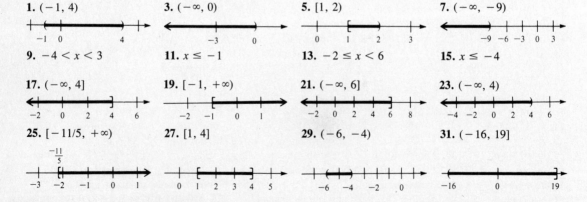

33. $[-3, 3]$

35. $(-\infty, -3] \cup [-1, +\infty)$

37. $[-2, 3]$

39. $(-\infty, 1/2) \cup (4, +\infty)$

41. $(-\infty, 0) \cup (0, +\infty)$

43. $((-5 - \sqrt{33})/2, (-5 + \sqrt{33})/2)$

45. $[1 - \sqrt{2}, 1 + \sqrt{2}]$

47. $[-2, 3/2] \cup [3, +\infty)$ **49.** $(-\infty, -2] \cup [0, 2]$ **51.** $(-2, 0) \cup (1/4, +\infty)$ **53.** 140 points
55. $[500, +\infty)$ **57.** $[45, +\infty)$ **59.** $(0, 5/4), (6, +\infty)$ **61.** For any x in $[0, 5/3) \cup (10, +\infty)$
63. 32°F to 86°F   **65.** $(-5, 3]$ **67.** $(-\infty, -2)$ **69.** $(-\infty, 6) \cup [15/2, +\infty)$
71. $(-\infty, 1) \cup (9/5, +\infty)$ **73.** $(-\infty, -3/2) \cup [-1/2, +\infty)$ **75.** $(-2, +\infty)$
77. $(0, 4/11) \cup (1/2, +\infty)$ **79.** $(-\infty, -2] \cup (1, 2)$ **81.** $(-\infty, 5)$ **83.** $[3/2, +\infty)$
85. $(5/2, +\infty)$ **87.** $[-8/3, 3/2] \cup (6, +\infty)$ **89.** If k is in $(-\infty, -4\sqrt{2}] \cup [4\sqrt{2} +\infty)$
91. If k is in $(-\infty, 0] \cup [8, +\infty)$

Section 3.7 (page 130)
1. $\{1, 3\}$ **3.** $\{-1/3, 1\}$ **5.** $\{2/3, 8/3\}$ **7.** $\{-6, 14\}$ **9.** $\{5/2, 7/2\}$ **11.** $\{-4/3, 2/9\}$
13. $\{-7/3, -1/7\}$ **15.** $\{-3/5, 11\}$ **17.** $\{-1/2\}$ **19.** $[-3, 3]$
21. $(-\infty, -1) \cup (1, +\infty)$ **23.** $\emptyset$ **25.** $[-10, 10]$ **27.** $(-4, -1)$
29. $(-\infty, -2/3) \cup (2, +\infty)$ **31.** $(-\infty, -8/3] \cup [2, +\infty)$ **33.** $(-3/2, 13/10)$
35. $|x - 2| \le 4$ **37.** $|z - 12| \ge 2$ **39.** $|k - 1| = 6$
41. If $|x - 2| \le .0004$, then $|y - 7| \le .00001$.   **43.** $m = 2; n = 20$

Chapter 3 Review Exercises (page 132)
1. $\{6\}$ **3.** $\{-11/3\}$ **5.** $\{-96/7\}$ **7.** $\{-2\}$ **9.** $\{13\}$ **11.** $x = -6b - a - 6$
13. $x = (6 - 3m)/(1 + 2k - km)$ **15.** $y = 6a/(4 - a)$ **17.** $C = (5/9)(F - 32)$
19. $j = n - nA/I$ **21.** $r_1 = kr_2/(r_2 - k)$ **23.** $L = V/(\pi r^2)$ **25.** $x = -4/(y^2 - 5y - 6p)$
27. \$500   **29.** 12 hr   **31.** $10 - 3i$ **33.** $-8 + 13i$ **35.** $19 + 17i$ **37.** 146
39. $7 - 24i$ **41.** $-30 - 40i$ **43.** $5/2 + (7/2)i$ **45.** $(11/5) + (2/5)i$ **47.** $-i$
49. $-i$ **51.** -1 **53.** $\{-7 \pm \sqrt{5}\}$ **55.** $\{5/2, -3\}$ **57.** $\{7, -3/2\}$ **59.** $\{3, -1/2\}$
61. $\{-2i \pm i\sqrt{5}\}$ **63.** -188; two complex solutions   **65.** 484; two rational solutions   **67.** 0;
one rational solution   **69.** 50 m by 225 m; or 112.5 m by 100 m   **71.** $\{\pm i, \pm 1/2\}$ **73.** $\{-2, 3\}$
75. $\{5/2, -15\}$ **77.** $\{63/2\}$ **79.** $\{3\}$ **81.** $\{-2, -1\}$ **83.** $\{1, -4\}$ **85.** $\{-1\}$
87. $(-7/13, +\infty)$ **89.** $(-\infty, 1]$ **91.** $(1, +\infty)$ **93.** $[4, 5]$ **95.** $[-4, 1]$
97. $(-2/3, 5/2)$ **99.** $(-\infty, -4] \cup [0, 4]$ **101.** $(-2, 0)$ **103.** $(-3, 1) \cup [7, +\infty)$
105. 0 to 30 (Remember: x must start at 0.)   **107.** $\{-1, 5\}$ **109.** $\{11/27, 25/27\}$
111. $\{4/3, -2/7\}$ **113.** $[-7, 7]$ **115.** $(-\infty, -3) \cup (3, +\infty)$ **117.** $[-6, -3]$
119. $(-2/7, 8/7)$ **121.** $(-\infty, -4) \cup (-2/3, +\infty)$

CHAPTER 4

Section 4.1 (page 140)

1–5.

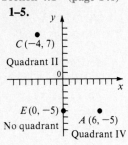

$C(-4, 7)$
Quadrant II
$E(0, -5)$
No quadrant
$A(6, -5)$
Quadrant IV

7.

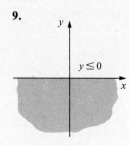

$x = 0$

9.

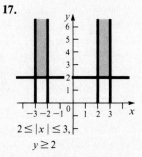

$y \leq 0$

11.

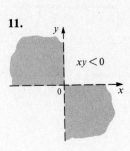

$xy < 0$

13.

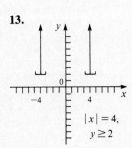

$|x| = 4$,
$y \geq 2$

15.

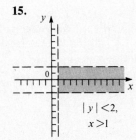

$|y| < 2$,
$x > 1$

17.

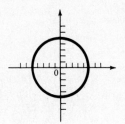

$2 \leq |x| \leq 3$,
$y \geq 2$

19. $8\sqrt{2}$; $(9, 3)$ **21.** $\sqrt{34}$; $(-11/2, -7/2)$ **23.** $\sqrt{133}$; $(2\sqrt{2}, 3\sqrt{5}/2)$ **25.** 7.616
27. 27.203 **29.** $(13, 10)$ **31.** $(-10, 11)$ **33.** Yes **35.** No **37.** No
39. Yes **41.** No **45.** $x^2 + y^2 = 25$ **47.** $(2 + \sqrt{7}, 2 + \sqrt{7})$, $(2 - \sqrt{7}, 2 - \sqrt{7})$

Section 4.2 (page 149)

1.

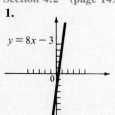

$y = 8x - 3$

3.

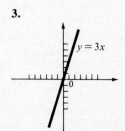

$y = 3x$

5.

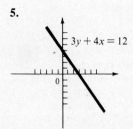

$3y + 4x = 12$

7.

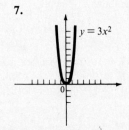

$y = 3x^2$

9.
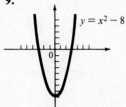
$y = x^2 - 8$

11.
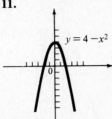
$y = 4 - x^2$

13.
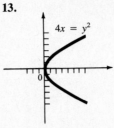
$4x = y^2$

15.
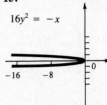
$16y^2 = -x$

17.
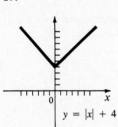
$y = |x| + 4$

19.
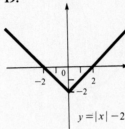
$y = |x| - 2$

21.
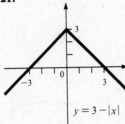
$y = 3 - |x|$

23.
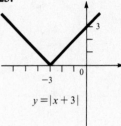
$y = |x + 3|$

25.
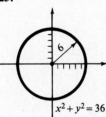
$x^2 + y^2 = 36$

27.
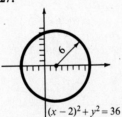
$(x - 2)^2 + y^2 = 36$

29.
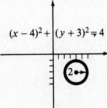
$(x - 4)^2 + (y + 3)^2 = 4$

31.
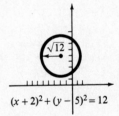
$(x + 2)^2 + (y - 5)^2 = 12$

33. $(x - 1)^2 + (y - 4)^2 = 9$ **35.** $(x + 8)^2 + (y - 6)^2 = 25$ **37.** $(x + 1)^2 + (y - 2)^2 = 25$
39. $(x + 3)^2 + (y + 2)^2 = 4$ **41.** $(-3, -4); r = 4$ **43.** $(6, -5); r = 6$
45. $(-4, 7); r = 0$ (a point)   **47.** $(0, 1); r = 7$ **49.** $(1.42, -.7); r = .8$
51. $(x + 2)^2 + (y + 3/2)^2 = 149/4$ **53.** Yes

55.
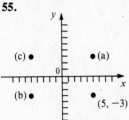

57.
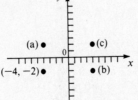

59.
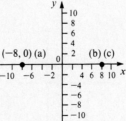

61.
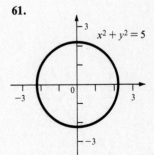
$x^2 + y^2 = 5$

63.

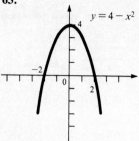

65.

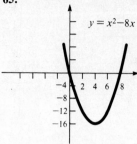

67.

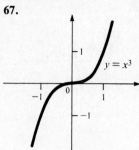

69.

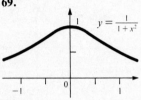

71.

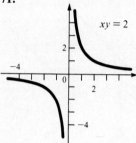

73. Reflected about the *x*-axis

Section 4.3   (page 154)
1. Linear   **3.** Linear   **5.** Not linear   **7.** Linear   **9.** Not linear   **11.** Not linear
13. Not linear   **15.** Not linear   **17.** Not linear
19.

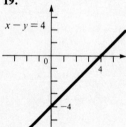

21.

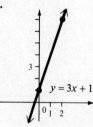

23.

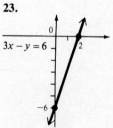

25.

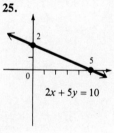

27.

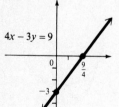

29.

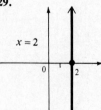

31.

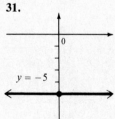

33.

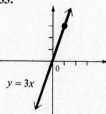

35.

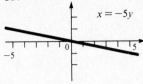

37.

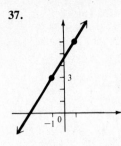

39.

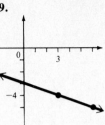

41.

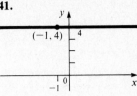

43.

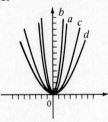

45. 1/5   **47.** 7/9   **49.** 3/5   **51.** Undefined slope
53. 2   **55.** 4   **57.** −3/4   **59.** 0   **61.** 0
63. .5785   **65.** Yes   **67.** No

69. (a) 16 (b) 11 (c) 6 (d) 8 (e) 4 (f) 0
(g) and (k) are graphed below (h) 0
(i) 40/3 (j) 80/3 (l) 8 (m) 6

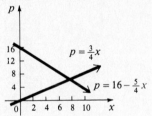

71. (a) See graph below (b) 125 (c) 50

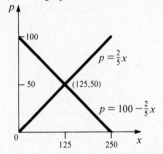

Section 4.4   (page 161)
1. $2x + y = 5$ **3.** $3x + 2y = -7$ **5.** $3x + 4y = 6$ **7.** $y = 2$ **9.** $x = -8$
11. $x + 3y = 10$ **13.** $-x + 4y = 13$ **15.** $3x + 4y = 12$ **17.** $2x - 3y = 6$
19. $x = -5$ **21.** $x = -6$ **23.** $5.081x + y = -4.69256$ **25.** $x + 3y = 11$
27. $2x - y = 9$ **29.** $x - y = 7$ **31.** $5x - 3y = -13$ **33.** $-2x + y = 4$
35. $x = -5$ **37.** No   **39.** (a) $-1/2$ (b) $-7/2$ **47.** $y = 640x + 1100; m = 640$
49. $y = -1000x + 40,000; m = -1000$ **51.** $y = 2.5x - 70; m = 2.5$

Section 4.5   (page 171)
1.

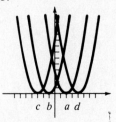

3.

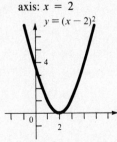

5. Vertex: (2, 0);
axis: $x = 2$

7. Vertex: $(-3, -4)$;
axis: $x = -3$

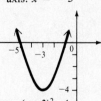

9. Vertex: $(-3, 2)$;
  axis: $x = -3$

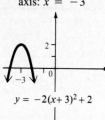

$y = -2(x+3)^2 + 2$

11. Vertex: $(-1, -3)$;
  axis: $x = -1$

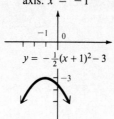

$y = -\frac{1}{2}(x+1)^2 - 3$

13. Vertex: $(1, 2)$;
  axis: $x = 1$

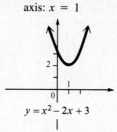

$y = x^2 - 2x + 3$

15. Vertex; $(-2, 6)$;
  axis: $x = -2$

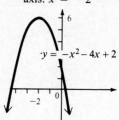

$y = -x^2 - 4x + 2$

17. Vertex: $(1, 3)$;
  axis: $x = 1$

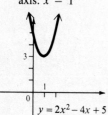

$y = 2x^2 - 4x + 5$

19.

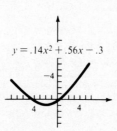

$y = .14x^2 + .56x - .3$

21.

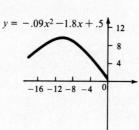

$y = -.09x^2 - 1.8x + .5$

23.

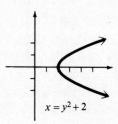

$x = y^2 + 2$

25.

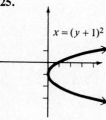

$x = (y+1)^2$

27.

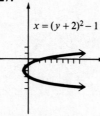

$x = (y+2)^2 - 1$

29.

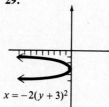

$x = -2(y+3)^2$

31.

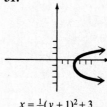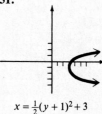

$x = \frac{1}{2}(y+1)^2 + 3$

33.

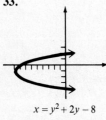

$x = y^2 + 2y - 8$

35.

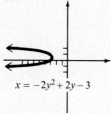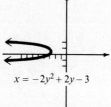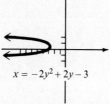

$x = -2y^2 + 2y - 3$

37. 80 by 160 ft   **39.** 300 sandwiches; 100
41. 5 in   **43.** 10, 10
45. (a) $x(500 - x) = 500x - x^2$
  (b)   (c) 250
             (d) $62,500

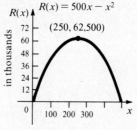

$R(x) = 500x - x^2$
(250, 62,500)

47.

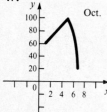

Oct.

49. $6\sqrt{3}$ cm   **51.** 25
53. $12y = x^2$
55. $-8x = y^2$
57. $4(y - 4) = (x - 3)^2$
59. $6(x + 33/6) = (y - 1)^2$

Section 4.6   (page 182)

1.

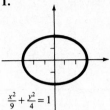

$$\frac{x^2}{9} + \frac{y^2}{4} = 1$$

3.

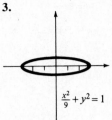

$$\frac{x^2}{9} + y^2 = 1$$

5.

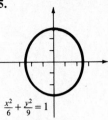

$$\frac{x^2}{6} + \frac{y^2}{9} = 1$$

7.

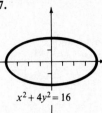

$$x^2 + 4y^2 = 16$$

9.

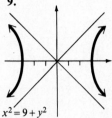

$$x^2 = 9 + y^2$$

11.

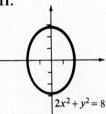

$$2x^2 + y^2 = 8$$

13.

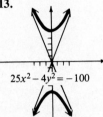

$$25x^2 - 4y^2 = -100$$

15.

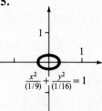

$$\frac{x^2}{(1/9)} + \frac{y^2}{(1/16)} = 1$$

17.

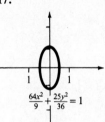

$$\frac{64x^2}{9} + \frac{25y^2}{36} = 1$$

19.

$$\frac{(x-1)^2}{9} + \frac{(y+3)^2}{25} = 1$$

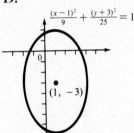

$$(1, -3)$$

21.

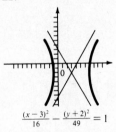

$$\frac{(x-3)^2}{16} - \frac{(y+2)^2}{49} = 1$$

23.

$$\frac{(y+1)^2}{25} - \frac{(x-3)^2}{36} = 1$$

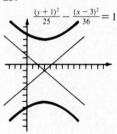

25.

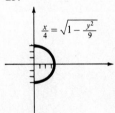

$$\frac{x}{4} = \sqrt{1 - \frac{y^2}{9}}$$

27.

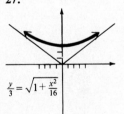

$$\frac{y}{3} = \sqrt{1 + \frac{x^2}{16}}$$

29.

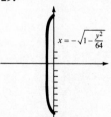

$$x = -\sqrt{1 - \frac{y^2}{64}}$$

31.

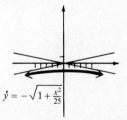

$$y = -\sqrt{1 + \frac{x^2}{25}}$$

33. Hyperbola

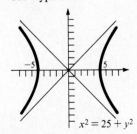

$$x^2 = 25 + y^2$$

35. Ellipse

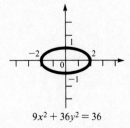

$$9x^2 + 36y^2 = 36$$

37. Circle

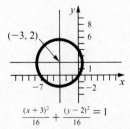

$$(-3, 2)$$

$$\frac{(x+3)^2}{16} + \frac{(y-2)^2}{16} = 1$$

39. Parabola

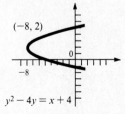

$$(-8, 2)$$

$$y^2 - 4y = x + 4$$

41. Empty set (no graph)

43. Circle

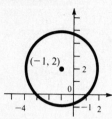

$$3x^2 + 6x + 3y^2 - 12y = 12$$

51. Empty set (no graph)

45. Parabola

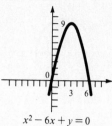

$$x^2 - 6x + y = 0$$

47. Hyperbola

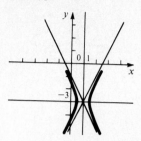

$$4x^2 - 8x - y^2 - 6y = 6$$

49. Ellipse

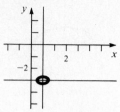

$$4x^2 - 8x + 9y^2 + 54y = -84$$

Section 4.7   (page 188)

1.

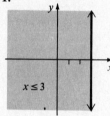

$x \le 3$

3.

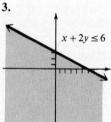

$x + 2y \le 6$

5.

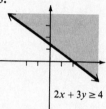

$2x + 3y \ge 4$

7.

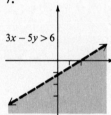

$3x - 5y > 6$

9.

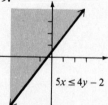

$5x \le 4y - 2$

11.

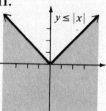

$y \le |x|$

13.

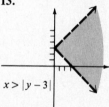

$x > |y - 3|$

15.

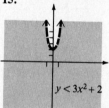

$y < 3x^2 + 2$

17.

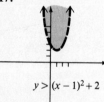

$y > (x - 1)^2 + 2$

19.

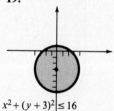

$x^2 + (y + 3)^2 \le 16$

21.

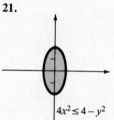

$4x^2 \le 4 - y^2$

23.

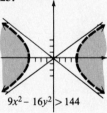

$9x^2 - 16y^2 > 144$

25.

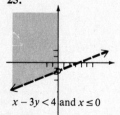

$x - 3y < 4$ and $x \le 0$

27.

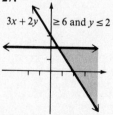

$3x + 2y \ge 6$ and $y \le 2$

29.

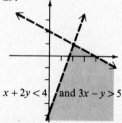

$x + 2y < 4$ and $3x - y > 5$

31.

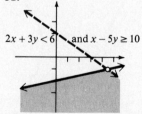

$2x + 3y < 6$ and $x - 5y \ge 10$

33.
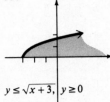
$y \le \sqrt{x+3},\ y \ge 0$

35.
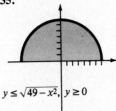
$y \le \sqrt{49-x^2},\ y \ge 0$

37.
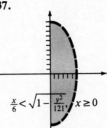
$\frac{x}{6} < \sqrt{1-\frac{y^2}{121}},\ x \ge 0$

39.
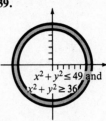
$x^2+y^2 \le 49$ and $x^2+y^2 \ge 36$

Chapter 4   Review Exercises   (page 190)
1. $\sqrt{85}$ **3.** 5   **5.** $-7;\ -1;\ 8;\ 23$ **7.** No such points exist

9.
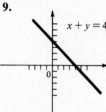
$x+y=4$

11.
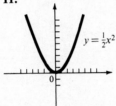
$y=\frac{1}{2}x^2$

13.
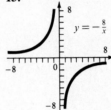
$y=-\frac{8}{x}$

15.
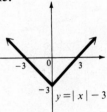
$y=|x|-3$

17. $(x+2)^2+(y-3)^2=25$ **19.** $(x+8)^2+(y-1)^2=289$ **21.** $(2,\ -3);\ r=1$
23. $(-7/2,\ -3/2);\ r=3\sqrt{6}/2$ **25.** Yes; yes; yes   **27.** No; no; no   **29.** Yes; no; no
31. Yes; yes; yes   **33.** 6/5   **35.** Undefined slope   **37.** 9/4   **39.** 1/5   **41.** 0
43.
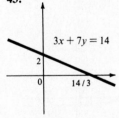
$3x+7y=14$

45.
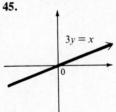
$3y=x$

47.
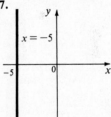
$x=-5$

49. $x+3y=10$ **51.** $2x+y=1$ **53.** $15x+30y=13$ **55.** $y=3/4$
57. $5x-8y=-40$ **59.** $y=-5$
61.
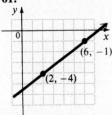
$(6,\ -1)$
$(2,\ -4)$

63.
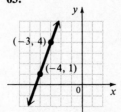
$(-3,\ 4)$
$(-4,\ 1)$

65.
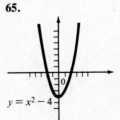
$y=x^2-4$

67.
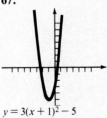
$y=3(x+1)^2-5$

69.

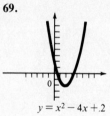

$$y = x^2 - 4x + .2$$

71. $(-3, -9)$; $x = -3$ **73.** $(7/2, -41/4)$; $x = 7/2$
75. $(-1, 4)$; $x = -1$ **77.** 5.5 and 5.5
79. A square, 45 m on a side

81. $(-2, 0)$; $y = 0$ **83.** $(2, -1)$; $y = -1$ **85.** $(29/4, 5/2)$; $y = 5/2$ **87.**

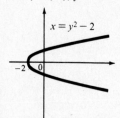

$$x = y^2 - 2$$

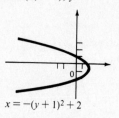

$$x = -(y + 1)^2 + 2$$

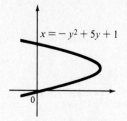

$$x = -y^2 + 5y + 1$$

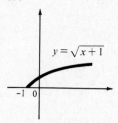

$$y = \sqrt{x + 1}$$

89.

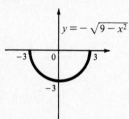

$$y = -\sqrt{9 - x^2}$$

91.

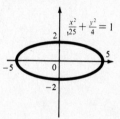

$$\frac{x^2}{25} + \frac{y^2}{4} = 1$$

93.

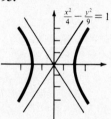

$$\frac{x^2}{4} - \frac{y^2}{9} = 1$$

95.

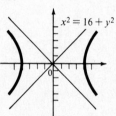

$$x^2 = 16 + y^2$$

97.

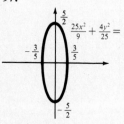

$$\frac{25x^2}{9} + \frac{4y^2}{25} = 1$$

99.

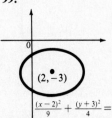

$$(2, -3)$$

$$\frac{(x - 2)^2}{9} + \frac{(y + 3)^2}{4} = 1$$

101.

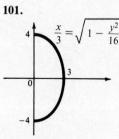

$$\frac{x}{3} = \sqrt{1 - \frac{y^2}{16}}$$

103.

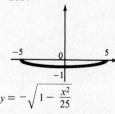

$$y = -\sqrt{1 - \frac{x^2}{25}}$$

105.

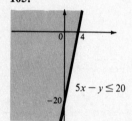

$$5x - y \leq 20$$

107.

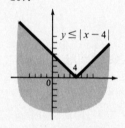

$$y \leq |x - 4|$$

109.

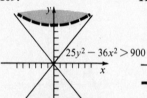

$$25y^2 - 36x^2 > 900$$

111.

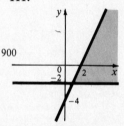

113.

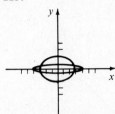

115.

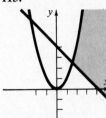

CHAPTER 5

Section 5.1   (page 197)

1. (a) 0   (b) 4   (c) 2   (d) 4   **3.** (a) -3 (b) -2 (c) 0   (d) 2   **5.** -1 **7.** -10
9. 4   **11.** $3a - 1$ **13.** 3   **15.** 72   **17.** $-6m - 1$ **19.** $15a - 7$
21. $25p^2 - 20p + 4$ **23.** $4p^2 + 3p - 1$ **25.** $3m^3 - m^2$ **27.** -66.8 **29.** -42.464
31. $(-\infty, +\infty); (-\infty, +\infty)$ **33.** $(-\infty, +\infty); [0, +\infty)$ **35.** $[-8, +\infty); [0, +\infty)$; function
37. $[-4, 4]; [0, 4]$ **39.** $[4, +\infty); [0, +\infty)$ **41.** $(-\infty, +\infty); (0, 1/5]$
43. $(-\infty, 1) \cup (1, 2) \cup (2, +\infty)$ **45.** $(-\infty, -1] \cup [5, +\infty); [0, +\infty)$ **47.** $(-\infty, +\infty); [0, +\infty)$
49. $[-5/2, +\infty); (-\infty, 0]$ **51.** $[-6, 6]; [0, 6]$ **53.** $(-\infty, -3] \cup [3, +\infty); (-\infty, 0]$
55. $[-5, 4]; [-2, 6]$ **57.** $(-\infty, +\infty); (-\infty, 12]$ **59.** $[-3, 4]; [-6, 8]$ **61.** 2   **63.** 16
65. 1/4   **67.** 2^m **69.** $2^{1/2} \approx 1.414$ **71.** (a) $x^2 + 2xh + h^2 - 4$ (b) $2xh + h^2$ (c) $2x + h$
73. (a) $6x + 6h + 2$ (b) $6h$ (c) 6   **75.** (a) $2x^3 + 6x^2h + 6xh^2 + 2h^3 + x^2 + 2xh + h^2$
(b) $6x^2h + 6xh^2 + 2h^3 + 2xh + h^2$ (c) $6x^2 + 6xh + 2h^2 + 2x + h$ **77.** $x = 25$
79. Break-even point is 45 units; do not produce.   **81.** Break-even point is -50; impossible to make a
profit here.

Section 5.2   (page 204)

For exercises 1–15, all are functions except 9 and 13.

1.

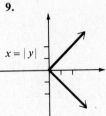

$y = |x + 1|$

3.

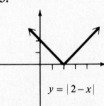

$y = |2 - x|$

5.

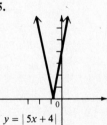

$y = |5x + 4|$

7.

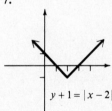

$y + 1 = |x - 2|$

9.

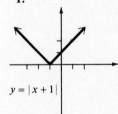

$x = |y|$

11.

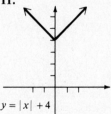

$y = |x| + 4$

13.

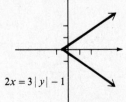

$2x = 3|y| - 1$

15.

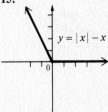

$y = |x| - x$

17.

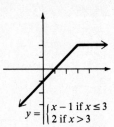

$y = \begin{cases} x - 1 \text{ if } x \le 3 \\ 2 \text{ if } x > 3 \end{cases}$

19.

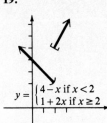

$y = \begin{cases} 4 - x \text{ if } x < 2 \\ 1 + 2x \text{ if } x \ge 2 \end{cases}$

21.

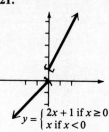

$y = \begin{cases} 2x + 1 \text{ if } x \ge 0 \\ x \text{ if } x < 0 \end{cases}$

23.

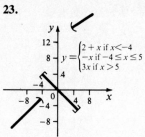

$y = \begin{cases} 2 + x \text{ if } x < -4 \\ -x \text{ if } -4 \le x \le 5 \\ 3x \text{ if } x > 5 \end{cases}$

25.

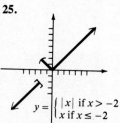

$y = \begin{cases} |x| \text{ if } x > -2 \\ x \text{ if } x \le -2 \end{cases}$

27.

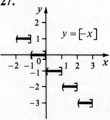

$y = [-x]$

29.

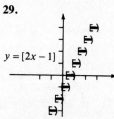

$y = [2x - 1]$

31.

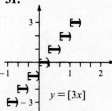

$y = [3x]$

33.

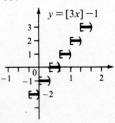

$y = [3x] - 1$

35. (a) \$11 (b) \$18
(c) \$32
(d)
(e) domain: $(0, +\infty)$ (at least in theory); range: $\{11, 18, 25, 32, 39, \ldots\}$

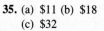

37. $- [-x]$ ounces

39.

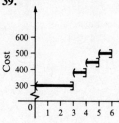

41. (a) 140 (b) 220 (c) 220 (d) 220 (e) 220 (f) 60
(g) 60 (h)

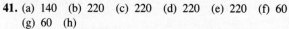

Section 5.3   **(page 212)**

1. (a)   (b) *y*-axis

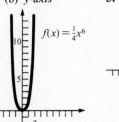

$$f(x) = \tfrac{1}{4}x^6$$

3. (a)   (b) Origin

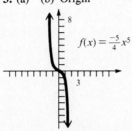

$$f(x) = \tfrac{-5}{4}x^5$$

5. (a)   (b) (0, 1)

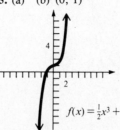

$$f(x) = \tfrac{1}{2}x^3 + 1$$

7. (a)   (b) (−1, 0)

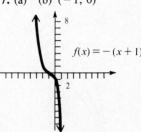

$$f(x) = -(x + 1)^3$$

9. (a)   (b) *x* = 1

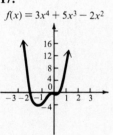

$$f(x) = (x - 1)^4 + 2$$

11.

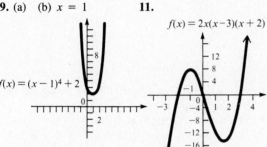

$$f(x) = 2x(x - 3)(x + 2)$$

13.

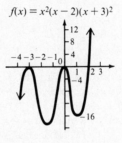

$$f(x) = x^2(x - 2)(x + 3)^2$$

15.

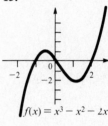

$$f(x) = x^3 - x^2 - 2x$$

17.

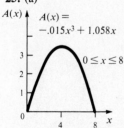

$$f(x) = 3x^4 + 5x^3 - 2x^2$$

19.

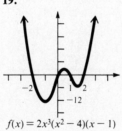

$$f(x) = 2x^3(x^2 - 4)(x - 1)$$

21.

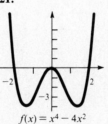

$$f(x) = x^4 - 4x^2$$

23.

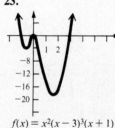

$$f(x) = x^2(x - 3)^3(x + 1)$$

25. (a)

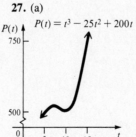

$$A(x) = -.015x^3 + 1.058x$$
$$0 \le x \le 8$$

27. (a)

$$P(t) = t^3 - 25t^2 + 200t$$

29. Odd   **31.** Even   **33.** Odd
35. Neither   **37.** Even   **39.** Neither
41. Maximum is 26.136 when *x* = −3.4; minimum
is 25 when *x* = −3.   **43.** Maximum is 1.048
when *x* = −.1; minimum is −5 when *x* = −1.
45. Maximum is 84 when *x* = −2; minimum is
−13 when *x* = −1.

(b) Between 4 and 5 hr,
closer to 5 hr   (c) From
about 1 hr to about 8 hr

(b) Increasing from *t* = 0
to *t* = 6 2/3 and from *t* = 10
on; decreasing from
t = 6 2/3 to *t* = 10

47.

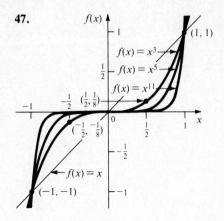

Section 5.4 (page 221)
1. Vertical asymptote: $x = 5$; horizontal asymptote: $y = 0$ **3.** Vertical asymptote: $x = 7/3$; horizontal asymptote: $y = 0$ **5.** Vertical asymptote: $x = -2$; horizontal asymptote: $y = -1$ **7.** Vertical asymptote: $x = -9/2$; horizontal asymptote: $y = 3/2$ **9.** Vertical asymptotes: $x = 3$, $x = 1$; horizontal asymptote: $y = 0$ **11.** Oblique asymptote: $y = x - 3$ **13.** Vertical asymptotes: $x = -2$, $x = 5/2$; horizontal asymptote: $y = 1/2$ **15.** No vertical asymptotes: horizontal asymptote: $y = 2$

17.

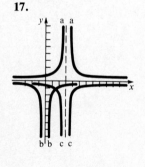

19.

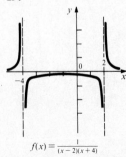

$$f(x) = \frac{1}{(x-2)(x+4)}$$

21.

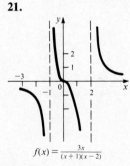

$$f(x) = \frac{3x}{(x+1)(x-2)}$$

23.

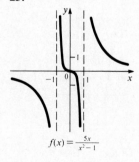

$$f(x) = \frac{5x}{x^2-1}$$

25.

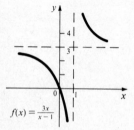

$$f(x) = \frac{3x}{x-1}$$

27.

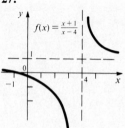

$$f(x) = \frac{x+1}{x-4}$$

29.

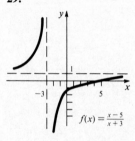

$$f(x) = \frac{x-5}{x+3}$$

31.

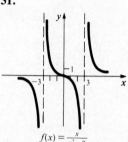

$$f(x) = \frac{x}{x^2-9}$$

33.

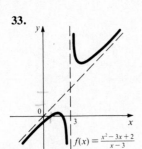

$f(x) = \frac{x^2 - 3x + 2}{x - 3}$

35.

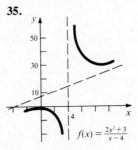

$f(x) = \frac{2x^2 + 3}{x - 4}$

37.

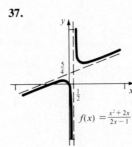

$f(x) = \frac{x^2 + 2x}{2x - 1}$

39.

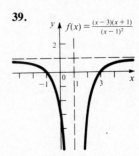

$f(x) = \frac{(x - 3)(x + 1)}{(x - 1)^2}$

41.

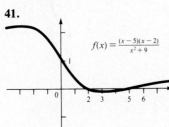

$f(x) = \frac{(x - 5)(x - 2)}{x^2 + 9}$

43.

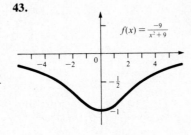

$f(x) = \frac{-9}{x^2 + 9}$

45. (a) 440, 400, 338, 259, 210, 176
(b)

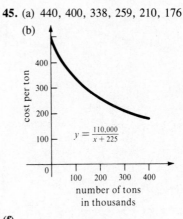

$y = \frac{110{,}000}{x + 225}$

cost per ton

number of tons in thousands

47. (a) (b) No

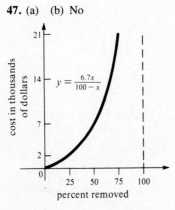

$y = \frac{6.7x}{100 - x}$

cost in thousands of dollars

percent removed

49. (a) \$65.5 tens of millions, or \$655,000,000 (b) \$64 tens of millions, or \$640,000,000 (c) \$60 tens of millions or \$600,000,000 (d) \$40 tens of millions or \$400,000,000 (e) \$0

(f)

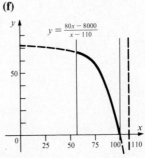

$y = \frac{80x - 8000}{x - 110}$

Section 5.5 (page 227)
1. $10x + 2$; $-2x - 4$; $24x^2 + 6x - 3$; $(4x - 1)/(6x + 3)$; all domains are $(-\infty, +\infty)$, except for f/g, which is $(-\infty, -1/2) \cup (-1/2, +\infty)$· **3.** $4x^2 - 4x + 1$; $2x^2 - 1$; $(3x^2 - 2x)(x^2 - 2x + 1)$; $(3x^2 - 2x)/(x^2 - 2x + 1)$; all domains are $(-\infty, +\infty)$, except for f/g, which is $(-\infty, -1) \cup (-1, +\infty)$
5. $\sqrt{2x + 5} + \sqrt{4x - 9}$; $\sqrt{2x + 5} - \sqrt{4x - 9}$; $\sqrt{(2x + 5)(4x - 9)}$; $\sqrt{(2x + 5)/(4x - 9)}$; all domains are $[9/4, +\infty)$, except f/g, which is $(9/4, +\infty)$ **7.** $5x^2 - 11x + 7$; $3x^2 - 11x - 3$; $(4x^2 - 11x + 2)(x^2 + 5)$; $(4x^2 - 11x + 2)/(x^2 + 5)$; all domains (including f/g) are $(-\infty, +\infty)$ **9.** 55
11. 1848 **13.** $-6/7$ **15.** $4m^2 + 6m + 1$ **17.** 1122 **19.** 97 **21.** $256k^2 + 48k + 2$
23. $24x + 4$; $24x + 35$ **25.** $-5x^2 + 20x + 18$; $-25x^2 - 10x + 6$ **27.** $-64x^3 + 2$; $-4x^3 + 8$
29. $1/x^2$; $1/x^2$ **31.** $\sqrt{8x^2 - 4}$; $8x + 10$ **33.** $x/(2 - 5x)$; $2(x - 5)$
35. $\sqrt{(x - 1)/x}$; $-1/\sqrt{x + 1}$ **49.** $18a^2 + 24a + 9$ **51.** $16\pi t^2$

Section 5.6   (page 234)

1. One-to-one   **3.** One-to-one   **5.** Not one-to-one   **7.** One-to-one   **9.** One-to-one
11. Not one-to-one   **13.** Not one-to-one   **15.** Not one-to-one   **17.** One-to-one   **19.** One-to-one   **21.** One-to-one   **23.** Not one-to-one   **25.** Inverses   **27.** Not inverses   **29.** Not inverses   **31.** Inverses   **33.** Inverses   **35.** Not inverses   **37.** Not inverses   **39.** Not inverses   **41.** Inverses   **43.** Not inverses

45.

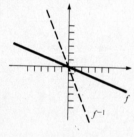

47.

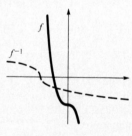

49.

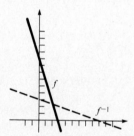

51. $f^{-1}(x) = (x + 5)/4$

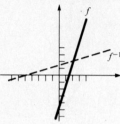

53. $f^{-1}(x) = -\dfrac{5}{2}x$

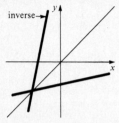

55. $f^{-1}(x) = \sqrt[3]{-x - 2}$

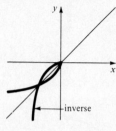

57. $f^{-1}(x) = (9 - x)/3$

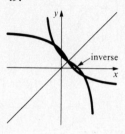

59. Not one-to-one

61. $f^{-1}(x) = 4/x$

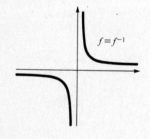

63. $f^{-1}(x) = (x + 5)/(3x + 6)$
Domain f:
$(-\infty, 1/3) \cup (1/3, +\infty)$
Domain f^{-1}:
$(-\infty, -2) \cup (-2, +\infty)$

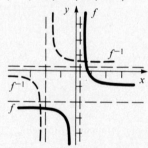

65. $f^{-1}(x) = x^2 - 6$
Domain: $[0, +\infty)$

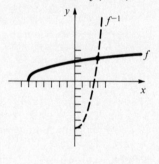

Section 5.7 (page 239)

1. $a = kb$ **3.** $x = k/y$ **5.** $r = kst$ **7.** $w = kx^2/y$ **9.** 220/7 **11.** 32/15
13. 18/125 **15.** 1075 **17.** 2304 ft **19.** .0444 ohms **21.** 140 kg per cm^2 **23.** \$1375
25. 8/9 metric tons **27.** 7500 lb **29.** 150 kg per m^2 **31.** 4,000,000 **33.** 1600 calls
35. About 4.94 **37.** About 7.4 km **39.** 263 lb

Chapter 5   Review Exercises   (page 243)

1. $(-\infty, +\infty)$ **3.** $(-\infty, +\infty)$ **5.** $(-\infty, +\infty)$ **7.** $[-7, 7]$ **9.** $(-\infty, +\infty)$
11. $(-\infty, 7) \cup (7, +\infty)$ **13.** -19 **15.** 2 **17.** 6 **19.** $-z^2 - 2z + 5$
21. $-4x - 2h + 4$ **23.** $-3x^2 - 3xh - h^2 + 4x + 2h$

25.

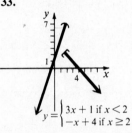

$y = -|x|$

27.

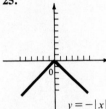

$y = -|x| - 2$

29.

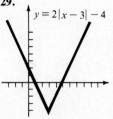

$y = 2|x - 3| - 4$

31.

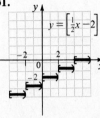

$y = \left[\frac{1}{2}x - 2\right]$

33.

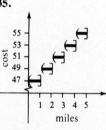

$y = \begin{cases} 3x + 1 \text{ if } x < 2 \\ -x + 4 \text{ if } x \geq 2 \end{cases}$

35.

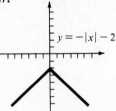

miles

37.

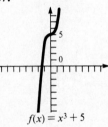

$f(x) = x^3 + 5$

39.

$f(x) = x^2(2x + 1)(x - 2)$

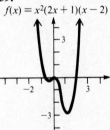

41.

$f(x) = 2x^3 + 13x^2 + 15x$

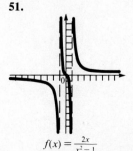

43. Neither
45. Odd

47.

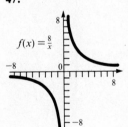

$f(x) = \frac{8}{x}$

49.

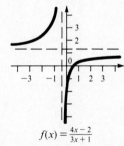

$f(x) = \frac{4x - 2}{3x + 1}$

51.

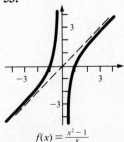

$f(x) = \frac{2x}{x^2 - 1}$

53.

$f(x) = \frac{x^2 - 1}{x}$

55. $4x^2 - 3x - 8$ **57.** 44   **59.** $16k^2 - 6k - 8$ **61.** $-23/4$ **63.** $(-\infty, +\infty)$
65. $\sqrt{x^2 - 2}$ **67.** $\sqrt{34}$ **69.** 1   **71.** Not one-to-one   **73.** Not one-to-one   **75.** Not
one-to-one   **77.** Not one-to-one
79. $f^{-1}(x) = (x - 3)/12$ **81.** $f^{-1}(x) = \sqrt[3]{x + 3}$ **83.** $f^{-1}(x) = \sqrt{25 - x^2}$; domain: $[0, 5]$

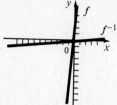
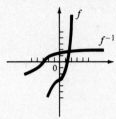
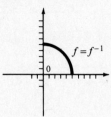

85. $m = kz^2$ **87.** $Y = (kMN^2)/X^3$ **89.** 27/2   **91.** 1372/729   **93.** 36 in

CHAPTER 6

Section 6.1   **(page 253)**
1. 8   **3.** 5   **5.** 4/9   **7.** $\sqrt{6}/3$
9. (a)   (b)   (c)   (d)

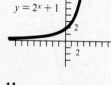
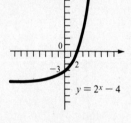
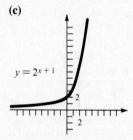
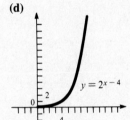

11.   **13.**   **15.**   **17.**

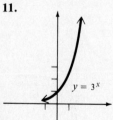
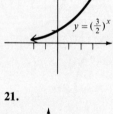
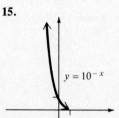
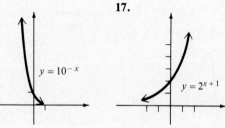

19.   **21.**   **23.** $\{1/2\}$ **25.** $\{-2\}$ **27.** $\{0\}$ **29.** $\{3\}$
31. $\{8\}$ **33.** $\{1/5\}$ **35.** $\{0\}$
37. $\{3/5\}$ **39.** $\{-2/3\}$

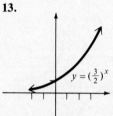
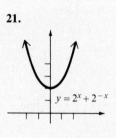

43.

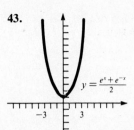

$$y = \frac{e^x + e^{-x}}{2}$$

45.

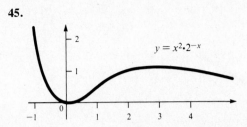

$$y = x^2 \cdot 2^{-x}$$

47.

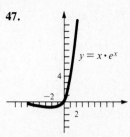

$$y = x \cdot e^x$$

49. (a) 500 g (b) 409 g (c) 335 g
(d) 184 g (e) $Q(t)$

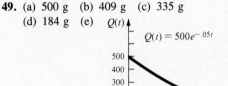

$$Q(t) = 500e^{-.05t}$$

51. (a) \$21,665.74 (b) \$29,836.49   **53.** \$3516.56
55. \$32,738.70   **57.** \$96,249.18   **59.** \$146,990.79
61. \$134,629.43   **63.** \$4054.96   **65.** \$8158.66
67. (a) \$4049.28 (b) \$4257.61 (c) \$4332.45
69. \$16,239.52   **71.** \$114,440.15
73. (a) About 2,690,000 (b) About 3,020,000
(c) About 9,600,000 (d) About 38,400,000   **75.** 2.718

Section 6.2 (page 261)
1. $\log_3 81 = 4$ **3.** $\log_{10} 10,000 = 4$ **5.** $\log_{1/2} 16 = -4$ **7.** $\log_{10} .0001 = -4$
9. $6^2 = 36$ **11.** $(\sqrt{3})^8 = 81$ **13.** $10^{-4} = .0001$ **15.** $m^n = k$ **17.** 2   **19.** 1
21. -3 **23.** $1/2$ **25.** $-1/6$ **27.** 9   **29.** 8   **31.** 9   **33.** {5}   **35.** {1/5}
37. $\log_3 2 - \log_3 5$ **39.** $\log_2 6 + \log_2 x - \log_2 y$ **41.** $1 + (1/2)\log_5 7 - \log_5 3$ **43.** Cannot
be simplified using the properties of logarithms   **45.** $\log_k p + 2 \cdot \log_k q - \log_k m$
47. $(1/2)(\log_m 5 + 3 \cdot \log_m r - 5 \cdot \log_m z)$ **49.** $\log_a (xy)/m$ **51.** $\log_m (a^2/b^6)$
53. $(\log_x (a^{3/2}b^{-4})$ or $\log_x (a^{3/2}/b^4)$ **55.** $\log_b [7x(x + 2)/8]$ **57.** $\log_a [(z - 1)^2(3z + 2)]$
59. $\log_5 (5^{1/3} m^{-1/3})$ or $\log_5(5^{1/3}m^{1/3})$
61. (a)

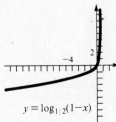

$$y = (\log_2 x) + 3$$

(b)

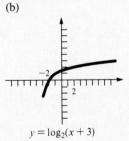

$$y = \log_2(x + 3)$$

(c)

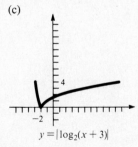

$$y = |\log_2(x + 3)|$$

63.

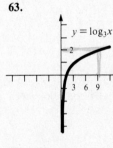

$$y = \log_3 x$$

65.

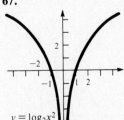

$$y = \log_{1/2}(1-x)$$

67.

$$y = \log_2 x^2$$

69.

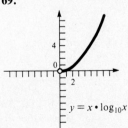

$$y = x \cdot \log_{10} x$$

71. .7781
73. .9542
75. 1.4771

77. (a) About 239
(b) About 477
(c) About 759 (d) $F(t)$

$$F(t) = 500 \log_{10}(2t + 2)$$

79. (a) 21 (b) 70 (c) 91 (d) 120 (e) 140
81. 398,000,000 I_0

Section 6.3 **(page 268)**
1. 1.3863   **3.** 2.8332   **5.** 5.8579   **7.** 3.8918   **9.** 3.7377   **11.** 11.3022   **13.** 1.43
15. .59   **17.** -1.59 **19.** .96   **21.** 1.89   **23.** -5.05
25.

$y = \ln x$

27.

$y = x \ln x$

29. 1   **31.** (a) 2.03 (b) 2.28 (c) 2.17 (d) 1.21   **33.** 378 ft   **35.** (a) 7,300,000;
77,300,000; 168,000,000; 175,000,000 (b) According to the given equation, the population can never increase
to 197,273,000.   **37.** 0.63   **39.** $21,665.74   **41.** $44,510.82   **43.** $25,424.98
45. $45,211.90   **47.** $80,466.31   **49.** (a) $32,590.96 (b) $32,599.00 (c) $32,599.35
51. $10,917.55   **53.** (a) 11% (b) 36% (c) 84%

Section 6.4 **(page 275)**
1. {1.631}   **3.** {1.069}   **5.** {$-.535$}   **7.** {$-.080$}   **9.** {.931}   **11.** {2.386}
13. {$-.123$}   **15.** $\emptyset$ **17.** {1.104}   **19.** {17.475}   **21.** {11}   **23.** {5}   **25.** {10}
27. {5}   **29.** {4}   **31.** $\emptyset$ **33.** {11}   **35.** {3}   **37.** $\emptyset$ **39.** {$-2, 2$}
41. {1, 10}   **43.** {2.423}   **45.** {$-.204$}   **47.** About 34.7 sec
49. (a) About 495 rabbits (b) About 2.75 mo   **51.** About 5600 yr   **53.** About 15,000 yr
55. About 3280 yr   **57.** 23 days   **59.** $t = (1000/k) \cdot \ln (P/P_0)$
61. $t = (-1/k) \cdot \log_{10} [(T - T_0)/(T_1 - T_0)]$ **63.** $I = I_0 \cdot 10^{d/10}$ **65.** $i = m[e^{(\ln A/P)/(nm)} - 1]$
67. (a) 315 (b) 229 (c) 142   **69.** (a) 611 million (b) 1007 million (c) 2028 million

Section 6.5 **(page 283)**
1. 2   **3.** 6   **5.** -4 **7.** -5 **9.** 2.9420   **11.** 4.1072   **13.** 0.8825
15. $6.9509 - 10$ or $.9509 - 4$ **17.** 0.5884   **19.** 4.8341   **21.** $6.5825 - 10$ or $.5825 - 4$
23. 34.4   **25.** .0747   **27.** 3124   **29.** .006367   **31.** 384   **33.** .206   **35.** 2.01
37. .735   **39.** 713   **41.** 890.5   **43.** 1.30   **45.** 15.57   **47.** 4.83   **49.** $-.02$
51. 5.62   **53.** 3.32   **55.** 1.79   **57.** {0.398}   **59.** {-0.192}   **61.** {-1.468}
63. {0.384}   **65.** 31.2 lb per ft^3 **67.** (a) About 350 yr (b) About 4000 yr (c) About 2300 yr
69. 38   **71.** 1,000,000   **73.** About 3160   **75.** 8.4   **77.** 13.5   **79.** 4×10^{-4}
81. 3.2×10^{-7} **83.** .00352

Chapter 6   Review Exercises   (page 286)

1. 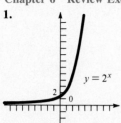   **3.**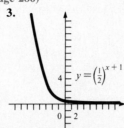

5. {5/3}   **7.** {3/2}   **9.** 800 g   **11.** $1790.19   **13.** About 12 yr   **15.** $\log_2 32 = 5$
17. $\log_{1/16} 1/2 = 1/4$ **19.** $\log_{10} 3 = .4771$ **21.** $10^{-3} = .001$ **23.** $10^{.537819} = 3.45$
25. $\log_3 m + \log_3 n - \log_3 5 - \log_3 r$ **27.** $2 \log_5 x + 4 \log_5 y + (1/5)(3 \log_5 m + \log_5 p)$
29. 3.555   **31.** 11.878   **33.** 2.255   **35.** 6.049   **37.** .8 m   **39.** 1.5 m
41. $1999.00   **43.** $43,988.40   **45.** $7797.47   **47.** $4272.85   **49.** $6002.84
51. $6289.05   **53.** $2663.54   **55.** $1293.68   **57.** $2346.33   **59.** About 7 yr
61. About 5.3 yr   **63.** $16,095.74   **65.** $580,792.63   **67.** {1.490}   **69.** {1.303}
71. {4}   **73.** {2}   **75.** 1.6590   **77.** 6.0899   **79.** $.7536 - 4$ **81.** 3150   **83.** .0446
85. 180   **87.** 5.97   **89.** 1.2785   **91.** $.9918 - 3$ **93.** .05582   **95.** 20.84
97. 1.433

CHAPTER 7

Section 7.1   (page 295)
1. 2×2; square   **3.** 3×4 **5.** 2×1; column   **7.** 4×1; column   **9.** 3×3; square
11. $x = 2$; $y = 4$; $z = 8$ **13.** $m = 8$; $n = -2$; $z = 2$; $y = 5$; $w = 6$
15. $x = -15$; $y = 5$; $k = 3$ **17.** $m = 6$; $n = -9$ **19.** $z = 18$; $r = 3$; $s = 3$; $p = 3$; $a = 3/4$

21. $\begin{bmatrix} 14 & -11 & -3 \\ -2 & 4 & -1 \end{bmatrix}$ **23.** $\begin{bmatrix} -12 & 16 \\ 4 & 2 \end{bmatrix}$ **25.** $\begin{bmatrix} 22 & -43 \\ 14 & -27 \\ -11 & 52 \end{bmatrix}$ **27.** Not possible

29. $\begin{bmatrix} 13 & 3 & 0 & -2 \\ 9 & -12 & 4 & 8 \\ 12 & -11 & -1 & 9 \end{bmatrix}$ **31.** $\begin{bmatrix} -12x + 8y & -x + y \\ x & 8x - y \end{bmatrix}$ **33.** $\begin{bmatrix} 10.874 & 17.011 \\ 12.609 & 15.573 \\ 8.892 & 14.517 \end{bmatrix}$

35. $\begin{bmatrix} -4 & 8 \\ 0 & 6 \end{bmatrix}$ **37.** $\begin{bmatrix} 8 & -16 \\ 0 & -12 \end{bmatrix}$ **39.** $\begin{bmatrix} 2 & 6 \\ -4 & 6 \end{bmatrix}$ **41.** $\begin{bmatrix} 60 & -10 \\ -44 & 9 \end{bmatrix}$

43. $\begin{bmatrix} 6 & 8 \\ -8 & 9 \end{bmatrix}$ **45.** $[18.2 \quad 19.0 \quad 17.6 \quad 18.5]$; $\begin{bmatrix} 18.2 \\ 19.0 \\ 17.6 \\ 18.5 \end{bmatrix}$ **47.** $\begin{bmatrix} 7 & 2 \\ 9 & 0 \\ 8 & 6 \end{bmatrix}$; $\begin{bmatrix} 7 & 9 & 8 \\ 2 & 0 & 6 \end{bmatrix}$

49. Yes, always true   **51.** Yes, always true   **53.** Yes, always true

Section 7.2   (page 304)
1. 2×2; 2×2 **3.** 4×4; 2×2 **5.** 3×2; BA cannot be found
7. AB cannot be found; 3×2 **9.** AB cannot be found; BA cannot be found.

11. $\begin{bmatrix} 13 \\ 25 \end{bmatrix}$ **13.** $\begin{bmatrix} -17 \\ -1 \end{bmatrix}$ **15.** $\begin{bmatrix} 17 & -10 \\ 1 & 2 \end{bmatrix}$ **17.** $\begin{bmatrix} -2 & 10 \\ 0 & 8 \end{bmatrix}$ **19.** $\begin{bmatrix} -2 & 5 & 0 \\ 6 & 6 & 1 \\ 12 & 2 & -3 \end{bmatrix}$

21. $[2 \quad 7 \quad -4]$ **23.** $\begin{bmatrix} 20 & 0 & 8 \\ 8 & 0 & 4 \end{bmatrix}$ **25.** Cannot be multiplied   **27.** $[-8]$

29. $\begin{bmatrix} -6 & 12 & 18 \\ 4 & -8 & -12 \\ -2 & 4 & 6 \end{bmatrix}$ **31.** $\begin{bmatrix} 11.7144 & 22.199 \\ 25.8752 & 53.5427 \end{bmatrix}$ **33.** $\begin{bmatrix} 16 & 22 \\ 7 & 19 \end{bmatrix}$

35. $\begin{bmatrix} -10 & 16 & 26 \\ 5 & 7 & 22 \end{bmatrix}$ **37.** No; no   **39.** Always true   **41.** Always true   **43.** Always true

45. Not true. $(A + B)^2 = (A + B)(A + B) = A^2 + AB + BA + B^2$. Since multiplication is not commutative, $AB + BA$ is not equal to $2AB$.   **47.** (a) $[3 \quad 1 \quad 2 \quad 4]$ (b) $[10.44 \quad 10.51 \quad 10.17]$

Section 7.3 (page 313)

1. $\begin{bmatrix} 2 & 4 \\ 0 & -1 \end{bmatrix}$ **3.** $\begin{bmatrix} 1 & 1/5 \\ 0 & 1 \end{bmatrix}$ **5.** $\begin{bmatrix} 1 & 5 & 6 \\ 0 & 13 & 11 \\ 4 & 7 & 0 \end{bmatrix}$ **7.** $\begin{bmatrix} -3 & 1 & -4 \\ 2 & 1 & 3 \\ -17 & 0 & -13 \end{bmatrix}$ **9.** Yes

11. No   **13.** Yes   **15.** No   **17.** Yes   **19.** $\begin{bmatrix} 0 & 1/2 \\ -1 & 1/2 \end{bmatrix}$ **21.** Does not exist

23. $\begin{bmatrix} 2 & 1 \\ -3/2 & -1/2 \end{bmatrix}$ **25.** $\begin{bmatrix} -2.5 & 5 \\ 12.5 & -15 \end{bmatrix}$ **27.** $\begin{bmatrix} -5 & 15 \\ 10 & -20 \end{bmatrix}$

29. $\begin{bmatrix} 1 & 0 & 0 \\ 0 & -1 & 0 \\ -1 & 0 & 1 \end{bmatrix}$ **31.** $\begin{bmatrix} -1/24 & 3/16 & -1/2 \\ 1/8 & -1/16 & 1/2 \\ 1/8 & -1/16 & -1/2 \end{bmatrix}$ **33.** $\begin{bmatrix} 15 & 4 & -5 \\ -12 & -3 & 4 \\ -4 & -1 & 1 \end{bmatrix}$

35. $\begin{bmatrix} 1/2 & 1/2 & -1/4 & 1/2 \\ -1 & 4 & -1/2 & -2 \\ -1/2 & 5/2 & -1/4 & -3/2 \\ 1/2 & -1/2 & 1/4 & 1/2 \end{bmatrix}$ **37.** $\begin{bmatrix} -10/3 & 5/9 & -10/9 \\ 20/3 & 5/9 & 80/9 \\ -5 & 5/6 & -20/3 \end{bmatrix}$

Section 7.4 (page 318)

1. -36 **3.** 7   **5.** 0   **7.** -26 **9.** -16 **11.** 0   **13.** $2x - 32$ **15.** $y^2 - 16$
17. $x^2 - y^2$ **19.** $-.57$ **21.** -1 **23.** 8   **25.** 17   **27.** 166   **29.** 0   **31.** 0
33. $4x$ **35.** $-.043804$ **37.** $\{-4\}$ **39.** $\{13\}$ **41.** $0, -5, 0$ **43.** $-6, 0, -6$
45. -88 **47.** 1   **49.** 5/2   **51.** 7   **53.** 8   **55.** $x + 3y - 11 = 0$

Section 7.5 (page 327)

1. Property 5   **3.** Property 1   **5.** By Property 4, multiply each element in the second row by $-1/4$. The determinant of the resulting matrix is 0, by Property 5.   **7.** Use Property 5 after multiplying each element in the third column by 1/2.   **9.** Use Property 5 after multiplying each element in the third column by 1/2.   **11.** Property 2   **13.** Property 3   **15.** Property 4   **17.** Property 6 (first row added to second row)   **19.** Property 6 (-2 times the elements of the first column added to the second column)   **21.** Property 6 (3 times the elements of the second row added to the third row)   **23.** Property 6
25. 0   **27.** 0   **29.** 5   **31.** 32   **33.** -32

Section 7.6 (page 329)

1. 3   **3.** 3   **5.** 21   **7.** $B = \begin{bmatrix} 0 & 2 & 3 \\ 2 & 0 & 4 \\ 3 & 4 & 0 \end{bmatrix}$ **9.** 12   **11.** 14

Chapter 7   Review Exercises   (page 331)

1. $m = -8; p = -7; x = 2; y = 9; z = 5$ **3.** $a = 5; x = 3/2; y = 0; z = 9$ **5.** $\begin{bmatrix} 0 & -2 & 7 \\ 6 & -1 & 10 \end{bmatrix}$

7. Cannot be done   **9.** Cannot be done   **11.** $\begin{bmatrix} 87 & 6.4 \\ 70 & 4.0 \\ 55 & 3.3 \end{bmatrix}; \begin{bmatrix} 87 & 70 & 55 \\ 6.4 & 4.0 & 3.3 \end{bmatrix}$ **13.** $\begin{bmatrix} 11 & 20 \\ 14 & 40 \end{bmatrix}$

15. $\begin{bmatrix} -3 \\ 10 \end{bmatrix}$ **17.** $\begin{bmatrix} -2 & 22 & 31 \\ 25 & 12 & 9 \end{bmatrix}$ **19.** Yes   **21.** No   **23.** $\begin{bmatrix} 3 & -1 \\ -5 & 2 \end{bmatrix}$

25. $\begin{bmatrix} 1/2 & 0 \\ 1/10 & 1/5 \end{bmatrix}$ **27.** $\begin{bmatrix} 2/3 & 0 & -1/3 \\ 1/3 & 0 & -2/3 \\ -2/3 & 1 & 1/3 \end{bmatrix}$ **29.** -25 **31.** -44 **33.** $\{-7/3\}$

35. Any number   **37.** One row is all zeros.   **39.** Multiply the elements of row 1 of the second matrix by 2.   **41.** In the first matrix, add to the elements of row 2 the results of multiplying each element in row 1 by 1.

CHAPTER 8

Section 8.1   (page 338)

1. $\{(6, 15)\}$ **3.** $\{(2, -3)\}$ **5.** $\{(4, 3)\}$ **7.** $\{(1, 2)\}$ **9.** $\{(4.35, -.986)\}$ (rounded)
11. $\{(3, 6)\}$ **13.** $\{(-1, 4)\}$ **15.** $\emptyset$ **17.** $\{(1, 3)\}$ **19.** $\{(4, -2)\}$
21. $\{(x, y)|4x - y = 9\}$ **23.** $\{(12, 6)\}$ **25.** $\{(4, 3)\}$ **27.** $\{(2, 2)\}$ **29.** $\{(1/5, 1)\}$
31. $\{(-4, -1)\}$ **33.** $\{(.38, -.47)\}$ **35.** $a = -3; b = -1$
37. Goats cost $30; sheep cost $25.   **39.** 32 days at $28   **41.** $18,000 at 9%; $12,000 at 10%
43. $x = 63; R = C = 346.5$ **45.** $x = 2100; R = C = 52,000$ **47.** $x = 370; R = C = 6900$
49. 32; 80   **51.** 28/3; 28   **53.** 5; 50/3

Section 8.2   (page 343)

1. $\{(1, 2, -1)\}$ **3.** $\{(2, 0, 3)\}$ **5.** $\emptyset$ **7.** $\{(1, 2, 3)\}$ **9.** $\{(-1, 2, 1)\}$ **11.** $\{(4, 1, 2)\}$
13. $\{(1/2, 2/3, -1)\}$ **15.** $\{(2, 4, 2)\}$ **17.** $\{(-1, 1, 1/3)\}$ **19.** $\{(.4, -.8, .7)\}$ **21.** $\emptyset$
23. $\{(x, -4x + 3, -3x + 4)\}$ **25.** $\{(x, -x + 15, -9x + 69)\}$
27. $\{(x, (-15 + x)/3, (24 + x)/3\}$ **29.** $a = 3/4; b = 1/4; c = -1/2$ **31.** $x^2 + y^2 + x - 7y = 0$
33. 6 ones; 9 fives; 15 twenties   **35.** 30 barrels each of $150 and $190 glue; 210 barrels of $120 glue
37. 15 cm; 12 cm; 6 cm   **39.** $50,000 at 10%; $10,000 at 9%; and $40,000 at 7.5%

Section 8.3   (page 351)

1. $\begin{bmatrix} 2 & 3 & | & 11 \\ 1 & 2 & | & 8 \end{bmatrix}$ **3.** $\begin{bmatrix} 1 & 5 & | & 6 \\ 1 & 2 & | & 8 \end{bmatrix}$ **5.** $\begin{bmatrix} 2 & 1 & 1 & | & 3 \\ 3 & -4 & 2 & | & -7 \\ 1 & 1 & 1 & | & 2 \end{bmatrix}$ **7.** $\begin{bmatrix} 1 & 1 & 0 & | & 2 \\ 0 & 2 & 1 & | & -4 \\ 0 & 0 & 1 & | & 2 \end{bmatrix}$

9. $\begin{bmatrix} 1 & 0 & 0 & | & 5 \\ 0 & 1 & 0 & | & -2 \\ 0 & 0 & 1 & | & 3 \end{bmatrix}$ **11.** $x = 2, y = 3$ **13.** $2x + y = 1, 3x - 2y = -9$

15. $x = 2, y = 3, z = -2$ **17.** $3x + 2y + z = 1, 2y + 4z = 22, -x - 2y + 3z = 15$
19. $\{(2, 3)\}$ **21.** $\{(-3, 0)\}$ **23.** $\{(7/2, -1)\}$ **25.** $\{(5/2 - 1)\}$ **27.** $\emptyset$
29. $\{(x, y)|6x - 3y = 1\}$ **31.** $\{(-2, 1, 3)\}$ **33.** $\{(-1, 23, 16)\}$ **35.** $\{(3, 2, -4)\}$
37. $\{(2, 1, -1)\}$ **39.** $\emptyset$ **41.** $\{(-1, 2, 5, 1)\}$ **43.** $\{(1.7, 2.4)\}$ **45.** $\{(.5, -.7, 6)\}$
47. $3000 at 8%; $1000 at 5%; $6000 at 9%   **49.** $1.25 for unleaded; $1.22 for regular; $1.23 for premium   **51.** $\{(z + 14/5, z - 12/5, z)\}$ **53.** $\{((12 - z)/7, (4z - 6)/7, z)\}$ **55.** $\emptyset$

57. (a) $x_3 + x_4 = 600$ (b) The system has no unique solution; equations $x_2 + x_3 = 700$ and $x_3 + x_4 = 600$ are dependent.   (c) $x_2 - x_4 = 100$ (d) $x_4 = 1000 - x_1$; 1000 (e) $x_4 = x_2 - 100$; 100
(f) $x_4 = 600 - x_3$; $600, 600$ (g) $600; 600; 700; 1000$

Section 8.4   (page 356)

1. $A = \begin{bmatrix} 1 & 1 \\ 2 & -1 \end{bmatrix};\ X = \begin{bmatrix} x \\ y \end{bmatrix};\ B = \begin{bmatrix} 8 \\ 4 \end{bmatrix}$ **3.** $A = \begin{bmatrix} 4 & 5 \\ 2 & 3 \end{bmatrix};\ X = \begin{bmatrix} x \\ y \end{bmatrix};\ B = \begin{bmatrix} 7 \\ 5 \end{bmatrix}$

5. $A = \begin{bmatrix} 1 & 1 & 1 \\ 2 & 1 & 3 \\ 5 & -2 & 2 \end{bmatrix};\ X = \begin{bmatrix} x \\ y \\ z \end{bmatrix};\ B = \begin{bmatrix} 9 \\ 17 \\ 16 \end{bmatrix}$ **7.** $\{(2, 3)\}$

9. $\{(-2, 4)\}$ **11.** $\{(4, -6)\}$ **13.** $\{(5/2, -1)\}$ **15.** $\{(4, 0)\}$ **17.** $\{(5/9, -11/3)\}$
19. $\{(x, y)\,|\,6x - 3y = 1\}$ **21.** $\{(1, 1, 2)\}$ **23.** $\{(1/4, 1/4, 13/4)\}$
25. $\{(12, -15/11, -65/11)\}$ **27.** $\{(1, 0, -1)\}$ **29.** $\{(6, 140, 90, -18)\}$ **31.** $\{(.3, .8)\}$
33. $\{(.3, .5, .3)\}$ **35.** 6 plates; 3 cups

Section 8.5   (page 361)

1. $\{(1, 1), (-2, 4)\}$ **3.** $\{(2, 1), (1/3, 4/9)\}$ **5.** $\{(2, 12), (-4, 0)\}$ **7.** $\{(2, -3), (-6/5, 17/5)\}$
9. $\{(-3/5, 7/5), (-1, 1)\}$ **11.** $\{(2, 2), (2, -2), (-2, 2), (-2, -2)\}$ **13.** $\{(0, 0)\}$
15. $\{(1, 0), (-1, 0)\}$ **17.** $\{(1, -1), (-1, 1), (1, 1), (-1, -1)\}$
19. $\{(3, 6), (-3, 6), (3, -6), (-3, -6)\}$ **21.** Same circles or $\{(x, y)\,|\,x^2 + y^2 = 10\}$ **23.** $\varnothing$
25. $\{(2, 3), (3, 2)\}$ **27.** $\{(-3, 5), (15/4, -4)\}$ **29.** $\{(4, -1/8), (-2, 1/4)\}$
31. $\{(3, 2), (-3, -2), (4, 3/2), (-4, -3/2)\}$ **33.** $\{(3, 5), (-3, -5)\}$
35. $\{(\sqrt{5}, 0), (-\sqrt{5}, 0), (\sqrt{5}, \sqrt{5}), (-\sqrt{5}, -\sqrt{5})\}$ **37.** $\{(3, -3), (3, 3)\}$
39. $\{((1 + \sqrt{13})/2, (-1 + \sqrt{13})/2), ((-1 - \sqrt{21})/2, (3 + \sqrt{21})/2)\}$ **41.** -2 and 8 **43.** Yes
45. 24 cm and 18 cm   **47.** $4 - \sqrt{3}$ or about 2.3 sec ($4 + \sqrt{3}$ makes the height of the disc negative)

Section 8.6   (page 366)

1.
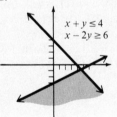
$x + y \leq 4$
$x - 2y \geq 6$

3.
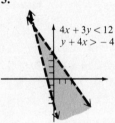
$4x + 3y < 12$
$y + 4x > -4$

5.
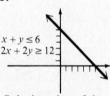
$x + y \leq 6$
$2x + 2y \geq 12$

Only the points of the line are included.

7.
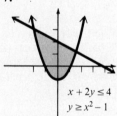
$x + 2y \leq 4$
$y \geq x^2 - 1$

9.
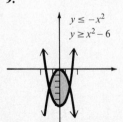
$y \leq -x^2$
$y \geq x^2 - 6$

11.
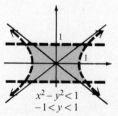
$x^2 - y^2 < 1$
$-1 < y < 1$

13.
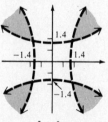
$2x^2 - y^2 > 4$
$2y^2 - x^2 > 4$

15.
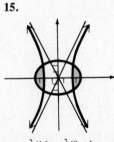
$x^2/16 + y^2/9 \leq 1$
$x^2/4 - y^2/16 \geq 1$

17.

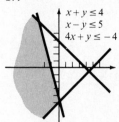

19. No solution

21.

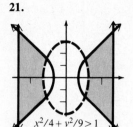

23.

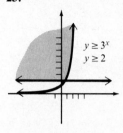

25.

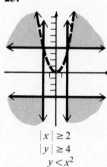

27.

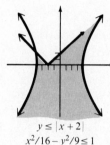

29.

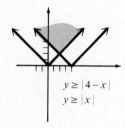

31. $(6/5, 6/5)$; $42/5$ **33.** $(17/3, 5)$; $49/3$ **35.** $112 with 4 pigs, 12 geese   **37.** 8 of #1; 3 of #2
39. 6.4 million gallons of gasoline and 3.2 million gallons of fuel oil, for maximum revenue of $16,960,000

Appendix Exercises   (page 372)
1. $\{(2, 2)\}$ **3.** $\{(2, -5)\}$ **5.** $\{(2, 0)\}$ **7.** $\{(2, 3)\}$ **9.** Can't use Cramer's rule, $D = 0$;
$\{(x, y) \mid 3x + 2y = 4\}$ **11.** Can't use Cramer's rule, $D = 0$; $\emptyset$ **13.** $\{(-1, 2, 1)\}$
15. $\{(-3, 4, 2)\}$ **17.** $\{(0, 0, -1)\}$ **19.** Can't use Cramer's rule, $D = 0$; $\emptyset$ **21.** Can't use
Cramer's rule, $D = 0$; $\{(x, (-13x + 10)/19, (4x + 32)/19)\}$ **23.** $\{(2, 3, 1)\}$ **25.** $\{(0, 4, 2)\}$
27. $\{(31/5, 19/10, -29/10)\}$ **29.** $\{(1, 0, -1, 2)\}$ **31.** $\{(-.2, -.8)\}$ **33.** $\{(.2, .5, -.2)\}$

Chapter 8   **Review Exercises**   (page 374)
1. $\{(5, 7)\}$ **3.** $\{(-5/4, -79/40)\}$ **5.** Dependent equations; $\{(x, y) \mid 5x - 2y = 11\}$
7. $\{(-1, 3)\}$ **9.** $\{(6, -12)\}$ **11.** $\{(2, 5)\}$ **13.** 10 at 25¢, 12 at 50¢
15. $10,000 at 6%, $20,000 at 7%, $20,000 at 9%   **17.** 17.5 g of 12 carat, 7.5 g of 22 carat
19. $\{(-1, 2, 3)\}$ **21.** $\{(1, 3, -1)\}$ **23.** 10 lb of $4.60 tea; 8 lb of $5.75 tea; 2 lb of $6.50 tea
25. 164 fives; 86 tens; 40 twenties   **27.** $\{(15z/11 + 6/11, 14z/11 - 1/11, z)\}$ **29.** $\{(-4, 6)\}$
31. $\{(-3, 2)\}$ **33.** $\{(0, 1, 0)\}$ **35.** $\{(2, 2)\}$ **37.** $\{(2, 1)\}$ **39.** $\{(-1, 0, 2)\}$
41. $\{(6z/11 + 5/11, 31z/11 + 2/11, z)\}$ **43.** $\{(-2, 3), (1, 0)\}$
45. $\{(2, 3), (-2, 3), (2, -3), (-2, -3)\}$ **47.** $\{(6, 2/3), (-4, -1)\}$
49. Yes; $\{((8 - 8\sqrt{41})/5, (16 + 4\sqrt{41})/5), ((8 + 8\sqrt{41})/5, (16 - 4\sqrt{41})/5)\}$
51. 10 and 6   or -10 and -6

53.

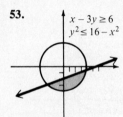

$$x - 3y \geq 6$$
$$y^2 \leq 16 - x^2$$

55. $2x + 3y/2 \leq 16;$
$3x + 2y/3 \leq 12;$
$x \geq 0; y \geq 0$

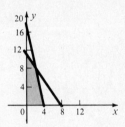

57. Maximum of 24 at $(0, 6)$ **59.** $\{(-2, 5)\}$ **61.** $\{(-4, 6, 2)\}$

CHAPTER 9

Section 9.1   (page 382)

1. $x^2 - 3x - 2$ **3.** $m^3 - m^2 - 6m$ **5.** $3x^2 + 4x + 3/(x - 5)$
7. $4m^2 - 4m + 1 + [-3/(m + 1)]$ **9.** $x^4 + x^3 + 2x - 1 + 3/(x + 2)$
11. $(1/3)x^2 - (1/9)x + 1/(x - 1/3)$ **13.** $9z^2 + 10z + 27 + 52/(z - 2)$ **15.** $y^2 + y + 1$
17. $x^3 + x^2 + x + 1$ **19.** $P(x) = (x - 1)(x^2 + 2x + 3) - 5$
21. $P(x) = (x + 2)(-x^2 + 4x - 8) + 20$ **23.** $P(x) = (x - 3)(x^3 + 2x + 5) + 20$
25. $P(x) = (x + 1)(3x^3 + x^2 - 11x + 11) + 4$ **27.** -8 **29.** 2   **31.** -6 **33.** $-6 - i$
35. $-2 - 3i$ **37.** Yes   **39.** Yes   **41.** Yes   **43.** Yes   **45.** No   **47.** Yes   **49.** No

Section 9.2   (page 388)

1. No   **3.** Yes   **5.** Yes   **7.** No   **9.** $P(x) = -(1/6)x^3 + (13/6)x + 2$
11. $P(x) = (-1/2)x^3 - (1/2)x^2 + x$ **13.** $P(x) = -10x^3 + 30x^2 - 10x + 30$
15. $P(x) = x^2 - 10x + 26$ **17.** $P(x) = x^3 - 4x^2 + 6x - 4$ **19.** $P(x) = x^3 - 3x^2 + x + 1$
21. $P(x) = x^4 - 6x^3 + 10x^2 + 2x - 15$ **23.** $P(x) = x^3 - 8x^2 + 22x - 20$
25. $P(x) = x^4 - 4x^3 + 5x^2 - 2x - 2$ **27.** $P(x) = x^4 - 10x^3 + 42x^2 - 82x + 65$
29. $P(x) = x^5 - 12x^4 + 74x^3 - 248x^2 + 445x - 500$ **31.** $P(x) = x^4 - 6x^3 + 17x^2 - 28x + 20$
33. $-1 + i, -1 - i$ **35.** $3, -2 - 2i$ **37.** $(-1 + i\sqrt{5})/2, (-1 - i\sqrt{5})/2$ **39.** $i, 2i, -2i$
41. $3, -2, 1 - 3i$ **43.** $2 - i, 1 + i, 1 - i$ **45.** $P(x) = (x - 2)(2x - 5)(x + 3)$
47. $P(x) = (x + 4)(3x - 1)(2x + 1)$ **49.** $P(x) = (x - 3i)(x + 4)(x + 3)$
51. $P(x) = (x - 1 - i)(2x - 1)(x + 3)$ **53.** Zeros are $2, -1, 3; P(x) = (x + 2)^2(x + 1)(x - 3)$
55. At $t = 3$ and $t = -2$

Section 9.3   (page 392)

1. $\pm 1, \pm 1/2, \pm 1/3, \pm 1/6$ **3.** $\pm 1, \pm 1/2, \pm 1/3, \pm 1/4, \pm 1/6, \pm 1/12, \pm 2, \pm 2/3$
5. $\pm 1, \pm 1/2, \pm 2, \pm 4, \pm 8$ **7.** $\pm 1, \pm 2, \pm 5, \pm 10, \pm 25, \pm 50$ **9.** $-1, -2, 5$
11. $2, -3, -5$ **13.** No rational zeros   **15.** $1, -2, -3, -5$
17. $-4, 3/2, -1/3; P(x) = (x + 4)(2x - 3)(3x + 1)$
19. $-3/2, -2/3, 1/2; P(x) = 2(x - 1/2)(3x + 2)(2x + 3) = (2x - 1)(3x + 2)(2x + 3)$
21. $1/2; P(x) = 2(x - 1/2)(x^2 + 4x + 8) = (2x - 1)(x^2 + 4x + 8)$
23. No rational zeros; prime   **25.** $-1, -2, -3, 4; P(x) = (x + 1)(x + 2)(x + 3)(x - 4)$
27. $-2, 2/3; P(x) = (x + 2)(3x - 2)(x^2 + 1)$ **29.** $1; P(x) = (x - 1)(x^4 + 4x^3 - x^2 - 12x - 12)$
31. $-2/3, -1, 3$ **33.** $1, -5/4$ **35.** 1

Section 9.4   (page 400)

17. (a) Positive: 1; negative: 2 or 0   (b) $-3, -1.4, 1.4$ **19.** (a) Positive: 2 or 0; negative: 1
(b) $-2, 1.6, 4.4$ **21.** (a) Positive: 1; negative: 0   (b) 1.5   **23.** (a) Positive: 3 or 1; negative: 1
(b) $1.1, -.7$ **25.** (a) Positive: 1; negative: 1   (b) $-1.5, 3.1$ **27.** $3.24, -1.24$
29. $-3.65, -.32, 1.65, 6.32$

31.

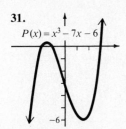

$P(x) = x^3 - 7x - 6$

33.

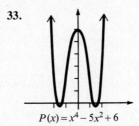

$P(x) = x^4 - 5x^2 + 6$

35.

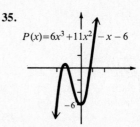

$P(x) = 6x^3 + 11x^2 - x - 6$

37.

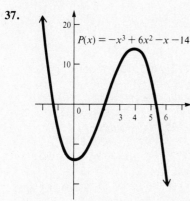

$P(x) = -x^3 + 6x^2 - x - 14$

Chapter 9   Review Exercises   (page 402)

1. $2x^2 - 3x + 2 + (-8)/(x + 1)$ **3.** $4m^2 + 9m + 18 + (-24)/(m - 2)$
5. $P(z) = (z + 2)(2z^2 + 11z + 8) + 4$ **7.** $P(m) = (m - 4)(2m^2 - 5m - 3) + 3$ **9.** 11
11. 28   **13.** $P(x) = x^3 - 10x^2 + 17x + 28$ **15.** $P(x) = x^4 - x^3 - 9x^2 + 7x + 14$
17. No   **19.** No   **21.** $P(x) = -2x^3 + 6x^2 + 12x - 16$
23. $P(x) = x^4 - 3x^2 - 4$ **25.** $P(x) = x^3 + x^2 - 4x + 6$
27. $3 + i, 3 - i, 2i, -2i; P(x) = (x - 3 - i)(x - 3 + i)(x - 2i)(x + 2i)$ **29.** 1/2, -1, 5
31. 4, $-1/2$, $-2/3$ **33.** None   **43.** -2.3, 4.6
45. $P(x) = 2x^3 - 11x^2 - 2x + 2$ **47.**

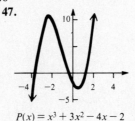

$P(x) = x^3 + 3x^2 - 4x - 2$

CHAPTER 10

Section 10.1   (page 407)
1. 10, 16, 22, 28, 34   **3.** 2, 4, 8, 16, 32   **5.** 1, -1, 1, -1, 1   **7.** 1/2, 4/5, 1, 8/7, 5/4
9. 4/3, 12/5, 20/7, 28/9, 36/11   **11.** -2, 8, -24, 64, -160 **13.** 2/3, 5/6, 10/11, 17/18, 26/27
15. x, x^2, x^3, x^4, x^5 **17.** 4, 9, 14, 19, 24, 29, 34, 39, 44, 49
19. -9, -13, -17, -21, -25, -29, -33, -37, -41, -45
21. 2, 4, 8, 16, 32, 64, 128, 256, 512, 1024
23. 1, 1, 2, 3, 5, 8, 13, 21, 34, 55   **25.** 35   **27.** 45   **29.** 29/20   **31.** 139/108   **33.** 3
35. 25/2   **37.** 89   **39.** 31   **41.** $x + x^2 + x^3 + x^4$ **43.** $x + 2x^2 + 3x^3 + 4x^4 + 5x^5$
45. 8   **47.** 29.0

Section 10.2   (page 413)
1. 4, 6, 8, 10, 12   **3.** 11, 9, 7, 5   **5.** 6, 2, -2, -6 **7.** -6, 0, 6, 12, 18, 24
9. Arithmetic; $d = 5$, $a_n = 7 + 5n$ **11.** Arithmetic; $d = -3$, $a_n = 21 - 3n$ **13.** Arithmetic;
$d = 7$, $a_n = -26 + 7n$ **15.** Arithmetic; $d = m$, $a_n = x + nm - m$ **17.** Arithmetic; $d = -m$,
$a_n = 2z + 2m - mn$ **19.** $a_8 = 19$; $a_n = 3 + 2n$ **21.** $a_8 = 7$; $a_n = n - 1$ **23.** $a_8 = -6$;
$a_n = 10 - 2n$ **25.** $a_8 = -3$; $a_n = -39 + 9n/2$ **27.** $a_8 = x + 21$; $a_n = x + 3n - 3$
29. $a_8 = 4m$; $a_n = mn - 4m$ **31.** 215   **33.** 125   **35.** 230   **37.** 205/2   **39.** 155/2
41. 16.745   **43.** 32.28   **45.** 18   **47.** 140   **49.** 570   **51.** -684 **53.** 500,500
55. 400   **57.** -78 **59.** $a_1 = 7$ **61.** $a_1 = 0$ **63.** $a_1 = 7$ **65.** $a_1 = 1$
67. 1281   **69.** 4680 bongs   **71.** 100 m; 550 m   **73.** 240 ft; 1024 ft   **75.** 170 m

Section 10.3   (page 419)
1. 2, 6, 18, 54   **3.** 1/2, 2, 8, 32   **5.** -2, 6, -18, 54   **7.** 3125, 625, 125, 25, 5, 1, 1/5
9. -1, 1, -1, 1, -1, 1   **11.** 3/2, 3, 6, 12, 24   **13.** $a_5 = 324$; $a_n = 4 \cdot 3^{n-1}$
15. $a_5 = -162$; $a_n = -2 \cdot 3^{n-1}$ **17.** $a_5 = -1875$; $a_n = -3(-5)^{n-1}$ **19.** $a_5 = 24$; $a_n = (3/2)2^{n-1}$
21. $a_5 = -256$; $a_n = -1(-4)^{n-1}$ **23.** Geometric; $r = 2$; $a_n = 6 \cdot 2^{n-1}$ **25.** Geometric; $r = 2$;
$a_n = (3/4)2^{n-1}$ **27.** Geometric; $r = -1/2$; $a_n = -4(-1/2)^{n-1}$ **29.** Not geometric
31. Geometric; $a_n = (9/4)2^{n-1}$ **33.** Geometric; $a_n = (-2/9)3^{n-1}$ **35.** 93   **37.** 33/4
39. 55/9   **41.** 124   **43.** 121/9   **45.** 860.95   **47.** -6.2737 **49.** 30   **51.** 60
53. 255/4   **55.** 406/625   **57.** 585/8   **59.** 120   **61.** 272/25
63. 2^{30} or $1,073,741,824   **65.** \$7,875,000   **67.** .00002%   **69.** $(1/32) \times 10^{15}$ molecules
71. \$12,487.56   **73.** \$40,634.48   **75.** $a_1 = 2$, $r = 3$ or $a_1 = -2$, $r = -3$
77. $a_1 = 128$, $r = 1/2$ or $a_1 = -128$, $r = -1/2$

Section 10.4   (page 424)
1. 2   **3.** 10   **5.** 1/2; sum would exist   **7.** 1.1   **9.** $-.9$; sum would exist   **11.** 64/3
13. 1000/9   **15.** 135   **17.** 512/3   **19.** 81   **21.** 3/2   **23.** 2   **25.** 1/5
27. Cannot be found   **29.** 1/3   **31.** Cannot be found   **33.** $-1/5$ **35.** 1/4   **37.** 1/9
39. Cannot be found   **41.** 5/9   **43.** 4/33   **45.** 31/99   **47.** 508/999   **49.** 311/900
51. 70 m   **53.** 1600   **55.** 12 m

Chapter 10   Review Exercises   (page 432)
1. 1/2, 2/3, 3/4, 4/5, 5/6   **3.** 8, 10, 12, 14, 16   **5.** 5, 2, -1, -4, -7 **7.** 5, 3, -2, -5, -3
9. 6, 2, -2, -6, -10 **11.** $3 - \sqrt{5}$, 4, $5 + \sqrt{5}$, $6 + 2\sqrt{5}$, $7 + 3\sqrt{5}$ **13.** 4, 8, 16, 32, 64
15. -3, 4, $-16/3$, 64/9, $-256/27$ **17.** $a_1 = -29$; $a_{20} = 66$ **19.** 20   **21.** $-x + 61$
23. 222   **25.** $84k$ **27.** 25   **29.** $x^5/125$ **31.** 15   **33.** $-10k$ **35.** 80
37. Cannot be found   **39.** 2/3   **41.** 512/999   **43.** 25/6   **45.** 690   **47.** 73/12
49. 50,005,000   **51.** 240   **53.** $r > 1$, so sum does not exist.   **55.** 64 cans   **57.** About 117
items

CHAPTER 11

Section 11.1   (page 438)
1. 5040   **3.** 720   **5.** 90   **7.** 336   **9.** 7   **11.** 1   **13.** 40,320   **15.** 120; 30,240
17. 30   **19.** 120   **21.** 4   **23.** 552   **25.** (a) $26^3 \cdot 10^3 = 17,576,000$ (b) 35,152,000
27. $26 \cdot 25 \cdot 24 \cdot 10 \cdot 9 \cdot 8 = 11,232,000$; $2(26 \cdot 25 \cdot 24 \cdot 10 \cdot 9 \cdot 8) = 22,464,000$

Section 11.2 (page 441)
1. 6   **3.** 56   **5.** 1365   **7.** 120   **9.** 14   **11.** 1   **13.** 27,405   **15.** 220; 220
17. 1326   **19.** 10   **21.** 28   **23.** (a) 84 (b) 10 (c) 40 (d) 28   **25.** $\binom{52}{5} = 2{,}598{,}960$

27. 4   **29.** $4 \cdot \binom{13}{5} = 5148$ **31.** $P(5, 5) = 120$ **33.** $\binom{12}{3} = 220$

Section 11.3 (page 448)
1. Let h = heads, t = tails. S = {h}   **3.** S = {hhh, hht, hth, thh, htt, tht, tth, ttt}   **5.** Let
c = correct, w = wrong. S = {ccc, ccw, cwc, wcc, wwc, wcw, cww, www}   **7.** (a) {hh, tt}, 1/2
(b) {hh, ht, th}, 3/4   **9.** (a) {2 and 4}, 1/10 (b) {1 and 3, 1 and 5, 3 and 5}, 3/10 (c) $\emptyset$, 0
(d) {1 and 2, 1 and 4, 2 and 3, 2 and 5, 3 and 4, 4 and 5}, 3/5   **11.** (a) 1/5 (b) 8/15 (c) 0
(d) 1 to 4 (e) 7 to 8   **13.** 1 to 4   **15.** 2/5   **17.** (a) 1/2 (b) 7/10 (c) 2/5   **19.** (a) 1/6
(b) 1/3 (c) 1/6   **21.** (a) 7/15 (b) 11/15 (c) 4/5

Section 11.4 (page 454)
1. $x^6 + 6x^5y + 15x^4y^2 + 20x^3y^3 + 15x^2y^4 + 6xy^5 + y^6$ **3.** $p^5 - 5p^4q + 10p^3q^2 - 10p^2q^3 + 5pq^4 - q^5$
5. $r^{10} + 5r^8s + 10r^6s^2 + 10r^4s^3 + 5r^2s^4 + s^5$ **7.** $p^4 + 8p^3q + 24p^2q^2 + 32pq^3 + 16q^4$
9. $2401p^4 + 2744p^3q + 1176p^2q^2 + 224pq^3 + 16q^4$
11. $729x^6 - 2916x^5y + 4860x^4y^2 + 4320x^3y^3 - 2160x^2y^4 - 576xy^5 + 64y^6$
13. $m^6/64 - 3m^5/16 + 15m^4/16 - 5m^3/2 + 15m^2/4 - 3m + 1$
15. $16p^4 + 32p^3q/3 + 8p^2q^2/3 + 8pq^3/27 + q^4/81$
17. $7920m^8p^4$ **19.** $126x^4y^5$ **21.** $180m^2n^8$ **23.** $4845p^8q^{16}$ **25.** $439{,}296x^{21}y^7$
27. 2.594   **29.** 245.937   **31.** 758.650   **35.** 2.112   **37.** .822

Chapter 11   Review Exercises (page 456)
1. 120   **3.** 1   **5.** 252   **7.** 7   **9.** 90   **11.** (a) 20 (b) 10
13. $26 \cdot 10 \cdot 10 \cdot 10 \cdot 26 \cdot 26 \cdot 26 = 456{,}976{,}000$; $26 \cdot 10 \cdot 9 \cdot 8 \cdot 25 \cdot 24 \cdot 23 = 258{,}336{,}000$
15. 4   **17.** 12   **19.** 4/13   **21.** 3/4   **23.** .86   **25.** $x^4 + 8x^3y + 24x^2y^2 + 32xy^3 + 16y^4$
27. $k^5/32 - 5k^4g/16 + 5k^3g^2/4 - 5k^2g^3/2 + 5kg^4/2 - g^5$ **29.** $25{,}344m^5n^{14}$
31. $-420 \cdot 3^{13}m^2n^{13} + 30 \cdot 3^{14}mn^{14} - 3^{15}n^{15}$ **33.** .868

Index